THE ORGANOMETALLIC CHEMISTRY OF THE TRANSITION METALS

THE ORGANOMETALLIC CHEMISTRY OF THE TRANSITION METALS

Fourth Edition

ROBERT H. CRABTREE

Yale University, New Haven, Connecticut

WILEY-INTERSCIENCE

A JOHN WILEY & SONS, INC., PUBLICATION

Published by John Wiley & Sons, Inc., Hoboken, New Jersey.
Published simultaneously in Canada.

No part of this publication may be reproduced, stored in a retrieval system, or transmitted in any form or by any means, electronic, mechanical, photocopying, recording, scanning, or otherwise, except as permitted under Section 107 or 108 of the 1976 United States Copyright Act, without either the prior written permission of the Publisher, or authorization through payment of the appropriate per-copy fee to the Copyright Clearance Center, Inc., 222 Rosewood Drive, Danvers, MA 01923, 978-750-8400, fax 978-646-8600, or on the web at www.copyright.com. Requests to the Publisher for permission should be addressed to the Permissions Department, John Wiley & Sons, Inc., 111 River Street, Hoboken, NJ 07030, (201) 748-6011, fax (201) 748-6008.

Limit of Liability/Disclaimer of Warranty: While the publisher and author have used their best efforts in preparing this book, they make no representations or warranties with respect to the accuracy or completeness of the contents of this book and specifically disclaim any implied warranties of merchantability or fitness for a particular purpose. No warranty may be created or extended by sales representatives or written sales materials. The advice and strategies contained herein may not be suitable for your situation. You should consult with a professional where appropriate. Neither the publisher nor author shall be liable for any loss of profit or any other commercial damages, including but not limited to special, incidental, consequential, or other damages.

For general information on our other products and services please contact our Customer Care Department within the U.S. at 877-762-2974, outside the U.S. at 317-572-3993 or fax 317-572-4002.

Wiley also publishes its books in a variety of electronic formats. Some content that appears in print, however, may not be available in electronic format.

Library of Congress Cataloging-in-Publication Data is available.

ISBN 0-471-66256-9

Printed in the United States of America.

10 9 8 7 6 5 4 3 2 1

CONTENTS

PREFACE

I would like to thank the many colleagues who kindly pointed out corrections, or contributed in some other way to this edition—Jack Faller, Ged Parkin, Robin Tanke, Joshua Telser, Fabiola Barrios-Landeros, Carole Velleca, Li Zeng, Guoan Du, Ipe Mavunkal, Xingwei Li, Marcetta Darensbourg, Greg Peters, Karen Goldberg, Odile Eisenstein, Eric Clot and Bruno Chaudret. I also thank UC Berkeley for hospitality while I was revising the book.

ROBERT H. CRABTREE

New Haven, Connecticut
January 2005

LIST OF ABBREVIATIONS

[]	Encloses complex molecules or ions
□	Vacant site or labile ligand
$1°, 2°, \ldots$	Primary, secondary, ...
A	Associative substitution (Section 4.4)
acac	Acetylacetone
AO	Atomic orbital
at.	Pressure in atmospheres
bipy	2,2′-Bipyridyl
Bu	Butyl
cata	Catalyst
CIDNP	Chemically induced dynamic nuclear polarization (Section 6.3)
CN	Coordination number
cod	1,5-Cyclooctadiene
coe	Cyclooctene
cot	Cyclooctatetraene
Cp, Cp*	C_5H_5, C_5Me_5
Cy	Cyclohexyl
∂^+	Partial positive charge
δ	Chemical shift (NMR)
Δ	Crystal field splitting (Section 1.4)
D	Dissociative substitution mechanism (Section 4.3)
d_σ, d_π	σ-Acceptor and π-donor metal orbitals (see Section 1.4)
diars	$Me_2AsCH_2CH_2AsMe_2$
dpe or dppe	$Ph_2PCH_2CH_2PPh_2$

dmf	Dimethylformamide
dmg	Dimethyl glyoximate
dmpe	$Me_2PCH_2CH_2PMe_2$
DMSO	Dimethyl sulfoxide
d^n	Electron configuration (Section 1.4)
η	Shows hapticity in π-bonding ligands (Section 2.1)
E, E^+	Generalized electrophile such as H^+
e	Electron, as in 18e rule
e.e.	Enantiomeric excess (Section 9.2)
en	$H_2NCH_2CH_2NH_2$
eq	Equivalent
Et	Ethyl
EPR	Electron paramagnetic resonance
eu	Entropy units
Fp	$(C_5H_5)(CO)_2Fe$
fac	Facial (stereochemistry)
Hal	Halogen
$HBpz_3$	Tris(pyrazolyl)borate
HOMO	Highest occupied molecular orbital
I	Nuclear spin
I	Intermediate substitution mechanism
IPR	Isotopic perturbation of resonance (Section 10.8)
IR	Infrared
κ	Shows hapticity in σ-bonding ligands (Section 2.1)
L	Generalized ligand, in particular a 2e ligand (L model for ligand binding is discussed in Section 2.1)
L_nM	Generalized metal fragment with n ligands
lin	linear
LUMO	Lowest unoccupied molecular orbital
μ	Descriptor for bridging (Section 1.1)
m-	Meta
Me	Methyl
mer	Meridional (stereochemistry)
m_r	Reduced mass
MO	Molecular orbital
ν	Frequency
nbd	Norbornadiene
NMR	Nuclear magnetic resonance (Sections 10.2–10.8)
NOE	Nuclear Overhauser effect (Section 10.7)
Np	Neopentyl
Nu, Nu^-	Generalized nucleophile, such as H^-
o-	Ortho
OAc	Acetate
oct	Octahedral (Table 2.5)
ofcot	Octafluorocyclooctadiene

OS	Oxidation state (Section 2.4)
p-	Para
Ph	Phenyl
py	Pyridine
RF	Radio frequency
SET	Single electron transfer (Section 8.6)
solv	Solvent
sq. py.	Square pyramidal (Table 2.5)
T_1	Spin-lattice relaxation time
tbe	t-BuCH$=$CH$_2$
thf	Tetrahydrofuran
triphos	MeC(CH$_2$PPh$_2$)$_3$
TBP or trig. bipy	Trigonal bipyramidal (Table 2.5)
TMEDA	Me$_2$NCH$_2$CH$_2$NMe$_2$
TMS	Trimethylsilyl
Ts	p-tolyl SO$_2$
VB	Valence bond
X	Generalized 1e anionic ligand (Section 2.1) (X$_2$ model for ligand binding is discussed on p. 126)

1

INTRODUCTION

Organometallic compounds, with their metal–carbon bonds (e.g., WMe_6), lie at the interface between classical organic and inorganic chemistry in dealing with the interaction between inorganic metal species and organic molecules. In the related metal–organic compound area, in contrast, the organic fragment is bound only by metal–heteroatom bonds [e.g., $Ti(OMe)_4$].

The organometallic field has provided a series of important conceptual insights, surprising structures, and useful catalysts both for industrial processes and for organic synthesis. Many catalysts are capable of very high levels of asymmetric induction in preferentially forming one enantiomer of a chiral product. The field is beginning to make links with biochemistry with the discovery of enzymes that carry out organometallic catalysis (e.g., acetyl CoA synthase). Ideas drawn from organometallic chemistry have helped interpret the chemistry of metal and metal oxide surfaces, both key actors in heterogeneous catalysis. The field is also creating links with the chemistry of materials because organometallic and metal–organic compounds are increasingly preferred as the precursors for depositing materials on various substrates via thermal decomposition of the metal compound. Nanoscience and nanotechnology are also benefiting with the use of such compounds as the most common precursors for nanoparticles. These small particles of a metal or alloy, with properties quite unlike the bulk material, are finding more and more useful applications in electronic, magnetic, or optical devices or in sensors.

Public concern for the environment has led to the rise of *green chemistry*, with the object of minimizing both energy use and chemical waste in industry

The Organometallic Chemistry of the Transition Metals, Fourth Edition, by Robert H. Crabtree
Copyright © 2005 John Wiley & Sons, Inc.

and commerce. One strategy is *atom economy* in which reactions are chosen that minimize the formation of by-products or unreacted starting materials. For example, rhodium or iridium-based catalysts directly convert MeOH and CO to MeCOOH with no significant by-products. Organometallic catalysis is likely to be a key contributor when climate change become severe enough to force government action to mandate the use of renewable fuels.

The presence of *d* electrons in their valence shell distinguishes the organometallic chemistry of the elements of groups 3–12 of the periodic table, the transition elements, from that of groups 1–2 and 12–18, the main-group elements. Group 12, and to some extent also group 3, often show greater resemblance to the main-group elements.

Transition metal ions can bind *ligands* (L) to give a coordination compound, or *complex* ML_n, as in the familiar aqua ions $[M(OH_2)_6]^{2+}$ (M = V, Cr, Mn, Fe, Co, or Ni). Organometallic chemistry is a subfield of coordination chemistry in which the complex contains an M−C or M−H bond [e.g., $Mo(CO)_6$]. Organometallic species tend to be more covalent, and the metal is often more reduced, than in other coordination compounds. Typical ligands that usually bind to metals in their lower oxidation states are CO, alkenes, and arenes, for example, $Mo(CO)_6$, $(C_6H_6)Cr(CO)_3$, or $Pt(C_2H_4)_3$.

In this chapter we review some fundamental ideas of coordination chemistry, which also apply to organometallic complexes.

1.1   WERNER COMPLEXES

Complexes in which the metal binds to noncarbon ligands have been known longest and are often called *classical* or *Werner complexes* such as $[Co(NH_3)_6]^{3+}$. The simplest metal–ligand bond is perhaps $L_nM−NH_3$, where an ammonia binds to a metal fragment. This fragment will usually also have other ligands, represented here by L_n. The bond consists of the lone pair of electrons present in free NH_3 that are donated to the metal to form the complex. The metal is a polyvalent Lewis acid capable of accepting the lone pairs of several ligands L, which act as Lewis bases.

Stereochemistry

The most common type of complex is ML_6, which adopts an octahedral coordination geometry (**1.1**) based on one of the Pythagorean regular solids. The ligands occupy the six vertices of the octahedron, which allows them to minimize their M−L bonding distances, while maximizing their L···L nonbonding distances. From the point of view of the coordination chemist, it is perhaps unfortunate that Pythagoras decided to name his solids after the number of faces (*octa* = eight) rather than the number of vertices. After ML_6, ML_4 and ML_5 are the next most common types. The solid and dashed wedges in **1.1** indicate bonds located in front of and behind the plane of the paper, respectively.

1.1 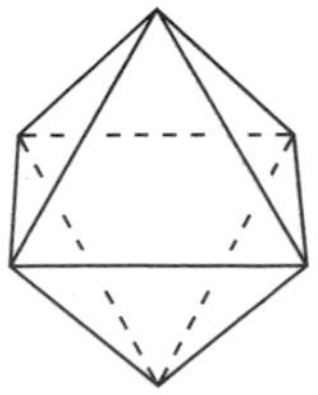   Octahedron

The assembly of metal and ligands that we call a *complex* may have a net ionic charge, in which case it is a complex ion (e.g., $[PtCl_4]^{2-}$). Together with the counterions, we have a complex salt (e.g., $K_2[PtCl_4]$). In some cases both the cation and the anion may be complex, as in the picturesquely named *Magnus' green salt* $[Pt(NH_3)_4][PtCl_4]$. Square brackets are used to enclose the individual complex molecules or ions where necessary to avoid ambiguity.

Those ligands that have a donor atom with more than one lone pair can donate one lone pair to each of two or more metal ions. This gives rise to polynuclear complexes, such as the orange crystalline compound **1.2** (L = PR_3). The bridging group is represented in formulas by using the Greek letter μ (pronounced "mu") as in $[Ru_2(\mu\text{-}Cl)_3(PR_3)_6]^+$. Note how **1.2** can be considered as two octahedral fragments sharing the face that contains the three chloride bridges.

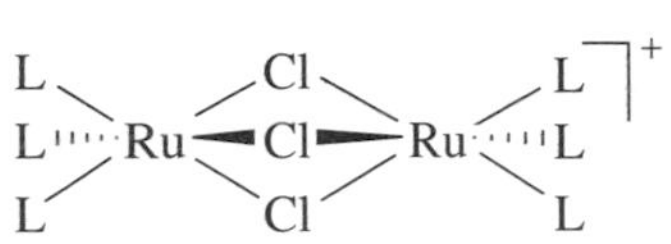

1.2

Chelate Effect

Other ligands can have more than one donor atom, each with its lone pair; an example is ethylenediamine ($NH_2CH_2CH_2NH_2$, often abbreviated "en"). Such ligands most commonly donate both lone pairs to the same metal to give a ring compound, known as a *chelate*, from the Greek word for "claw" (**1.3**). Chelate ligands may be bidentate, such as ethylenediamine, or polydentate, such as **1.4** and **1.5**.

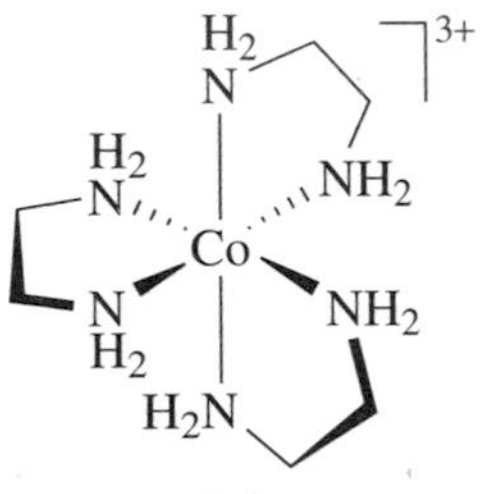

1.3

The early Russian investigator Chugaev first drew attention to the fact that chelating ligands are much less easily displaced from a complex than are monodentate ligands of the same type. The reason is illustrated in Eq. 1.1:

$$[M(NH_3)_6]^{n+} + 3en \longrightarrow [M(en)_3]^{n+} + 6NH_3 \qquad (1.1)$$

Formation of the tris chelate releases six NH_3 molecules so that the total number of particles increases from four to seven. This creates entropy and so favors the chelate form. Each chelate ring usually leads to an additional factor of about 10^5 in the equilibrium constant for reactions such as Eq. 1.1. Equilibrium constants for complex formation are usually called *formation constants*; the higher the value, the more stable the complex.

Chelation not only makes the complex more stable but also forces the donor atoms to take up adjacent or cis sites in the resulting complex. Polydentate chelating ligands with three or more donor atoms also exist. Macrocyclic ligands, such as **1.4** and **1.5** confer an additional increment in the formation constant (the macrocyclic effect); they tend to be given rather lugubrious trivial names, such as *cryptates* (**1.4**) and *sepulchrates* (**1.5**).[1]

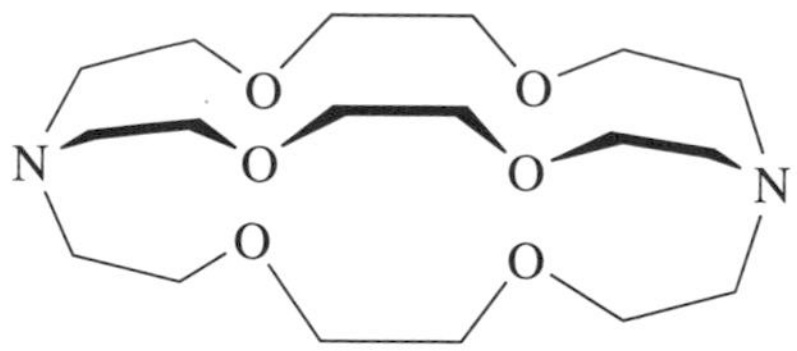

1.4 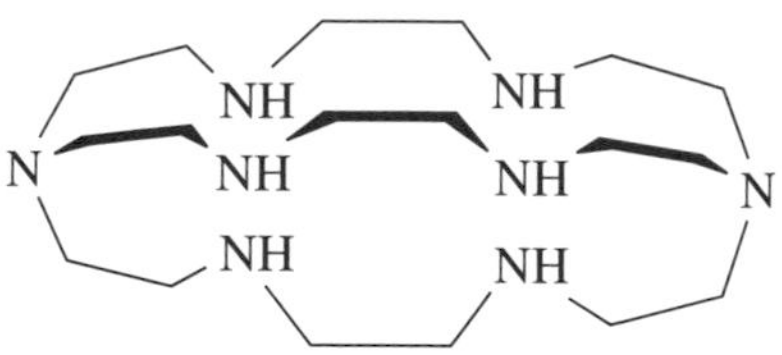   **1.5**

Werner Complexes

Alfred Werner developed the modern picture of coordination complexes in the 20 years that followed 1893, when, as a young scientist, he proposed that in the well-known cobalt ammines (ammonia complexes) the metal ion is surrounded by six ligands in an octahedral array as in **1.6** and **1.7**. In doing so, he was

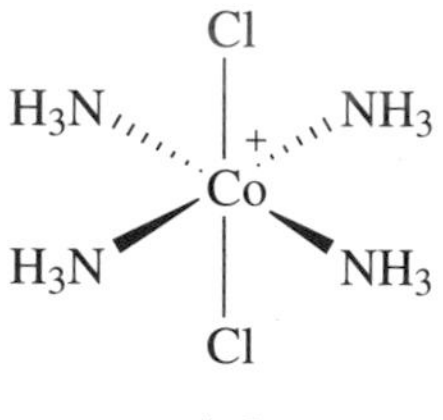

1.6 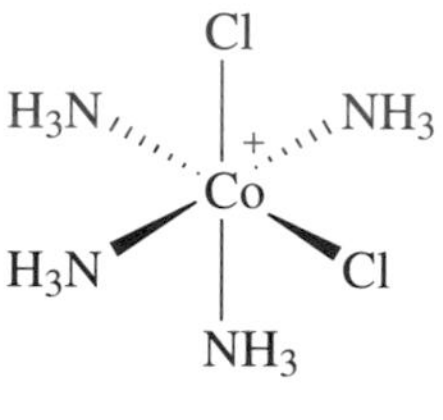   **1.7**

opposing all the major figures in the field, who held that the ligands were bound to one another in chains, and that only the ends of the chains were bound to the metal as in **1.8** and **1.9**. Jørgensen, who led the traditionalists against the

$$
\begin{array}{l}
\text{Co}\!\!<\!\!\overset{\textstyle \text{Cl}}{\text{Cl}} \\
\quad\text{NH}_2\!-\!\text{NH}_2\!-\!\text{NH}_2\!-\!\text{NH}_2\!-\!\text{Cl}
\end{array}
$$

1.8

$$
\begin{array}{l}
\text{Co}\!\!<\!\!\overset{\textstyle \text{Cl}}{\text{Cl}} \\
\quad\text{NH}_2\!-\!\text{NH}_2\!-\!\text{NH}_2\!-\!\text{NH}_2\!-\!\text{Cl}
\end{array}
$$

1.9

Werner insurgency, was not willing to accept that a trivalent metal, Co^{3+}, could form bonds to six groups; in the chain theory, there were never more than three bonds to Co. Each time Werner came up with what he believed to be proof for his theory, Jørgensen would find a way of interpreting the chain theory to fit the new facts. For example, coordination theory predicts that there should be two isomers of $[Co(NH_3)_4Cl_2]^+$ (**1.6** and **1.7**). Up to that time, only a green one had ever been found. We now call this the *trans isomer* (**1.6**) because the two Cl ligands occupy opposite vertices of the octahedron. According to Werner's theory, there should also have been a second isomer, **1.7** (cis), in which the Cl ligands occupy adjacent vertices. Changing the anionic ligand, Werner was able to obtain both green and purple isomers of the nitrite complex $[Co(NH_3)_4(NO_2)_2]^+$. Jørgensen quite reasonably (but wrongly) countered this finding by arguing that the nitrite ligands in the two isomers were simply bound in a different way (*linkage isomers*), via N in one case ($Co-NO_2$) and O ($Co-ONO$) in the other. Werner then showed that there were two isomers of $[Co(en)_2Cl_2]^+$, one green and one purple, in a case where no linkage isomerism was possible. Jørgensen brushed this observation aside by invoking the two chain isomers **1.8** and **1.9** in which the topology of the chains differ.

In 1907, Werner finally succeeded in making the elusive purple isomer of $[Co(NH_3)_4Cl_2]^+$ by an ingenious route (Eq. 1.2) via the carbonate $[Co(NH_3)_4(O_2CO)]$ in which two oxygens of the chelating dianion are necessarily cis. Treatment with HCl at $0°C$ liberates CO_2 and gives the cis dichloride. Jorgensen, receiving a sample of this purple cis complex by mail, conceded defeat.

$$
\begin{array}{ccc}
\left[\begin{array}{c}
\overset{\textstyle \text{O}}{\underset{}{\parallel}} \\
\text{C} \\
\text{O}\diagdown\ \ \diagup\text{O} \\
\text{H}_3\text{N}_{\prime\prime\prime\prime}\ \big|\ _{\prime\prime\prime\prime}\text{O} \\
\text{Co} \\
\text{H}_3\text{N}\diagup\ \big|\ \diagdown\text{NH}_3 \\
\text{NH}_3
\end{array}\right]^{+}
& \xrightarrow{\ \text{HCl}\ } &
\left[\begin{array}{c}
\text{Cl} \\
\text{H}_3\text{N}_{\prime\prime\prime\prime}\ \big|\ _{\prime\prime\prime\prime}\text{Cl} \\
\text{Co} \\
\text{H}_3\text{N}\diagup\ \big|\ \diagdown\text{NH}_3 \\
\text{NH}_3
\end{array}\right]^{+}
\end{array}
\qquad (1.2)
$$

1.10   **1.11**

Finally, Werner resolved optical isomers of some of his compounds of the general type $[Co(en)_2X_2]^{2+}$ (**1.10** and **1.11**). Only an octahedral array can account for the optical isomerism of these complexes. Even this point was challenged on the grounds that only organic compounds can be optically active, and so the optical activity must reside in the organic ligands. Werner responded by resolving a complex (**1.12**) containing only inorganic elements. This species has the extraordinarily high specific rotation of 36,000° and required 1000 recrystallizations to resolve. Werner won the chemistry Nobel Prize for this work in 1913.

1.12

1.2   THE TRANS EFFECT

We now move from complexes of tripositive cobalt, often termed "Co(III) compounds," where the III refers to the +3 oxidation state (Section 2.4) of the central metal, to the case of Pt(II). In the 1920s, Chernaev discovered that certain ligands, L^t, facilitate the departure of a second ligand, L, trans to the first, and their replacement or *substitution*, by an external ligand. Ligands, L^t, that are more effective at this labilization are said to have a higher *trans effect*. We consider in detail how this happens on page 109, for the moment we need only note that the effect is most clearly marked in substitution in Pt(II), and that the highest trans effect ligands form either unusually strong σ bonds, such as $L^t = H^-$, Me^-, or $SnCl_3^-$, or unusually strong π bonds, such as $L^t = CO$, C_2H_4, and thiourea [$(NH_2)_2CS$, a ligand often represented as "tu"].

The same ligands also weaken the trans M–L bonds, as shown by a lengthening of the M–L distances found by X-ray crystallography or by some spectroscopic measure, such as M,L coupling constant in the nuclear magnetic resonance (NMR) spectroscopy (Section 10.4), or the ν(M–L) stretching frequency in the IR (infrared) spectrum (Section 10.9). A change in the ground-state thermodynamic properties, such as these, is usually termed the *trans influence* to distinguish it from the parallel effect on the properties of the transition state for the substitution reaction, which is the trans effect proper, and refers to differences in *rates* of substitution and is therefore a result of a change in the energy difference between the ground state and transition state for the substitution.

Note that Pt(II) adopts a coordination geometry different from that of Co(III). The ligands in these Pt complexes lie at the corners of a square with the metal at the center. This is called the *square planar geometry* (**1.13**).

$$L_{\prime\prime\prime\prime_{\prime}} \qquad _{\prime\prime\prime}L$$
$$Pt$$
$$L \qquad L$$

1.13

An important application of the trans effect is the synthesis of specific isomers of coordination compounds. Equations 1.3 and 1.4 show how the cis and trans isomers of $Pt(NH_3)_2Cl_2$ can be prepared selectively by taking advantage of the trans effect order $Cl > NH_3$, so $L^t = Cl$. This example is also of practical interest because the cis isomer is an important antitumor drug, but the trans isomer is ineffective. In each case the first step of the substitution can give only one isomer. In Eq. 1.3, the cis isomer is formed in the second step because the Cl trans to Cl is more labile than the Cl trans to the lower trans effect ligand, ammonia. On the other hand, in Eq. 1.4, the first Cl to substitute labilizes the ammonia trans to itself to give the trans dichloride as final product.

$$\begin{bmatrix} Cl & Cl \\ & Pt & \\ Cl & Cl \end{bmatrix}^{2-} \xrightarrow{NH_3} \begin{bmatrix} Cl & NH_3 \\ & Pt & \\ Cl & Cl \end{bmatrix}^- \xrightarrow{NH_3} \begin{matrix} Cl & NH_3 \\ & Pt & \\ Cl & NH_3 \end{matrix} \qquad (1.3)$$

$$\begin{bmatrix} H_3N & NH_3 \\ & Pt & \\ H_3N & NH_3 \end{bmatrix}^{2+} \xrightarrow{Cl^-} \begin{bmatrix} H_3N & Cl \\ & Pt & \\ H_3N & NH_3 \end{bmatrix}^+ \xrightarrow{Cl^-} \begin{matrix} H_3N & Cl \\ & Pt & \\ Cl & NH_3 \end{matrix} \qquad (1.4)$$

A trans effect series for a typical Pt(II) system is given below. The order can change somewhat for different metals and oxidation states.

$$OH^- < NH_3 < Cl^- < Br^- < CN^-, CO, C_2H_4, CH_3^- < I^- < PR_3 < H^-$$

$\leftarrow$ low trans effect   high trans effect $\rightarrow$

1.3   SOFT VERSUS HARD LIGANDS

Table 1.1 shows formation constants for different metal ion (acid)–halide ligand (base) combinations,[2] where large positive numbers mean strong binding. The series of halide ions starts with F^-, termed *hard* because it is small, difficult to polarize, and forms predominantly ionic bonds. It binds best to a hard cation, H^+, which is also small and difficult to polarize. This hard–hard combination is therefore a good one.

The halide series ends with I^-, termed *soft* because it is large, easy to polarize, and forms predominantly covalent bonds. It binds best to a soft cation, Hg^{2+}, which is also large and easy to polarize. In this context, high polarizability means that electrons from each partner readily engage in covalent bonding. The Hg^{2+}/I^- soft–soft combination is therefore a very good one—by far the best in the table—and dominated by covalent bonding.[3]

Soft bases have lone pairs on atoms of the second or later row of the periodic table (e.g., Cl^-, Br^-, PPh_3) or have double or triple bonds (e.g., CN^-, C_2H_4, benzene). Soft acids can also come from the second or later row of the periodic table (e.g., Hg^{2+}) or contain atoms that are relatively electropositive (e.g., BH_3) or are metals in a low (≤ 2) oxidation state [e.g., Ni(0), Re(I), Pt(II), Ti(II)]. An important part of organometallic chemistry is dominated by soft–soft interactions (e.g., metal carbonyl, alkene, and arene chemistry).

TABLE 1.1   Hard and Soft Acids and Bases: Some Formation Constants[a]

Metal Ion (Acid)	Ligand (Base)			
	F^- (Hard)	Cl^-	Br^-	I^- (Soft)
H^+ (hard)	3	−7	−9	−9.5
Zn^{2+}	0.7	−0.2	−0.6	−1.3
Cu^{2+}	1.2	0.05	−0.03	—
Hg^{2+} (soft)	1.03	6.74	8.94	12.87

[a]The values are the negative logarithms of the equilibrium constant for $[M.aq]^{n+} + X^- \rightleftharpoons [MX.aq]^{(n-1)+}$ and show how H^+ and Zn^{2+} are hard acids, forming stronger complexes with F^- than with Cl^-, Br^-, or I^-. Cu^{2+} is a borderline case, and Hg^{2+} is a very soft acid, forming much stronger complexes with the more polarizable halide ions.

- High-trans-effect ligands labilize the ligand located opposite to themselves.
- Hard ligands have first-row donors and no multiple bonds (e.g., NH_3).
- Soft ligands have second- or later-row donors and/or multiple bonds (e.g., PH_3 or CO).

1.4   THE CRYSTAL FIELD

An important advance in understanding the spectra, structure, and magnetism of transition metal complexes is provided by the *crystal field* model. The idea is to find out how the *d* orbitals of the transition metal are affected by the presence of the ligands. To do this, we make the simplest possible assumption about the ligands—they act as negative charges. For Cl^- as a ligand, we just think of the net negative charge on the ion; for NH_3, we think of the lone pair on nitrogen acting as a local concentration of negative charge. If we imagine the metal ion isolated in space, then the *d* orbitals are *degenerate* (have the same energy). As the ligands L approach the metal from the six octahedral directions $\pm x$, $\pm y$, and $\pm z$, the *d* orbitals take the form shown in Fig. 1.1. Those *d* orbitals that point toward the L groups ($d_{x^2-y^2}$ and d_{z^2}) are destabilized by the negative charge of the ligands and move to higher energy. Those that point away from L (d_{xy}, d_{yz}, and d_{xz}) are less destabilized.

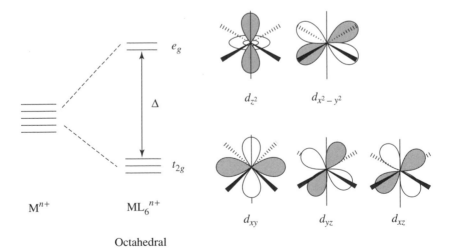

FIGURE 1.1   Effect on the *d* orbitals of bringing up six ligands along the $\pm x$, $\pm y$, and $\pm z$ directions. In this figure, shading represents the symmetry (not the occupation) of the *d* orbitals; shaded parts have the same sign of ψ.

The pair of orbitals that are most strongly destabilized are often identified by their symmetry label, e_g, or simply as d_σ, because they point along the M–L σ-bonding directions. The three more stable orbitals have the label t_{2g}, or simply d_π; these point away from the ligand directions but can form π bonds with the ligands. The magnitude of the energy difference between the d_σ and d_π set, usually called the *crystal field splitting*, and labeled Δ (or sometimes $10\,Dq$) depends on the value of the effective negative charge and therefore on the nature of the ligands. Higher Δ leads to stronger M–L bonds.

High Spin Versus Low Spin

Cobalt, which is in group 9 of the periodic table, has the electron configuration $[Ar]4s^2 3d^7$ in the free atom, with nine valence electrons. Once the atom forms a complex, however, the d orbitals become more stable as a result of metal–ligand bonding, and the electron configuration becomes $[Ar]4s^0 3d^9$ for the case of a Co(0) complex, or $[Ar]3s^0 4d^6$ for Co(III), usually shortened to d^9 and d^6, respectively. This picture explains why Co^{3+}, the metal ion Werner studied, has such a strong preference for the octahedral geometry. With its d^6 configuration, six electrons just fill the three low-lying d_π orbitals of the crystal field diagram and leave the d_σ empty. This is a particularly stable arrangement, and other d^6 metals, Mo(0), Re(I), Fe(II), Ir(III), and Pt(IV) also show a very strong preference for the octahedral geometry. Indeed, low spin d^6 is by far the commonest type of metal complex in organometallic chemistry. In spite of the high tendency to spin-pair the electrons in the d^6 configuration (to give the *low-spin* form $t_{2g}^6 e_g^0$), if the ligand field splitting is small enough, then the electrons may occasionally rearrange to give the *high-spin* form $t_{2g}^4 e_g^2$. In the high-spin form all the unpaired spins are aligned, as prescribed for the free ion by Hund's rule. This is shown in Fig. 1.2. The factor that favors the high-spin form is the fact that fewer electrons are paired up in the same orbitals and so the electron–electron repulsions are reduced. On the other hand, if Δ becomes large enough, then the energy gained by dropping from the e_g to the t_{2g} level will be

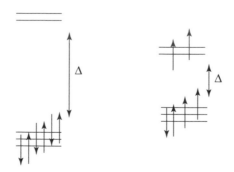

FIGURE 1.2 In a d^6 metal ion, both low- and high-spin complexes are possible depending on the value of Δ. A high Δ leads to the low-spin form.

sufficient to drive the electrons into pairing up. The spin state of the complex can usually be determined by measuring the magnetic moment of the complex. This is done by weighing a sample of the complex in a magnetic field gradient. In the low-spin form of a d^6 ion, the molecule is *diamagnetic*, that is, it is very weakly repelled by the field. This behavior is exactly the same as that found for the vast majority of organic compounds, which are also spin-paired. On the other hand, the high-spin form is *paramagnetic*, in which case it is attracted into the field because there are unpaired electrons. The complex does not itself form a permanent magnet as does a piece of iron or nickel (this property is called *ferromagnetism*) because the spins are not aligned in the crystal in the absence of an external field, but they do respond to the external field by lining up together when we measure the magnetic moment.

Although the great majority of organometallic complexes are diamagnetic, because Δ is usually large in these complexes, we should not lose sight of the possibility that any given complex or reaction intermediate may be paramagnetic. This will always be the case for molecules such as d^5 $V(CO)_6$, which have an uneven number of electrons. For molecules with an even number of electrons, a high-spin configuration is more likely for the first row metals, where Δ tends to be smaller than in the later rows. Sometimes the low- and high-spin isomers have almost exactly the same energy. Each state can now be populated, and the relative populations of the two states vary with temperature; this happens for $Fe(dpe)_2Cl_2$, for example.

Inert Versus Labile Coordination

In an octahedral d^7 ion we are obliged to place one electron in the higher-energy (less stable) d_σ level to give the configuration $t_{2g}^6 e_g^1$, to make the complex paramagnetic (Fig. 1.3). The net stabilization, the *crystal field stabilization energy* (CFSE) of such a system will also be less than for d^6 (low spin), where we can put all the electrons into the more stable t_{2g} level. This is reflected in the chemistry of octahedral d^7 ions [e.g., Co(II)], which are more reactive than their d^6 analogs. For example, they undergo ligand dissociation much more readily. The reason

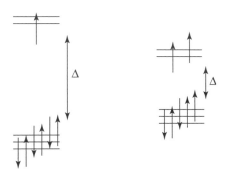

FIGURE 1.3 A d^7 octahedral ion is paramagnetic even in the low-spin form.

is that the d_σ levels are M−L σ-antibonding in character (Section 1.5). Werner studied Co(III) because the ligands tend to stay put. This is why Co(III) and other low-spin d^6 ions are often referred to as *coordinatively inert*; d^3 ions such as Cr(III) are also coordination inert because the t_{2g} level is now exactly half-filled, another favorable situation. On the other hand, Co(II) and other non-d^6 and -d^3 ions can be *coordinatively labile*. The second- and third-row transition metals form much more inert complexes because of their higher Δ and CFSE.

Low- Versus High-Field Ligands

The colors of transition metal ions often arise from the absorption of light that corresponds to the d_π−d_σ energy gap, Δ. The spectrum of the complex can then give a direct measure of this gap and, therefore, of the crystal field strength of the ligands. So-called *high-field ligands* such as CO and C_2H_4 give rise to a large value of Δ. *Low-field ligands*, such as H_2O or NH_3, can give such a low Δ that the spin pairing is lost and even the d^6 configuration can become paramagnetic (Fig. 1.2, right side).

The *spectrochemical series* of ligands, which lists the common ligands in order of increasing Δ, allows us to see the general trend that π-donor ligands such as halide or H_2O tend to be weak-field and π-acceptor ligands such as CO tend to be strong-field ligands as discussed in Section 1.6. These π effects are not the whole story, however, because H, which has no π-donor or acceptor properties at all, is nevertheless a very strong field ligand, probably because of the very strong M−H σ bonds it forms.

$$I^- < Br^- < Cl^- < F^- < H_2O < NH_3 < PPh_3 < CO, H < SnCl_3^-$$

$\leftarrow$ low Δ high Δ $\rightarrow$

$\leftarrow$ π donor π acceptor/strong σ donor $\rightarrow$

Hydrides and carbonyls therefore have very strong M−L bonds (L = H, CO) and have a very strong tendency to give diamagnetic complexes. High-field ligands, such as high-trans-effect ligands, tend to form strong σ and/or π bonds, but the precise order is significantly different in the two series.

Odd Versus Even d^n Configurations

If a molecule has an odd number of electrons, not all of them can be paired up. An odd d^n configuration, such as d^7 (e.g., [Re(CO)$_3$(PCy$_3$)$_2$]), therefore, guarantees paramagnetism if we are dealing with a mononuclear complex—one containing only a single metal atom. In dinuclear complexes, the odd electrons on each metal may pair up, however, as in the diamagnetic d^7−d^7 dimer, [(OC)$_5$Re−Re(CO)$_5$]. Complexes with an even d^n configuration can be diamagnetic or paramagnetic depending on whether they are high or low spin, but low-spin diamagnetic complexes are much more common in organometallic chemistry because the most commonly encountered ligands are high field.

Other Geometries

In 4 coordination, two geometries are common, tetrahedral and square planar, for which the crystal field splitting patterns are shown in Fig. 1.4. For the same ligand set, the tetrahedral splitting parameter is smaller than that for the octahedral geometry by a factor of $\frac{2}{3}$ because we now have only four ligands, not six, and so the chance of having a high-spin species is greater. The ordering of the levels is also reversed; three increase and only two decrease in energy. This is because the d_{xy}, d_{yz}, and d_{xz} orbitals now point toward and the $d_{x^2-y^2}$ and d_{z^2} orbitals away from the ligands. The d^{10} ions [e.g., Zn(II), Pt(0), Cu(I)] are often tetrahedral. The square planar splitting pattern is also shown. This geometry tends to be adopted by diamagnetic d^8 ions such as Au(III), Ni(II), Pd(II) or Pt(II), and Rh(I) or Ir(I); it is also common for paramagnetic d^9, such as Cu(II).

For a given geometry and ligand set, metal ions tend to have different values of Δ. For example, first-row metals and metals in a low oxidation state tend to have low Δ, while second- and third-row metals and metals in a high oxidation state tend to have high Δ. The trend is illustrated by the *spectrochemical series* of metal ions in order of increasing Δ.

$$Mn^{2+} < V^{2+} < Co^{2+} < Fe^{2+} < Ni^{2+} < Fe^{3+} < Co^{3+} < Mn^{4+}$$

$$< Rh^{3+} < Ru^{3+} < Pd^{4+} < Ir^{3+} < Pt^{4+}$$

$\leftarrow$ low Δ high Δ $\rightarrow$

$\leftarrow$ low valent, first row   high valent, third row $\rightarrow$

Third-row metals therefore tend to form stronger M−L bonds and more thermally stable complexes and are also more likely to give diamagnetic complexes. Comparison of the same metal and ligand set in different oxidation states is complicated by the fact that low oxidation states are usually accessible only with strong-field ligands that tend to give a high Δ (see the spectrochemical series of ligands on page 12).

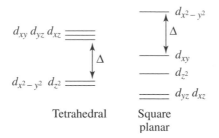

FIGURE 1.4 Crystal field splitting patterns for the common 4-coordinate geometries: tetrahedral and square planar. For the square planar arrangement, the z axis is conventionally taken to be perpendicular to the square plane.

This is why third-row metals tend to be used when isolation of stable compounds is the aim. When catalysis is the goal (Chapter 9), the intermediates involved have to be reactive and therefore relatively less stable, and first- or second-row metals are sometimes preferred.

Isoconfigurational Ions

Transition metals tend to be treated as a group rather than as individual elements. One reason is that d^n ions of the same configuration (e.g., $n = 6$) show important similarities independent of the identity of the element. This means that d^6 Co(III) is closer in properties to d^6 Fe(II) than to d^7 Co(II). The variable valency of the transition metals leads to many cases of isoconfigurational ions.

1.5   THE LIGAND FIELD

The crystal field picture gives a useful qualitative understanding, but, once having established what to expect, we turn to the more sophisticated *ligand field* model, really a conventional molecular orbital, or MO, picture for accurate electronic structure calculations. In this model (Fig. 1.5), we consider the s, the three p, and the five d orbitals of the valence shell of the isolated ion as well as the six lone pair orbitals of a set of pure σ-donor ligands in an octahedron around the metal. Six of the metal orbitals, the s, the three p, and the two d_σ, which we will call the dsp_σ set, find symmetry matches in the six ligand lone-pair orbitals. In combining the six metal orbitals with the six ligand orbitals, we make a bonding set of six (the M$-$L σ bonds) that are stabilized, and an antibonding set of six (the M$-$L σ^* levels) that are destabilized when the six L groups approach to bonding distance. The remaining three d orbitals, the d_π set, do not overlap with the ligand orbitals, and remain nonbonding. In a d^6 ion, we have 6e (six electrons) from Co^{3+} and 12e from the ligands, giving 18e in all. This means that all the levels up to and including the d_π set are filled, and the M$-$L σ^* levels remain unfilled. Note that we can identify the familiar crystal field splitting pattern in the d_π and two of the M$-$L σ^* levels. The Δ splitting will increase as the strength of the M$-$L σ bonds increase. The bond strength is the analog of the effective charge in the crystal field model. In the ligand field picture, high-field ligands are ones that form strong σ bonds. We can now see that a d_σ orbital of the crystal field picture is an M$-$L σ-antibonding orbital.

The L lone pairs start out in free L as pure ligand electrons but become bonding electron pairs shared between L and M when the M$-$L σ bonds are formed; these are the 6 lowest orbitals in Fig. 1.5 and are always completely filled (12 electrons). Each M$-$L σ-bonding MO is formed by the combination of the ligand lone pair, L(σ), with M(d_σ) and has both metal and ligand character, but L(σ) predominates. Any MO will more closely resemble the parent atomic orbital that lies closest in energy to it, and L(σ) almost always lies below M(d_σ) and therefore closer to the M$-$L σ-bonding orbitals. This means that electrons

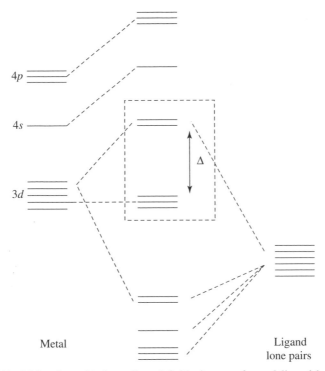

FIGURE 1.5 Molecular orbital, or ligand field picture, of metal ligand bonding in an octahedral ML_6 complex. The box contains the d orbitals.

that were purely L lone pairs in the free ligand gain some metal character in the complex; in other words, the $L(\sigma)$ lone pairs are partially transferred to the metal. As L becomes more basic, the energy of the $L(\sigma)$ orbital increases, and the extent of electron transfer will increase. An orbital that is higher in energy will appear higher in the MO diagram and will tend to occupy a larger volume of space, and any electrons in it will tend to be less stable and more available for chemical bonding or removal by ionization.

Ligands are generally *nucleophilic* because they have available (high-lying) electron lone pairs. The metal ion is *electrophilic* because it has available (low-lying) empty d orbitals. The nucleophilic ligands, which are lone-pair donors, attack the electrophilic metal, an acceptor for lone pairs, to give the metal complex. Metal ions can accept multiple lone pairs so that the complex formed is not just ML but ML_n $(n = 2–9)$.

1.6   BACK BONDING

Ligands such as NH_3 are good σ donors but are not significant π acceptors. CO, in contrast, is an example of a good π acceptor. Such π-*acid* ligands are of very great importance in organometallic chemistry. They tend to be very high

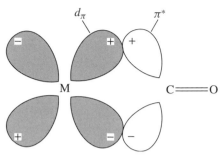

FIGURE 1.6 Overlap between a filled metal d_π orbital and an empty CO π^* orbital to give the π component of the M—CO bond. The shading refers to occupancy of the orbitals and the $+$ and $-$ signs, to the symmetry. The M—CO σ bond is formed by the donation of a lone pair on C into an empty d_σ orbital on the metal (not shown).

field ligands and form strong M—L bonds. All have empty orbitals of the right symmetry to overlap with a filled d_π orbital of the metal. In the case of CO, this orbital is the CO π^*. Figure 1.6 shows how overlap takes place to form the M—C π bond. It may seem paradoxical that an antibonding orbital such as the $\pi^*(CO)$ can form a bond, but this orbital is antibonding only with respect to C and O and can still be bonding with respect to M and C.

We can make the ligand field diagram of Fig. 1.5 appropriate for the case of $W(CO)_6$ by including the π^* levels of CO (Fig. 1.7). The d_π set of levels still find no match with the six $CO(\sigma)$ orbitals, which are lone pairs on C. They do interact strongly with the empty CO π^* levels. Since the Md_π set are filled in this d^6 complex, the d_π electrons that were metal centered now spend some of their time on the ligands: This means that the metal has donated some electron density to the ligands. This *back bonding* is a key feature of M—L bonds where L is unsaturated (i.e., has multiple bonds). Note that this can only happen in d^2 or higher configurations; a d^0 ion such as Ti^{4+} cannot back bond and seldom forms stable carbonyl complexes.

As antibonding orbitals, the CO π^* levels are high in energy, but they are able to stabilize the d_π set as shown in Fig. 1.7. This has two important consequences: (1) The ligand field splitting parameter Δ rises, explaining why π-bonding ligands have such a strong ligand field; and (2) back bonding allows electron density on the metal as it makes its way back to the ligands. This, in turn, allows low-valent or zero-valent metals to form complexes. Such metals are in a reduced state and already have a high electron density. (They are said to be very *basic* or *electron rich*.) They cannot accept further electrons from pure σ donors; this is why $W(NH_3)_6$ is not a stable compound. By back bonding, the metal can get rid of some of this excess electron density. In $W(CO)_6$ back bonding is so effective that the compound is air stable and relatively unreactive; the CO groups have so stabilized the electrons that they have no tendency to be abstracted by air as an oxidant. In $W(PMe_3)_6$, in contrast, back bonding is inefficient and the compound exists but is very air sensitive and reactive.

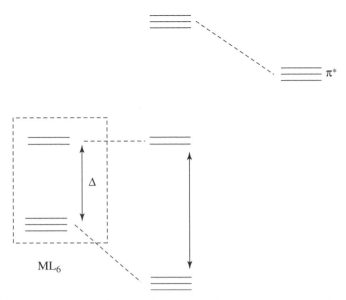

FIGURE 1.7 Effect of "turning on" the π interaction between a π-acceptor ligand and the metal. The unoccupied, and relatively unstable π^* orbitals of the ligand are shown on the right. Their effect is to stabilize the filled d_π orbitals of the complex and so increase Δ. In $W(CO)_6$, the lowest three orbitals are filled.

Spectroscopic and theoretical studies show that for CO this π back donation is usually comparable to or greater than the CO-to-metal electron donation in the σ bond. One of the most direct arguments is structural. The M=C bond in metal carbonyls is usually substantially shorter than an M—C single bond. This is easiest to test when both types of bond are present in the same complex, such as $CpMo(CO)_3Me$, where M—C is 2.38 Å, and M=CO is 1.99 Å. We have to remember that a putative M—CO single bond would be shorter than 2.38 Å by about 0.07 Å, to allow for the higher s character (and therefore shorter bond length) of the sp hybrid on CO compared to the sp^3 hybrid of the methyl group. The remaining shortening of 0.32 Å is still substantial.

To confirm that it really is the π^* orbital of CO that is involved in the back bonding, we turn to IR spectroscopy. If CO were bound to the metal by its carbon lone pair, nonbonding with respect to CO, then the $\nu(CO)$ frequency in the complex would differ very little from that in free CO. The compound BH_3, which is as pure as a σ acceptor as will bind to CO, shows a slight shift of $\nu(CO)$ to higher energy: free CO, 2149 cm^{-1}; H_3B—CO, 2178 cm^{-1}. Metal complexes, in contrast, show $\nu(CO)$ coordination shifts of hundreds of wavenumbers to lower energy, consistent with the weakening of the C—O bond that would be expected if the π^* orbital were being filled [e.g., $Cr(CO)_6$, $\nu(CO) = 2000$ cm^{-1}]. Not only is there a coordination shift, but the shift is larger in cases where we would expect stronger back donation and vice versa. A net positive charge raises $\nu(CO)$, and a net negative charge lowers it [e.g., $V(CO)_6^-$, 1860 cm^{-1};

$Mn(CO)_6{}^+$, 2090 cm^{-1}]. The effect of replacing three π-acceptor COs by the three pure σ-donor nitrogens of the tren ligand ($H_2NCH_2CH_2NHCH_2CH_2NH_2$) is almost as great as changing the net ionic charge by one unit [e.g., $Cr(tren)(CO)_3$, 1880 cm^{-1}]. This makes $v(CO)$ a good indicator of how electron rich a metal is, and it often correlates well with other ways of estimating nucleophilic character, such as the ease of removing an electron.[4]

Series of compounds such as $V(CO)_6{}^-$, $Cr(CO)_6$, and $Mn(CO)_6{}^+$ are said to be *isoelectronic complexes* because they have the same number of electrons distributed in very similar structures. Isoelectronic ligands are CO and NO$^+$ or CO and CN$^-$, for example. Strictly speaking, CO and CS are not isoelectronic, but as the difference between O and S lies in the number of core levels, while the valence shell is the same, the term *isoelectronic* is often extended to cover such pairs. A comparison of isoelectronic complexes or ligands can be useful in making analogies and pointing out contrasts.[5]

The dipole moments of a variety of coordination compounds show that the bond moments of the M−L bonds of most σ-donor ligands are about 4 D, with the donor atom positive. In contrast, metal carbonyls show an M−C bond moment that is essentially zero because the M→L back donation and L→M direct donation, together with CO polarization (Section 2.6), cancel out. Formation of the M−CO bond weakens the C−O bond relative to free CO. This will still lead to a stable complex as long as the energy gained from the M−C bond exceeds the loss in C−O. Bond weakening in L on binding is a very common feature in many M−L systems.

Frontier Orbitals

The picture for CO holds with slight modifications for a whole series of π acceptor (or soft) ligands, such as alkenes, alkynes, arenes, carbenes, carbynes, NO, N_2, and PF$_3$. Each has a filled orbital that acts as a σ donor and an empty orbital that acts as a π acceptor. These orbitals are almost always the highest occupied (*HOMO*) and lowest unoccupied molecular orbitals (*LUMO*) of L, respectively. The HOMO of L is a donor to the LUMO of the metal, which is normally d_σ. The LUMO of the ligand accepts back donation from a filled d_π orbital of the metal. The HOMO and LUMO of each fragment, the so-called *frontier orbitals*, nearly always dominate the bonding. This is because strong interactions between orbitals require not only that the overlap between the orbitals be large but also that the energy separation be small. The HOMO of each fragment, M and L, is usually closest in energy to the LUMO of the partner fragment than to any other vacant orbital of the partner. Strong bonding is expected if the HOMO–LUMO gap of both partners is small. A small HOMO–LUMO gap usually makes a ligand soft because it is a good π acceptor, and a d^6 metal soft because it is a good π donor.

π-Donor Ligands

Ligands such as OR$^-$, F$^-$, and Cl$^-$ are π donors as a result of the lone pairs that are left after one lone pair has formed the M−L σ bond. Instead of stabilizing the

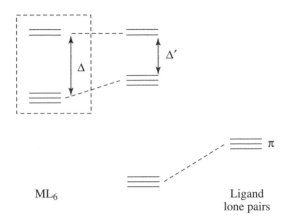

FIGURE 1.8 Effect of "turning on" the π interaction between a π-donor ligand and the metal. The occupied, and relatively stable, lone-pair (π) orbitals of the ligand are shown on the right. Their effect is to destabilize the filled d_π orbitals of the complex and so decrease Δ. This is effectively a repulsion between two lone pairs, one on the metal and the other on the ligand.

d_π electrons of a d^6 ion as does a π acceptor, these electrons are now destabilized by what is effectively a repulsion between two filled orbitals. This lowers Δ, as shown in Fig. 1.8, and leads to a weaker M−L bond than in the π-acceptor case (e.g., $CoF_6{}^{3-}$). Lone pairs on electronegative atoms such as Cl and O are much more stable than the $M(d_\pi)$ level, and this is why they are lower in Fig. 1.8 than are the π^* orbitals in Fig. 1.7. If the metal has empty d_π orbitals, as in the d^0 ion Ti^{4+}, π donation from the ligand to the metal d_π orbitals now leads to stronger metal–ligand bonding; d^0 metals therefore form particularly strong bonds with π-donor ligands [e.g., $W(OMe)_6$, $[TiF_6]^{2-}$].

- Coordination inert cases include d^6 octahedral low spin and d^3 octahedral.
- In the commonest geometry, octahedral, d orbitals split into a 3-below-2 pattern.
- The splitting varies as: 3rd row > 2nd row > 1st row metal and high-field > low-field ligand.
- Ligands with back bonding (and also hydride) are soft and high field.

1.7   ELECTRONEUTRALITY

In 1948 Pauling proposed the powerful *electroneutrality principle*. This says that the atoms in molecules arrange themselves so that their net charges fall within rather narrow limits, from about $+1$ to -1 overall. In fact, the range for any given element is likely to be narrower than this, and tends toward a preferred charge, which differs according to the electronegativity of the element concerned. The nonmetals, such as C, N, or O, tend to be closer to -1, and the metals, such as Li, Mg, and Fe, tend to be closer to $+1$. This implies that as far as electroneutrality arguments go, an element will bond best to other elements that have complementary preferred charges. In this way, each can satisfy the other. An electropositive element prefers an electronegative one, as in the compounds NaCl and TiO_2, and elements with an intermediate electronegativity tend to prefer each other, as in HgS and Au metal. An isolated Co^{3+} ion is not a electroneutral species, as it has an excessively high positive charge. In its compounds it will therefore seek good electron donors as ligands, such as O^{2-} in Co_2O_3, or NH_3, in the ammine (NH_3) complexes. On the other hand, an isolated W(0) atom is too electron rich for its electronegativity, so it will prefer net electron-attracting ligands such as CO that can remove electron density.

Trends with Oxidation State

There is a deeper reason why the d orbitals of transition metals are available for back donation only in electron-rich complexes. Co(III), for example, has a filled d_π level, but Co(III) does not bind CO because the d_π orbital is too low in energy and therefore not sufficiently basic. The reason is that the s, p, and d orbitals respond differently to a change in the charge on the metal. If the metal is in a high oxidation state, like Co(III), then there are electron "holes" in the valence shell compared with the neutral atom. This means that the valence shell of the ion is positive with respect to the situation in the atom. Since d orbitals tend to have their maximum electron density far away from the nucleus (because they have two planar nodes or planes of zero electron density that pass through the nucleus), p orbitals reach their maximum somewhat closer to the nucleus (one planar node), and s orbitals reach their maximum at the nucleus (no planar nodes), the orbitals will be less sensitive to the $3+$ change in the net charge that took place on going from Co(0) to Co(III), in the order $d > p > s$. In other words, the d orbitals will be much more strongly stabilized than the others on going from the atom to the ion. This is why the atomic electron configuration for the transition metals involves s-orbital occupation (e.g., Co, $d^7 s^2$), but the configuration of the ion is d^6, not $d^4 s^2$. On the other hand, the more electron rich (i.e., the more reduced, or low oxidation state) the metal complex, the less positive will be the charge on the metal. This will destabilize the d orbitals and make them more available for back donation.

Periodic Trends

We also alter the orbital energies as we go from left to right in the transition series. For each step to the right, a proton is added to the nucleus. This extra positive charge stabilizes all the orbitals. The earlier metals are more electropositive because it is easier to remove electrons from their less stable energy levels. The sensitivity of the orbitals to this change is different from what we saw above. This time the order is $d \sim s > p$ because the s orbital, having a maximum electron density at the nucleus, is more stabilized by the extra protons that we add for each step to the right in the periodic table, than are the p orbitals, which have a planar node at the nucleus. The d orbitals are stabilized because of their lower principal quantum number (e.g., $3d$ versus $4s$ and $4p$ for Fe). The special property of the transition metals is that all three types of orbital are in the valence shell and have similar energies so they are neither too stable nor too unstable to contribute significantly to the bonding. Metal carbonyls, for example, are most stable for groups 4–10 because CO requires d-orbital participation to bind effectively.

There is a large difference between a d^0 state and a d^2 state, both common in the early transition metals [e.g., d^0 Ti(IV) and a d^2 Ti(II)]. The d^0 oxidation state cannot back bond because it lacks d electrons, while a d^2 state often has an exceptionally high back-bonding power because early in the transition series the d orbitals are relatively unstable for the reasons mentioned above. The d^0 Ti(IV) species $(C_5H_5)_2TiCl_2$ therefore does not react with CO at all, while the corresponding d^2 Ti(II) fragment, $(C_5H_5)_2Ti$, forms a very stable monocarbonyl, $(C_5H_5)_2Ti(CO)$, with a very low $\nu(CO)$, indicating very strong back bonding.

Finally, as we go down a group from the first-row transition element to the second row, the outer valence electrons become more and more *shielded* from the nucleus by the extra shell of electrons that has been added. They are therefore more easily lost, and the heavier element will be the more basic and more electronegative, and high oxidation states will be more stable. This trend also extends to the third row, but as the f electrons that were added to build up the lanthanide elements are not as effective as s, p, or even d electrons in shielding the valence electrons from the nucleus, there is a smaller change on going from the second- to the third-row elements than was the case for moving from the first row to the second. Compare, for example, Cr(VI) in Na_2CrO_4 and Mn(VII) in $KMnO_4$; both are powerful oxidizing agents, with their stable analogs in the second and third rows, Na_2MoO_4, Na_2WO_4, and $KReO_4$, which are only very weakly oxidizing. Similarly, the increase in covalent radii is larger on going from the first to the second row than it is on going from the second to the third. This is termed the *lanthanide contraction*.

Ionic compounds with excessively high positive or negative net ionic charges are not normally formed. The great majority of compounds are neutral, net charges of ± 1 are not uncommon, but net ionic charges of ± 2 or greater are increasingly rare unless there is some special reason to expect them, such as the presence of several metals to share the ionic charge.

1.8   TYPES OF LIGAND

Most ligands form the M—L σ bond by using a lone pair, that is, a pair of electrons that are nonbonding in the free ligand. For ligands such as PR_3 or pyridine, these lone pairs are often the HOMO and the most basic electrons in the molecule. Classical Werner coordination complexes always involve lone-pair donor ligands.

There are two other types of ligand found in organometallic compounds, π and σ, of which C_2H_4 and H_2 are typical examples.

π Complexes

Ethylene has no lone pairs, yet it binds strongly to low-valent metals. In this case the HOMO is the C=C π bond, and it is these electrons that form the M–L σ bond, as shown in Fig. 1.9a, hence the term π-complex. The arrow marked "1" represents the π-bonding electron pair of ethylene being donated to the metal. There is also a back-bonding component (marked "2") where the π^* orbital of ethylene plays the role of acceptor. Since the C=C π bond lies both above and below the molecular plane, the metal has to bind out of the C_2H_4 plane, where the electrons are. This type of binding is represented as $(\eta^2\text{-}C_2H_4)$ (pronounced "eta–two ethylene") where η represents the *hapticity* of the ligand, defined as the number of atoms in the ligand bonded to the metal.

σ Complexes

Molecular hydrogen has neither a lone pair nor a π bond, yet it also binds as an intact molecule to metals in such complexes as $[W(\eta^2\text{-}H_2)(CO)_3L_2]$. The only available electron pair is the H–H σ bond, and this becomes the donor ("3" in Fig. 1.9b). Back donation in this case ("4" in Fig. 1.9b) is accepted by the H_2 σ^* orbital. The metal binds side-on to H_2 to maximize σ–d_σ overlap. Related σ *complexes*[6] are formed with C–H, Si–H, B–H, and M–H bonds. In general, the basicity of electron pairs decreases in the following order: lone pairs > π-bonding pairs > σ-bonding pairs, because being part of a bond stabilizes electrons. The usual order of binding ability is therefore as follows: lone-pair donor > π donor > σ donor.

M–L Bonding

For lone-pair donors the M–L π bond can have 2e and be attractive, as we saw for M–CO (M = d^6 metal, Figs. 1.6 and 1.7) or 4e and be repulsive, as is

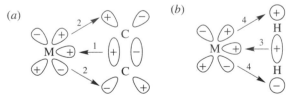

FIGURE 1.9 (a) Bonding of a π-bond donor, ethylene, to a metal. The arrow labeled "1" represents electron donation from the filled C=C π bond to the empty d_σ orbital on the metal; "2" represents the back donation from the filled M(d_π) orbital to the empty C=C π^*. (b) Bonding of a σ-bond donor, hydrogen, to a metal. The label "3" represents electron donation from the filled H–H σ bond to the empty d_σ orbital on the metal, and "4" represents the back donation from the filled M(d_π) orbital to the empty H–H σ^*. Only one of the four lobes of the d_σ orbital is shown.

the case for $M-F^-$ ($M = d^6$ metal, Fig. 1.8). For σ and π donors, the $M-L$ π bond is nearly always attractive because if it were not, L would not bind strongly enough to form an isolable complex. In the π-bond case, an $M(d_\pi)$ electron pair is donated to an empty antibonding orbital of the ligand, usually a π^* for π-bond donors and a σ^* for σ-bond donors (Fig. 1.9b). In the case of a π ligand such as ethylene, this back bonding weakens the $C=C$ π bond but does not break it because C_2H_4 is still held together by strong $C-C$ and $C-H$ σ bonds that are not involved in $M-L$ bond formation. The $C=C$ distance of 1.32 Å in free ethylene is lengthened only to 1.35–1.5 Å in the complex. PF_3 is unusual because it is a strong π acceptor even though it has no multiple bonds; in Section 4.2 we see that PF σ^* orbital plays the role of ligand LUMO.

For σ donors such as H_2,[6] or an alkane,[7] forming the $M-L$ σ bond partially depletes the $H-H$ σ bond because electrons that were fully engaged in keeping the two H atoms together in free H_2 are now also delocalized over the metal (hence the name *two-electron, three-center bond* for this interaction). Back bonding into the $H-H$ σ^* causes additional weakening or even breaking of the $H-H$ σ bond because the σ^* is antibonding with respect to $H-H$. Free H_2 has an $H-H$ distance of 0.74 Å, but the $H-H$ distances in H_2 complexes go all the way from 0.82 to 1.5 Å. Eventually the $H-H$ bond breaks and a dihydride is formed (Eq. 1.5). This is the *oxidative addition reaction* (see Chapter 6). Formation of a σ complex can be thought of as an incomplete oxidative addition. Table 1.2 classifies common ligands by the nature of the $M-L$ σ and π bonds. Both σ and π bonds bind side-on to metals when they act as ligands.

$$L_nM + H_2 \rightleftharpoons L_nM-\overset{\displaystyle H}{\underset{\displaystyle H}{|}} \rightleftharpoons L_nM\overset{\displaystyle H}{\underset{\displaystyle H}{<}} \qquad (1.5)$$

$$\sigma \text{ complex} \qquad\qquad \text{oxidative addition product}$$

Ambidentate Ligands

Some ligands have several alternate types of electron pair available for bonding. For example, aldehydes (**1.14**) have the $C=O$ π bond and lone pairs on the oxygen. When they act as π-bond donors, aldehydes bind side-on (**1.15**) like ethylene, when they act as lone-pair donors, they bind end-on (**1.16**). Equilibria such as Eq. 1.6 [R = aryl; $L_nM = CpRe(NO)PPh_3^+$] are possible, as Gladysz has shown.[8a] The more sterically demanding π-bound form (**1.15**) is favored for unhindered metal complexes; **1.15** also involves back donation and so is also favored by more electron-donor metal fragments and more electron-acceptor R groups. Alkenes have both a $C=C$ π bond and $C-H$ σ bonds. Gladysz[8b] has

TABLE 1.2 Types of Ligand[a]

	Strong π Acceptor	Weak π Bonding	Strong π Donor
Lone-pair donor	CO PF$_3$	CH$_3{}^-$ H^{-c}	CR$_2{}^-$ OR$^-$
	CR$_2{}^{+b}$	NH$_3$	F$^-$
π-Bonding electron	C$_2$F$_4$	C$_2$H$_4$	
pair donor	O$_2$	RCHO^d	
σ-Bonding electron	Oxidative	R$_3$Si$-$H, H$_2$	
pair donor	addition^e	R$_3$C$-$H	

[a]Ligands are listed in approximate order of π-donor/acceptor power, with acceptors to the left.
[b]CH$_2{}^+$ and CH$_2{}^-$ refer to Fischer and Schrock carbenes of Chapter 11.
[c]Ligands like this are considered here as anions rather than radicals.
[d]Can also bind as a lone-pair donor (Eq. 1.6).
[e]Oxidative addition occurs when σ-bond donors bind very strongly (Eq. 1.5).

also shown how metals can move from one face of a C=C bond to the other via intermediate σ binding to the C$-$H bond (Eq. 1.7).

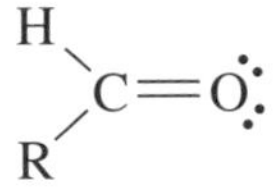

1.14

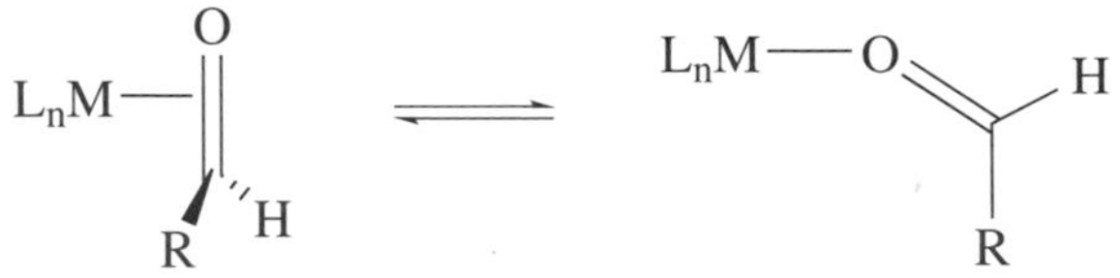

1.15   **1.16**

(1.6)

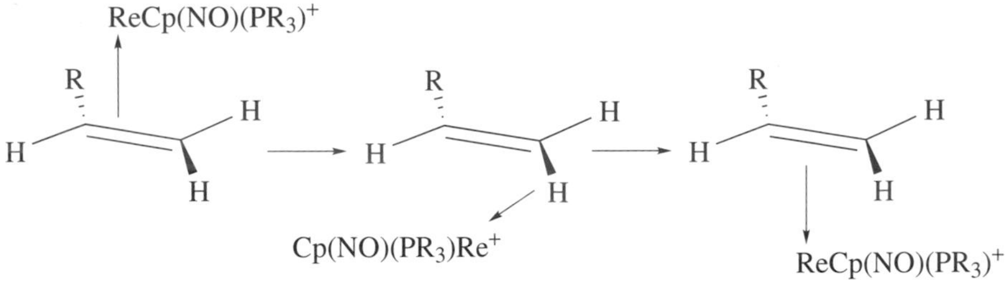

(1.7)

The $\{(NH_3)_5Os^{II}\}^{2+}$ fragment in Eq. 1.8 is a strong π donor because NH$_3$ is strongly σ donor but not a π-acceptor ligand. The metal is electron rich in spite of the 2+ ionic charge, and it prefers to bind to a π acceptor an aromatic C=C bond of aniline. Oxidation to Os^{III} causes a sharp falloff in π-donor power because the extra positive charge stabilizes the d orbitals, and the complex rearranges to the N-bound aniline form.[9] This illustrates how the electronic character of a metal can be altered by changing the ligand set and oxidation state; soft Os(II) binds to the soft C=C bond and hard Os(III) binds to the hard NH_2 group.

$$(1.8)$$

Spectator Versus Actor Ligands

Spectator ligands remain unchanged during chemical transformations. Actor ligands dissociate or undergo some chemical conversion. For example, there is a very extensive chemistry of $[CpFe(CO)_2X]$ and $[CpFe(CO)_2L]^+$ (Cp = cyclopentadienyl; X = anion; L = neutral ligand) where the $\{CpFe(CO)_2\}$ fragment remains intact. The role of these ligands is to impart solubility in organic solvents, prevent departure of the metal, and influence the electronic and steric properties of the complex so as to favor the desired goal. An important part of the art of organometallic chemistry is to pick suitable spectator ligand sets to facilitate certain types of reaction. Apparently small changes in ligand can entirely change the chemistry. For example, PPh_3 is an exceptionally useful ligand with tens of thousands of complexes known while apparently similar compounds NPh_3, $BiPh_3$, and $P(C_6F_5)_3$ appear to be of very little use as ligands. One aspect of the ligand is the nature of the donor atom, so an N donor such as NPh_3 is likely to be very different from a P donor such as PPh_3. Another factor is the nature of the substituents, so that the strongly electron-withdrawing C_6F_5 substituents in $P(C_6F_5)_3$ appear to completely deactivate the lone pair from being able to take part in coordinate bonding. The strong effect of the steric factor is shown by the difference between PMe_3 and $P(C_6H_{11})_3$; up to five or even six of the smaller PMe_3 ligands are easily able to bind to a typical metal to give stable complexes, while only two or at most three of the bulky $P(C_6H_{11})_3$ ligands can normally bind to a single metal at the same time.

One role of spectator ligands is to block certain sites, say of an octahedron, to leave a specific set of sites available for the actor ligands so the desired chemistry can occur. These spectator ligands are commonly polydentate with the donor atoms arranged in specific patterns. A small sample of such ligands is shown in Fig. 1.10. The tridentate ligands can bind to an octahedron either

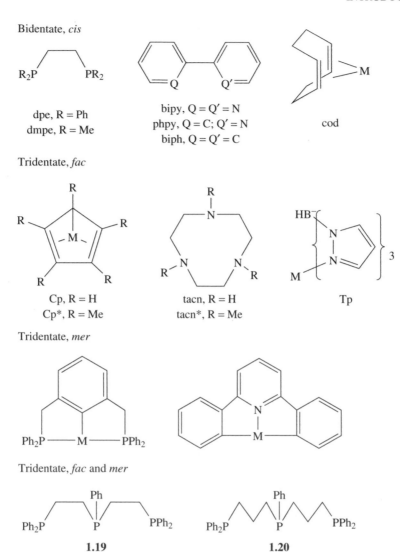

Bidentate, *cis*

dpe, R = Ph
dmpe, R = Me

bipy, Q = Q′ = N
phpy, Q = C; Q′ = N
biph, Q = Q′ = C

cod

Tridentate, *fac*

Cp, R = H
Cp*, R = Me

tacn, R = H
tacn*, R = Me

Tp

Tridentate, *mer*

Tridentate, *fac* and *mer*

1.19

1.20

FIGURE 1.10   Selection of common ligands with different binding preferences. Bidentate trans-binding ligands are extremely rare. The metal is shown where the binding mode might otherwise be unclear. Cp and Cp* can formally be considered as facial tridentate ligands (see Chapter 5).

in a *mer* (meridonal) fashion **1.18** (pincer ligands) or *fac* (facial) **1.17**, or in some cases, in both ways. The choice of ligand is still something of an art because subtle stereoelectronic effects, still not fully understood, can play an important role. Ligands **1.19** and **1.20** impart substantially different properties to their complexes in spite of their apparent similarity, probably as a result of the greater flexibility of the three-carbon linker in **1.20**.

L
L ⤬
L

L
L ⤬
L

1.17
fac

1.18
mer

- Ligands donate via their highest occupied molecular orbital (HOMO) and accept back bonding via their lowest unoccupied molecular orbital (LUMO).
- Metal–ligand bond strengths tend to increase as the ligand donor orbital changes: σ bond $< \pi$ bond $<$ lone pair.
- Changes in the ligand set can greatly change the chemistry at the metal.

REFERENCES

1. A. M. Sargeson, *Pure Appl. Chem.* **56**, 1603, 1984.
2. S. Ahrland, J. Chatt, and N. R. Davies, *Chem. Soc. Revs.* **12**, 265, 1958.
3. C. E. Housecroft and A. G. Sharpe, *Inorganic Chemistry*, Pearson, Edinburgh, 2005, Chaps. 19–20.
4. A. D. Hunter, V. Mozol, and S. D. Tsai, *Organometallics* **11**, 2251, 1992.
5. A. J. Ashe, H. Yang, X. D. Fang, and J. W. Kampf, *Organometallics* **21**, 4578, 2002; S. Y. Liu, M. M. C. Lo, and G. C. Fu, *Angew. Chem. Int. Ed.* **41**, 174, 2002.
6. G. J. Kubas, *Metal Dihydrogen and σ- Bond Complexes*, Kluwer/Plenum, New York, 2001; R. H. Crabtree, *Angew. Chem. Int. Ed.* **32**, 789, 1993.
7. S. Geftakis and G. E. Ball, *J. Am. Chem. Soc.*, **120**, 9953, 1998; C. Hall and R. N. Perutz, *Chem. Rev.* **96**, 3125, 1996; D. W. Lee and C. M. Jensen, *J. Am. Chem. Soc.* **118**, 8749, 1996.
8. (a) N. Q. Mendez, J. W. Seyler, A. M. Serif, and J. A. Gladysz, *J. Am. Chem. Soc.* **115**, 2323, 1993; (b) T. S. Peng and J. A. Gladysz, *J. Am. Chem. Soc.* **114**, 4174, 1992.
9. H. Taube, *Pure Appl. Chem.* **63**, 651, 1991.

PROBLEMS

1. How many isomers would you expect for a complex with the empirical formula $Pt(NH_3)_2Cl_2$?

2. Predict the structure of $[Me_3Pt(\mu_3\text{-I})]_4$. The arrangement of the Pt and I atoms is often considered to be analogous to that of the vertices in one of the Pythagorean regular solids; which one do you think it is?

3. Why is $R_2PCH_2CH_2PR_2$ so much better as a chelating ligand than $R_2PCH_2PR_2$? Why is H_2O a lower-field ligand than NH_3?

4. How would you design a synthesis of the complex *trans*-$[PtCl_2(NH_3)(tu)]$, (the trans descriptor refers to the fact a pair of identical ligands, Cl in this case, is mutually trans), given that the trans effect order is tu > Cl > NH_3 [tu = $(H_2N)_2CS$]?

5. Consider the two complexes $MeTiCl_3$ and $(CO)_5W(thf)$. Predict the order of reactivity in each case toward the following sets of ligands: NMe_3, PMe_3, CO.

6. How could you distinguish between a square planar and a tetrahedral structure in a nickel(II) complex of which you have a pure sample, without using crystallography?

7. You have a set of different ligands of the PR_3 type and a large supply of $(CO)_5W(thf)$ with which to make a series of complexes $(CO)_5W(PR_3)$. How could you estimate the relative ordering of the electron-donor power of the different PR_3 ligands?

8. The stability of metal carbonyl complexes falls off markedly as we go to the right of group 10 in the periodic table. For example, copper forms only a few weakly bound complexes with CO. Why is this? What oxidation state, of the ones commonly available to copper, would you think form the strongest CO complexes?

9. Low-oxidation-state complexes are often air sensitive (i.e., they react with the oxygen in the air), but are rarely water sensitive. Why do you think this is so?

10. $MnCp_2$ is high spin, while $MnCp_2^*$ ($Cp^* = \eta^5$-C_5Me_5) is low spin. How many unpaired electrons does each metal have, and which ligand has the stronger ligand field?

11. Make up a problem on the subject matter of this chapter and provide an answer. This is a good thing for you to do for subsequent chapters as well. It gives you an idea of topics and issues on which to base questions and will therefore guide you in studying for tests.

2

GENERAL PROPERTIES OF ORGANOMETALLIC COMPLEXES

Organometallic chemistry is concerned with the metal–carbon bond, of which the simplest is the M–C single bond of metal alkyls. As σ-bonding ligands, alkyls are closely related to the ligands found in coordination compounds, such as Cl, H_2O, and NH_3. A larger class of organometallic ligands (CO, C_2H_4) are soft and can π bond. The structures of some typical organometallic compounds in later chapters of this book show many examples of such π-bonding ligands as butadiene, benzene, cyclopentadienyl (C_5H_5 or Cp), and allyl. There are several differences between complexes of these ligands and coordination compounds containing Cl^-, H_2O, and NH_3. The metals are more electron rich, in the sense that the metal bears a greater negative charge in the organometallic complex. The M–L bonds are much more covalent and often have a substantial π component. The metal d orbitals are higher in energy and by back donation perturb the electronic structure of the ligands much more than is the case for coordination compounds. The organometallic ligands can be polarized and therefore activated toward chemical reactions, σ and π bonds in the ligands can be weakened or broken, and chemical bonds can be made or broken within and between different ligands. This rich pattern of reactions is characteristic of organometallic chemistry.

In this chapter, we look at the 18-electron rule and at the ionic and covalent models that are commonly used for electron counting. We then examine the ways in which binding to the metal can perturb the chemical character of a ligand, an effect that lies at the heart of organometallic chemistry.

2.1 THE 18-ELECTRON RULE

The 18e rule[1] is a way to help us decide whether a given d-block transition metal organometallic complex is likely to be stable. Not all the organic formulas we can write down correspond to stable species. For example, CH_5 requires a 5-valent carbon and is therefore not stable. Stable compounds, such as CH_4, have the noble gas octet, and so carbon can be thought of as following an 8e rule. This corresponds to carbon using its s and three p orbitals to form four filled bonding orbitals and four unfilled antibonding orbitals. On the covalent model, we can consider that of the eight electrons required to fill the bonding orbitals, four come from carbon and one each comes from the four H substituents. We can therefore think of each H atom as being a 1e ligand to carbon.

To assign a formal oxidation state to carbon in an organic molecule, we impose an ionic model by artificially dissecting it into ions. Each electron pair in any bond is assigned to the most electronegative of the two atoms or groups that constitute the bond. For methane, this dissection gives $C^{4-} + 4H^+$, with carbon as the more electronegative element. This makes methane an 8e compound with an oxidation state of -4, usually written C(-IV). Note that the net electron count always remains the same, whether we adopt the covalent (4e {C atom} $+ 4 \times$ 1e {4H atoms} = 8e) or ionic (8e{C^{4-}ion} $+ 4 \times$ 0e{$4H^+$ions} = 8e) model.

The 18e rule, which applies to many low-valent transition metal complexes, follows a similar line of reasoning. The metal now has one s, and three p orbitals, as before, but now also five d orbitals. We need 18e to fill all nine orbitals; some come from the metal, the rest from the ligands. Only a limited number of combinations of metal and ligand give an 18e count. Figure 1.5 shows that 18e fills the molecular orbital (MO) diagram of the complex ML_6 up to the d_π level, and leaves the M—L antibonding d_σ^* orbitals empty. The resulting configuration is analogous to the closed shell present in the group 18 elements and is therefore called the *noble gas configuration*. Each atomic orbital (AO) on the metal that remains nonbonding will clearly give rise to one MO in the complex; each AO that interacts with a ligand orbital will give rise to one bonding MO, which will be filled in the complex, and one antibonding MO, which will normally be empty. Our nine metal orbitals therefore give rise to nine low-lying orbitals in the complex, and to fill these we need 18 electrons.

Table 2.1 shows how the first-row carbonyls mostly follow the 18e rule. Each metal contributes the same number of electrons as its group number, and each CO contributes 2e from its lone pair; π back bonding makes no difference to the electron count for the metal. In the free atom, it had pairs of d_π electrons for back bonding; in the complex it still has them, now delocalized over metal and ligands.

In cases where we start with an odd number of electrons on the metal, we can never reach an even number, 18, by adding 2e ligands such as CO. In each case the system resolves this problem in a different way. In $V(CO)_6$, the complex is 17e but is easily reduced to the 18e anion $V(CO)_6^-$. Unlike $V(CO)_6$, the $Mn(CO)_5$ fragment, also 17e, does dimerize, probably because, as

TABLE 2.1 First-Row Carbonyls

$V(CO)_6$	17e; 18e $V(CO)_6^-$ also stable
$Cr(CO)_6$	Octahedral
$(CO)_5Mn-Mn(CO)_5$	M—M bond contributes 1e to each metal; all the CO groups are terminal
$Fe(CO)_5$	Trigonal bipyramidal
$(CO)_3Co(\mu\text{-}CO)_2Co(CO)_3$	μ-CO contributes 1e to each metal, and there is also an M—M bond
$Ni(CO)_4$	Tetrahedral

a 5-coordinate species, there is more space available to make the M—M bond. This completes the noble gas configuration for each metal because the unpaired electron in each fragment is shared with the other in forming the bond, much as the 7e methyl radical dimerizes to give the 8e compound, ethane. In the 17e fragment $Co(CO)_4$, dimerization also takes place via a metal–metal bond, but a pair of COs also move into bridging positions. This makes no difference in the electron count because the bridging CO is a 1e ligand to each metal, so an M—M bond is still required to attain 18e. The even-electron metals are able to achieve 18e without M—M bond formation, and in each case they do so by binding the appropriate number of COs; the odd-electron metals need to form M—M bonds.

Ionic Versus Covalent Model

Unfortunately, there are two conventions for counting electrons: the ionic and covalent models, both of which have roughly equal numbers of supporters. Both methods lead to *exactly the same net result*; they differ only in the way that the electrons are considered as "coming from" the metal or from the ligands. Take $HMn(CO)_5$: We can adopt the covalent model and argue that the H atom, a 1e ligand, is coordinated to a 17e $Mn(CO)_5$ fragment. On the other hand, on the ionic model, we can consider the complex as being an anionic 2e H^- ligand coordinated to a cationic 16e $Mn(CO)_5^+$ fragment. The reason is that H is more electronegative than Mn and so is formally assigned the bonding electron pair when we dissect the complex. Fortunately, no one has yet suggested counting the molecule as arising from a 0e H^+ ligand and an 18e $Mn(CO)_5^-$ anion; ironically, protonation of the anion is the most common preparative method for this hydride.

These different ways of assigning electrons are simply models. Since all bonds between dissimilar elements have at least some ionic and some covalent character, each model reflects a facet of the truth. The covalent model is probably more appropriate for the majority of low-valent transition metal complexes, especially with the unsaturated ligands we will be studying. On the other hand, the ionic model is more appropriate for high-valent complexes with N, O, or Cl ligands, such as are found in coordination chemistry or in the organometallic chemistry described in Chapter 15. In classical coordination chemistry, the oxidation state model played a dominant role because the oxidation state of the types of compound studied could almost always be unambiguously defined. The rise of the covalent model has paralleled the growth in importance of organometallic compounds, which tend to involve more covalent M$-$L bonds and for which oxidation states cannot always be unambiguously defined (see Section 2.4). We have therefore preferred the covalent model as being most appropriate for the majority of the compounds with which we will be concerned. It is important to be conversant with both models, however, because each can be found in the literature without any indication as to which is being used, so you should practice counting under the other convention after you are happy with the first. We will also refer to any special implications of using one or other model as necessary.

Electron Counts for Common Ligands and Hapticity

In Table 2.2 we see some of the common ligands and their electron counts on the two models. The symbol L is commonly used to signify a neutral ligand, which can be a lone-pair donor, such as CO or NH_3, a π-bond donor, such as C_2H_4, or a σ-bond donor such as H_2, which are all 2e ligands on both models. The symbol X refers to ligands such as H, Cl, or Me, which are 1e X ligands on the covalent model and 2e X^- ligands on the ionic model. In the covalent model we regard them as 1e X$\cdot$ radicals bonding to the neutral metal atom; in the ionic model, we regard them as 2e X^- anions bonding to the M^+ cation. Green[2] has developed a useful extension of this nomenclature by which more complicated ligands can be classified. For example, benzene (2.1) can be considered as a combination of three C=C ligands, and therefore as L_3.* The allyl group can be considered as a

TABLE 2.2　Common Ligands and Their Electron Counts

Ligand	Type	Covalent Model	Ionic Model
Me, Cl, Ph, Cl, η^1-allyl, NO (bent)[a]	X	1e	2e
Lone-pair donors: CO, NH_3	L	2e	2e
π-Bond donors: C_2H_4	L	2e	2e
σ-Bond donors: (H_2)	L	2e	2e
M$-$Cl (bridging)	L	2e	2e
η^3-Allyl, κ^2-acetate	LX	3e	4e
NO (linear)[a]		3e	2e[a]
η^4-Butadiene	L_2[b]	4e	4e
=O (oxo)	X_2	4e	2e
η^5-Cp	L_2X	5e	6e
η^6-Benzene	L_3	6e	6e

[a]Linear NO is considered as NO^+ on the ionic model; see Section 4.1.
[b]The alternative LX_2 structure sometimes adopted gives the same electron count.

*Undergraduates will need to become familiar with organic "line notation," in which only C$-$C bonds are shown and enough H groups must be added to each C to make it 4-valent. For example, 2.6 represents $MCH_2CH=CH_2$.

combination of an alkyl and a C=C group. The two resonance forms **2.2** and **2.3** show how we can consider allyl groups in which all three carbons are bound to the metal as LX ligands. This can also be represented in the delocalized form as **2.4**. In such a case, the *hapticity* of the ligand, the number of ligand atoms bound to the metal, is three and so **2.5**, referred to as "bis-π-allyl nickel" in the older literature, is now known as bis-η^3-allyl nickel, or [Ni(η^3-C$_3$H$_5$)$_2$]. Occasionally the letter "h" is used instead of η, and sometimes η is used without a superscript as a synonym for the older form, π; such things tend to be frowned on. The electron count of the η^3 form of the allyl group is 3e on the covalent model and 4e on the ionic model, as suggested by the LX label. The advantage of the LX label is that those who follow the covalent model will translate LX as meaning a 3e ligand, and the devotees of the ionic model will translate LX as meaning a 4e ligand. The Greek letter κ (kappa) is normally used instead of η when describing ligands that bind via heteroatoms, such as κ^2-acetate.

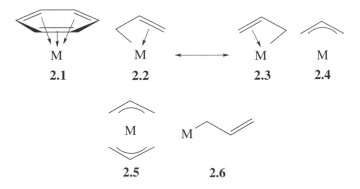

The allyl group can also bind in another way (**2.6**). Since only one carbon is now bound to the metal, this is the η^1-allyl, or σ-allyl, form. In this bonding mode, the allyl behaves as an X-type ligand, like a methyl group, and is therefore a 1e ligand on the covalent model and a 2e ligand on the ionic model. Some examples of electron counting are shown in Fig. 2.1. Note the dissection of **2.7**–**2.12** (Fig. 2.1) into atoms and radicals in the covalent model and into ions in the ionic model.

Bridging ligands are very common and are prefixed by the symbol μ. Bridging CO ligands are usually counted as shown in Table 2.1. On the ionic model, a bridging Cl$^-$ donates a pair of electrons to each of two L$_n$M$^+$ groups. On the covalent model, we first form L$_n$M−Cl, the Cl of which carries a lone pair, which is donated to the second metal in forming the bridge. An L$_n$MCl group is effectively acting as a ligand to the second metal. If ML$_n$ = M$'$L$_n$, then the two bonds to Cl are indistinguishable by resonance between **2.13** and **2.14**:

$$L_nM-\overset{\cdot\cdot}{Cl}: \;+\; \overset{+}{M'L_n} \longrightarrow L_nM\overset{Cl}{\diagdown}\overset{+}{M'L_n} \longleftrightarrow \overset{+}{L_nM}\overset{Cl}{\diagup}M'L_n$$

$$\textbf{2.13} \qquad\qquad\qquad\qquad \textbf{2.14}$$

$$(2.1)$$

Ionic Model				Covalent Model	
$C_5H_5^-$	6e			$C_5H_5\cdot$	5e
$C_5H_5^-$	6e	Fe		$C_5H_5\cdot$	5e
Fe^{2+}	6e			Fe	8e
	18e				18e
		2.7			
Mo^{4+}	2e			Mo	6e
$4 \times H^-$	8e	$MoH_4(PR_3)_4$		$4 \times H\cdot$	4e
$4 \times PR_3$	8e	**2.8**		$4 \times PR_3$	8e
	18e				18e
Ni^{2+}	8e			Ni	10e
$2 \times C_3H_5^-$	8e	Ni		$2 \times C_3H_5\cdot$	6e
	16e				16e
		2.9			
Mo	6e			Mo	6e
$2 \times C_6H_6$	12e	Mo		$2 \times C_6H_6$	12e
	18e				18e
		2.10			
$2 \times Cl^-$	4e		Cl	$2 \times Cl$	2e
Ti^{4+}	0e	Ti	Cl	Ti	4e
$2 \times C_5H_5^-$	12e			$2 \times C_5H_5\cdot$	10e
	16e				16e
		2.11			
Co^{3+}	6e			Co	9e
$2 \times C_5H_5^-$	12e	Co		$2 \times C_5H_5\cdot$	10e
	18e			Positive charge[a]	−1e
		2.12			18e

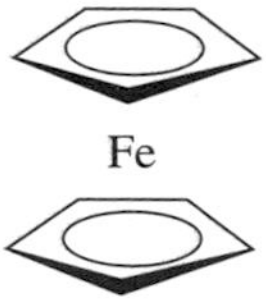

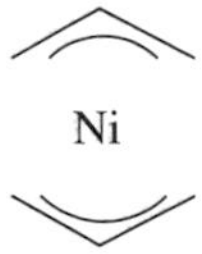

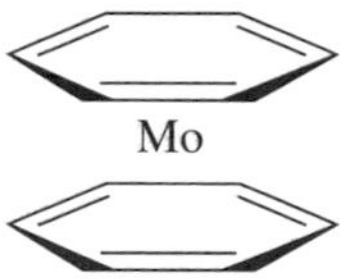

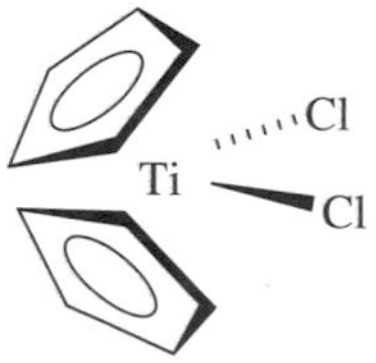

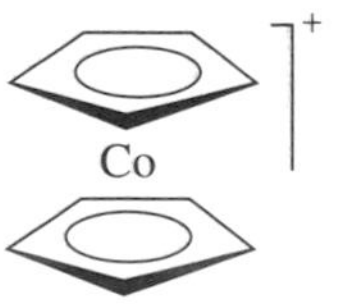

[a]To account for the positive ionic charge on the complex as a whole; for anions, the net charge is added to the total.

FIGURE 2.1 Electron counting on the covalent and ionic models.

For electron counting purposes, we can consider that the chlorine atom is a 1e donor to M and M−Cl is a 2e donor to M' via its lone pair (or, on the ionic model, that Cl^- is a 2e donor to each metal via two lone pairs). A triply bridging Cl would donate 1e to the first and 2e each to the other two metals on the covalent model. The same usually holds true for other X-type ligands, such as halide, $-SR$, $-OR$, or $-PR_2$. A bridging carbonyl is like a ketone from the point of view of electron counting; it is a 1e donor to each metal. (This is true for both models because users of the ionic model regard CO as a neutral ligand even when bridging.) Other ligands of the same type are bridging methylene, $M-CH_2-M$, and bridging oxo, $M-O-M$, which are 1e ligands to each metal on the covalent model and 2e ligands on the ionic model.

As shown in **2.13** and **2.14**, we often write $M-X$ to signify the covalent bond, but $L \rightarrow M$ for the coordinate bond, as an indication that both electrons are regarded as "coming from" the ligand L.

For complex ions, we have to adjust for the net ionic charge in making the electron count. For example, $CoCp_2^+$ (**2.12** in Fig. 2.1) is counted on the covalent model as follows. The neutral Co atom has 9e because it is group 9; from Table 2.2, the two neutral Cp groups add 10e; the net ionic charge is 1+, so one electron has been removed to make the cation. The electron (e) count is therefore $9 + 10 - 1 = 18e$. Electron counting can be summarized by Eq. 2.2, which shows the electron count for a generalized complex $[MX_aL_b]^{c+}$, where N is the group number of the metal (and therefore the number of electrons in the neutral M atom), a and b are the numbers of ligands, and c is the net ionic charge:

$$\text{e count (covalent model)} = N + a + 2b - c \tag{2.2}$$

When we use the ionic model for electron counting, we first have to calculate the oxidation state of the metal (Section 2.4). The oxidation state is the ionic charge left on the metal after removal of the ligands, taking care to assign the electron pairs in the $M-L$ bonds to the more electronegative atom in each case. (If two atoms have the same electronegativity, one electron is assigned to each; see also Section 2.4.) For $CoCp_2^+$, we must remove two Cp's as Cp^- ions (as C is more electronegative than Co); this leaves Co^{3+}, which has a d^6 configuration. This means that $CoCp_2^+$ has $6 + (2 \times 6) = 18$ electrons. For the general case of $[MX_aL_b]^{c+}$, this procedure leaves the metal as $M^{(c+a)+}$, and therefore the metal is in the oxidation state $(c + a)$, and has $N - c - a$ electrons. We now have to add 2e for each X^-, and 2e for each L in putting the complex back together:

$$\text{e count (ionic model)} = N - a - c + 2a + 2b = N + a + 2b - c \tag{2.3}$$

You will see that this reduces to Eq. 2.2 and so the two methods of electron counting are equivalent and always give the same result.

2.2   LIMITATIONS OF THE 18-ELECTRON RULE

There are many cases in which the electron count for a stable complex is not 18; examples are $MeTiCl_3$, 8e; Me_2NbCl_3, 10e; WMe_6, 12e; $Pt(PCy_3)_2$, 14e; $[M(H_2O)_6]^{2+}$ (M = V, 15e; Cr, 16e; Mn, 17e; Fe, 18e), $CoCp_2$, 19e; and $NiCp_2$,

20e. For the 18e rule to be useful, we need to be able to predict when it will be obeyed and when it will not.

The rule works best for hydrides and carbonyls because these are sterically small, high-field ligands. Because they are small, as many generally bind as are required to achieve 18e. With high-field ligands, Δ for the complex will be large. This means that the d_σ^* orbitals that would be filled if the metal had more than 18e are high in energy and therefore poor acceptors. On the other hand, the d_π orbitals that would have to give up electrons if the molecule had less than 18e and are low in energy because of π bonding by CO (or, in the case of H, because of the very strong σ bond and the absence of repulsive π interactions with lone pairs). The d_π level is therefore a good acceptor, and to be stable, a complex must have this level filled (otherwise the electrophilic metal will gain electrons by binding more CO, or the solvent or some functional group in the ligands until the 18e configuration is attained).

Conversely, the rule works least well for high-valent metals with weak-field ligands. In the hexaaqua ions $[M(H_2O)_6]^{2+}$ (M = V, Cr, Mn, Fe, Co, Ni), the structure is the same whatever the electron count of the metal and so must be dictated by the fact that six H_2O's fit well around a metal ion. H_2O has two lone pairs, one of which it uses to form a σ bond. This leaves one remaining on the ligand, which acts as a π donor to the metal and so lowers Δ; H_2O is therefore a weak-field ligand. If Δ is small, then the tendency to adopt the 18e configuration is also small because it is easy to add electrons to the low-lying d_σ^* or to remove them from the high-lying d_π.

An important class of complexes follow a 16e, rather than an 18e, rule because one of the nine orbitals is very high lying and is usually empty. This can happen for the d^8 metals of groups 8–11 (Table 2.3). Group 8 shows the least and group 11 the highest tendency to become 16e. When these metals are 16e, they normally adopt the square planar geometry, but large distortions can occur.[3] Some examples of 16e complexes of this sort are $RhClL_3$, $IrCl(CO)L_2$, $PdCl_2L_2$, and $[PtCl_4]^{2-}$, $[AuMe_4]^-$ (L = 3° phosphine).

TABLE 2.3 The d^8 Metals that can Adopt a 16e Square Planar Configuration

	Group		
8	9	10	11
Fe(0)[a]	Co(I)[b]	Ni(II)	Cu(III)[c]
Ru(0)[a]	Rh(I)[b]	Pd(II)	—
Os(0)[a]	Ir(I)[b]	Pt(II)	Au(III)

[a]These metals prefer 18e to 16e.
[b]The 16e configuration is more often seen, but 18e complexes are common.
[c]A rare oxidation state.

The smaller metal clusters, such as $Os_3(CO)_{12}$, often obey the 18e rule for each metal, but for clusters of six metals or more, there can be deviations for which special cluster counting rules have been devised (Chapter 13). The rule is not useful for main-group elements, such as $ZnMe_2$, 14e; $MeHg(bipy)^+$, 16e; $[I(py)_2]^+$, 20e; $[SbF_6]^-$, 22e; and IF_7, 24e, where no particular electron count is favored. The lanthanides and actinides have seven f orbitals to fill before they even start on the d orbitals, and so they are essentially never able to bind a sufficient number of ligands to raise the electron count to the $s^2 p^6 d^{10} f^{14}$, or 32e configuration of the appropriate noble gas; some examples are $U(cot)_2$, 22e, and Cp_2LuMe, 28e. The stoichiometry of an f block complex tends to be decided by steric saturation of the space around the metal. Paramagnetic complexes [e.g., $V(CO)_6$, 17e; Cp_2Fe^+, 17e; Cp_2Ni, 20e] generally do not obey the 18e rule,[4] but many of these have reactions in which they attain an 18e configuration, for example, the 19e $CpFe(\eta^6\text{-arene})$ is a powerful 1e reductant.[5]

Complexes of d^0 metals can pose special problems. Many such complexes have electron counts below 18e (e.g., $TiMe_4$, $CpWOCl_3$). An ambiguity often arises when the ligands have additional π-type lone pairs that can—at least in principle—be donated into empty metal d_π orbitals as shown in Fig. 1.8. For example, $W(OMe)_6$ is apparently a 12e species, but each oxygen has two π-type lone pairs for a total of 24 additional electrons that could be donated to the metal. Almost any even electron count could therefore be assigned and for this reason electron counting is less useful in discussions of early metal and d^0 organometallic chemistry.

2.3 ELECTRON COUNTING IN REACTIONS

It is often useful to consider changes in the electron count of a metal during a reaction. For example, an 18e complex might be reluctant to add a 2e ligand, such as PPh_3, without first losing a 2e ligand or rearranging in some way to generate a 2e vacancy at the metal. The 20e intermediate (or transition state) that would be involved if an extra ligand were to bind, is likely to be less stable than the 16e intermediate (or transition state) involved in the loss of a ligand. If all the ligands originally present are firmly bound, as in $FeCp_2$, then we do not expect a 2e reagent, such as a phosphine, to bind. On the other hand, H^+ is a zero-electron (0e) reagent, and can react with an 18e species, such as ferrocene (Eq. 2.4). This protonation also illustrates the electron-rich (basic) character of the metal common for organometallic compounds, but not seen for aqua complexes and other coordination compounds.

$$Cp_2Fe + H^+ = [Cp_2FeH]^+ \qquad (2.4)$$

Because H^- is a 2e reagent like PPh_3, we would not expect H^- to attack the metal in ferrocene. Note that this result is the same whether we use the ionic or covalent

TABLE 2.4   Reagent Electron Counts

0e	1e	2e	3e	4e
H^+	$H\cdot^a$	$H^-(LiAlH_4)^b$	NO	$C_3H_5^-(C_3H_5MgBr)$
$Me^+(MeI)$	$Me\cdot^a$	$Me^-(LiMe)$		Butadiene
$Br^+(Br_2)^c$		PPh_3, NO^+		NO^-
		Cl^-, CO, H_2		

^{a}These species are unstable and so they are invoked as reactive intermediates in mechanistic schemes, rather than used as reagents in the usual way.
^{b}The reagents in parentheses are the ones most commonly used as a source of the species in question.
^{c}Br_2 can also be a source of $Br\cdot$, a 1e reagent, as well as of Br^+, depending on conditions.

model. The reagents on the left-hand side of Eq. 2.4 are already separated for us, on any model, H^+ is 0e and Cp_2Fe is 18e. Ironically, neither model applied to $[Cp_2FeH]^+$ gives the dissection shown on the left-hand side of Eq. 2.4. We will therefore speak of H^+ and H^- as 0e and 2e *reagents*, respectively, even though H is a 1e ligand (ionic model: 2e) to make the distinction clear.

In terms of electron counting, any X ligand that bears a negative charge, as in Cl^-, is a 2e reagent, like PPh_3. Table 2.4 shows the effect of net charges on some other reagents. This table also tells us about possible isoelectronic replacements of one ligand by another. So, for example, an X^- group can replace an L ligand without a change in the electron count.

$$W(CO)_5(thf) + Cl^- = [W(CO)_5Cl]^- \tag{2.5}$$

The reaction of Eq. 2.6 turns a 1e alkyl group into a 2e alkene group. To retain the 18e configuration, the complex must become positively charged, which implies that the H must be lost as H^- and that an electrophilic reagent (such as Ph_3C^+) must be used. In this way the 18e rule helps us pick the right reagent.

$$Cp(CO)_2Fe-CH(CH_3)_2 + Ph_3C^+$$
$$= [Cp(CO)_2Fe(\eta^2\text{-}CHMe{=}CH_2)]^+ + Ph_3CH \tag{2.6}$$

As you look at the equations in the pages to come, become familiar with electron counting of stable complexes and with counting the ligands that are gained or lost in reactions.

- Many compounds have 18 valence electrons so counting electrons is a vital skill.
- Ionic and covalent models give the same electron count (Eq. 2.2).
- Some d^8 metals prefer 16 electron square planar geometries (Table 2.3).

2.4 OXIDATION STATE

The oxidation state of a metal in a complex is simply the charge that the metal would have on the ionic model. In practice, all we have to do for a neutral complex is to count the number of X ligands. For example, Cp_2Fe has two L_2X ligands and so can be represented as MX_2L_4; this means that the oxidation state (OS) is 2+, so Cp_2Fe is said to be Fe(II). For a complex ion, we need also to take account of the net charge as shown for $[MX_aL_b]^{c+}$ in Eq. 2.7. For example, Cp_2Fe^+ is Fe(III), and $[W(CO)_5]^{2-}$ is W(-II). Once we have the oxidation state, we can immediately obtain the corresponding d^n configuration. This is simply the number of d electrons that would be present in the free metal ion that corresponds to the oxidation state we have assigned. For Cp_2Fe^+ the OS is Fe(III), which corresponds to the Fe^{3+} ion. The iron atom, which is in group 8, has 8e, and so the ion has $8 - 3 = 5e$. Cp_2Fe^+ is therefore said to be a d^5 complex. Equation 2.8 gives the value of n in a general form. The significance of the d^n configuration is that it tells us how to fill up the crystal field diagrams we saw in Section 1.4. For example, the odd number for Cp_2Fe^+ implies paramagnetism because in a mononuclear complex we cannot pair five electrons whatever the d-orbital splitting.

$$OS = c + a \tag{2.7}$$

$$n = N - (c + a) = N - c - a \tag{2.8}$$

Many organometallic compounds have low or intermediate formal oxidation states. High oxidation states are now gaining more attention and in Chapters 11 and 15, we look at these interesting species in detail. Back donation is severely reduced in higher oxidation states because (1) there are fewer (or no) nonbonding d electrons available and (2) the increased partial positive charge present on the metal in the high-oxidation-state complex strongly stabilizes the d levels so that any electrons they contain become less available. Those high-valent species that do exist generally come from the third-row metals. The extra shielding provided by the f electrons added in building up the lanthanides makes the outer electrons of the third-row metals less tightly bound and therefore more available. High oxidation states can be accessible if the ligands are small and non-π-bonding like H or Me, however, as in the d^0 species WMe_6 and $ReH_7(dpe)_2$.

It is often useful to refer to the oxidation state and d^n configuration, but they are a formal classification only and do not allow us to deduce the real partial charge present on the metal. It is therefore important not to read too much into oxidation states and d^n configurations. Organometallic complexes are not ionic, and so an Fe(II) complex, such as ferrocene, does not contain an Fe^{2+} ion. Similarly, WH_6L_3, in spite of being W(VI), is certainly closer to $W(CO)_6$ in terms of the real charge on the metal than to WO_3. In real terms, the hexahydride may even be more reduced and more electron rich than the W(0) carbonyl. CO groups are excellent π acceptors, so the metal in $W(CO)_6$ has a much lower electron density than a free W(0) atom; on the other hand, the W−H bond in

WH_6L_3 is only weakly polar, and so the polyhydride has a much higher electron density than the W^{6+} suggested by its W(VI) oxidation state (which assumes a dissection: $W^+ H^-$). For this reason, the term *formal oxidation state* is often used for the value of OS as given by Eq. 2.7.

Ambiguous Oxidation States

More problematic are cases in which even the formal oxidation state is ambiguous and cannot be specified. Any organometallic fragment that has several resonance forms that contribute to a comparable extent to the real structure can be affected. For example, this is the case for the resonance forms **2.15** and **2.16** in butadiene complexes. One structure is L_2 (or π_2), the other LX_2 (or $\pi\sigma_2$).* The binding of butadiene as **2.15** leaves the oxidation state of the metal unchanged, but as **2.16** it becomes more positive by two units. On the covalent model, each gives exactly the same electron count: 4e. On the ionic model, the count changes by 2e (**2.15**, 4e; **2.16**, 6e), but this is compensated by a 2e "oxidation" of the metal. Any given complex has a structure that is intermediate between the extremes defined by **2.15** and **2.16**; we never see two distinct forms of the same complex, one like **2.15**, one like **2.16**. Note that the electron count remains the same for all resonance forms of a complex. In the case of W(butadiene)$_3$, we can attribute any even oxidation between W(0) and W(VI) to the molecule by counting one or more of the ligands as LX_2, rather than L_2. To avoid misunderstandings it is therefore necessary to specify the resonance form to which the formal oxidation state applies. For neutral ligands like butadiene, the neutral L_2 form is generally used because this is the stable form of the ligand in the free state. Yet structural studies show that the ligand often more closely resembles **2.16** than **2.15**. Clearly, we can place no reliance on the formal oxidation state to tell us about the real charge on the metal in W(butadiene)$_3$. We will see later (e.g., Section 4.2) several ways in which we can learn something about the real charge. In spite of its ambiguities, the oxidation state convention is almost universally used in classifying organometallic complexes.

2.15   **2.16**

Another type of oxidation state ambiguity occurs in cases where an electron is localized on an easily reduced ligand, rather than on the metal. Such is the case for the green paramagnetic species shown, RuBr(CO)(PPh$_3$)$_2$L, where the complex appears to be 19e Ru(I) but electron paramagnetic resonance (EPR) data shows that the 19th electron is in fact located on the organic diazopyridine ligand, L, which is reduced to the L•$^-$ organic radical anion; the metal is in fact 18e Ru(II).[6]

---

*We prefer the LX notation because it holds for all types of ligands, including carbenes and nitrosyls where a $\pi\sigma$ notation does not apply.

Maximum Oxidation State

The oxidation state of a complex can never be higher than the group number of the transition metal involved. Titanium can have no higher oxidation state than Ti(IV), for example, because Ti has only four valence electrons with which to form bonds and $TiMe_6$ therefore cannot exist.

2.5 COORDINATION NUMBER AND GEOMETRY

The coordination number (CN) of a complex is easily defined in cases in which the ligands are all monodentate; it is simply the number of ligands present [e.g., $[PtCl_4]^{2-}$, CN = 4, $W(CO)_6$, CN = 6]. A useful generalization is that the coordination number cannot exceed 9 for the transition metals. This is because the metal only has 9 valence orbitals, and each ligand needs its own orbital. In most cases the CN is less than 9, and some of the 9 orbitals will either be lone pairs on the metal or engaged in back bonding.

Each coordination number has one or more coordination geometries associated with it. Table 2.5 lists some examples. In order to reach the maximum coordination number of 9, we need relatively small ligands (e.g., $[ReH_9]^{2-}$). Coordination numbers lower than 4 tend to be found with bulky ligands, which cannot bind in greater number without prohibitive steric interference between the ligands [e.g., $Pt(PCy_3)_2$].

TABLE 2.5   Some Common Coordination Numbers and Geometries

2	linear	—M—	$(Me_3SiCH_2)_2Mn$
3	trigonal	—M	$Al(mesityl)_3$
	T-shaped	—M—	$Rh(PPh_3)_3^+$
4	square planar	—M—	$RhCl(CO)(PPh_3)_2$
	tetrahedral	M	$Ni(CO)_4$

(*continued overleaf*)

TABLE 2.5   (*continued*)

5	trigonal bipyramidal		$Fe(CO)_5$
	square pyramidal		$Co(CNPh)_5{}^{2+}$
6	octahedral		$Mo(CO)_6$
7	capped octahedron		$ReH(PR_3)_3(MeCN)_3{}^{2+}$
	pentagonal bipyramid		$IrH_5(PPh_3)_2$
8	dodecahedral[a]		$MoH_4(PR_3)_4$
	square antiprism		$TaF_8{}^{3-}$
9	tricapped[b] trigonal prism		$ReH_9{}^{2-}$

[a]The smaller ligands tend to go to the less hindered A sites. Two A and two B sites each lie on a plane containing the metal. One such plane is shown dotted; the other lies at right angles to the first. [b]The tricapped trigonal prism is shown as viewed along its threefold axis. The vertices of the triangles are the axial ligand positions. The equatorial M—L bonds are shown explicitly.

Unfortunately, the definition of coordination number and geometry is less clear-cut for organometallic species, such as Cp_2Fe. Is this molecule 2-coordinate (there are two ligands), 6-coordinate (there are six electron pairs involved in metal–ligand bonding), or 10-coordinate (the 10 C atoms are all within bonding distance of the metal)? Most often, it is the second definition that is used, which is equivalent to counting up the number of lone pairs provided by the ligands on the ionic model. We use this as the CN in what follows.

Equations 2.9–2.12 summarize the different counting rules as applied to our generalized d^n transition metal complex $[MX_aL_b]^{c+}$, where N is the group number. In Eq. 2.9, the CN cannot exceed 9:

$$\text{Coordination number:} \quad CN = a + b \leq 9 \tag{2.9}$$

$$\text{Electron count:} \quad N + a + 2b - c = 18 \tag{2.10}$$

$$\text{Oxidation state:} \quad OS = a + c \leq N \tag{2.11}$$

$$d^n \text{ configuration:} \quad d^n = d^{(N-OS)} = d^{(N-a-c)} \tag{2.12}$$

d^n Configuration and Geometry

The d^n configuration of the metal is a good guide to the preferred geometry adopted, as indicated in Table 2.6, because of the ligand field effects specific to each configuration. The d^0, d^5 (hs), and d^{10} configurations are special because they have the same number of electrons (zero, one, or two, respectively) in each d orbital. This symmetric electron distribution means there are no ligand field effects and the ligand positions are sterically determined. The standard model for predicting geometries in main-group chemistry, VSEPR (valence shell electron pair repulsion), works reliably only when ligand field effects are absent. In transition metal systems, this means only for d^0, d^5 (hs), and d^{10} cases where the d electrons are not considered. For example, in d^{10} PtL_4, we consider only the four L lone pairs, which, in accordance with VSEPR, are arranged in a tetrahedral geometry.

Steric Effects and Geometry

Large ligands favor low coordination numbers [e.g., $Pt(PCy_3)_2$]. These ligands also favor distortions from electronically preferred geometries. For example,

TABLE 2.6 Common Geometries with Typical d^n Configuration

Coordination Number	Geometry	d^n Configuration	Example
3	T-shaped	d^8	$[Rh(PPh_3)_3]^+$
4	Tetrahedral	d^0, d^5 (hs), d^{10}	$Pd(PPh_3)_4$
4	Square planar	d^8	$[RhCl(PPh_3)_3]$
5	Trigonal bipyramidal	d^8, $\{d^6\}^a$	$[Fe(CO)_5]$
6	Octahedral	d^0, d^3, d^5 (ls), d^6	$[Mn(CO)_6]^+$
8	Dodecahedral	d^2	$WH_4(PMePh_2)_4$
9	TTP^b	d^0	$[ReH_9]^{2-}$

$^a\{d^6\}$ means that a distorted version of this geometry occurs for this d configuration (see Section 4.3).
^{b}Tricapped trigonal prism. hs = high spin; ls = low spin.

$[CuBr_4]^{2-}$, $[Ni(CN)_4]^{2-}$, and $[PtI_4]^{2-}$ electronically prefer square planar, but steric effects cause a distortion toward the less hindered tetrahedral geometry.

Generalizing the 18e Rule

We can now generalize the 18e rule for complexes of any coordination number, n. Figure 2.2 shows the situation for a complex ML_n for $n = 4–9$ where there are n M—L σ-bonding orbitals and $(9 - n)$ nonbonding d orbitals. The value of n appropriate for this situation is the CN defined in Eq. 2.11, so for $[Ni(\eta^3$-allyl)2]$ (**2.9**), $n = 4$; for Cp_2Fe (**2.7**) or $[Mo(\eta^6$-$C_6H_6)_2]$ (**2.10**), $n = 6$; and for Cp_2TiCl_2 (**2.11**) or $MoH_4(PR_3)_4$ (**2.8**), $n = 8$. Filling the bonding and nonbonding levels—a total of nine orbitals—requires 18 electrons. Normally the antibonding orbitals are empty. In Fig 2.2, each type of orbital—bonding, nonbonding, and antibonding—is represented by a thick horizontal line, although in reality each group is spread out in a pattern that depends on the exact geometry and ligand set. Figure 1.4 shows the nonbonding orbitals for tetrahedral and square planar geometries, for example.

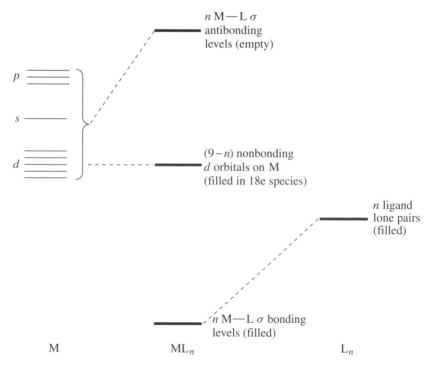

FIGURE 2.2   Schematic description of the bonding in a metal complex ML_n ($n = 4–9$), showing how the molecular orbitals of the complex can be considered as dividing into three classes: bonding (always filled), nonbonding (filled in 18e complexes), and antibonding (almost always empty). Each thick horizontal line represents several orbitals somewhat spread out in energy, depending on the exact nature of the complex.

Net Ionic Charge

The net ionic charge, $c+$ in $[MX_aL_b]^{c+}$, has a strong influence on the chemistry of a complex. A positive charge increases its electrophilic character (e.g., $[Ru(\eta^6\text{-}C_6H_6)]^{2+}$) and a negative charge increases its nucleophilic character (e.g., $[Fe(CO)_4]^{2-}$). It also affects back bonding to the ligands (Section 2.7, Table 2.9).

2.6 EFFECTS OF COMPLEXATION

The chemical character of many ligands is profoundly modified on binding to the metal. For the typical range of metal fragments L_nM, there is a smooth gradation of properties from strongly σ acceptor to strongly π basic. A typical unsaturated ligand Q is depleted of charge and made more electrophilic by a σ-acceptor L_nM fragment, but made to accept electrons and therefore become more nucleophilic for a π-basic L_nM fragment. As an example, free benzene is very resistant to attack by nucleophiles but reacts with electrophiles. In the complex $(C_6H_6)Cr(CO)_3$, in contrast, the $Cr(CO)_3$ fragment is a good acceptor by virtue of its three CO ligands and so depletes the electron density on the aromatic ring. This makes it susceptible to nucleophilic attack but resistant to electrophilic attack. A factor that increases the electrophilic character of the ligands is a net positive charge on the complex, such as $[Ru(\eta^6\text{-}C_6H_6)_2]^{2+}$. On the other hand, both Cp groups and phosphines are strong donors, and so the acetyl **2.17** in Eq. 2.13 is very largely in the carbene (see Chapter 11) form **2.18**. It is subject to electrophilic attack to give **2.19**:

(2.13)

Polarization

A third important situation occurs if the metal fragment is somewhere in the middle of the range of electronic properties mentioned above and is both a σ acceptor and a π donor. It might be thought that the unsaturated ligand would differ little in its chemical character from the situation in the free state. In fact, the ligand can still be strongly activated by polarization. This is because the σ donation from the ligand to the metal usually depletes the electron density of one atom or set of atoms in the ligand, but π back donation from the metal raises the electron density on a different set of atoms. For example, in the case of molecular nitrogen, N_2, σ donation to the metal comes from a lone pair on the nitrogen directly bonded to the metal. The back bonding from the metal goes into a π^* orbital that is delocalized over both nitrogens. This means that the nitrogen directly bound to the metal tends to become positively charged, and the terminal nitrogen negatively charged on binding:

$$M\!-\!\overset{\partial+}{N}\!\equiv\!\overset{\partial-}{N}$$

2.20

This polarization activates the coordinated N_2 toward chemical reactions, such as protonation at the terminal nitrogen and nucleophilic attack at the vicinal nitrogen; the free ligand is, of course, nonpolar and notably unreactive. The general situation is summarized in Table 2.7. If a ligand is normally reactive toward, say, nucleophiles, we can deactivate it by binding to a nucleophilic metal. The metal can then be thought of as acting as a protecting group. A ligand that is inert toward nucleophilic attack can be activated by binding to an electrophilic metal.

Paradoxically, stronger binding does not always lead to stronger ligand activation. An excellent example is coordinated H_2 (Section 1.8), a ligand that is enormously acidified on binding. The pK_a of free H_2 is near 35 but that for bound H_2 often lies in the range 0–20 with the more weakly bound ligands at the lower end of the range (i.e., most acidified).

Free $\neq$ Bound

The bound form of a given ligand is usually very different in properties compared to the same ligand in the free state. A knowledge of the behavior of organic

TABLE 2.7   Effect of Electronic Character of Metal Fragment on Tendency for Attached Ligand to Undergo Nucleophilic or Electrophilic Attack

Character of Free Ligand	Character of ML_n Fragment[a]		
	σ Acid	Polarizing	π Base
Susceptible to electrophilic attack	Suppresses susceptibility	May enhance susceptibility	Enhances
Susceptible to nucleophilic attack	Enhances susceptibility	May enhance susceptibility	Suppresses
Unreactive	May allow nu. attack	May allow both nu. and el. attack	May allow el. attack

[a] Abbreviations: nu. = nucleophilic; el. = electrophilic.

carbenes, dienes, or other species can be misleading in trying to understand the chemistry of their complexes. For example, a notable feature of diene chemistry is their reaction with dienophiles in the Diels–Alder reaction. Dienes coordinated in the η^4 fashion do not give this reaction. In a sense, we can consider that the complex is already a Diels–Alder adduct, with the metal as the dienophile.

The properties of the metal ions as well as those of the ligands are both altered on complex formation. For example, Co(III) is very strongly oxidizing in a simple compound such as the acetate, which will even oxidize hydrocarbons. We know from Werner's work that almost all of this oxidizing power can be quenched by binding six ammonias to the Co(III) ion. The resulting $[Co(NH_3)_6]^{3+}$ ion lacks the severe electron deficiency of the acetate complex because of the presence of six strong σ-donor ligands. Conversely, molybdenum atoms are strongly reducing, yet $Mo(CO)_6$ is an air-stable compound with only modest reducing properties because CO removes electron density from the metal by back donation.

Finally, it is important to remember that donor and acceptor are relative terms. If we take a complex $L_nM—H$, in which the hydride ligand bears no strong positive or negative charge, then we can consider the complex as arising from $L_nM^+ + H^-$, $L_nM\cdot + H\cdot$, or $L_nM^- + H^+$. We would have to regard H^- as a strong donor to L_nM^+, H^+ as a strong acceptor from L_nM^-, and $H\cdot$ as being neither with respect to $L_nM\cdot$. Normally the ionic model is assumed and the first type of dissection is implied.

2.7   DIFFERENCES BETWEEN METALS

Changing the metal has an important effect on the properties of the resulting complexes. So great are the differences that it is not unusual for a single research group to confine itself to one part of the periodic table. As we move from left to right, the electronegativity of the elements increases substantially. This means that the orbitals in which the electrons are located start out relatively high in energy and fall steadily as we go to the right. Table 2.8 shows the Pauling electronegativities of the transition elements. The early transition metals are electropositive and so readily lose all their valence electrons. These elements are therefore often found in the highest permissible oxidation state, such as d^0 Zr(IV) and Ta(V). Lower oxidation states, such as d^2 Zr(II) and Ta(III), are very easily oxidized because the two d electrons are in an orbital of relatively high energy and, therefore, are easily lost. These systems can be very air sensitive. Not only are these electrons easily lost to an oxidizing agent but also have a strong tendency to be lost to the π^* orbitals of an unsaturated ligand in back donation. This makes d^2 early metal ions very π basic and able to bind π ligands strongly with the effects we saw in Section 2.6. Ligands such as CO, C_6H_6, and C_2H_4, which require back bonding for stability, will tend to bind only weakly, if at all, to d^0 metals.

Late metals, in contrast, are relatively electronegative, so they tend to retain their valence electrons. The low oxidation states, such as d^8 Pd(II), tend to be stable, and the higher ones, such as d^6 Pd(IV), often find ways to return to Pd(II);

TABLE 2.8   Pauling Electronegativities of the Transition Elements[a]

Sc	Ti	V	Cr	Mn	Fe	Co	Ni	Cu
1.3	1.5	1.6	1.6	1.6	1.8	1.9	1.9	1.9
Y	Zr	Nb	Mo	Tc	Ru	Rh	Pd	Ag
1.2	1.3	1.6	2.1	1.9	2.2	2.3	2.2	1.9
La	Hf	Ta	W	Re	Os	Ir	Pt	Au
1.1	1.3	1.5	2.3	1.9	2.2	2.2	2.3	2.5

[a]Lanthanides and actinides: 1.1–1.3. The electronegativities of important ligand atoms are H, 2.2; C, 2.5; N, 3.0; O, 3.4; F, 4; Si, 1.9; P, 2.2; S, 2.6; Cl, 3.1; Br, 2.9; I, 2.6. Effective electronegativities of all elements are altered by their substituents, for example, the electronegativities estimated for an alkyl C, a vinyl C, and a propynyl C are 2.5, 2.75, and 3.3, respectively.

that is, they are oxidizing. Back donation is not so marked as with the early metals, and so any unsaturated ligand attached to the weak π-donor Pd(II) will accumulate a positive charge. As we see later (page 213), this makes the ligand subject to attack by nucleophiles Nu^- and is the basis for important applications in organic synthesis.

Table 2.9 shows that several types of changes all cause an increase in $\nu(CO)$ values of metal carbonyls, corresponding to a reduction in the basicity of the

TABLE 2.9   Effects of Changing Metal, Net Charge, and Ligands on π Basicity of Metal, as Measured by $\nu(CO)$ Values (cm^{-1}) of the Highest Frequency Band in IR Spectrum

		Changing Metal Across the Periodic Table				
$V(CO)_6$	$Cr(CO)_6$	$Mn_2(CO)_{10}$	$Fe(CO)_5$		$Co_2(CO)_8$	$Ni(CO)_4$
1976	2000	2013(av)[a]	2023(av)[a]		2044(av)[b]	2057
	$Cr(CO)_4$		$Fe(CO)_4$			$Ni(CO)_4$
	1938[c]		1995[c]			2057
		Changing Metal down the Periodic Table				
$[Cr(CO)_6]$		$[Mo(CO)_6]$	$W(CO)_6$			
2000		2004	1998			
		Changing Net Ionic Charge in an Isoelectronic Series				
$[Ti(CO)_6]^{2-}$		$[V(CO)_6]^-$	$Cr(CO)_6$		$[Mn(CO)_6]^+$	
1747[d]		1860[d]	2000		2090	
		Replacing π- Acceptor CO Groups by Non-π-Acceptor Amines				
$[Mn(CO)_6]^+$		$[(MeNH_2)Mn(CO)_5]^+$	$[(en)Mn(CO)_4]^+$		$[(tren)Mn(CO)_3]^+$	
2090		2043(av)	2000(av)		1960	

[a]Average of several bands.
[b]Of isomer without bridging CO groups.
[c]Unstable species observed in matrix; the advantage of this series is that it keeps the number of COs the same for each metal.
[d]The band positions may be slightly lowered by coordination of the CO oxygens to the counterions.
en = $H_2NCH_2CH_2 NH_2$; tren = $H_2NCH_2CH_2NHCH_2CH_2 NH_2$.

metal and in the strength of back bonding to CO: (1) making the net ionic charge one unit more positive, (2) replacing one CO by a pure σ-donor amine ligand, and (3) moving to the right by one periodic group. All three changes seem to have approximately equal effects. The first series, going from $[Ti(CO)_6]^{2-}$ to $[Fe(CO)_6]^{2+}$, involves changes of metal as well as of ionic charge, but comparison with the series $Cr(CO)_4$ to $Ni(CO)_4$ suggests that about one half of the total effect is due to the change of metal and the other half to the change in ion charge.

First-row metals have lower M–L bond strengths and crystal field splittings compared with their second- and third-row analogs. They are more likely to undergo 1e redox changes rather than the 2e changes often associated with the second and third rows. Finally, the first-row metals do not attain high oxidation states so easily as the second and especially the third row. Mn(V), (VI), and (VII) (e.g., MnO_4^-) are rare and usually highly oxidizing; Re(V) and (VII) are not unusual and the complexes are not strongly oxidizing.

- Oxidation state (Section 2.4) is a useful classification but can be ambiguous.
- Metal complexes with similar d^n configurations have similar chemistry.
- The splitting pattern of the d orbitals changes with geometry.
- The chemistry of many ligands changes profoundly on complexation.
- Ion charge, ligand set, and metal all affect the basicity of the metal (Table 2.9).

2.8   OUTER-SPHERE COORDINATION

The distinction between inner and outer spheres was first introduced by Alfred Werner. The part of a complex that is bound together by covalent or coordinate bonds, and placed within square brackets in the formula such as Cp_2Fe^+ in $[Cp_2Fe]Cl$, is considered as the inner sphere of the complex. The outer coordination sphere consists of solvent and any counterions that may be bound to the inner-sphere ligands by intermolecular forces such as hydrogen bonding, ion pairing, or dipole–dipole forces. Taube's[6] description of electron transfer between metal complexes made extensive use of the idea of outer-sphere interaction of the reacting metal complexes, but this fascinating chemistry involves classical Werner complexes and so is not described here.

The outer-sphere ligands, such as solvent, are often weakly bound, and the outer-sphere complex (OSC) normally has a variable structure and little more than a transient existence. In certain cases,[7] however, a much stronger interaction seems to be possible. A new type of hydrogen bond has been found between the hydridic hydride of a metal complex and the protonic hydrogen of an OH or NH group, having an estimated OS bond strength of 4–6 kcal/mol.[8] The resulting *dihydrogen bond* is illustrated by the neutron diffraction structure of

[ReH$_5$(PPh$_3$)$_3$]C$_8$H$_6$N—H (**2.21**) with its N—H···H-Re dihydrogen bond. For an idea of the significance of the H···H distance of 1.73 Å in **2.22**, we can compare it with a normal H···H contact, normally no less than the sum of the van der Waals radii of two hydrogen atoms, 2.4 Å.

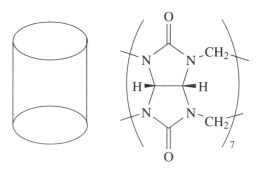

2.21

In a potential medical application,[9] the toxicity of some binuclear Pt(II) antitumor compounds was much reduced, without affecting the beneficial biological activity, by encapsulation in cucurbit[7]uril (**2.22**) a molecule having a cylindrical cavity. NMR spectroscopy showed that the encapsulated Pt(II) complex exchanges only slowly on the NMR timescale with free Pt(II) complex.

cylindrical cavity
2.22

REFERENCES

1. C. Tolman, *Chem. Soc. Rev.* **1**, 337, 1972.

2. M. L. H. Green, *J. Organomet. Chem.* **500**, 127, 1995.

3. M. Ogasawara, S. A. Macgregor, W. E. Streib, K. Folting, O. Eisenstein, and K. G. Caulton, *J. Am. Chem. Soc.* **117**, 8869, 1995.

4. P. Hamon, L. Toupet, J.-R. Hamon, and C. Lapinte, *Chem. Commun.* 341, 1994.

5. D. Astruc, *Acc. Chem. Res.* **24**, 36, 1991.

6. V. G. Poulopoulou and H. Taube, *Inorg. Chim. Acta* **319**, 123, 2001.

7. K. Zamaraev, *New J. Chem.* **18**, 3, 1994.

8. R. H. Crabtree, P. E. M. Siegbahn, O. Eisenstein, A. L. Rheingold, and T. F. Koetzle, *Acc. Chem. Res.* **29**, 349, 1996, and references cited.

9. N. J. Wheate, A. I. Day, R. J. Blanch, A. P. Arnold, C. Cullinane, and J. G. Collins, *Chem. Comm.* 1424, 2004.

PROBLEMS

Answering Problems

It is important that any intermediate you suggest in an organometallic reaction be reasonable. Does it have an appropriate electron count, coordination number, and oxidation state? If it is the only known Rh(V) carbonyl, it may be open to criticism. Check that the organic fragment is also reasonable. Sometimes students write diagrams without stopping to consider that their structure contains 5-valent carbon. Indicate the hapticity of each ligand.

1. Give the electron counts, formal oxidation states, and d^n configurations of the following: $[Pt(NH_3)_4]^{2+}$, $PtCl_2(NH_3)_2$, $PtCl_4^{2-}$, $(\eta^5\text{-}C_5H_5)_2Ni$, $[(R_3P)_3 Ru(\mu\text{-}Cl)_3Ru(PR_3)_3]^+$, ReH_9^{2-}, $CpIrMe_4$, $TaMe_5$, $(\eta^5\text{-}C_5H_5)_2TiCl_2$, and $MeReO_3$.

2. A complex is found to correspond to the empirical formula $(CO)_3ReCl$. How could it attain the 18e configuration without requiring any additional ligands?

3. How could a complex of empirical formula $Cr(CO)_3(C_6H_5)_2$ attain the 18e configuration?

4. A complex $Ti(\eta^2\text{-}MeN{=}CH{-}CH{=}NMe)_2$ is found to be chelated via nitrogen. What oxidation state should we assign to Ti? Is any alternative assignment possible?

5. Count the valence electrons in the complexes shown in Problem 1, but using a different model (ionic or covalent) from the one you used originally.

6. Given the existence of $(CO)_5Mn{-}Mn(CO)_5$, deduce the electron counting rule that applies to M$-$M bonds. Verify that the same holds for $Os_3(CO)_{12}$, which contains three Os$-$Os bonds and only terminal CO groups. What structure do you think is most likely for $Rh_4(CO)_{12}$?

7. Show how the valence electron count for the carbon atom in $CH_3NH_3^+$ can be evaluated considering the molecule as an ammonia complex. Can the methylene carbon in $CH_2{=}C{=}O$ be treated in a similar way?

8. Water has two lone pairs. Decide whether both or only one of these should normally be counted, given that the following typical complexes exist: $IrH_2(H_2O)_2(PPh_3)_2^+$, $(\eta^6\text{-}C_6H_6)Os(H_2O)_3^{2+}$.

9. Acetone can bind in an η^2 (via C and O) and an η^1 fashion (via O). Would you expect the electron count to be the same or different in the two forms?

What kind of metal fragments would you expect would be most likely to bind acetone as (a) an η^1 and (b) an η^2 ligand? Would either binding mode be expected to enhance the tendency of the carbonyl carbon to undergo nucleophilic attack?

10. Predict the hapticity of each Cp ring in $Cp_2W(CO)_2$, and of each "triphos" in $[Pd\{(PPh_2CH_2CH_2)_3CPh\}_2]^{2+}$.

11. Assign the oxidation states, d^n configurations, and electron counts for the two species shown below, which are in equilibrium in solution. Use both the covalent and ionic models.

$$W(\eta^2\text{-}H_2)(CO)_3(PR_3)_2 \rightleftharpoons W(H)_2(CO)_3(PR_3)_2$$

3

METAL ALKYLS, ARYLS, AND HYDRIDES AND RELATED σ-BONDED LIGANDS

Metal alkyls and aryls are perhaps the simplest organometallic species. Yet transition metal examples remained very rare until the principles governing their stability were understood in the 1960s and 1970s. These principles make a useful starting point for our study of alkyls because they introduce some of the most important organometallic reactions, which we will go on to study in more detail in later chapters.

After alkyls, we move on to metal hydrides and dihydrogen complexes, another area with important implications for later discussions.

3.1 TRANSITION METAL ALKYLS AND ARYLS

The story of metal alkyls starts with the main-group elements, particularly Li, Mg, Zn, Hg, As, and Al, and alkyls of some of these elements are still very widely useful. Indeed, it is scarcely possible to carry out an organic synthesis of any complexity without using an alkyllithium reagent. Organometallic chemistry began in a dramatic way. Working in a Parisian military pharmacy in 1757, Cadet made the appallingly evil-smelling cacodyl oxide (Greek: $kakos$ = stinking), later shown by Bunsen[1] to be $Me_2As-O-AsMe_2$, from As_2O_3 and $KOOCCH_3$. In 1848, an attempt by Edward Frankland[2] to prepare free ethyl radicals by reaction of ethyl iodide with metallic zinc instead gave a colorless liquid that proved to be diethylzinc. Because arsenic is a semimetal, $ZnEt_2$ was the first unambiguous case of a molecular compound with a metal–carbon bond and Frankland is often

The Organometallic Chemistry of the Transition Metals, Fourth Edition, by Robert H. Crabtree
Copyright © 2005 John Wiley & Sons, Inc.

considered a founder of organometallic chemistry. It was only with Victor Grignard's discovery (1900) of the alkylmagnesium halide reagents, RMgX, however, that organometallic chemistry began to make a major impact through its application in organic synthesis. The later development of organolithium reagents is associated with Schlenk (1914) and Ziegler (1930). Ziegler was also instrumental in showing the utility of organoaluminum reagents.

Metal Alkyls as Stabilized Carbanions

Grignard reagents, RMgX, provided the first source of nucleophilic alkyl groups, $R^{\delta-}$, to complement the electrophilic alkyl groups, $R^{\delta+}$, available from the alkyl halides. Metal alkyls result from combining an alkyl anion with a metal cation. In combining, the alkyl anion is stabilized to a different extent depending on the electronegativity of the metal concerned. Alkyls of the electropositive elements of groups 1–2, as well as Al and Zn, are sometimes called *polar organometallics* because the alkyl anion is only weakly stabilized and retains much of the strongly nucleophilic and basic character of the free anion. Polar alkyls all react with traces of humidity to hydrolyze the M—C bond to form M—OH and release RH. Air oxidation also occurs very readily, and so polar organometallics must be protected from both air and water. Alkyls of the early transition metals, such as Ti or Zr, can also be very air and water sensitive, but as we move to less electropositive metals by moving to the right and down the periodic table, the compounds become much less reactive, until we reach Hg, where the Hg—C bond is so stable that $[\text{Me—Hg}]^+$ cation is indefinitely stable in aqueous sulfuric acid in air. As we go from the essentially ionic $NaCH_3$, to the highly polar covalent Li and Mg species, to the essentially covalent late metal alkyls, the reactivity falls steadily along the series, showing the effect of changing metal (Fig. 3.1).

The inherent stability of the R fragment plays a role, too. As an sp^3 ion, CH_3^- is the most reactive. As we move to sp^2 $C_6H_5^-$ and even more to sp $RC\equiv C^-$, where the lone pair of the anion is stabilized by being in an orbital with more s character, the intrinsic reactivity falls off. The same trend makes the acidity of the hydrocarbons increase as we go from CH_4 ($pK_a = \sim50$) to C_6H_6 ($pK_a = \sim43$) and to $RC\equiv CH$ ($pK_a = \sim25$), so the anion from the latter is the most stable and least reactive.

Following the successful syntheses of main-group alkyls, many attempts were made to form transition metal alkyls. Pope and Peachey's Me_3PtI (1909) was an early but isolated example of a d-block metal alkyl. Attempts during the 1920s through 1940s to make further examples of d-block alkyls all failed. This was especially puzzling because by then almost every nontransition element had been shown to form stable alkyls. These failures led to the view that transition metal–carbon bonds were unusually weak; for a long time after that, few serious attempts were made to look for them. In fact, we now know that such M—C bonds are strong (30–65 kcal/mol is typical). It is the existence of several easy decomposition pathways that makes many transition metal alkyls unstable. Kinetics, not thermodynamics, was to blame for the synthetic failures. This is

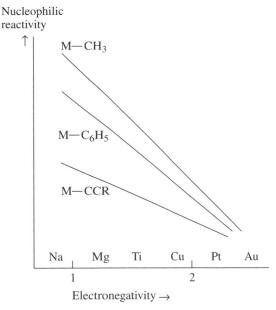

FIGURE 3.1 Schematic diagram showing qualitatively how the nucleophilic reactivity of main-group and transition metal alkyls to protons or air oxidation depends on the alkyl itself and the electronegativity of the metal.

fortunate because it is easier to manipulate the system to block decomposition pathways than it is to increase the bond strength. In order to be able to design stable alkyls, we must look at some of these pathways to see how they can be inhibited. This example of the historical evolution of our ideas implies that just as some of the early assumptions in this area proved to be wrong, some of our ideas today will probably turn out to be wrong, too—the problem is we do not know which ones!

β Elimination

The major decomposition pathway for alkyls is *β elimination*[3] (Eq. 3.1), which converts a metal alkyl into a hydridometal alkene complex. We study it in detail in Section 7.4. For the moment we need only note that this very common mechanistic type can occur whenever

1. The β carbon of the alkyl bears a hydrogen substituent.
2. The M—C—C—H unit can take up a roughly coplanar conformation,[3b] which brings the β hydrogen close to the metal.
3. There is a vacant site on the metal, symbolized here as □, cis to the alkyl.
4. The reaction is much more rapid for d^2 and higher metals than for d^0 and main-group alkyls.

Requirements 1 and 3 arise because it is the β hydrogen of the alkyl that is transferred to the metal to give the product hydride. The geometry of the situation means that a cis site is required on the metal and a coplanar M$-$C$-$C$-$H arrangement in the ligand. The elimination is believed to be concerted; that is, C$-$H bond breaking and M$-$C and M$-$H bond making happen at the same time.

$$\text{(3.1)}$$

3.1

The term "vacant site" of requirement 3 needs some clarification. It does *not* simply mean that there should be a gap in the coordination sphere large enough to accommodate the incoming ligand. There must also be an empty orbital ready to accept the β-H, or more exactly, the pair of electrons that constitutes the β-C$-$H bond. The electron count of the product alkene hydride is 2e more than that of the alkyl starting material. An 18e alkyl is much more reluctant to β-eliminate via a 20e intermediate than is a 16e alkyl, which can go via an 18e alkene hydride. Even if the alkene subsequently dissociates, which is often the case, we still have to stabilize the transition state leading to the alkene hydride intermediate if we want the reaction to be fast. An 18e alkyl, on the other hand, is said to be *coordinatively saturated*. By this we mean that an empty orbital is not available. Some 18e alkyls do β-eliminate, but detailed mechanistic study often shows that the prior dissociation of some ligand is required in the rate-determining step.

Main-group alkyls can also β-eliminate (e.g., Eq. 3.2), but this usually happens much more slowly. The reason for this difference is believed to be the greater ability of d-block metals to stabilize the transition states involved that resemble agostic alkyl complexes such as **3.7**.

$$[(\text{EtMeCH})_3\text{Al}]_2 \xrightarrow{\text{reflux}} [(\text{EtMeCH})_2\text{Al}(\mu\text{-H})]_2 + \text{butene} \qquad \text{(3.2)}$$

Stable Alkyls

To have a kinetically stable alkyl, we must block the β-elimination pathway for decomposition. This can happen for

1. Alkyls that have no β hydrogen:

$$\text{WMe}_6 \qquad \text{Ti}(\text{CH}_2\text{Ph})_4 \qquad \text{W}(\text{CH}_2\text{SiMe}_3)_6 \qquad \text{TaCl}_2(\text{CH}_2\text{CMe}_3)_3$$

$$\text{C}_2\text{F}_5\text{Mn}(\text{CO})_5 \qquad \text{LAuCF}_2\text{CF}_2\text{Me} \qquad \text{Pt}(\text{C}\equiv\text{CCF}_3)_2\text{L}_2$$

$$\text{Pt}(\text{CH}_2\text{COMe})\text{Cl}(\text{NH}_3)_2$$

2. Alkyls for which the β hydrogen is unable to approach the metal as a result of the geometry or bulk of the ligand:

$$PtH(C\equiv CH)L_2 \qquad PdPh_2L_2 \qquad Cr(CMe_3)_4 \qquad Cr(CHMe_2)_4$$

$$CpL'_3MoCH=CHCMe_3$$

3. Alkyls in which the M–C–C–H unit cannot become *syn*-coplanar:[3c]

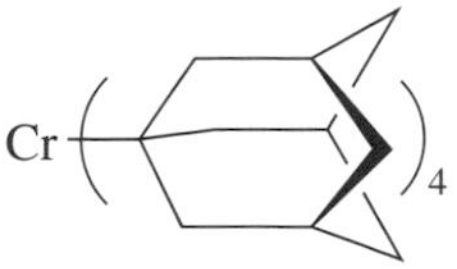

[Cr(1-adamantyl)₄] 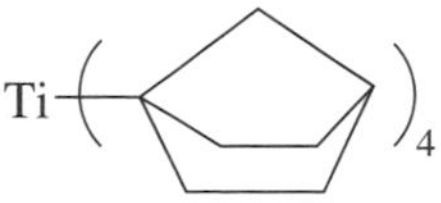   Ti(6-norbornyl)₄   L₂Pt(CH₂)₃

The first two would give "forbidden" anti-Bredt olefins if they were to β-eliminate.

4. An 18-electron species with firmly bound ligands, which will not dissociate to generate a vacant site:

$$Cp(CO)_2FeCH_2CH_3, \qquad Cp(CO)_3MoCH_2CH_3,$$
$$\textbf{3.2} \qquad\qquad\qquad \textbf{3.3}$$

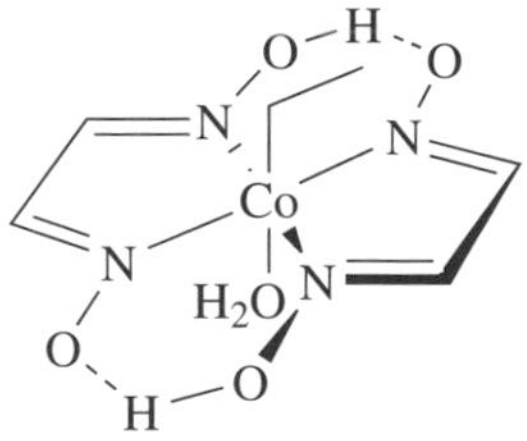

$$Cp(CO)IrPrH, \qquad [Cr(H_2O)_5Et]^{2+}$$
$$\textbf{3.5} \qquad\qquad \textbf{3.6}$$

3.4

5. Some d^0 alkyls:

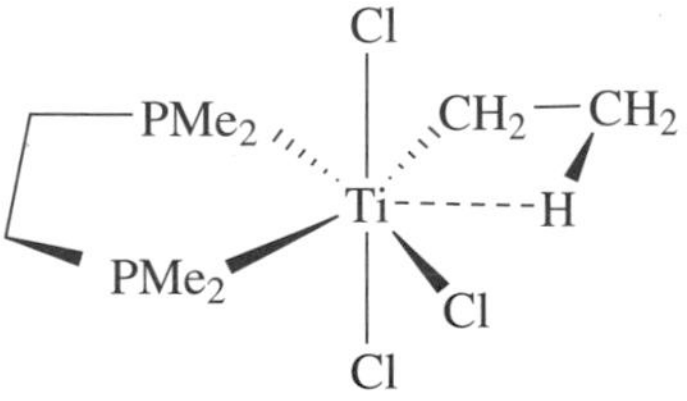

3.7

Some of these cases call for special comment. WMe_6, like WH_6, has a trigonal prismatic structure **3.8**,[4a] not the octahedral structure usually found for ML_6

species. Albright and Eisenstein[4b] had previously predicted that $d^0\,MX_6$ species would be trigonal prismatic where X is not a π donor. Methyl compounds are especially numerous, and the small size of this ligand allows the formation of polyalkyls. Often, substitution with electron-withdrawing or bulky groups (e.g., $-CH_2Ph$, $-CH_2SiMe_3$) also gives stable alkyls. The vinyl and phenyl groups both have β hydrogens, but they do not β-eliminate easily. One reason may be that the β hydrogens are further from the metal in these sp^2-hybridized systems with $120°$ angles at carbon, less favorable for delivery of the β-H than in the sp^3 ethyl group $(109°)$. In addition, as is the case for other electronegative alkyl groups, the phenyl and vinyl groups have stronger M−C bonds than does the ethyl group.

3.8

The *iso*-propyl and *tert*-butyl chromium complexes are unusual. Presumably, their steric bulk prevents the β-C−H bond from reaching the metal. These structures seem to be sterically saturated. The examples containing noncoplanar M−C−C−H groups mainly involve cyclic alkyls, in which the rigidity of the ring system holds the M−C−C−H dihedral angle near $60°$ and far away from the value of $0°$ required for β elimination. The fourth group includes those systems with no vacant site (**3.2**, **3.3**, and **3.5**) and others that have such a site, but not cis to the alkyl (**3.4**, assuming that the aqua ligand can dissociate). Compound **3.6** is not an 18e species, but as a d^3 Cr(III) complex it is coordination inert (see Section 1.4).

Agostic Alkyls

Rarer are those species in which all the criteria appear to be favorable but in which β elimination still does not occur. In some of these (e.g., **3.7**) the β-C−H bond is bound to the metal in a way that suggests that the alkyl is beginning the approach to the transition state for β elimination, but the reaction has been arrested along the way. These *agostic* alkyls can be detected by X-ray or neutron crystal structural work and by the high-field shift of the agostic H in the proton NMR. The lowering of the J(C,H) and ν(CH) in the NMR and IR spectra, respectively, on binding is symptomatic of the reduced C−H bond order in the agostic system.[5] The reason that β elimination does not occur in **3.7** is that the d^0 Ti has no electron density to back donate into the σ^* orbital of the C−H bond. This back donation breaks the C−H bond in the β-elimination reaction, much as happens in oxidative addition (see Eq. 1.5). Agostic binding of C−H bonds also provides a way to stabilize coordinatively unsaturated species. They are also found in transition states for reactions such as alkene insertion/β elimination either by experiment (see Fig. 12.4) or in theoretical work.[6]

We saw earlier that we need a 2e vacant site (an empty d orbital) on the metal for β elimination. Now we see that we also need an available electron pair (a filled d orbital) for breaking the C—H bond. There is a very close analogy between these requirements and those for binding a soft ligand such as CO. Both processes require a metal that is both σ acidic and π basic. In the case of CO, binding leads to a reduction in the CO bond order. In the case of the β-C—H bond of an alkyl group, this binding can reduce the C—H bond order to zero, by cleavage to give the alkene hydride complex. Alternatively, if the metal is a good σ acid but a poor π base, an agostic system may be the result, and the C—H bond is only weakened, not completely broken. Many of the characteristic reactions of organometallic chemistry require both σ-acid and π-base bifunctional character. This is why transition metals, with their *partly* filled d orbitals, give these reactions so readily.

Halide Elimination

β Elimination of halide can also occur. Early transition metals, such as Ti, the lanthanides, and the actinides do not tend to form stable fluoroalkyls because the very high M—F bond strengths of these elements encourages β elimination of the halide. The late transition metals have weaker M—F bonds and do form stable fluoroalkyls. Not only do these ligands lack β-Hs, but the M—C bond strengths are very high, as is also true for other alkyls MCH_2X, where X is any electronegative group. CF_3, like PF_3, can also act as a π acceptor via the σ^* orbitals of the C—F bond (see Section 4.2), which also makes the M—C bond stronger for the π-basic late metals. The C_6F_5 group forms extremely stable aryls with the late transition metals in which an aryl π^* orbital acts as electron acceptor.[7]

Reductive Elimination

A second very common decomposition pathway for metal alkyls is *reductive elimination* ("red. elim." in Eq. 3.3).[8] This leads to a decrease by two units in both the electron count and the formal oxidation state. (This is why the reaction is labeled "reductive.") We study it in detail in Chapter 6. In principle, it is available to all complexes, even if they are d^0 or 18e, provided a stable oxidation state exists two units more reduced than the oxidation state in the starting alkyl. In fact, in many instances reductive elimination is not observed, for example, if X in **3.9** is a halogen. The reason is that for alkyl halides, the position of equilibrium for Eq. 3.3 usually lies well over to the side of **3.9**; in other words, **3.9** is usually more stable thermodynamically. Some examples of the loss of alkyl halide are known, however.

$$L_nM(Me)X \xrightarrow{\text{red. elim.}} L_nM + MeX \qquad (3.3)$$

$$\text{\textbf{3.9}, 18e} \qquad\qquad \text{\textbf{3.10}, 16e}$$

On the other hand, when X = H, the reaction is usually both kinetically facile and thermodynamically favorable, so isolable alkyl hydrides are rare. Where X = CH$_3$, the thermodynamics still favor elimination, but the reaction is generally much slower kinetically. It is often the case that reactions involving a hydrogen atom are much faster than those involving any other element; this is because H carries no electrons other than bonding electrons, and these are in a $1s$ orbital, which is capable of making and breaking bonds in any direction in the transition state. The sp^3 orbital of the CH$_3$ fragment is directed in space, and so there can often be poorer orbital overlap in the transition state.

Stability from Bulky Substituents

Bulky ligands provide a general strategy for stabilizing many different classes of organometallic complex. Associative decomposition pathways for alkyls, such as by reaction with the solvent or with another molecule of the complex, can also be important, especially for 16e metals. These can often be suppressed with bulky coligands. For example, square planar Ni(II) alkyls are vulnerable to attack along the z direction perpendicular to the plane. The o-tolyl complex **3.11**, in which this approach is blocked, is more stable than the analogous diphenyl, **3.12**, for example. This steric factor has made the use of bulky alkyl groups, such as neopentyl (CH$_2$CMe$_3$) or trimethylsilylmethyl (CH$_2$SiMe$_3$) common in organometallic chemistry.

3.11   **3.12**

Where β elimination cannot occur for the reasons discussed above, α elimination sometimes takes over. This leads to the formation of species called *carbenes*, which have M=C double bonds. The first step in the thermal decomposition of Ti(CH$_2t$-Bu)$_4$ is known to be α elimination to Ti(=CHt-Bu)(CH$_2t$-Bu)$_2$. Similarly, attempts to prepare Ta(CH$_2t$-Bu)$_5$ led to formation of the carbene complex, t-BuCH=Ta(CH$_2t$-Bu)$_3$. Carbenes and α elimination are discussed in Sections 11.1 and 7.4.

Where a heteroatom such as N or O is present to activate the adjacent C$-$H bonds for reaction, double C$-$H bond cleavage can occur at the same carbon.[9] In Eq. 3.4, the first cleavage, an oxidative addition, and the second, an α elimination, can be observed stepwise for R = H. Even for the ArNEt$_2$ analog (R = CH$_3$), where there is a choice between α elimination and β elimination in the second step, the product still comes exclusively from α elimination. In Eq. 3.5, the carbene is again formed but the hydrogen produced is now trapped by half of the Ru(II) as a dihydrogen complex.

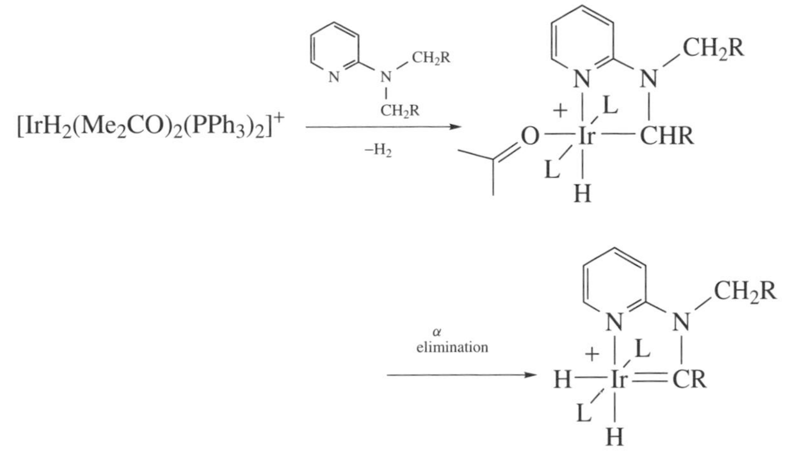

$$[IrH_2(Me_2CO)_2(PPh_3)_2]^+ \xrightarrow{-H_2}$$

(3.4)

$$\xrightarrow[\text{elimination}]{\alpha}$$

$$(L = PPh_3; R = H \text{ or } CH_3)$$

$$[RuHCl(PiPr_3)_2]_2 \longrightarrow [RuH(H_2)Cl(PiPr_3)_2] + $$

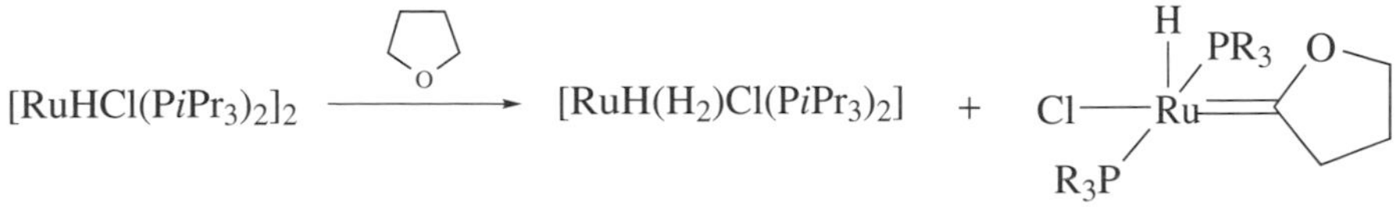

(3.5)

The α heteroatom may stabilize the alkyl by allowing back donation into the C-X σ^* to be discussed in Chapter 4 (Fig. 4.3), while the carbene is additionally stabilized by X to C(p_π) donation (see **11.1**).

Preparation of Metal Alkyls

The chief methods for the synthesis of alkyls involve (1) an R^- reagent, (2) an R^+ reagent, (3) oxidative addition, and (4) insertion. Typical examples of these are shown in Eqs. 3.6–3.15:

1. From an R^- reagent (nucleophilic attack on the metal):

$$WCl_6 \xrightarrow{LiMe} WMe_6 + LiCl \qquad (3.6)$$

$$NbCl_5 \xrightarrow{ZnMe_2} NbMe_2Cl_3 + ZnCl_2 \qquad (3.7)$$

2. From an R^+ reagent (electrophilic attack on the metal):

$$Mn(CO)_5^- \xrightarrow{MeI} MeMn(CO)_5 + I^- \qquad (3.8)$$

$$\text{CpFe(CO)}_2^- \xrightarrow{\text{Ph}_2\text{I}^+} \text{Cp}_2\text{Fe(CO)}_2\text{Ph} + \text{PhI} \qquad (3.9)$$

$$[\text{Mn(CO)}_5]^- \xrightarrow{\text{CF}_3\text{COCl}} \text{CF}_3\text{COMn(CO)}_5 \xrightarrow{-\text{CO}} \text{CF}_3\text{Mn(CO)}_5 \quad (3.10)$$

3. By oxidative addition:

$$\text{IrCl(CO)L}_2 \xrightarrow{\text{MeI}} \text{MeIrICl(CO)L}_2 \qquad (3.11)$$

$$\text{PtL}_4 \xrightarrow{\text{MeI}} \text{MePtIL}_2 \qquad (3.12)$$

$$(\text{L} = \text{PPh}_3)$$

$$\text{Cr}^{2+}(\text{aq}) \xrightarrow{\text{MeI}} \text{CrMe(H}_2\text{O)}_5^{2+} + \text{CrI(H}_2\text{O)}_5^{2+} \qquad (3.13)$$

4. By insertion:

$$\text{PtHCl(PEt}_3)_2 + \text{C}_2\text{H}_4 \longrightarrow \text{PtEtCl(PEt}_3)_2 \qquad (3.14)$$

$$\text{Cp(CO)}_3\text{MoH} \xrightarrow{\text{CH}_2\text{N}_2} \text{Cp(CO)}_3\text{MoCH}_3 \qquad (3.15)$$

A Grignard or organolithium reagent usually reacts with a metal halide or a cationic metal complex to give an alkyl, often by nucleophilic attack on the metal. Alternatively (case 2), a sufficiently nucleophilic metal can undergo electrophilic attack. Both these pathways have direct analogies in reactions that make bonds to carbon or nitrogen in organic chemistry (e.g., the reaction of MeLi with Me$_2$CO or of NMe$_3$ with MeI). Transfer of an alkyl group from one metal, such as Zn, Mg, or Li, to another, such as a transition metal, is called *transmetalation*. In Eq. 3.10, we use the fact that acyl complexes can often be persuaded to lose CO (Section 7.1). This is very convenient in this case because reagents that donate CF$_3^+$ are not readily available; CF$_3$I, for example, has a δ^-CF$_3$ group and a δ^+ I.

Oxidative Addition[10]

With the third general method of making alkyls, we encounter the very important *oxidative addition* reaction, which we study in detail in Chapter 6. This term is used any time we find that an X—Y bond has been broken by the insertion of a metal fragment L$_n$M into the X—Y bond. X and Y can be any one of a large number of groups, some of which are shown in Eq. 3.16:

$$
\begin{array}{c}
X \\
| \\
Y
\end{array}
\quad + \quad ML_n
\quad \longrightarrow \quad
\begin{array}{c}
X \\
\diagdown \\
Y \diagup
\end{array}
ML_n
$$

OS = 0   OS = 2

16e   18e

CN = n CN = $n + 2$

$(XY = H_2, \ R_3C—H, \ Cl—H, \ RCO—Cl, \ Cl—Cl, \ Me—I, \ R_3Si—H)$

(3.16)

Certain L_nM fragments are often considered carbenelike because there is an analogy between their insertion into $X—Y$ bonds and the insertion of an organic carbene, such as CH_2, into a C–H, Si–H, or O–H bond (Eq. 3.17). In Section 13.2, we will see how the *isolobal principle* allows us to understand the orbital analogy between the two systems. There are several mechanisms for oxidative addition (Chapter 6). For the moment we need only note that the overall process fits a general pattern in which the oxidation state, the coordination number, and the electron count all rise by two units. This means that a metal fragment of oxidation state n can normally give an oxidative addition only if it also has a stable oxidation state of $(n + 2)$, can tolerate an increase in its coordination number by 2, and can accept two more electrons. This last condition requires that the metal fragment be 16e or less. An 18e complex can still undergo the reaction, provided at least one 2e ligand (e.g., PPh_3 or Cl^-) is lost first. Oxidative addition is simply the reverse of the reductive elimination reaction that we saw in Section 3.1.

$$
\begin{array}{c}
X \\
| \\
Y
\end{array}
\quad + \quad :CH_2
\quad \longrightarrow \quad
\begin{array}{c}
X \\
\diagdown \\
Y \diagup
\end{array}
CH_2
$$

6e   8e   (3.17)

CN = 2   CN = 4

$(XY = R_3C—H, \ R_3Si—H, \ RCO_2—H, \ RO—H)$

A special case of oxidative addition is cyclometalation, in which a C–H bond in a ligand oxidatively adds to a metal to give a ring. Because of this ring formation, the reaction can be highly selective, for example, only one of the nine distinct CH bonds in benzoquinoline is cleaved when cyclometalation of Eq. 3.18 occurs. This kind of selectivity has been used in catalytic tritiation (Chapter 9) and in the Murai reaction[10] (page 428), in which aromatic ketones first undergo selective cyclometalation and the resulting aryl group is then functionalized in subsequent steps.

$$(3.18)$$

$$(3.19)$$

The third example of oxidative addition (Eq. 3.13) is a binuclear variant appropriate to those metals (usually from the first row) that prefer to change their oxidation state, coordination number, and electron count by one unit rather than two.

Insertion

The fourth general route, *insertion* (studied in detail in Chapter 7), is particularly important because it allows us to make an alkyl from an alkene and a metal hydride. We shall see in Chapter 9 how this sequence can lead to a whole series of catalytic transformations of alkenes, such as hydrogenation with H_2 to give alkanes, hydroformylation with H_2 and CO to give aldehydes, and hydrocyanation with HCN to give nitriles. Such catalytic reactions are among the most important applications of organometallic chemistry. Olefin insertion is the reverse of the β-elimination reaction of Section 3.1. Since we insisted earlier on the kinetic instability of alkyls having β-H substituents, it might seem inconsistent that we can make alkyls of this type in this way. In practice, it is not unusual to find that only a small equilibrium concentration of the alkyl may be formed in such an insertion. This is enough to enable a catalytic reaction to proceed if the alkyl is rapidly trapped in some way. For example, in catalytic hydrogenation, the alkyl is trapped by reductive elimination with a second hydride to give the product alkane. On the other hand, if the alkene is a fluorocarbon, then the product of insertion is a fluoroalkyl, and these are often very stable thermodynamically.[11] Compare the reversibility of C_2H_4 insertion with the irreversible formation of the

C_2F_4 insertion product in Eq. 3.19. The reason is the high M—C bond strength in the latter cases, as discussed in Section 3.1.

Another way to trap the alkylmetal complex is to fill the vacant site that opens up on the metal in the insertion with another ligand:

$$(3.20)$$

Although oxidative addition can be seen as an insertion of L_nM into X—Y, the term "insertion" in organometallic chemistry is reserved for the insertion of a ligand into an M—X bond (Sections 7.1–7.3).

One final route to alkyls is the attack of a nucleophile on a metal alkene complex, discussed in Chapter 5. This route is more useful for the synthesis of metal vinyls from alkyne complexes; vinyls are also formed from alkyne insertion into M—H bonds:

$$[LFeH(H_2)]^+ + HC{\equiv}CR \longrightarrow [LFe(H_2)(CH{=}CHR)]^+ \quad (3.21)$$

$$\{L = P(CH_2CH_2PPh_2)_3\}$$

Bridging Alkyls and Related Ligands

Alkyls can also be bridging ligands. In the case of main-group elements, such as Al, this seems to happen by a 2e, three-center bond involving only the metals and carbon [e.g., $Me_4Al_2(\mu{-}Me)_2$, **3.13**]. On the other hand, the transition metals tend to prefer to bridge by an agostic C—H bond (e.g., **3.14**). A number of remarkable bridges have also been found that involve an essentially planar methyl with the two d^0 metals coordinated each side of the plane (**3.15**).[4b]

3.13   **3.14**   **3.15**

Alkylidenes, Alkylidynes, and Carbides

The carbon atom of the alkylidene $(CR_2)^9$ group is able to form two normal covalencies, one to each metal (e.g., **3.16**). The alkylidene can also act as a terminal ligand, in which case it forms a double bond to the metal [e.g., $Cp_2^*Ta(=CH_2)Me$], which gives it a distinctive chemistry discussed in Chapter 11. Alkylidynes, CR, can bridge to three or to two metals or act as a terminal group with an $M\equiv C$ triple bond (e.g., **3.17**, **3.18**, and **3.19**). Finally, a carbon atom can bridge four metals as in $C(HgOAc)_4$ (**3.20**) but is more commonly found in *metal clusters* (Chapter 13), which are complexes that contain two or more metal–metal bonds. In the example shown (**3.21**), carbon is 6-coordinate!

3.16   **3.17**   **3.18**   **3.19**

3.20   **3.21**

Metalacycles

Cyclic dialkyls $M(CH_2)_n$ are *metalacycles*.[12] Metalacyclopropanes ($n = 2$) are more usefully thought of as metal–alkene complexes, but the higher homologs do indeed behave like dialkyls and have certain characteristic properties, such as the following interesting rearrangements:

(3.22)

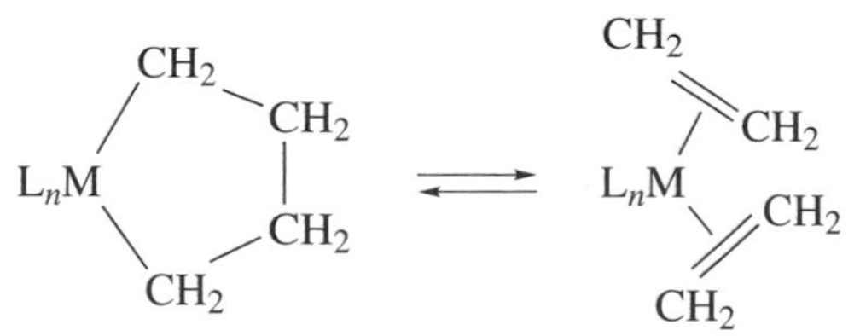

$$L_nM \overset{CH_2}{\underset{CH_2}{\diagup}} \begin{matrix} CH_2 \\ \vert \\ CH_2 \end{matrix} \rightleftharpoons L_nM \overset{CH_2}{\underset{CH_2}{\diagup}} \begin{matrix} CH_2 \\ \vert \\ CH_2 \end{matrix} \qquad (3.23)$$

We look at these reactions in detail in Sections 11.3 and 6.7, respectively. For the moment we need only note that the β-C$-$H of these cyclic dialkyls is held away from the metal and so is not available for β elimination. The β-C$-$C bond *is* held close to the metal, however, and so the rearrangements in Eq. 3.22 and 3.23 are really β eliminations involving a C$-$C, rather than a C$-$H, bond. The reaction of Eq. 3.22 is of particular significance because it is the key step of an important catalytic reaction, alkene metathesis, which converts propene to butene and ethylene (Chapter 11). The anion of [Li(tmeda)]$_2$[(CH$_2$)$_4$Pt(CH$_2$)$_4$] contains two tetramethylene rings bound to square planar Pt(II) and is thermally rather stable.[13] Cyclic diaryls can be very stable; an example[14] is shown below:

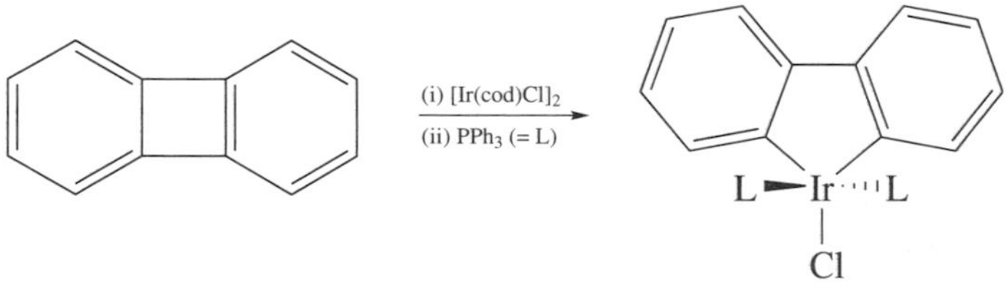

$$\text{(i) [Ir(cod)Cl]}_2 \quad \text{(ii) PPh}_3 \text{ (= L)} \qquad (3.24)$$

It would be very hard to be certain that such an unusual structure was correct without the X-ray crystallographic data shown in Fig. 3.2. This shows the compound is a 5-coordinate monomer with a distorted Y geometry (see Eq. 4.30). The atom positions are uncertain as a result of thermal motion and experimental error, so probability ellipsoids are used to represent the atoms. There is a 50% probability that the atom is located within its ellipsoid. The atoms furthest from the central heavy atom tend to show larger ellipsoids, probably as a result of greater thermal librational motion.

Aryl, vinyl, and acyl ligands have empty π^* orbitals that can accept electron density from the metal, and these also form strong M$-$C bonds. Pentahalophenyl ligands are exceptionally stable and strongly bound.[15] Vinyls and acyls also have an alternative η^2-bonding mode[16] shown as **3.22** and **3.23**, when the electron

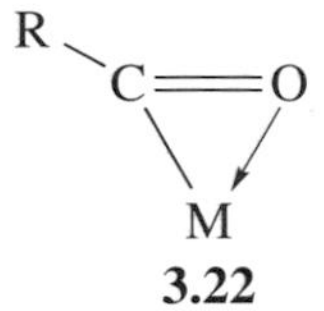

$$\begin{matrix} R \\ \diagdown \\ C = O \\ \diagdown \diagup \\ M \end{matrix}$$

3.22

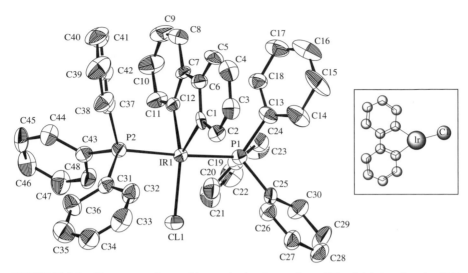

FIGURE 3.2   X-ray crystallographic results for the product of Eq. 3.24 showing the 50% probability ellipsoids for each atom. Hydrogen atoms, poorly located by X-ray methods, are omitted for clarity. Inset shows the unusual distortion in more detail. [Reproduced from Ref. 14a with permission.]

count goes from 1e to 3e. The η^2 forms are probable intermediates in the cis/trans isomerization of metal vinyl complexes (Eq. 3.25).[17]

$$\text{L}_n\text{M} \diagup \!\!\!\!\diagdown_{\text{R}} \;\rightleftharpoons\; \text{L}_n\text{M} \overset{\diagup}{\underset{\text{H} \quad \text{R}}{\diagdown}} \;\rightleftharpoons\; \text{L}_n\text{M} \diagup\!\!\!\!\diagdown \diagup^{\text{R}} \qquad (3.25)$$

3.23

- Alkyls can decompose by β elimination if an accessible C—H is β ($\text{M}-\text{CR}_2-\text{CR}_2-\text{H}$) to an unsaturated ($\leq$ 16 electron) metal (Eq. 3.1).
- Reductive elimination of RMX forms RX and extrudes M (Eq. 3.3).
- Oxidative addition of RX to M forms RMX (Eq. 3.16).
- Insertion of $\text{R}_2\text{C}=\text{CR}_2$ into M—H gives $\text{M}-\text{CR}_2-\text{CR}_2-\text{H}$ (Eq. 3.20).

3.2   RELATED σ-BONDED LIGANDS

Group 14 Elements

The closest noncarbon analog of the metal alkyl is the metal silyl $\text{M}-\text{SiR}_3$ (R = alkyl, aryl, or OH).[18] Trimethylsilyl transition metal complexes are much more numerous than are complexes of the t-butyl group, stable examples of which are rare. The most important reasons for this are probably that β elimination involving Me_3Si is inhibited by the relative instability of Si=C double bonds. The silyl complex is also less sterically congested than the CMe_3 group because the M−Si bond is much longer than M−C. Finally, M−SiR_3 bonds are strong because of the same π interaction we discuss for M−PR_3 bonds in Section 4.2. Similar SnR_3 complexes are also known; an important class consists of $SnCl_3$ complexes. Polystannyl derivatives, such as $[Pt(SnCl_3)_3(cod)]^-$, are possible in this case. Many poly(trichlorostannyl) complexes are catalytically active, perhaps because the very high trans influence of this group helps labilize other ligands and so create sites for substrate binding.

Groups 15–17

On moving to the right of C in the periodic table, we encounter the dialkylamido, alkoxo,[19] and fluoro ligands. Examples are $[Mo(NMe_2)_4]$, $[W(NMe_2)_6]$, $[(PhO)_3Mo\equiv Mo(OPh)_3]$, $Zr(OtBu)_4$, and Cp_2TiF_2. Their most important feature is the presence on the heteroatom of one (−NR_2), two (−OR), or three (−F) lone pairs. In a late transition metal complex, which is 18e and so has filled d orbitals, these lone pairs only weaken the M−X bond by repulsion of the filled metal orbitals (**3.24**; shading denotes filled orbitals; see also Fig. 1.8). In the case of an early metal, in contrast, the complex is often d^0, and has less than 18e. There are therefore empty d_π orbitals available, which can accept electron density from the lone pairs of X and so strengthen the M−X bond (**3.25**). The early metals are therefore said to be *oxophilic* or *fluorophilic*. This effect is just one example of a general difference between the early and the late metals. As electropositive elements, the early metals are more often seen in high oxidation states. In these states they seek to attract electron density from the ligands, so hard, π-donor ligands such as NR_2, OR, or F are favored. The late metals, which are more electronegative and have more d electrons available, tend to prefer lower oxidation states and soft π-acceptor ligands such as CO (**3.26**); amide, alkoxo, and fluoro complexes of the late metals are known, however, especially in situations such as **3.27**, where the 16e metal can accept some of the heteroatom lone-pair electron density.[20]

There are interesting structural consequences of this type of binding, especially in early metal complexes and with bulky alkoxides. The M−O−R angles tend to be larger than the usual tetrahedral angle. There are even cases where the angle is essentially 180°. The oxygen rehybridizes from sp^3 to sp^2 or even sp so as to put one or both of the lone pairs in p orbitals, which makes them more available for overlap with empty metal d orbitals; this in turn makes the M−O−R angle open to 120° (**3.28**) or 180° (**3.29**).

In many cases intermediate angles are also seen. The alkoxide needs to be bulky or else it can otherwise simply bridge to a second metal center, which achieves the same object of transferring electron density from the alkoxide to the metal without the necessity of rehybridizing; bulkiness strongly inhibits this

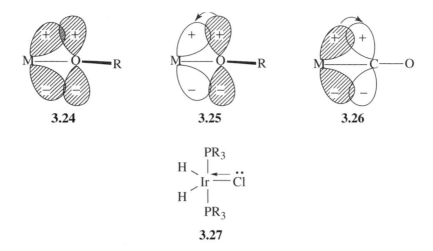

bridging. A linear alkoxide can be considered as donating both of its lone pairs to the metal. As such it is now a 5e (ionic model: 6e) donor. A sufficiently bulky alkoxide of this type can give complexes reminiscent of the corresponding cyclopentadienyls (e.g., **3.30** resembles Cp$_2$NbX$_3$).[21]

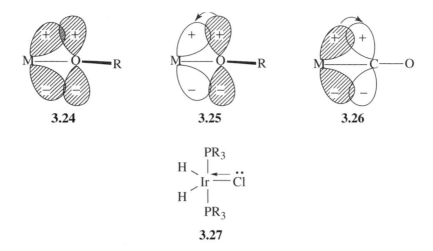

3.30

(Trp = trypticyl)

In dialkylamido ligands, NR_2, the lone pair is very basic and so the ligand often adopts a planar conformation, which puts the lone pair in a p orbital from which it can be donated to the metal. This resembles the situation in the planar NR_2 group found in organic amides, $RCONR_2$, where the π^* of the RCO group plays the role of acceptor.

M—OR or M—NR_2, although they lack M—C bonds, show certain similarities to alkyls; β elimination can still occur as shown in Eq. 3.26, but instead of an alkene, a ketone, aldehyde, or mine is formed. This reaction has the important consequence that alcohols can act as reducing agents for metal complexes, especially in the presence of a base. The base converts the coordinated alcohol to the alkoxide, which can then β-eliminate. The alkoxides **3.31** and **3.32** are particularly stable ligands because they lack β hydrogens.

3.31   **3.32**

$$L_nM \quad \begin{matrix} R \\ | \\ X-CH \\ | \quad \backslash \\ H \end{matrix} \longrightarrow L_nM-H + X=CH \begin{matrix} R \\ | \\ \end{matrix} \qquad (3.26)$$

$$(X = O, NR)$$

The heavier elements of groups 15–17 also give σ-bonded complexes, but the ligands $-PR_2$, $-SR$, and $-Cl$ have a much higher tendency to bridge than do their first-row analogs. This has been a serious problem in developing the chemistry of thiolate complexes, a particularly important area because cysteine thiolate is the soft ligand in metalloenzymes, the catalysts of biology.

Groups 12–13

Moving to the left of C, the boryl group, $-BR_2$, has an empty p orbital and so is able to accept back-bonding electrons from a late transition metal. M—X bonds where X is itself a metal have special properties considered in Chapter 13. In the case where X is $Au(PPh_3)$, the ligand is small enough to form polyaurated derivatives such as **3.33**, which show some resemblance to polyhydrides. As well as being bonded to the central metal, the gold atoms are also mutually bonded to give a *metal cluster* (Chapter 13).

3.33

3.3 METAL HYDRIDE COMPLEXES

The M—H bond plays a very important role in organometallic chemistry because metal hydrides[22] can undergo insertion with a wide variety of unsaturated compounds to give stable species or reaction intermediates containing M—C bonds. These are not only synthetically useful but many of the catalytic reactions we study later involve hydride insertion as the key step.

Hieber was the first to report a metal hydride complex with the discovery of $H_2Fe(CO)_4$ in 1931. His claim that this compound contains an Fe—H bond remained controversial for many years; Sidgwick, in 1950, regarded it as having the structure $(CO)_2Fe(COH)_2$. Only with the discovery of Cp_2ReH, $PtHCl(PR_3)_2$, and the striking polyhydride $K_2[ReH_9]$ in the period 1955–1964, did the reality of the M—H bond as a normal covalency become widely accepted. The discovery of molecular hydrogen complexes in 1984 stimulated intense activity, which continues today. For such a simple ligand, H has a remarkably rich chemistry.

Characterization

Hydrides are usually detected by ^1H NMR because they resonate to high field of $SiMe_4$ in a region $(0-60\delta)$ normally free of other ligand resonances. They couple with the metal, where this has $\frac{1}{2}$ spin, and with cis $(J = 15-30$ Hz$)$ and trans $(J = 90-150$ Hz$)$ phosphines, which is often useful for determining the stereochemistry of the complex. Inequivalent hydrides also couple with each other $(J = 1-10$ Hz$)$. IR studies show M—H stretching frequencies in the range $1500-2200$ cm^{-1}, but the intensities are often weak, and so the method is not entirely reliable. Hydrides, especially paramagnetic hydrides can be very difficult to characterize.[23]

Crystallographic studies are problematic because the hydride is such a poor scatterer of X rays. Hydrides may not be detected or may not be distinguishable with certainty from random electron density maxima in the neighborhood of the metal. Since X rays are scattered by electron density, not by the atomic nuclei, it is the M—H bonding electrons that are detected; these lie between the two nuclei, so that X-ray methods systematically underestimate the true M—H internuclear distance by approximately 0.1 Å. The best data for detecting hydrides are obtained at low temperatures (to reduce thermal motion) and at low angles (because hydride tends to give low angle scattering). Neutron diffraction detects the proton itself, which scatters neutrons relatively efficiently, so accurate distances can be obtained, but much larger crystals (1 vs. 0.01 mm^3) are usually needed for neutron work.

Synthesis

The main synthetic routes to hydrides are shown in Eqs. 3.27– 3.33:

1. By protonation:

$$[Fe(CO)_4]^{2-} \xrightarrow{\text{H}^+} [HFe(CO)_4]^- \xrightarrow{\text{H}^+} H_2Fe(CO)_4 \qquad (3.27)$$

$$Cp_2WH_2 \xrightarrow{+\text{H}^+} [Cp_2WH_3]^+ \qquad (3.28)$$

2. From hydride donors:

$$WCl_6 + LiBEt_3H + PR_3 \longrightarrow WH_6(PR_3)_3 \qquad (3.29)$$

3. From H$_2$:

$$IrCl(CO)(PPh_3)_2 \xrightarrow{\text{H}_2} IrH_2Cl(CO)(PR_3)_2 \qquad (3.30)$$

$$WMe_6 + PMe_2Ph \xrightarrow{\text{H}_2} WH_6(PMe_2Ph)_3 \qquad (3.31)$$

4. From a ligand:

$$RuCl_2(PPh_3)_3 + KOCHMe_2 + PPh_3 \longrightarrow$$
$$RuH_2(PPh_3)_4 + Me_2CO + KCl \qquad (3.32)$$

$$Cr(CO)_6 + OH^- \longrightarrow [Cr(CO)_5(COOH)]^- \xrightarrow{-CO_2}$$

$$[CrH(CO)_5]^- \xrightarrow{Cr(CO)_6, -CO} [(CO)_5Cr-H-Cr(CO)_5]^- \qquad (3.33)$$

Protonation requires a basic metal complex, but the action of a main-group hydride on a metal halide is very general. The third route, oxidative addition, requires a metal that can undergo this reaction but is of particular importance in catalysis. The reaction of hydrogen with the d^0 alkyl WMe$_6$ cannot go via oxidative addition because that would cause the W to exceed its maximum permitted oxidation state of 6. This type of reaction is called σ-*bond metathesis* (Eq. 3.31). Finally, hydrides are formed by the β elimination of a variety of groups.

Reactions

Hydrides are kinetically very reactive species and undergo a wide variety of transformations; some of the more significant are shown in Eqs. 3.34–3.37. Hydride transfer and insertion are closely related; the former implies that a hydridic hydride is attacking an electrophilic substrate.

1. Deprotonation:

$$WH_6(PMe_3)_3 + NaH \longrightarrow Na[WH_5(PMe_3)_3] + H_2 \qquad (3.34)$$

2. Hydride transfer and insertion:

$$Cp_2^*ZrH_2 + CH_2O \longrightarrow Cp_2^*Zr(OMe)_2 \qquad (3.35)$$

$$Cp_2ZrHCl + RCH=CH_2 \longrightarrow Cp_2ZrCl(CH_2-CH_2R) \qquad (3.36)$$

3. H atom transfer:

$$[Co(CN)_5H]^{3-} + PhCH=CHCOOH \longrightarrow$$
$$[Co(CN)_5]^{3-} + PhC\cdot H-CH_2COOH \qquad (3.37)$$

Several carbonyl hydrides, such as $HCo(CO)_4$, are quite strong acids because the CO groups are able to delocalize the negative charge of the corresponding metal anion. When bound to the more electropositive early metals, the hydrogen tends to carry a significant negative charge, and these hydrides tend to be the most reactive toward transfer of H^- to an electrophilic substrate such as an aldehyde or ketone (Eq. 3.35). The later metals impart much less negative charge to the hydride (the hydride may even be positively charged in some cases), so that the word *hydride* should not be taken literally. Protonation of a hydride with loss of H_2 is a common method to open up a coordination site; for example, $IrH_5(PCy_3)_2$ reacts with HBF_4 in MeCN to give $[IrH_2(MeCN)_2(PCy_3)_2]^+$.

The reactivity of a hydride may depend strongly on the nature of the reaction partner. For example, $CpW(CO)_3H$ has been shown to be an H^+ donor toward simple bases, an H· donor toward styrene, and an H^- donor to a carbonium ion.[24] Many hydrides react with excess CCl_4 to give $CHCl_3$ and the metal chloride, a reaction that has been used to detect metal hydride complexes.

Bridging Hydrides

Hydrides have a high tendency to bridge two or more metals.[25] This bridge is bent and resembles a BHB bridge in boranes but is quite unlike the linear hydrogen bond, which results from an electrostatic attraction. It can be thought of as a σ-bond complex (Section 1.8) in which M—H as the donor binds to

M′ as the acceptor. Note the difference between this situation, where the pair of electrons in the M−H bond is the donor, and that for other bridging ligands in which a lone pair on the bridging atom acts as the donor to the second metal (M−Cl:→M). One way of counting electrons in this system is to consider M−H as a 2e donor to M′. For example, in $[(CO)_5Cr-H-Cr(CO)_5]^-$, we can put the charge on one Cr and regard that Cr as the M−H donor to obtain $[(CO)_5Cr^- -H \rightarrow Cr(CO)_5]$. Both Cr are now 18e species. The three-center bonding of a σ-bond complex implies the presence of M−M bonding. (In Section 13.1 we look at another convention for counting bridging hydrides.) The same idea can be applied synthetically; for example, L_nM-H often reacts with 16e $M'L'_n$ [or the equivalent system stabilized by a labile solvent $[(solv)M'L'_n]$] to give the bridged species $L_nM-H-M'L'_n$. Subsequent rearrangement to give multiply bridged systems, such as $[L_2HIr(\mu\text{-}H)_3IrHL_2]^+$ or $[H_2L_2Re(\mu\text{-}H)_4ReH_2L_2]$ (L = PPh$_3$), commonly occurs, however.

3.4 σ COMPLEXES

Sigma complexes[26,27] (Section 1.8) contain ligands of type X−H that bind by donation of the X−H σ-bonding electrons in a 2e 3-center bond to the metal (**3.34**). They are considered neutral 2e donor, L-type ligands. Examples are known where X = H, Si, Sn, B, P, but at least one H normally has to be present in the ligand. The H atom has a small atomic radius and carries no lone pairs or other substituents, allowing the hydrogen end of the X−H bond to approach close to the metal and so allow the filled M d_π orbital to back-bond relatively strongly only into the lobe of the X−H σ* orbital that is located on the H atom (compare the strictly side-on case in Fig. 1.9b). This leads to a canted side-on structure (**3.35**), where the H is much closer to the metal than the X group.

Back donation into the X−H σ* level is an essential part of the bonding because pure Lewis acids such as AlMe$_3$ or BF$_3$ do not form isolable H$_2$ or HX σ complexes. On the other hand, very strong back donation breaks the X−H bond in an oxidative addition, for example, to give a dihydride (**3.36**). There are even a few cases where complexes of types **3.36** and **3.37** are in equilibrium (e.g., Eq 3.38):

$$[W(H)_2(CO)_3(PCy_3)_2] \rightleftharpoons [W(H_2)(CO)_3(PCy_3)_2] \qquad (3.38)$$

3.34   3.35   3.36   3.37

H_2 Complexes and Nonclassical Hydrides[26,27]

The most important ligand of the σ type is dihydrogen, H_2, and the first dihydrogen complexes (**3.37**) were discovered by Kubas;[26] many others are now known (**3.38–3.41**):

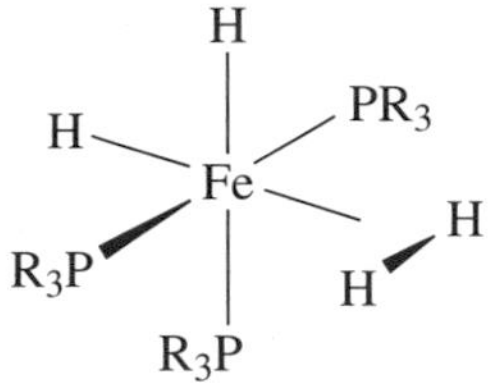

3.38

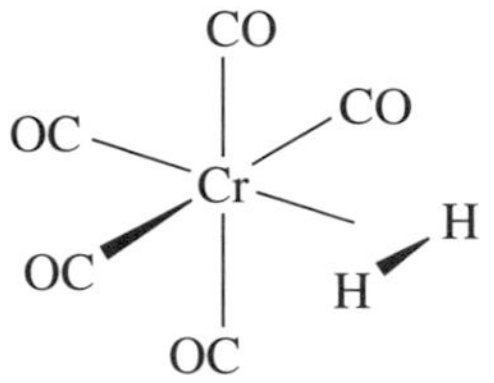

3.39

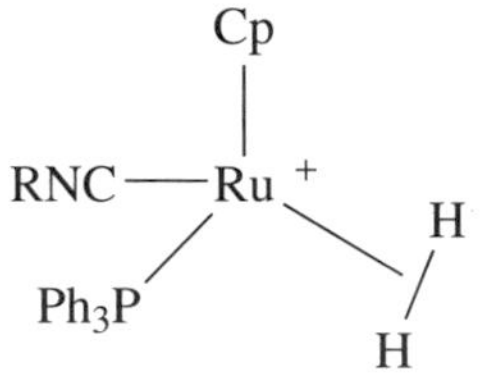

3.40

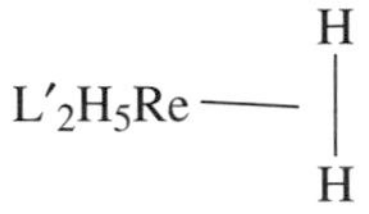

3.41

$(L' = P(o\text{-tol})_3)$

Free H_2 is a very weak acid ($pK_a = 35$) but binding as a σ complex makes it a very much better acid ($pK_a = 0$–20). This degree of acidification is very remarkable considering that ligands such as H_2O that bind via a lone pair are minimally acidified (by 2–4 pK_a units) on binding to the same metal sites. The reason is that while OH^- is bound only a little better than OH_2 to these metal fragments, H^- ion is bound very much better than H_2. For any acid, AH, incremental stabilization of the A^- product of acid dissociation relative to AH itself on binding translates into an acidification of the AH molecule. This explains why acidification is greatest when H_2 binds to a cationic fragment (better able to stabilize H^-) having weak back-bonding power (so H_2 itself binds weakly). Deprotonation of coordinated H_2 provides a route for the heterolytic activation of hydrogen: H^+ is abstracted by a base and H^- is retained by the metal. The following equilibrium (Eq. 3.39) illustrates an intramolecular version of this reaction. Remarkably, the position of equilibrium depends on the location of the counterion in the ion pair, which is in turn fixed by the steric size of L.[28]

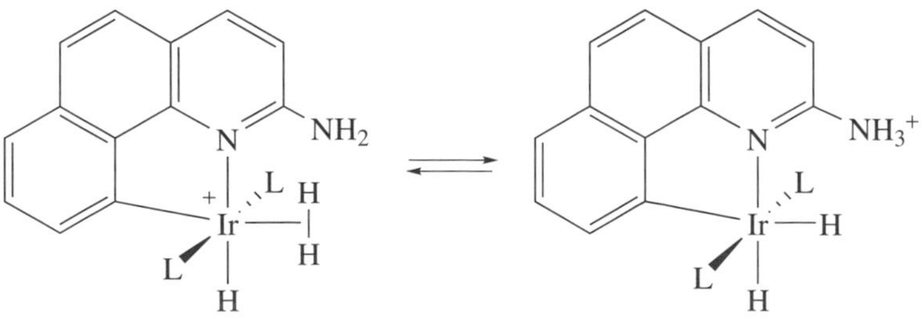

(3.39)

Complexes with H−H bonds are often called *nonclassical* hydrides.[27] By the bonding model of Fig. 1.9b, we expect that more π-basic metals will tend to split the H_2 and form a classical dihydride **3.36**, while less π-basic metals will tend to form the dihydrogen complex, **3.37**. Morris and co-workers[29] have shown how increasing the electron density at the metal favors the dihydride complex **3.36** by looking at the IR stretching frequency of the corresponding N_2 complex: the lower $v(N_2)$ the more π basic the site, and the more the dihydride **3.36** is favored. Since π basicity rises as we go down the periodic table, this accounts for the difference in structure between the nonclassical tetrahydride, $M(H_2)H_2(PR_3)_3$, where M is Fe or Ru, and the classical Os analog $OsH_4(PR_3)_3$. The role of a positive charge in reducing the basicity of a metal center is illustrated in the equation below in which a classical pentahydride is protonated to give a bis(dihydrogen) dihydride cation.[27]

$$Ir^VH_5(PCy_3)_2 + H^+ \longrightarrow [Ir^{III}(H_2)_2H_2(PCy_3)_2]^+$$

Coordinated dihydrogen can often be deprotonated with base;[27] for $[Ir(H_2)_2H_2(PCy_3)_2]^+$ this happens even with NEt_3. In $[CpRe(NO)(CO)(H_2)]^+$, the H_2 ligand has a pK_a of −2.5, making it a strong acid.[30] Formation of an H_2 complex can be a good way to activate it heterolytically, in which case H^- is retained by the metal and H^+ is released. Several H_2 complexes can both exchange with free H_2 or D_2 and exchange with solvent protons and thus can catalyze isotope exchange between gas-phase D_2 and solvent protons.[31]

$Cp^*FeH(dppe)$ shows faster protonation at the Fe−H bond, so that $[Cp^*Fe^{II}(H_2)(dppe)]^+$ is obtained at −80°C; on warming above −40°C, the complex irreversibly converts to the classical form $[Cp^*Fe^{IV}(H)_2(dppe)]^+$. The Fe−H is the better kinetic base (faster protonation), but the Fe is the better thermodynamic base (dihydride more stable).[32]

Characterization

Dihydrogen complexes have been characterized by X-ray, or, much better, neutron diffraction. An IR absorption at 2300–2900 cm^{-1} is assigned to the H−H stretch, but it is not always seen. The H_2 resonance appears in the range 0 to -10δ in the ^1H NMR and is often broad. The presence of an H−H(D) bond is shown by the H,D coupling constant of 20–34 Hz in the ^1H NMR spectrum of the H−D analog. This compares with a value of 43 Hz for free HD and ∼1 Hz for classical H−M−D species.

The H,D coupling depends on the H−H bond order, and the H⋯H distance in an H_2 complex can even be reliably estimated from the Morris equation:[29]

$$d_{HH} (Å) = 1.42 - 0.0167\{J_{HD} (Hz)\} \tag{3.40}$$

Stretched H_2 complexes with H−H distances above 1 Å are less common; for example, d(H−H) is 1.36 Å (*n* diffraction) in **3.41**. They are difficult to distinguish from classical hydrides other than by neutron diffraction or H, D coupling.

Dihydrogen Bonding

In standard hydrogen bonding, a weak acid such as an N−H or O−H bond binds with 5−10 kcal/mol bond strength to a weak base, typically a nitrogen or oxygen lone pair to give structures such as O−H···O or N−H···N, where the dotted bond represents the weak hydrogen bonding interaction. These are of critical importance in biology. Since M−H can protonate to give M(H$_2$), the M−H bond must be considered as a weak base. An interesting consequence of this basic character is that an M−H bond can take part as the weak base component of a hydrogen bond, to give structures such as N−H···H−M or O−H···H−M. Now two hydrogens are involved in the bond, so it has been called the dihydrogen bond. It can also be seen as a proton−hydride interaction because the N or O carries a protonic hydrogen while the M carries a hydridic hydrogen. The bond strength is not very much different than in the conventional hydrogen bond, and the H···H distance is typically 1.8 Å, much shorter than the sum of the van der Waals radii of two hydrogens, 2.4 Å. The N−H or O−H acid approaches the M−H base in a side-on manner because the proton has to get close to the pair of electrons in the M−H bond that constitute the weak base. Examples are known of complexes with intramolecular (**3.42**) and intermolecular (**3.43**) dihydrogen bonds.[33]

3.42   **3.43**

Agostic Species

Sigma complexes of C−H bonds are also known, but the binding is not as strong as in the H$_2$ case and examples of the L$_n$M(alkane) type are still rare[34] because they dissociate too easily. Where a ligand is already firmly bound to the metal, a C−H bond of that ligand can much more commonly bind to the metal, if the metal fragment has 16 or fewer valence electrons and can accept the 2e donor C−H bond as an additional ligand. These are agostic complexes in which C−H σ binding occurs as part of a chelating system. The case of an agostic alkyl (**3.7**) has already been mentioned, but many other types of ligands behave in this way. The case of **3.44** is interesting because of the fast, reversible oxidative addition/reductive elimination, detected by NMR (*fluxionality*).

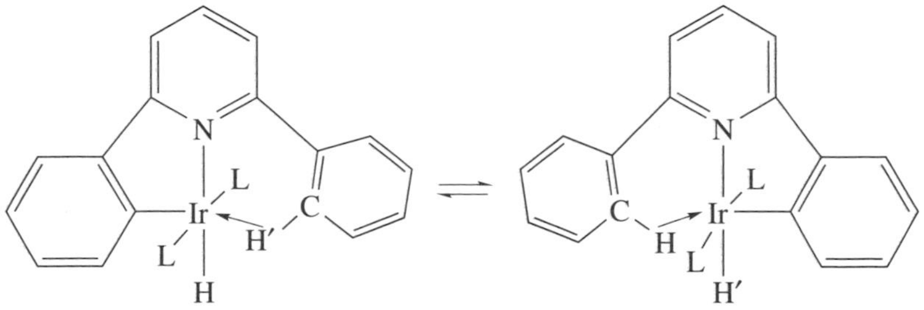

3.44

$$(3.41)$$

Other σ Complexes

Sigma complexes are now also known or suspected for X—M, where X = Si, Sn, B, P, S.[26] While atoms in organic compounds are either bonded or nonbonded, inorganic compounds can have bond orders between 0 and 1. That is why many structures have dotted lines, indicating partial bonds, for example, Eq. 3.42.[35]

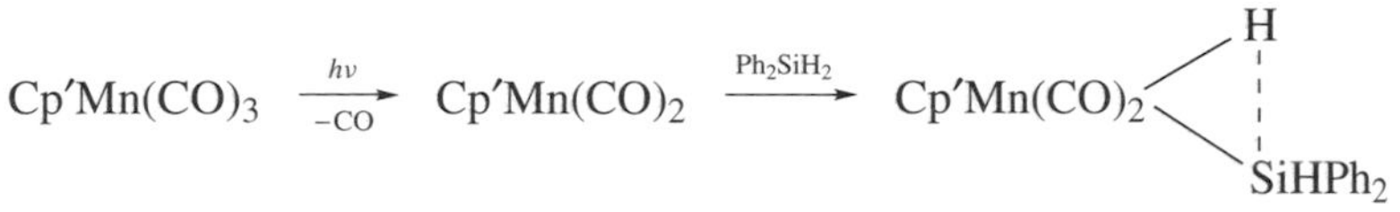

$$(3.42)$$

$$(Cp' = \eta^5\text{-MeC}_5\text{H}_4)$$

- Hydrides M—H are key reactive species.
- M—H can protonate to give complexes M(H—H), an example of a σ complex, where a σ bond is bound but not broken.

3.5 BOND STRENGTHS FOR CLASSICAL σ-BONDING LIGANDS

Classical σ-bonding ligands such as H, CH_3, and Cl form strong M—X bonds with metals. Bond strengths or bond dissociation energies (BDEs) are defined as the energy required to break the M—X bond homolytically, that is, by:

$$L_nM-X \longrightarrow L_nM\bullet + X\bullet \qquad (3.43)$$

Bond strengths can be useful guides in predicting whether proposed steps in catalytic cycles are energetically reasonable. For example, oxidative addition of a C—F bond to a metal would require that the necessary loss of the large C—F bond energy of $\sim$120 kcal/mol be compensated by the formation of sufficiently strong M—C and M—F bonds. It is much more difficult to determine BDEs in organometallic chemistry than it is for organic compounds because only the latter usually burn cleanly to give defined products, and calorimetry is therefore possible. Instead, a number of other methods have been developed. For example, Fig. 3.3 illustrates a thermodynamic cycle that has proved useful for studies on metal hydrides. It relies on our ability to measure all the other steps in the cycle except the one involving the M—H BDE, and therefore to estimate the BDE by Hess's law. By measuring the acid dissociation constant of the hydride and the potential required for oxidizing the conjugate base, the metal anion, the ΔG values corresponding to steps b and c can be estimated from Eqs. 3.44–3.45.

$$\Delta G = -RT \ln K \qquad (3.44)$$

$$\Delta G = \frac{RT}{F} \ln E_0 \qquad (3.45)$$

The H^+/H_2 potential gives ΔG for step d, leaving the bond strength of H_2 and the solvation energy of H•, which are both known. The only unknown is now the M—H BDE. Methods useful for M—C BDEs are discussed in Section 16.2.

Typical data[36] for M—X BDEs of various types are shown in Fig. 3.4, in which the M—X BDE is plotted against the H—X BDE. The good correlation between the two set of figures is rather surprising. The only significant deviation is the case of L_nM—H, which is normally stronger than L_nM—CH$_3$ by 15–25 kcal/mol even though Me—H and H—H have almost the same BDEs. Labinger and Bercaw[37] have discussed this problem in some detail.

In organic chemistry it is a useful approximation to say that the same type of bond will have a very similar bond strength wherever it occurs. In organometallic compounds this seems not to be generally true.[38] The activation energy for phosphine loss from Cp*Ru(PMe$_3$)$_2$X (**3.45**) is a measure of the M—P bond strength,

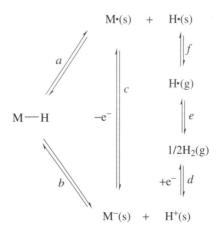

FIGURE 3.3 Thermodynamic cycle involved in one method of determining the M—H bond strength (s = solution, g = gas).

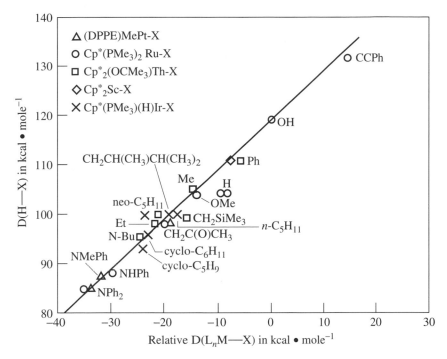

FIGURE 3.4 Relative bond energies $D(L_nM-X)$ versus the HX bond energy $D(H-X)$ showing the good correlation obtained. Reproduced from Ref. 22 with permission.

TABLE 3.1 M−P Bond Strength Differences (kcal/mol) in Cp*(Me₃P)XRu(−PMe₃) as a Function of the Nature of X

		σ-Donor Ligands			
H	>7	−C≡CPh	+2	CH_3	0^a
CH_2Ph	−2	Ph	−3	CH_2SiMe_3	−6
		π-Donor Ligands			
CH_3	0^a	Cl	−7	OH	−11
NHPh	−12	Ph	−3		

^{a}Zero by definition: this non-π-donor ligand is taken as a reference point for all the compounds studied.

because the incoming ligand is believed not to bond significantly to the metal in the transition state where the PMe₃ is almost completely lost (D mechanism; see Section 4.3), so the barrier to the process is essentially equal to the M−P BDE. If the M−P BDE were constant, the activation energy would not change as X changes. Table 3.1 shows that for a series of σ-bonding ligands, the activation energy differences (and therefore M−P BDE differences) relative to the X = Me compound vary widely depending on the steric size of the ligand. Organic compounds, with 4-coordinate carbon, do not normally have strong intersubstituent repulsions. In contrast, metal ions in organometallic compounds often have much higher coordination numbers. For example, in **3.45**, 8 atoms are directly bonded to the metal. Intraligand repulsions are therefore common and relief of these repulsions on ligand dissociation favors ligand loss and makes the M−P bond strength sensitive to changes in steric bulk.

3.45

The barrier for PMe$_3$ loss is also affected when X is a π-donor ligand because X is then capable of stabilizing the 16e Cp*Ru(PMe$_3$)X fragment by π-electron donation from X to Ru (as illustrated in **3.46**). Relative to the non-π-bonding X = Me case, the barrier to PMe$_3$ loss is lowered by the presence of a π-donor X, to an extent that roughly corresponds with the π-donor power of X. This electronic effect is comparable in importance to the steric effect discussed above. All this may mean that no one set of BDE values is likely to be generally applicable in organometallic chemistry. Indeed, M−CO bond energies all the way from 22 to 84 kcal/mol have been reported[39] for an extensive series of metal carbonyls.

3.46

REFERENCES

1. L. Cockroft, *Chem. Br.* **35**(4), 49, 1999.

2. (a) C. Russell, *Chem. Br.* **35**(9), 43, 1999. (b) Frankland is also significant in having helped Darwin "immeasurably" in his biological research[2c] by providing advice on chemical aspects. (c) J. Browne, *Charles Darwin*, Knopf, New York, Vol. 2, p. 409.

3. G. M. Whitesides et al.: (a) *J. Am. Chem. Soc.* **98**, 6521, 1976; (b) *J. Am. Chem. Soc.* **94**, 5258, 1972; (c) J. A. Mata, C. Incarvito, and R. H. Crabtree, *Chem. Comm.* 184, 2003.

4. (a) A. Haaland et al., *J. Am. Chem. Soc.* **112**, 4547, 1990; X. F. Wang and L. Andrews *J. Am. Chem. Soc.* **124**, 5636, 2002; (b) T. A. Albright, O. Eisenstein, et al., *Inorg. Chem.* **28**, 1611, 1989.

5. (a) M. Brookhart and M. L. H. Green, *J. Organomet. Chem.* **250**, 395, 1983; M. L. H. Green, *Prog. Inorg. Chem.* **36**, 1, 1988.

6. S. Niu and M. B. Hall, *Chem. Rev.* **100**, 353, 2000.

7. M. Thornberry et al., *Organometallics* **19**, 5352, 2000.

8. C. J. Elsevier, *Coord. Chem. Rev.* **186**, 809, 1999.

9. E. Clot, J. Chen, D. H. Lee, S. Y. Sung, L. N. Appelhans, J. W. Faller, R. H. Crabtree, and O. Eisenstein, *J. Am. Chem Soc.*, **126**, 8795, 2004; G. Ferrando-Miguel, J. N. Coalter, H. Gérard, J. C. Huffman, O. Eisenstein, K. G. Caulton, *New J. Chem.* **26**, 687, 2002.

10. H. Kurosawa and A. Yamamoto, *Fundamentals of Molecular Catalysis*, Elsevier, Amsterdam, 2003.

11. T. J. Marks, *Bonding Energetics in Organometallic Compounds*, ACS Symposium Series, 1990.

12. J. Campora and E. Carmona, *Coord. Chem. Rev.* **195**, 207, 1999.

13. H.-O. Fröhlich et al., *Angew Chem. Int. Ed.* **32**, 387, 1993.

14. (a) O. Eisenstein, R. H. Crabtree, et al., *Organometallics*, **14**, 1168, 1995; (b) G. Ujaque et al., *New J. Chem.* **22**, 1493, 1998.

15. R. Usón et al., *Adv. Organomet. Chem.* **28**, 219, 1988.

16. L. D. Durtee and I. P. Rothwell, *Chem. Rev.* **88**, 1059, 1988.

17. R. H. Crabtree, *New. J. Chem.* **27**, 771, 2003.

18. T. D. Tilley, in *The Chemistry of Organosilicon Compounds*, S. Patai, ed., Wiley, Chichester, 1989, Vol. 2, p. 1415; W. R. Roper, L. J. Wright, et al., *J. Am. Chem. Soc.* **114**, 9682, 1992.

19. K. G. Caulton and L. G. Hubert-Pfalzgraf, *Chem. Rev.* **90**, 969, 1990.

20. A. C. Cooper, O. Eisenstein, and K. G. Caulton, *New J. Chem.* **22**, 307, 1998.

21. P. T. Wolczanski et al., *J. Am. Chem. Soc.* **119**, 247, 1997; *J. Organomet. Chem.* **591**, 194, 1999.

22. A. Dedieu, *Transition Metal Hydrides*, VCH, New York, 1992.

23. J. L. Kersten, A. L. Rheingold, K. H. Theopold, and C. P. Casey, *Angew. Chem.* **31**, 1341, 1992.

24. R. M. Bullock et al., *Comments Inorg. Chem.* **12**, 1, 1991; *J. Am. Chem. Soc.* **112**, 6886, 1990; R. M. Bullock, J. R. Norton, et al., *Angew. Chem. Int. Ed.* **31**, 1233, 1992.

25. R. Bau, R. G. Teller, S. W. Kirtley, and T. F. Koetzle, *Acc. Chem. Res.* **12**, 176, 1979.

26. G. J. Kubas, *Metal Dihydrogen and σ-Bond Complexes*, Kluwer, New York, 2001.

27. D. M. Heinekey et al., *J. Am. Chem. Soc.* **127**, 850, 2005; R. H. Crabtree, *Angew. Chem. Int. Ed.* **32**, 789, 1993.

28. K. Gruet, E. Clot, O. Eisenstein, D.-H. Lee, B. P. Patel, and R. H. Crabtree, *New J. Chem.* **27**, 80, 2003.

29. R. H. Morris et al., *Coord. Chem. Rev.* **121**, 155, 1992; S. Sabo-Etienne and B. Chaudret, *Coord. Chem. Rev.* **180**, 381, 1998.

30. D. M. Heinekey et al., *J. Am. Chem. Soc.* **112**, 5166, 1990, and references cited therein.

31. A. C. Albeniz, D. M. Heinekey, and R. H. Crabtree, *Inorg. Chem.* **30**, 3632, 1991.

32. C. Lapinte et al., *J. Organomet. Chem.* **428**, 49, 1992.

33. R. H. Crabtree, *Science* **282**, 2000, 1998; R. Custelcean and J. E. Jackson, *Chem. Rev.* **101**, 1963, 2001.

34. C. A. Reed, P. D. Boyd et al., *J. Am. Chem. Soc.* **119**, 3633, 1997; S. Geftakis and G. E. Ball, *J. Am. Chem. Soc.* **120**, 9953, 1998.

35. U. Schubert, *Adv. Organomet. Chem.* **30**, 1, 1990.

36. H. E. Bryndza, J. Bercaw et al., *J. Am. Chem. Soc.* **109**, 1444, 1987.

37. J. Labinger and J. E. Bercaw, *Organometallics* **7**, 926, 1988.

38. H. E. Bryndza, J. E. Bercaw et al., *Organometallics* **8**, 379, 1989.

39. K. Krogh-Jespersen, A. S. Goldman, S. P. Nolan et al., *J. Am. Chem. Soc.* **120**, 9256, 1998.

PROBLEMS

1. [Pt(Ph$_3$P)$_2$(RC≡CR)] reacts with HCl to give **3.47**. Propose a mechanism for this process to account for the fact that the H in the product vinyl is endo with respect to the metal, as shown in **3.47**.

$$(Ph_3P)_2ClPt - \overset{\displaystyle R}{\underset{\displaystyle H}{\bigvee}} - R$$

3.47

2. In which direction would you expect a late transition metal hydride to undergo insertion with CH$_2$=CF$_2$ to give the most stable alkyl product?

3. Suggest an efficient method for preparing IrMe$_3$L$_3$ from IrClL$_3$, LiMe, and MeCl.

4. Propose three alkoxides, which should be as different in structure as possible, that you would examine in trying to make a series of stable metal derivatives, say, of the type Mo(OR)$_6$. Would you expect CpFe(CO)$_2$(OR) to be linear or bent at O? Explain.

5. What is the metal electron count for H$_2$Fe(CO)$_4$ and ReH$_9{}^{2-}$? Would the electron count be changed if any of these species had a nonclassical structure?

6. Ligands of type X−Y only give 2e three-center bonds to transition metals if X = H and Y lack lone pairs. Why do you think this is so? (*Hint:* Consider possible alternative structures if X and Y are nonhydrogen groups.)

7. Reductive eliminations can sometimes be encouraged to take place by oxidizing the metal. Why do you think this is so?

8. Given that the HOMO of a d^8 square planar complex is the d_{z^2} orbital, predict which rotational conformer of the aryl groups in NiPh$_2$L$_2$ will be (a) electronically and (b) sterically favored.

9. Give the electron counts, oxidation states, and d^n configurations in the following: $L_3Ru(\mu\text{-}CH_2)_3RuL_3$, $[(CO)_5Cr(\mu\text{-}H)Cr(CO)_5]^-$, and WMe_6.

10. $Me_2CHMgBr$ reacts with $IrClL_3$ to give $IrHL_3$. How can this be explained, and what is the organic product formed?

11. Certain 16e metal hydrides catalytically convert free 1-butene to free 2-butene. Propose a plausible mechanism, using the symbol $[M]-H$ to represent the catalyst. Would an 18e metal hydride be able to carry out this reaction?

4

CARBONYLS, PHOSPHINE COMPLEXES, AND LIGAND SUBSTITUTION REACTIONS

We first examine how CO, phosphines, and related species act as ligands, then look at ways in which one ligand can replace another in a *substitution* reaction:

$$L_nM-L + L' = L_nM-L' + L \tag{4.1}$$

This has been studied in most detail for the case of the substitution of CO groups in metal carbonyls by a variety of other ligands, such as tertiary phosphines, PR_3. The principles involved will be important later, for example, in catalysis.

4.1 METAL COMPLEXES OF CO, RNC, CS, AND NO

A chance 1884 observation by Ludwig Mond led to an important advance in the nickel refining industry. When he found his nickel valves were being eaten away by CO, he deliberately heated Ni powder in a CO stream to form a volatile compound, $Ni(CO)_4$, the first metal carbonyl. The Mond refining process was based on the fact that the carbonyl can be decomposed to give pure nickel by further heating. Lord Kelvin was so impressed by this result that he remarked that Mond "gave wings to nickel."

Unlike a simple alkyl, CO is an unsaturated ligand, by virtue of the C—O multiple bond. As we saw in Section 1.6, such ligands are soft because they are capable of accepting metal d_π electrons by back bonding; that is, these ligands

The Organometallic Chemistry of the Transition Metals, Fourth Edition, by Robert H. Crabtree
Copyright © 2005 John Wiley & Sons, Inc.

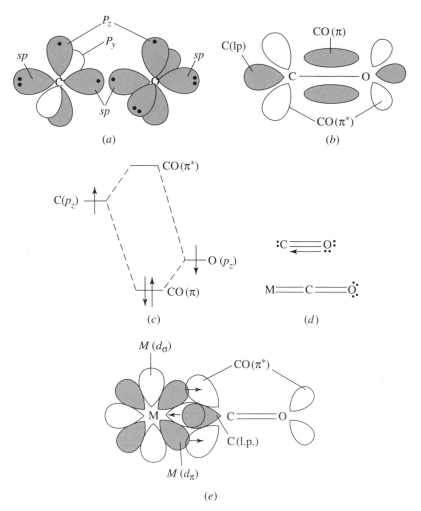

FIGURE 4.1 Electronic structure of CO and carbonyl complexes. Shading represents occupied orbitals (*a*) and (*b*) building up CO from C and O, each atom having two *p* orbitals and two *sp* hybrids. In (*a*), the dots represent the electrons occupying each orbital in the C and O atoms. In (*b*), only one of the two mutually perpendicular sets of π orbitals is shown. (*c*) An MO diagram showing a π bond of CO. (*d*) Valence bond representations of CO and the MCO fragment. (*e*) An MO picture of the MCO fragment. Again, only one of the two mutually perpendicular sets of π orbitals is shown.

are π acceptors. This contrasts to hard ligands, which are σ donors, and often π donors, too (e.g., H_2O, alkoxides). CO can act as a spectator or an actor ligand.

As we saw in Section 1.6, we look first at the frontier orbitals of M and L because these usually dominate the M—L bonding. The electronic structure of free CO is shown in Fig. 4.1*a* and 4.1*b*. We start with both the C and the O *sp*-hybridized. The singly occupied *sp* and p_z orbitals on each atom form a σ and a π bond, respectively. This leaves the carbon p_y orbital empty, and the oxygen p_y orbital doubly occupied, and so the second π bond is formed only after we have formed a dative bond by transfer of the lone pair of $O(p_y)$ electrons into the empty $C(p_y)$ orbital. This transfer leads to a C^--O^+ polarization of the molecule, which is almost exactly canceled out by a partial C^+-O^- polarization of all three bonding orbitals because of the higher electronegativity of oxygen. The free CO molecule therefore has a net dipole moment very close to zero. In Fig. 4.1c the reason for the polarization of the π_z orbital is shown in MO terms. An orbital is always polarized so as to favor the AO that is closest in energy and so the C—O π MO has more O than C character. The valence bond picture of CO and one form of the MCO system is shown in Fig. 4.1d.

It is not surprising that the metal binds to C, not O, because the ligand HOMO is the C, not the O lone pair; this is because O is more electronegative and so its orbitals have lower energy. In addition, the $CO(\pi^*)$ LUMO is polarized toward C, and so M—CO π overlap will also be optimal at C not O. Figure 4.1e shows how the CO HOMO, the carbon lone pair, donates electrons to the metal LUMO, the empty $M(d_\sigma)$ orbital, and metal HOMO, the filled $M(d_\pi)$ orbital, back donates to the CO LUMO. While the former removes electron density from C, the latter increases electron density at both C and O because $CO(\pi^*)$ has both C and O character. The result is that C becomes more positive on coordination, and O becomes more negative. This translates into a polarization of the CO on binding.

This metal-induced polarization chemically activates the CO ligand. It makes the carbon more sensitive to nucleophilic and the oxygen more sensitive to electrophilic attack. The polarization will be modulated by the effect of the other ligands on the metal and by the net charge on the complex. In $L_nM(CO)$, the CO carbon becomes particularly ∂^+ in character if the L groups are good π acids or if the complex is cationic [e.g., $Mo(CO)_6$ or $[Mn(CO)_6]^+$], because the CO-to-metal σ-donor electron transfer will be enhanced at the expense of the metal to CO back donation. If the L groups are good donors or the complex is anionic [e.g., $Cp_2W(CO)$ or $[W(CO)_5]^{2-}$], back donation will be encouraged, the CO carbon will lose its pronounced ∂^+ charge, but the CO oxygen will become significantly ∂^-. The range can be represented in valence bond terms as **4.1**,* the extreme in which CO acts as a pure σ donor, through **4.2** and **4.3**, the extreme in which both the π_x^* and π_y^* are both fully engaged in back bonding. Neither extreme is reached in practice, but each can be considered to contribute differently to the real structure according to the circumstances. In general, polarization effects are of great importance in determining the reactivity of unsaturated ligands, and the same sort of effects we have seen for CO will be repeated for the others, with nuances in each case depending on the chemical character of the particular ligand. Note that, on the covalent model, the electron count of CO in **4.1–4.3** is 2e. The same e count applies to all true resonance forms.

We can tell where any particular CO lies on the continuum between **4.1** and **4.3**, by looking at the IR spectrum. Because **4.3** has a lower C=O bond order than **4.1**,

*The + and − in **4.1–4.3** are *formal charges* and do not necessarily reflect the real charge, which is shown here by ∂^+ or ∂^- signs.

$$M^- \longleftarrow C^{\partial+} \equiv O^+ \qquad M=C=O \qquad M^+ \equiv C-O^-$$

$$\textbf{4.1} \qquad\qquad\qquad \textbf{4.2} \qquad\qquad\qquad \textbf{4.3}$$

the greater the contribution of **4.3** to the real structure, the lower the observed CO stretching frequency, $\nu(CO)$; the normal range is $1820–2150$ cm^{-1}. The MO picture leads to a similar conclusion. As the metal to CO π^* back bonding becomes more important, we populate an orbital that is antibonding with respect to the C=O bond, and so we lengthen and weaken the CO bond. In a metal carbonyl, the M−C π bond is made at the expense of the C=O π bond. The high intensity of the CO stretching bands, also partly a result of polarization on binding, means that IR spectroscopy is extremely useful. From the band position, we can tell how good the metal is as a π base. From the number and pattern of the bands, we can tell the number and stereochemistry of the COs present (see Chapter 10).

Carbonyls bound to very poor π-donor metals, where **4.1** is the predominant contributor to the bonding, have very high $\nu(CO)$ bands as a result of weak back donation. When these appear to high energy of the 2143 cm^{-1} band of free CO, the complexes are sometimes called *nonclassical carbonyls*.[1a] Even d^0 species can bind CO, for example, the nonclassical, formally d^0 Zr(IV) carbonyl complexes, $[Cp_2^*Zr(\kappa^2\text{-}S_2)(CO)]$, prepared from reaction of d^2 $[Cp_2^*Zr(CO)_2]$ with S_8 at $80°C$, has a $\nu(CO)$ stretching frequency of 2057 cm^{-1}.[1b] One of the most extreme weak π-donor examples is $[Ir(CO)_6]^{3+}$ with $\nu(CO)$ bands at 2254, 2276, and 2295 cm^{-1}. The X-ray structure of the related complex $[IrCl(CO)_5]^{2+}$ shows the long M−C $[2.02(2)\text{Å}]$ and short C−O $[1.08(2)\text{Å}]$ distances expected from structure **4.1**.[1c] The highest oxidation state carbonyl known is *trans*-$[OsO_2(CO)_4]^{2+}$ with $\nu(CO) = 2253$ cm^{-1}.[1c] Carbonyls with exceptionally low $\nu(CO)$ frequencies are found for negative oxidation states (e.g., $[Ti(CO)_6]^{2-}$; $\nu(CO) = 1747$ cm^{-1}) or where a single CO is accompanied by non-π-acceptor ligands (e.g., $[ReCl(CO)(PMe_3)_4]$; $\nu(CO) = 1820$ cm^{-1}); these show short M−C and long C−O bonds.

Although **4.1–4.3** represent three ideal structures in the bonding range possible for CO, no one structure can be said to perfectly represent the situation for any particular case. There is therefore considerable looseness in the way carbonyls are represented in organometallic structures. Often, M−CO or M−C=O are used. Whatever picture is chosen for graphical representation, the bonding picture discussed above still applies.

Preparations of CO Complexes

Typical examples are shown in Eqs. 4.2–4.7:

1. From CO:

$$Fe \xrightarrow{\text{CO, 200 atm. 200°}} Fe(CO)_5 \qquad\qquad (4.2)^{2a}$$

$$IrCl(cod)L_2 + CO \rightarrow IrCl(CO)L_2 \underset{}{\overset{CO}{\rightleftharpoons}} IrCl(CO)_2L_2 \quad (4.3)^{2b}$$

$$(L = PMe_3)$$

2. From CO and a reducing agent (reductive carbonylation):

$$NiSO_4 + CO + S_2O_4{}^{2-} = Ni(CO)_4 \quad (4.4)^3$$

$$Re_2O_7 + 17CO \longrightarrow (CO)_5Re-Re(CO)_5 + 7CO_2 \quad (4.5)$$

$$Cr(CO)_4(tmeda) \overset{Na}{\longrightarrow} Na_4[Cr(CO)_4] \quad (4.6)^4$$
$$\textbf{4.3}$$

$$(tmeda = Me_2NCH_2CH_2NMe_2)$$

3. From a reactive organic carbonyl compound:

$$RhClL_3 + RCXO \overset{oxidative\ addition}{\longrightarrow}$$

$$\{XRhCl(COR)L_3\} \overset{retro\ migratory\ insertion}{\longrightarrow}$$

$$\{XRhCl(CO)RL_2\} \overset{reductive\ elimination}{\longrightarrow} RX + RhCl(CO)L_2 \quad (4.7)^5$$

$$(L = PPh_3; X = H\ or\ Cl)$$

The first method requires that the metal already be in a reduced state because only π-basic metals can bind CO. If a high-oxidation-state complex is the starting material, then we need to reduce it first as shown in the second method. Equation 4.5 illustrates the high tendency of CO groups to stabilize M−M bonds; not only are COs small ligands but they also leave the metal atom with a net charge similar to that in the bulk metal. In this case the product has no bridging carbonyls, and the dimer is held together by the M−M bond only. Equation 4.6 shows the ability of CO to stabilize polyanionic species by acting as a strong π acceptor and delocalizing the negative charge over the CO oxygens. $Na_4[Cr(CO)_4]$ has the extraordinarily low $\nu(CO)$ of 1462 cm^{-1}, the extremely high anionic charge on the complex, and ion pairing of Na$^+$ to the carbonyl oxygen contribute to the lowering by favoring the M≡C−ONa resonance form, which is related to **4.3**.

The third route involves abstraction of CO from an organic compound. This can happen for aldehydes, alcohols, and even CO$_2$ (see Eq. 12.20). In the example shown in Eq. 4.7, the reaction requires three steps; the second step is the reverse of migratory insertion. The success of the reaction in any given instance relies in part on the thermodynamic stability of the final metal carbonyl product, which is greater for a low-valent metal. Note that the first step in the case of an aldehyde is oxidative addition of the aldehyde C−H bond. It is much more difficult for the metal to break into a C−C bond so ketones, R_2CO, are usually resistant to this reaction.

Since COs are small and strongly held ligands, as many will usually bind as are required to achieve coordinative saturation. This means that metal carbonyls, in common with metal hydrides, show a strong preference for the 18e configuration.

Reactions of Metal Carbonyls

Typical reactions are shown in Eqs. 4.8–4.13. All of these depend on the polarization of the CO on binding, and so change in importance as the coligands and net charge change. For example, types 1 and 3 are promoted by the electrophilicity of the CO carbon and type 2 by nucleophilicity at CO oxygen.

1. Nucleophilic attack at carbon:

$$L_nM-CO \xrightarrow{\ Nu^-\ } L_nM{=}C\overset{\displaystyle Nu}{\underset{\displaystyle O^-}{\Big\backslash}} \tag{4.8}$$

$$(CO)_5Mo(CO) \xrightarrow{\ LiMe\ } (CO)_5Mo{=}C\overset{\displaystyle Me}{\underset{\displaystyle OLi}{\Big\backslash}} \xrightarrow{\ MeI\ } (CO)_5Mo{=}C\overset{\displaystyle Me}{\underset{\displaystyle OMe}{\Big\backslash}} \tag{4.9}$$

$$(CO)_5Mo(CO) \xrightarrow{\ Me_3N^+-O^-\ } (CO)_5Mo{=}C\overset{\displaystyle O-\overset{+}{N}Me_3}{\underset{\displaystyle O^-}{\Big\backslash}} \tag{4.10}$$

$$(CO)_5Mo-\square + CO_2 + NMe_3 \longleftarrow (CO)_5Mo^-{-}C\overset{\displaystyle O-\overset{+}{N}Me_3}{\underset{\displaystyle O}{\Big\Vert}}$$

These reactions give carbenes (Chapter 11) or carbenelike intermediates. The reaction of Eq. 4.10 is particularly important because it is one of the rare ways in which the tightly bound CO can be removed to generate an open site at the metal. In this way a ligand L′, which would normally not be sufficiently strongly binding to replace the CO, can now do so.

$$[Cp(NO)(PPh_3)ReCO]^+ \xrightarrow{\ LiBHEt_3\ } Cp(NO)(PPh_3)Re(CHO) \tag{4.11}$$

This reaction (Eq. 4.11) produces the unusual formyl ligand, which is important in CO reduction to MeOH (Section 12.3). It is stable in this case because the 18e complex provides no empty site for rearrangement to a hydridocarbonyl complex.

2. Electrophilic attack at oxygen:

$$Cl(PR_3)_4Re-CO \xrightarrow{AlMe_3} [Cl(PR_3)_4Re-CO{\rightarrow}AlMe_3] \qquad (4.12)$$

Protonation of this Re carbonyl occurs at the metal, as is most often the case, but the bulkier acid, $AlMe_3$, prefers to bind at the CO oxygen.

3. Finally, there is the migratory insertion reaction that we looked at in Section 3.3:

$$MeMn(CO)_5 \xrightarrow{PMe_3} (MeCO)Mn(CO)_4(PMe_3) \qquad (4.13)$$

Bridging CO Groups

CO has a high tendency to bridge two metals (e.g., **4.4** $\rightleftharpoons$ **4.5**):

$$(4.14)$$

The electron count remains unchanged on going from **4.4** to **4.5**. The 15e CpFe(CO) fragment is completed in **4.4** by an M−M bond, counted as a 1e contributor to each metal, and a terminal CO counting as 2e. In **4.5**, on the other hand, we count 1e from each of the two bridging CO (μ^2-CO) groups and 1c from the M−M bond. The bridging CO is not entirely ketonelike because an M−M bond seems almost always to accompany a CO bridge. The CO stretching frequency in the IR spectrum falls to $1720–1850$ cm^{-1} on bridging. Consistent with the idea of a nucleophilic attack by a second metal, a bridging CO is more basic at O than the terminal ligand. A good illustration of this is the fact that a Lewis acid can bind more strongly to the oxygen of a bridging CO and so displace the equilibrium of Eq. 4.15 toward **4.6**. Similar $[CpM(CO)_x]_2$ species are known for many different metals.[6a]

$$(4.15)$$

Cotton[6b] studied the *semibridging carbonyl* in which the CO is neither fully terminal nor fully bridging but intermediate between the two. This is one of the many cases in organometallic chemistry where a stable species is intermediate in character between two bonding types and shows us a "stopped action" view of the conversion of one to the other. An example is **4.7** in which you can see that each semibridging CO is bending in response to the second metal atom being close by.

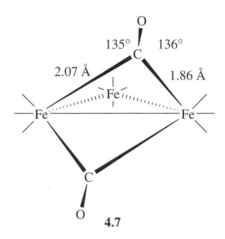

4.7

Triply and even quadruply bridging CO groups are also known in metal cluster compounds, for example, $(Cp^*Co)_3(\mu^3\text{-}CO)_2$ (**4.8**). These have CO stretching frequencies in the range of $1600–1730\ cm^{-1}$. PdSiO, a very unstable molecule seen only at low temperatures, is the only SiO complex known.[7]

O
‖
C
Cp*Co ———— CoCp*
            CoCp*
C
‖
O

4.8

Isonitriles

Many 2e ligands closely resemble CO. Replacement of the CO oxygen with the related, but less electronegative, fragment RN gives isonitrile, RNC, a ligand that is a significantly better electron donor than CO. It stabilizes more cationic and higher-oxidation-state complexes than does CO [e.g., $[Pt(CNPh)_4]^{2+}$], for which in many cases no CO analog is known, but tends to bridge less readily than does CO. It is also more sensitive to nucleophilic attack at carbon to give aminocarbenes (Eq. 11.3) and has a higher tendency for migratory insertion. Unlike the situation for CO, the CN stretching vibration in isonitrile complexes is often lower than in the free ligand. The C lone pair is nearly nonbonding with respect to CO (i.e., does not contribute to the CO bond) for carbonyls but is much more antibonding with respect to CN in isonitriles. Depletion of electron density in this lone pair by donation to the metal therefore has little effect on $v(CO)$ but raises $v(CN)$. Back bonding lowers both $v(CO)$ and $v(CN)$. Depending on the balance of σ versus π bonding, $v(CN)$ is raised for weak π-donor metals, such as Pt(II), and lowered for strong π-donor metals, such as Ni(O). Cases such as NbCl(CO)(CNR)(dmpe)$_2$ have been found in which back bonding to an isonitrile is so strong that this normally linear ligand becomes bent at N (129°–144°), indicating that the resonance form **4.9** has become dominant. The M−C bond is also unusually short (2.05 Å compared to 2.32 Å for an Nb−C single bond) in the bent isonitrile case, and the $v(CN)$ is unusually low (1750 cm^{-1} compared to ~ 2100 cm^{-1} for the linear type), again consistent with the structure **4.9**.[8] The appalling stench of volatile isonitriles may be a result of their binding to a metal ion acting as a receptor in the human nose.[9]

$$M=C=\overset{\cdot\cdot}{N}\diagdown_{R}$$

4.9

Thiocarbonyls

CS is not stable above $-160°C$ in the free state, but a number of complexes are known, such as RhCl(CS)(PPh$_3$) (Eq. 4.16) and Cp(CO)Ru(μ^2-CS)$_2$RuCp(CO), but so far no "pure" or *homoleptic* examples of M(CS)$_n$. They are usually made from CS$_2$ or by conversion of a CO to a CS group. Perhaps because of the lower tendency of the second-row elements such as S to form double bonds, the M$^+\equiv$C−S$^-$ form analogous to **4.3** is more important for MCS than MCO: the MC bond therefore tends to be short and CS is a better π acceptor than CO. Perhaps for this reason, CO and not CS tends to be substituted in a mixed carbonyl-thiocarbonyl complex.

$$\text{RhCl(PPh}_3)_3 \xrightarrow{\text{CS}_2} \textit{trans}\text{-RhCl(CS)(PPh}_3)_2 + \text{SPPh}_3 \qquad (4.16)^{10}$$

Typical $v(CS)$ ranges for CS complexes are 1273 cm^{-1} for free CS, 1040–1080 cm^{-1} for M$_3$(μ_3-CS), 1100–1160 cm^{-1} for M$_2$(μ_2-CS), and 1160–1410 cm^{-1} for M−CS.[11]

Nitrosyls[12]

Free NO is a stable free radical because the ON—NO bond in the dimer is very weak. In a surprising development, NO was found to be important in biological signaling having a biosynthetic pathway and specialized sensor proteins.[13] It forms an extensive series of diamagnetic nitrosyl complexes by binding to odd-electron metal fragments. As an alternative to using free NO for the synthesis of nitrosyl complexes, NO^+, available as the salt $NOBF_4$, is isoelectronic with CO and can often replace CO in a substitution reaction. In the majority of nitrosyl complexes, the MNO unit is linear, and in such cases, the NO is usually considered as behaving as the 2e donor NO^+ on the ionic model and as a 3e ligand on the covalent model. NO^+ is isoelectronic with CO and thus binds in a linear fashion. Replacing a CO by an NO^+ means that the complex will bear an extra positive (or one less negative) charge. This increases the reactivity of the system toward nucleophiles and is a standard strategy for activating an otherwise unreactive complex for such a reaction (e.g., Eq. 4.17).[14]

$$(4.17)$$

(Nu = enamine or PhMgBr)

We can mentally construct NO from CO by adding an extra proton (and a neutron) to the carbon nucleus to give us NO^+, and a single electron to the π^* orbital to account for the extra valence electron of N versus C. We look first at the ionic model (Fig. 4.2). In bringing $CpMo(CO)_2$ and NO together to form $CpMo(CO)_2(lin\text{-}NO)$, we first remove the unpaired electron from NO to give NO^+ and place this electron on Mo, which gives it a zero oxidation state in this case. Binding of NO^+ as a 2e donor to $CpMo(CO)_2^-$, a 16e fragment, gives an 18e configuration. On the other hand, the 17e fragment, $[Co(diars)_2X]^+$, binds NO to give a complex with a bent nitrosyl structure. In this case, we first carry out an electron transfer from the metal to NO to get the 16e fragment $[Co(diars)_2X]^{2+}$ and NO^-; the NO^- is then a 2e ligand to bring the total electron count to 18. The formal oxidation state of the metal is obtained by considering a linear NO as NO^+ and a bent NO as NO^-, for example $Cr(lin\text{-}NO)_4$ is formally Cr(-IV) with the tetrahedral geometry appropriate for d^{10}. The conversion of a linear to a bent NO is considered to lead to an increase in the formal oxidation state by two units (e.g., Eq. 4.18). Raising the electron density on a metal will encourage the linear-to-bent conversion because in the bent NO a pair of electrons originally assigned to the complex becomes a lone pair on nitrogen; in the language of the ionic model, the electron-rich metal reduces the NO^+ to NO^-. For example,

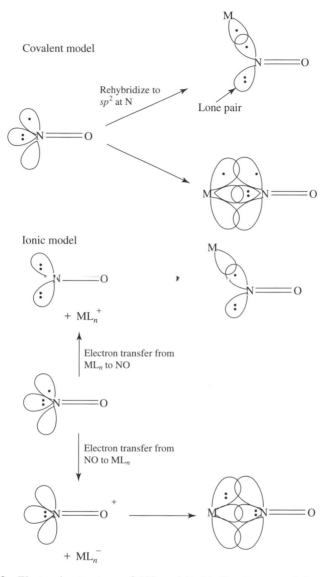

FIGURE 4.2 Electronic structure of NO and its binding to a metal fragment on the covalent and ionic models.

the Fe(III) center in the oxidized form of myoglobin, an iron protein found in muscle, forms a linear NO complex, but on reduction to Fe(II) the NO switches to the bent form.[15]

On the covalent model, a linear NO is a 3e ligand. In this case there is no need to rehybridize. The metal has a singly occupied d_π orbital, which binds with the singly occupied NO(π^*) to give an M—N π bond, and the N(lp) (lone pair) donates to the empty $M(d_\sigma)$ in the normal way to give the σ bond. A bent NO is a 1e X ligand such as a chlorine atom, but as the electron is in a π^* orbital in free NO, the N has to rehybridize to put this electron in an sp^2 orbital pointing toward the metal in order to bind.

A 17e L_nM fragment can bond to NO to give only a bent 18e nitrosyl complex, while a 15e L_nM fragment can give either an 18e linear or a 16e bent complex. The 16e bent NO complexes are not uncommon. Some complexes have both bent and linear NO: for example, $ClL_2Ir(lin\text{-}NO)(bent\text{-}NO)$. Equations 4.18 and 4.19 show examples where the linear and bent nitrosyl isomers are in equilibrium.[16,17] For the Co case, the linear complex has $\nu(NO)$ at 1750 cm^{-1} and the bent NO has $\nu(NO)$ at 1650 cm^{-1}; unfortunately, the typical $\nu(NO)$ ranges for the two structural types overlap. These equilibria also show that it is not always possible to decide whether an NO is linear or bent by finding out which structure leads to an 18e configuration. Only if a linear structure would give a 20e configuration, as in **4.10** in Eq. 4.20, can we safely assign a bent structure.

$$CoCl_2L_2(lin\text{-}NO) \rightleftharpoons CoCl_2L_2(bent\text{-}NO) \qquad (4.18)^{16a}$$
$$\text{18e, Co(I)} \qquad\qquad \text{16e, Co(III)}$$

$$(o\text{-}C_6H_4O_2)_2L_2Ir(lin\text{-}NO) \rightleftharpoons (o\text{-}C_6H_4O_2)_2L_2Ir(bent\text{-}NO) \quad (4.19)^{16b}$$
$$\text{18e, Ir(I)} \qquad\qquad \text{16e, Ir(III)}$$

$$(L = PPh_3)$$

$$[Co(lin\text{-}NO)(diars)_2]^{2+} + X^- \longrightarrow [CoX(bent\text{-}NO)(diars)_2]^+ \qquad (4.20)$$
$$\text{18e, Co(I)} \qquad\qquad\qquad \textbf{4.10}, \text{18e, Co(III)}$$

The discovery that NO and CO are important messenger molecules in the mammalian brain and exert their effect by binding to metalloprotein receptors will certainly provoke increased interest in the area.[17]

Typical nitrosyls, together with some preparative routes, are shown in Eqs. 4.21–4.26. The first two cases show linear–bent equilibria. Equation 4.21 shows that NO, unlike most ligands, can replace all the COs in a metal carbonyl to give a homoleptic nitrosyl. The last two cases show the use of the stable cation NO$^+$ (isoelectronic with CO) in synthesis. NO$^+$ is a powerful 1e oxidizing agent and it is even capable of oxidizing many bulk metals (Eq. 4.25). The resulting higher-oxidation-state ions cannot usually bind NO, however.

$$Cr(CO)_6 + NO + h\nu = Cr(lin\text{-}NO)_4 \qquad (4.21)^{18}$$

$$Mn(CO)_5I + NO = Mn(lin\text{-}NO)_3(CO) \qquad (4.22)$$

$$IrH_5(PR_3)_2 + NO$$

$$= (R_3P)(lin\text{-}NO)_2Ir-Ir(lin\text{-}NO)_2(PR_3) \qquad (4.23)^{19}$$

$$(toluene)Cr(CO)_3 + NO^+ + MeCN = trans\text{-}[Cr(lin\text{-}NO)_2(MeCN)_4]^{2+}$$

$$(4.24)^{20}$$

$$Pd + 2NO^+ + MeCN = [Pd(MeCN)_4]^{2+} + 2NO \qquad (4.25)$$

Like CO, coordinated NO can give the migratory insertion reaction:

$$[CpCo(NO)]^- \xrightarrow{\text{RI}} [CpCoR(NO)] \xrightarrow{\text{PPh}_3} [CpCo(NOR)PPh_3] \qquad (4.26)^{21}$$

Cyanide

Cyanide ion, CN^-, is gaining importance as an ionic CO analog.[22] It has been found as ligand for the active-site iron in a number of hydrogenases (Chapter 16). Its complexes date back to alchemical times. Diesbach, a Berlin draper, boiled beef blood in a basic medium to obtain the dye, Prussian blue, still in common use. It was later shown to be a coordination polymer containing $Fe^{II}-C\equiv N-Fe^{III}$ units; note how the softer Fe(II) binds the softer C end of cyanide. This can claim to be considered both the first organometallic and the first coordination compound.

Dinitrogen

Dinitrogen (N_2) is a ligand of great importance in connection with biological nitrogen fixation (conversion to ammonia), discussed in Section 16.3.[23] It binds to metals much less strongly than CO because it is both a weaker σ donor and a weaker π acceptor.

- Back bonding to CO strengthens the $M-C$ but weakens the $C-O$ bond lowering $\nu(CO)$ in the IR spectrum.
- $M-CO$ is subject to nucleophilic attack at C (Eq. 4.8) particularly when the metal is incapable of strong back bonding.

4.2 PHOSPHINES AND RELATED LIGANDS

Tertiary phosphines, PR_3, are important because they constitute one of the few series of ligands in which electronic and steric properties can be altered in a systematic and predictable way over a very wide range by varying R. They also stabilize an exceptionally wide variety of ligands of interest to the organometallic chemist as their phosphine complexes $(R_3P)_nM-L$. Phosphines are more commonly spectator than actor ligands.

Structure and Bonding

Like NH_3, phosphines have a lone pair on the central atom that can be donated to a metal. Unlike NH_3, they are also π acids, to an extent that depends on the nature of the R groups present on the PR_3 ligand. For alkyl phosphines, the π acidity is weak; aryl, dialkylamino, and alkoxy groups are successively

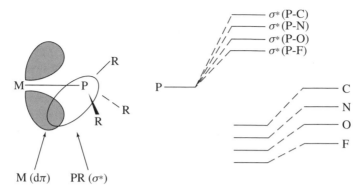

FIGURE 4.3 Empty P−R σ^* orbital plays the role of acceptor in metal complexes of PR$_3$. As the atom attached to phosphorus becomes more electronegative, the empty P−X σ^* orbital becomes more stable and so moves to lower energy and becomes a better acceptor from the metal. Shading represents orbital occupation.

more effective in promoting π acidity. In the extreme case of PF$_3$, the π acidity becomes as great as that found for CO.

In the case of CO the π^* orbital accepts electrons from the metal. The σ^* orbitals of the P−R bonds play the role of acceptor in PR$_3$.[24] Figure 4.3 shows the MO picture. Whenever the R group becomes more electronegative, the orbital that the R fragment uses to bond to phosphorus becomes more stable (lower in energy). This implies that the σ^* orbital of the P−R bond also becomes more stable. At the same time, the phosphorus contribution to σ^* increases, and so the size of the σ^* lobe that points toward the metal increases (the larger the energy gap between two atomic orbitals, the more the more stable atomic orbital contributes to σ, and the least stable to σ^*). Both of these factors make the empty σ^* more accessible for back donation. The final order of increasing π-acid character is

$$\text{PMe}_3 \approx \text{P(NR}_2)_3 < \text{PAr}_3 < \text{P(OMe)}_3 < \text{P(OAr)}_3 < \text{PCl}_3 < \text{CO} \approx \text{PF}_3$$

P(NR$_2$)$_3$ is a better donor than it should be based on the argument of Fig. 4.3, probably because the basic N lone pairs compete with the metal d_π orbitals in donating to PR σ^*.

Occupation of the P−R σ^* by back bonding from the metal also implies that the P−R bonds should lengthen slightly on binding. In practice, this is masked by a simultaneous shortening of the P−R bond due to donation of the P lone pair to the metal, and the consequent decrease in P(lone pair)−R(bonding pair) repulsions. To eliminate this complication, Orpen[24a] has compared the crystal structures of pairs of complexes, such as $[(\eta^3\text{-C}_8\text{H}_{13})\text{Fe}\{\text{P(OMe)}_3\}_3]^{n+}$, where $n = 0$ or 1. The M−P σ bonds are similar in both cases, but the cationic iron in the oxidized complex is less π basic and so back-donates less to the phosphite; this leads to a longer Fe−P distance (difference: $+0.015 \pm 0.003$ Å), and a shorter

P—O distance (-0.021 ± 0.003 Å). Once again, as in the case of CO, the M—L π bond is made at the expense of a bond in the ligand, but this time it is a σ, not a π, bond.

Further evidence for the π-acceptor character of phosphines comes from the diamagnetism of the octahedral d^2 species, *trans*-TiMe$_2$(dmpe)$_2$ (**4.11**). In order to be diamagnetic, the three d_π orbitals have to split as shown in **4.12**. For this to happen, either the axial ligands have to be π donors or the equatorial ligands have to be π acceptors. Since $-CH_3$ was shown not to be a significant π donor, the dmpe must be an acceptor. In Fig. 1.7, six π-acceptor ligands caused all three d_π orbitals to drop in energy; here four π acceptors in the xy plane ($2 \times$ dmpe) cause the d_{xy} orbital to be lowered below d_{xz} and d_{yz} to give **4.12**.[25] Note that Ti(II) is a very strong π donor, as we saw in Section 2.7, and so the situation is very favorable for detecting M—(PR$_3$) π bonding.

4.11   **4.12**

Tolman Electronic Parameter and Cone Angle

The electronic effect of various PR$_3$ ligands can be adjusted by changing the R group as, quantified by Tolman,[26] who compared the $v(CO)$ frequencies of a series of complexes of the type LNi(CO)$_3$, containing different PR$_3$ ligands. The stronger donor phosphines increase the electron density on Ni, which passes some of this increase along to the COs by back donation. This, in turn, lowers $v(CO)$ as shown in Fig. 4.4. Particularly for chelating ligands, the $v(CO)$ frequencies of L$_2$Mo(CO)$_4$ can also be used for this purpose because v_{Mo} correlates with v_{Ni}.[27]

The second important feature of PR$_3$ as a ligand is the variable steric size, which can be adjusted by changing R. COs are so small that as many can bind as are needed to achieve 18e. In contrast, the same is rarely true for phosphines, where only a certain number of phosphines can fit around the metal. This can be a great advantage in that by using bulky PR$_3$ ligands, we can favor forming low-coordinate metals or we can leave room for small but weakly binding ligands, which would be excluded by a direct competition with a smaller ligand such as PMe$_3$ or CO. The usual maximum number of phosphines that can bind to a single metal is two for PCy$_3$ or P(*i*-Pr)$_3$, three or four for PPh$_3$, four for PMe$_2$Ph, and five or six for PMe$_3$. Examples of stable complexes showing these principles at work are Pt(PCy$_3$)$_2$ and [Rh(PPh$_3$)$_3$]$^+$, both coordinatively unsaturated species that are stabilized by bulky phosphines, and W(PMe$_3$)$_6$, a rare case of a hexakis–phosphine complex.

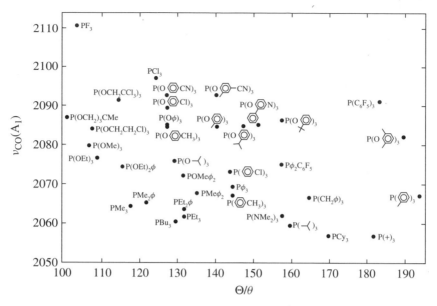

FIGURE 4.4 Electronic and steric effects of common P-donor ligands plotted on a map according to Tolman (v in cm^{-1}, θ in degrees). Reproduced from Ref. 26 with permission of the American Chemical Society.)

Tolman has also quantified the steric effects of phosphines with his *cone angle*. This is obtained by taking a space-filling model of the M(PR$_3$) group, folding back the R substituents as far as they will go, and measuring the angle of the cone that will just contain all of the ligand, when the apex of the cone is at the metal (**4.13**). Although the procedure may look rather approximate, the angles obtained have been very successful in rationalizing the behavior of a wide variety of complexes. The results of these studies also appear on Fig. 4.4 with the electronic parameters.

4.13

An important part of organometallic chemistry consists in varying the steric and electronic nature of the ligand environment of a complex to promote whatever properties are desired: activity or selectivity in homogeneous catalysis, reversible binding of a ligand, facile decomposition, or high stability. A key feature of the PR_3 series of ligands is that we can relatively easily change electronic effects without changing steric effects [e.g., by moving from PBu_3 to $P(O^iPr)_3$] or change steric effects without changing electronic effects [e.g., by moving from PMe_3 to $P(o\text{-tolyl})_3$]. One outcome of increasing the ligand electron donor strength, for example, might be to perturb an oxidative addition/reductive elimination equilibrium in favor of the oxidative addition product. Likewise, increasing the steric bulk is expected to favor low-coordination-number species. We can therefore expect the chemistry of a phosphine-containing complex to vary with the position of the phosphine in the Tolman map.

Bite Angle[28]

Chelate ligands such as the $Ph_2P(CH_2)_nPPh_2$ series usually enforce a cis arrangement of the two phosphorus atoms as well as discouraging ligand dissociation by the chelate effect. Different chelates differ in their preferred *bite angle* (P–M–P angle), but many such ligands are very flexible and have a wide range of accessible bite angles. Very rigid diphosphines are also available, such as the phenoxathiin shown.[28b] In rare cases, PR_3 groups can bridge.[29]

Computational Tolman Parameters

$LNi(CO)_3$ species are not available for all classes of ligands, and to extend the Tolman parameters to other classes the v (CO) values of $LNi(CO)_3$ have been predicted computationally.[30a]

Tolman steric parameter have also been calculated,[30b] but here it can be a problem to compare cone-shaped PR_3 ligands with fan-shaped aryls or carbenes.

Phosphine Analogues

Complexes of AsR_3 and SbR_3 are known but tend to be less useful than PR_3 complexes, in part because the Q–R bond is more readily cleaved for the heavier elements. N-heterocyclic carbenes (e.g., **11.6**) have been proposed as an electronic and steric analog of phosphines (Section 11.1).

- Phosphines are so useful because they are electronically and sterically tunable (Figure 4.4).

4.3 DISSOCIATIVE SUBSTITUTION

The reactions of phosphines with metal carbonyls, investigated by Basolo,[31a] form the basis for our understanding of organometallic substitution reactions in general. The phosphine is usually refluxed with the carbonyl in an organic solvent, such as ethanol or toluene. One can distinguish two extreme mechanisms for substitution, one dissociative,[31] labeled D, and the other associative, labeled A. Intermediate cases are often labeled I: I_a if closer to A and I_d if closer to D.[31b]

Kinetics

The dissociative extreme involves a slow initial loss of a CO to generate a vacant site at the metal, which is trapped by the incoming ligand L. In general, a dissociative step precedes an associative step. Because the rate-determining step is dissociation of CO, the reaction is usually independent of the concentration of L, and the rate is the same for any of a series of different L ligands. This leads to a simple rate equation:

$$\text{Rate} = k_1[\text{complex}] \tag{4.27}$$

$$L_nM-CO \underset{+CO, k_{-1}}{\overset{-CO, k_1}{\rightleftharpoons}} L_nM-\square \overset{+L', k_2}{\longrightarrow} L_nM-L' \tag{4.28}$$

In some cases, the back reaction, k_{-1}, becomes important, in which case the intermediate, $L_nM-\square$, partitions between the forward and back reactions.[31] Increasing the concentration of L does now have an effect on the rate because k_2 now competes with k_{-1}. The rate equation derived for Eq. 4.28 is shown in Eq. 4.29. It reduces to Eq. 4.27, if the concentration of CO, and therefore the rate of the back reaction, is negligible.

The overall rate is usually controlled by the rate at which the leaving ligand dissociates. Ligands that bind less well to the metal dissociate faster than does CO. For example, $Cr(CO)_5L$ shows faster rates of substitution of L in the order $L = CO < Ph_3As < py$. For similar ligands, say, phosphines, the larger the cone angle, the faster the dissociation:[31b]

$$\text{Rate} = \frac{k_1k_2[\text{L}][\text{complex}]}{k_{-1}[\text{CO}] + k_2[\text{L}]} \tag{4.29}$$

This mechanism tends to be observed for 18e carbonyls. The alternative, initial associative attack of a phosphine would generate a 20e species. While it is not forbidden to have a 20e transition state (after all, $NiCp_2$ is a stable 20e species), the 16e intermediate of Eq. 4.28 provides a lower-energy path in many cases. This is reminiscent of the S_N1 mechanism of substitution in alkyl halides where halide dissociates. The activation enthalpy required for the reaction is normally close to the M−CO bond strength because this bond is largely broken in going to the transition state. $\Delta S^{\ddagger}$ is usually positive and in the range 10–15 eu (entropy units), as expected for a dissociative process in which the transition state is less ordered.

Stereochemistry of Dissociative Substitution

A dissociative substitution of a d^6 ML_6 complex may go with retention or loss of the starting stereochemistry depending on the behavior of the d^6 ML_5 intermediate formed after initial dissociation of L. Unlike the d^8 ML_5 situation, where a trigonal bipyramid (TBP) is preferred, a d^6 ML_5 species is unstable in a TBP geometry and tends to undergo a distortion. Figure 4.5 shows why this is so.[32] The pure TBP geometry requires that two electrons occupy the two highest filled orbitals. Hund's rule predicts a triplet paramagnetic ground state for such a situation. The distortion may take place in one of two ways, either to the square pyramidal (SP) geometry or to the distorted TBP (DTBP) geometry. In either case, the system is stabilized because the two electrons can pair up and occupy the lower-lying orbital. In the SP and DTBP structures, the equatorial ligands form the letters T and Y, respectively. An SP geometry is favored when L′ is a high-trans-effect ligand such as H and a DTBP geometry when L′ is a π donor such as Cl. If the SP geometry (**4.14**) is preferred for the intermediate in Eq. 4.30, the incoming ligand can simply replace the leaving group and we may have retention of stereochemistry. On the other hand, if the DTBP geometry (**4.15**) is favored, inversion of the stereochemistry is more probable. Complications can occur because **4.14** and **4.15** can both be fluxional, in which case unexpected products can be obtained. Crystal structures of the rare stable examples of d^6 ML_5 species show SP, DTBP, or even intermediate[32] geometries, but never pure TBP.

$$(4.30)$$

square pyramidal
or T geometry

4.14

distorted
trigonal pyramidal
or Y geometry

4.15

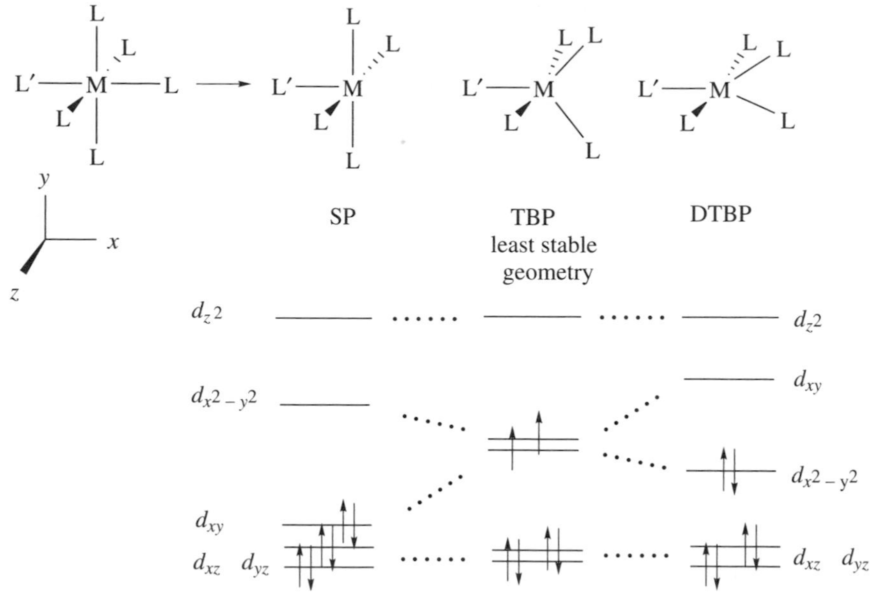

FIGURE 4.5 Crystal field basis for the distortion of the d^6 ML_5 intermediate formed after initial dissociation of L from a d^6 ML_6 complex in dissociative substitution. Pure TBP (LML = 120°) is the least stable geometry and distortion occurs to DTBP (LML = 75°) if L′ is a π donor or to SP (LML = 180°) if L′ is a high-trans-effect ligand.

Electronic and Steric Factors

The dissociative mechanism tends to be most favored in TBP d^8, followed by d^{10} tetrahedral and then d^6 octahedral. For example, d^8 $CO_2(CO)_8$ has a half-life for CO dissociation of a few tens of minutes at 0°, but for d^6 $Mn_2(CO)_{10}$ at room temperature the half-life is about 10 years! This order is consistent with the relative stabilities of the stereochemistries of the starting material and of the intermediates in each case, as predicted by crystal field arguments (Section 1.4). Substitution rates tend to follow the order 3rd row < 2nd row > 1st row.[18] For example, at 50°, the rate constants for CO dissociation in $M(CO)_5$ are Fe 6×10^{-11}, Ru 3×10^{-3}, and Os 5×10^{-8} s^{-1}. The rate for Fe is exceptionally slow, perhaps because $Fe(CO)_4$, but not the Ru or Os analog, has a high-spin ground state having low stability, leading to a higher activation energy for CO loss.

Whereas 18e organometallic complexes are usually diamagnetic, non-18e intermediates may have more than one possible spin state, such as singlet ($\downarrow\uparrow$) and triplet ($\uparrow\uparrow$) for $M(CO)_4$ (M = Fe, Ru, Os). Different spin states are isomers with different structures and different potential-energy surfaces; the reaction pathways favored by the various possible spin states can in principle differ greatly. Transitions between spin states are generally thought to be very fast, but data are sparse. This is an aspect of transition metal chemistry that is still far from well understood.[33]

Phosphines do not replace all the carbonyls in a complex, even in a case where the particular phosphine is sterically small enough to do so. The reaction of $Mo(CO)_6$ with a monodentate alkylphosphine never proceeds further than the *fac*-$Mo(CO)_3L_3$ stage. This is in part because the phosphines are much more electron donating than the carbonyls they replace. The remaining COs therefore benefit from increased back donation and are more tightly held in consequence. The fac stereochemistry (**4.16**), in which the PR_3 ligands occupy a face of the octahedron, is preferred electronically to the mer arrangement (**4.17**), in which the ligands occupy a meridian. This is because the COs have a higher trans effect than do the phosphines, and so substitution continues until there are no COs trans to a CO. The mer arrangement is less sterically encumbered, however, and is seen for bulky L groups.

fac	mer
4.16	**4.17**

Dissociation of a ligand is accelerated for bulky ligands. We shall see in Section 9.4 how this affects the dissociation of a phosphite from NiL_4 in a key step in olefin hydrocyanation, an important catalytic reaction. The degree of dissociation can be predicted from the appropriate cone angles, and the bulky phosphite $P(O\text{-}o\text{-tolyl})_3$ gives one of the very best catalysts. Triphenylphosphine is very useful in a wide variety of catalysts for the same reason.

Dissociation can sometimes be encouraged in various ways. For example, a chloride ligand can often be substituted in the presence of Ag^+ because AgCl is precipitated. Tl^+ is used in cases where Ag^+ oxidizes the complex and is therefore unsatisfactory. Protonation can also be used to remove ligands such as alkyl or hydride groups. Weakly bound solvents are often useful ligands synthetically because they can be readily displaced. As a π donor, thf is a poor ligand for W(0), and $W(CO)_5(thf)$ readily reacts with a wide range of ligands L to give $W(CO)_5L$.

Substitution of halide for alkyl or hydride is often carried out with RMgX or $LiAlH_4$. Cyclopentadienyls may be prepared from CpNa or CpTl, in which case the insoluble TlCl precipitates and helps drive the reaction.

Certain types of ligand are more likely to dissociate than others. The chelate effect means that polydentate ligands will dissociate less easily, for example. Carbon-donor ligands of the L_n type, such as η^6-C_6H_6 (L_3 type) or CO (L), will tend to dissociate more easily than L_nX ligands such as η^5-Cp (L_2X) or Me (X). This is because L_n ligands tend to be stable in the free state, but L_nX ligands would have to dissociate as radicals or ions, which is usually less favorable. Among non-carbon-donor ligands, the anions or cations can be very stable in solution (e.g., H^+ or Cl^-) and may well dissociate in a polar solvent. The electronic configuration of the metal is also important: substitution-inert d^6 octahedral complexes are much less likely to dissociate a ligand than are substitution-labile d^8 TBP metals, as we saw in Section 1.4.

Redox catalysis of substitution[31b] is covered in Section 4.5.

Hard Ligands for High-Valent Metals

Amines, hard and incapable of back donation, have a very limited ability to bind to low-valent metals, but there are a greater number of complexes in midrange oxidation states, for example, [$PtCl_2(NH_3)_2$]. Chelating amines, often deprotonated, have proved very useful for favoring high oxidation states; the π lone pair of a deprotonated R_2N ligand makes it a π donor, appropriate for a d^0 metal. In Eq. 4.31, the W(IV) starting material has such a high tendency to achieve W(VI) that it dehydrogenates and rearranges ethylene to extrude H_2 to give an ethylidyne (triyl or X_3) ligand and also allows trapping of the $M^+\equiv C-O^-$ (X_3, **4.3**) bonding mode of CO.[34]

(4.31)

Alkoxide and halides, classic ligands for d^0 metals, also have π lone pairs but being more electronegative are somewhat less effective π donors (see Section 3.2 and **3.46**).

4.4   ASSOCIATIVE MECHANISM

Kinetics

The slow step in associative substitution[35] is the attack of the incoming ligand L' on the complex to form an intermediate that rapidly expels one of the original ligands L. In general, an associative step precedes a dissociative step.

$$L_n M \xrightarrow{\ +L',\, k_1\ } L_n M{-}L' \xrightarrow{\ -L,\ \text{fast}\ } L_{n-1}M{-}L' \qquad (4.32)$$

The rate of the overall process is now controlled by the rate at which the incoming ligand can attack the metal in the slow step, and so L' appears in the rate equation:

$$\text{Rate} = k_1[L'][\text{complex}] \qquad (4.33)$$

This mechanism is often adopted by 16e complexes because the intermediate is now 18e, and so can usually provide a lower energy route than the 14e intermediate that would be formed in dissociative substitution. The reaction is analogous to the nucleophilic attack of OH^- on a $C{=}O$ in ester hydrolysis, for example. The entropy of activation is negative ($\Delta S^{\ddagger} = -10$ to -15 eu), as one might expect for the more ordered transition state required.[35]

Origin of the Trans Effect

The classic examples of the associative mechanism are shown by 16e, d^8 square planar species, such as complexes of Pt(II), Pd(II), and Rh(I). The 18e intermediate is a trigonal bipyramid with the incoming ligand in the equatorial plane (**4.18**). By microscopic reversibility, if the entering ligand occupies an equatorial site, the departing ligand must leave from an equatorial site. Loss of an equatorial ligand gives a stable square, planar species, but loss of an axial ligand would leave a much less favorable tetrahedral fragment. This has important consequences for the stereochemistry of the product and provides a simple rationale for the trans effect (Section 1.2). In Eq. 4.34, the incoming ligand is labeled L^i, the departing ligand L^d. We need to postulate that L^t, the ligand of highest trans effect, has the highest tendency to occupy the equatorial sites in the intermediate. This will ensure that the ligand L^d, trans to L^t, will also be in an equatorial site. Now, either L^t or L^d may be lost to give the final product; since L^t, as a good π-bonding ligand, is likely to be firmly bound, L^d, as the most labile ligand in the equatorial plane, is forced to leave. This is equivalent to saying that L^d is labilized by the trans effect of L^t. Good π-acid ligands are high in the trans effect series because they find the more π-basic equatorial sites in the TBP intermediate more congenial—the metal is a better π donor to these sites. Hydrogen also has a high trans effect, in part because of the lack of lone pairs, such as would be found for Cl^-, for example, minimize the ligand—metal (d_π) repulsions.

$$(4.34)$$

4.18

Other Factors

It is not uncommon for the solvent, present as it is in such high molarity, to act as L^i and expel L^d to give a solvated 4-coordinate intermediate. This intermediate can then undergo a second associative substitution with the ultimate ligand to give the final product. Substitutions of one halide for another on Pd and Pt(II) can follow this route:[36]

$$L_2MCl_2 \xrightarrow{+solv.\ slow} [L_2M(solv)Cl]^+ \xrightarrow{+Br^-,\ fast} L_2MBrCl \qquad (4.35)$$

It is easy to imagine that, because it is cationic, the solvated intermediate would be much more susceptible to Br^- attack than the starting complex. Because the solvent concentration cannot normally be varied without introducing rate changes due to solvent effects, the [solv] term does not usually appear in the experimental rate equation, which therefore has the form

$$Rate = k_s[complex] + k_a[complex][L'] \qquad (4.36)$$

where the first term refers to the solvent-assisted associative route, and the second to the direct associative reaction, which will become relatively more important as less strongly ligating solvents are used. If k_a is zero, this type of reaction can wrongly appear to be dissociative.

Ligand Rearrangements

Eighteen-electron complexes can also undergo associative substitution. Such complexes usually contain a ligand capable of rearranging and accepting an extra pair of electrons, so that the metal can avoid a 20e configuration. Nitrosyls, with their bent to linear rearrangements, can do this. For example, $Mn(CO)_4(NO)$ shows a second-order rate law and a negative entropy of activation, $\Delta S^{\ddagger}$, consistent with this mechanism:

$$(CO)_4Mn(lin\text{-}NO) \xrightarrow{\text{L. slow}} (CO)_4LMn(bent\text{-}NO) \xrightarrow{-CO.\ \text{fast}} (CO)_3LMn(lin\text{-}NO)$$

$$(4.37)^{37}$$

$$\text{Rate} = k_a[\text{complex}][\text{L}]$$

Indenyl complexes undergo associative substitution much faster than their Cp analogs. This is a result of the indenyl slipping from an η^5 to an η^3 structure. This is favorable for the indenyl group because the fused benzo ring regains its full aromatic stabilization energy as the 8 and 9 carbons dissociate from the metal and participate fully in the aromaticity of the benzo ring. These arguments have been strengthened by the isolation of several stable complexes with an η^3, or even an η^1 indenyl group, formed by the attack of a ligand on an η^1 indenyl complex. Having an indenyl is not required, $CpRh(CO)_2$ undergoes associative substitution, and the unsubstituted Cp is assumed to slip.[38] Several other ligands are capable of rearranging in a similar way; some examples are shown in Eqs. 4.38–4.42:

$$(4.38)$$

$$(4.39)$$

$$(4.40)$$

$$(4.41)$$

$$(4.42)$$

- Unsaturated ($\leq$ 16 electron) complexes can give associative substitution where the incoming ligand L_i initially binds to the metal; the rate depends on L_i.

- Saturated (18 electron) complexes can give dissociative substitution where a ligand has to be lost before L_i can bind to the metal; the rate is independent of L_i.

- Saturated complexes can give associative substitution if a ligand can rearrange to make the metal unsaturated (Eq. 4.38).

4.5 REDOX EFFECTS, THE I MECHANISM, AND REARRANGEMENTS IN SUBSTITUTION

Odd-electron species formed in redox-initiated substitution are more difficult to study and are often transients rather than stable compounds.[39]

17e and 19e Species

Astruc et al. note that 17e species such as $[Cp^*Fe(C_6Me_6)]$ $[SbF_6]_2$ can be powerful one-electron oxidants, and 19e species such as Cp_2Co can be powerful reductants.[40] As one might expect for a complex with an electron in an $M-L$ σ^* orbital, 19e species[41] tend to be much more dissociatively labile than their 18e counterparts. This means that substitution of 18e species may be catalyzed by reduction. For example, $Fe(CO)_5$ can be substituted with electrochemical catalysis as shown in Eq. 4.43, where $[Fe(CO)_5]\cdot^-$ is the chain carrier in the catalytic cycle:

$$Fe(CO)_5 \xrightarrow{+e^-} [Fe(CO)_5]\cdot^- \xrightarrow{-CO} [Fe(CO)_4]\cdot^-$$

$$\xrightarrow{+L} [Fe(CO)_4L]\cdot^- \xrightarrow{Fe(CO)_5} [Fe(CO)_5]\cdot^- + Fe(CO)_4L \quad (4.43)^{42}$$

The substitution of $[(ArH)Mn(CO)_3]^+$ by PPh_3 to give $[(ArH)Mn(CO)_2L]^+$ is catalyzed in the same way.[43]

Although the green-black 17e complex $V(CO)_6$ is stable, many 17e species,[44] such as $Mn(CO)_5\cdot$[45] and $Co(CO)_4\cdot$,[46] are isolable[47] only in matrices at low temperature or are transient intermediates at room temperature. These and other 17e species also undergo very rapid substitution, but usually by an associative pathway.[17] $V(CO)_6$, for example,[48] undergoes second-order (associative) ligand exchange at room temperature, while the 18e $[V(CO)_6]^-$ does not substitute or lose CO even in molten PPh_3. This means that substitution in an 18e species can often be catalyzed by oxidation. The presence of air is sometimes enough to cause substitution to occur, which can lead to irreproducibility or to problems in interpreting the rate. Electrochemically oxidizing $CpMn(CO)_2(MeCN)$ in the presence of PPh_3 leads to the substitution of the acetonitrile not in just one but in as many as 250 molecules of the complex per 1e abstracted.[49] The chain reaction of Eqs. 4.44 and 4.45 accounts for this result because the product radical reoxidizes the starting material, and the cycle can be repeated.

$$CpMn(CO)_2(MeCN) \xrightarrow{-e} [CpMn(CO)_2(MeCN)]\cdot^+$$

$$\xrightarrow{+L} [CpMn(CO)_2L]\cdot^+ \qquad (4.44)$$

$$CpMn(CO)_2(MeCN) + [CpMn(CO)_2L]\cdot^+ \longrightarrow$$

$$[CpMn(CO)_2(MeCN)]\cdot^+ + CpMn(CO)_2L \quad (4.45)$$

Alternatively, a trace of a free radical, Q, can abstract a 1e ligand from the metal, and the substitution can be catalyzed by a chain reaction such as is shown in Eq. 4.46. The last step regenerates the chain carrier $(CO)_nM\cdot$:

$$(CO)_nMX \xrightarrow{Q} (CO)_nM\cdot \xrightarrow{L} (CO)_{(n-1)}LM\cdot$$

$$\xrightarrow{(CO)_nMX} (CO)_{(n-1)}LMX + (CO)_nM\cdot \quad (4.46)$$

Note that Eqs. 4.43–4.46 all involve 17e/19e interconversions, while the previous examples of A and D mechanisms in diamagnetic molecules (e.g., Eqs. 4.28 and 4.32) involved 16e/18e interconversions.

While most 19e species are reactive transients, some are stable enough to isolate. Tyler[50] has isolated $(\eta^5\text{-}Ph_4C_5H)Mo(CO)_2L_2$ [L_2 = 2,3-bis(diphenylphosphino)maleic anhydride] and Astruc[51] $CpFe(\eta^6\text{-arene})$ as stable 19e species. Mössbauer and EPR (electron paramagnetic resonance) data for the Fe(I) species suggested the 19th electron is largely located on the metal; the X-ray crystal structure shows that all 11 carbons of both rings are coordinated, but the Fe$-$C(Cp) distances are 0.1 Å longer than in analogous 18e species. Sometimes the 19th electron is largely ligand based, as in $CoCp_2$.[51a] The addition of a salt such as $NaPF_6$ can completely change the outcome of a substitution reaction to give ionic products instead of the neutral ones formed in the absence of a salt. This effect has so far been studied for 19e species,[51b] but it could be useful in other types of substitution.

The Interchange Mechanism

There is evidence that certain soft nucleophiles show a second-order, associative component for their substitution even in cases such as $Mo(CO)_6$, where it is not obvious how the molecule can rearrange to avoid being 20e when the L^i binds. We have seen that 20e intermediates are unlikely, but a 20e transition state seems to be possible. An intermediate is a species that has to survive as an independent entity, if only briefly. The lifetime of a transition state, on the other hand, is comparable with a molecular vibration, or about 10^{-13} s. It is necessarily an unstable entity, and 20e transition states are not uncommon. In such cases that although both L^i and L^d bind simultaneously to the metal, they do so more weakly than they would in a more stable 18e intermediate. This is called the *interchange mechanism of substitution* and is designated I. The I mechanisms are further divided into I_a, in which L^i and L^d bind more strongly to the metal in the transition state, and I_d, in which they bind more weakly.[52] Experimentally, it is not easy to distinguish an I_a from an A mechanism because the evidence for I_a is essentially negative: the absence of a detectable intermediate. In spite of the great sophistication of modern methods of detection of transient intermediates,[53] it may be that we do not see one and will therefore take an A mechanism to be I_a. This problem is fully discussed in a review by Darensbourg.[31b]

Rearrangements of Coordinatively Unsaturated Species

When an 18e complex loses a ligand, it is common for one of the remaining ligands to rearrange so as to fill the vacant site created. This is simply the reverse of the processes we saw in Eqs. 4.38–4.42. For example, an acetate might chelate as shown in Eq. 4.47. The rearrangement product may be stable, in which case it may be observed directly, or it may be unstable, and an incoming ligand L^i may displace it. The closest analogy in organic chemistry is neighboring group participation (Eq. 4.48):

$$\text{(4.47)}$$

$$\text{(4.48)}$$

This stabilization of what would otherwise be coordinatively unsaturated intermediates can accelerate substitution reactions. In addition, species that appear from their stoichiometry to be coordinatively unsaturated intermediates may not in fact be what they seem. For example, on heating $Mo(N_2)_2(PMe_2Ph)_4$, N_2 is lost and $Mo(PMe_2Ph)_4$ (**4.19**) is formed:

$$\text{(4.49)}$$

4.19

Complex **4.19** might seem to have an electron count of 14e, but in fact it has rearranged to an 18e complex in which one of the phosphines binds via an η^6 arene ring, not via phosphorus at all. Other common ways that apparently 18e species can rearrange is by dimerization via a potentially bridging ligand (Eq. 4.51), via an agostic ligand (Eq. 4.53), or by the process known as *cyclometallation* (e.g., Eqs. 4.50 and 4.52); this is simply the oxidative addition of a C–H bond in a ligand to the metal:

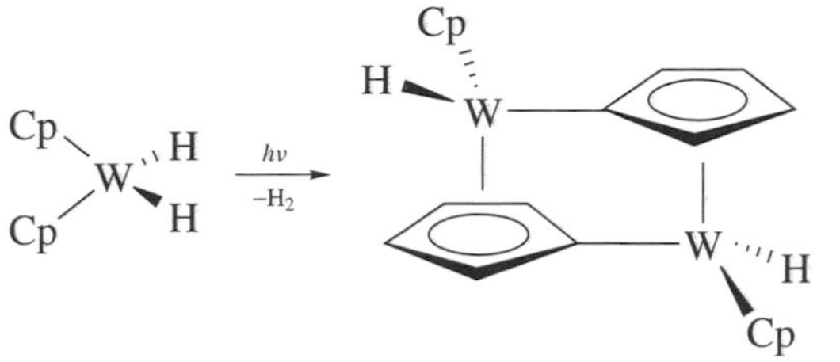

(4.50)

$$\text{ReCl(N}_2)(\text{PMe}_2\text{Ph})_4 \xrightarrow[\text{-N}_2]{\text{heat}} (\text{PMe}_2\text{Ph})_4\text{Re} \overset{\text{Cl}}{\underset{\text{Cl}}{\cdots}} \text{Re}(\text{PMe}_2\text{Ph})_4 \qquad (4.51)$$

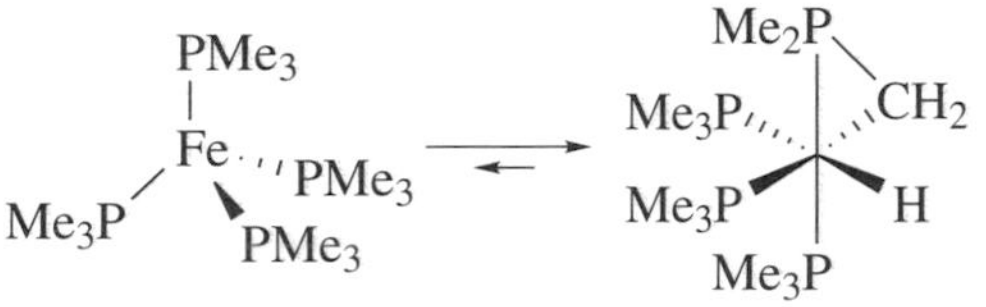

(4.52)

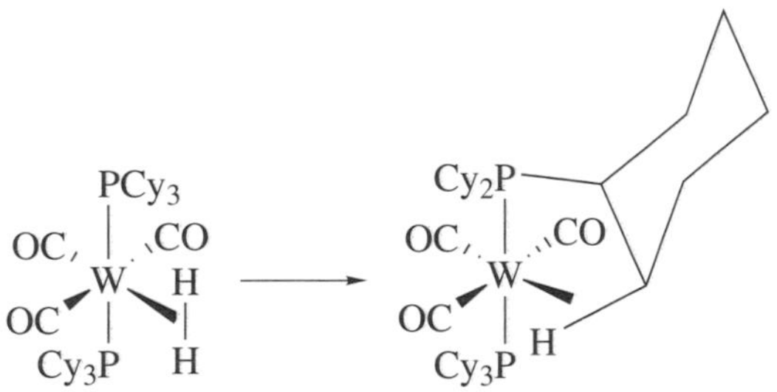

(4.53)

yellow   purple

There are also cases in which there are reasons to believe that apparently highly coordinatively unsaturated species are authentic, for example, Cp_2^*ScMe, $Cr(CH_2Ph)_4$, $Pt(PCy_3)_2$, or $[Rh(PPh_3)_3]^+$. It is always hard to rule out weak interactions with the ligands and solvent in solution, however. In such cases steric bulk of the ligands may play a role in stabilizing the compound.

4.6 PHOTOCHEMICAL SUBSTITUTION

Photochemical reactions can occur when light is absorbed by a compound. In this process, an electron is promoted and the ground-state electronic configuration is changed to that of one of the excited states. Even the longer-lived of these states survive for only 10^{-6}–10^{-9} s, and so if any photochemistry is to occur, the excited state must react very quickly and bimolecular steps are usually too slow to contribute. If a molecule of product is formed for every photon absorbed, the quantum yield, Φ, is said to be unity. Otherwise the electron falls back to the ground state and the compound either emits light (luminescence) or is heated up thermally; in this case, chemistry does not occur and Φ for product formation will normally be less than unity.

Carbonyls

Substitution reactions of carbonyls, such as $W(CO)_6$, are accelerated by ultraviolet (UV) or, for colored carbonyls, by visible light. For example, on irradiation in thf as solvent the pentacarbonyl $W(CO)_5(thf)$ is obtained. This is a useful synthetic intermediate because it reacts with a variety of ligands L to give $W(CO)_5L$ cleanly by rapid thermal substitution, rather than more highly substituted species, such as *fac*-$W(CO)_3L_3$, which are obtained from $W(CO)_6$ and L on heating. The most reasonable mechanism for such reactions is the photon-induced promotion of a d_π electron into a d_σ level, which is M–L σ-antibonding in character, and so dissociative substitution is more rapid in the excited state. Knowing the UV–visible spectrum of the starting material is useful in designing the experiment. The complex must absorb at the irradiation wavelength to be used, but if the product also absorbs, then subsequent photochemistry may lower the yield. The buildup of highly absorbing decomposition products can stop the photochemistry by absorbing all the light.

The photolysis of $W(CO)_5L$ can lead either to loss of L or of a CO group cis to L, according to the wavelength used. This result can be understood[54] in terms of the crystal field diagram for the complex, shown in Fig. 4.6. Since the symmetry is lower than octahedral because of the presence of L, both the d_σ and the d_π levels split up in a characteristic pattern. The L ligand, conventionally placed on the z axis, is usually a lower-field ligand than CO and so the d_{z^2} orbital is stabilized with respect to the $d_{x^2-y^2}$. As we saw in Section 1.5, these are really M–L σ^* orbitals, $d_{x^2-y^2}(\sigma_{xy}^*)$ playing this role for ligands in the xy plane, and $d_{z^2}(\sigma_z^*)$ for the ligands along the z axis. This means that irradiation at ν_1 tends to populate the σ_z^*, which will labilize the L ligand because it lies on the z axis. Irradiation at ν_2 will tend to populate σ_{xy}^*, and so one of the cis COs will be labilized because they lie in the xy plane, cis to L. Where L is pyridine, the appropriate wavelengths are ~400 nm (ν_1) and <250 nm (ν_2), respectively. The method has often been used to synthesize *cis*-$Mo(CO)_4L_2$ complexes.

$W(CO)_4(phen)$ has near UV and visible absorbtions at 366 and 546 nm. The first corresponds to promotion of a d_π electron to the d_σ level and is referred to as a ligand field (LF) band. The 546-nm band is a *metal-to-ligand charge transfer* (or MLCT) band and corresponds to promoting a metal d_π electron to a π^* level of the dipy ligand; the excited state therefore contains a 17e metal and a reduced ligand $W\bullet^+(CO)_4(phen\bullet^-)$. Irradiation in either band leads to substitution by PPh_3, for example, to give $W(CO)_3(PPh_3)(phen)$.

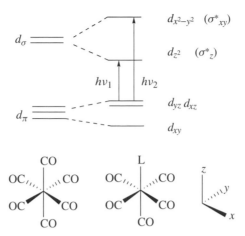

FIGURE 4.6 Crystal field basis for the selectivity observed in the photolysis of $M(CO)_5L$ complexes. Irradiation at a frequency ν_1 raises an electron from the filled d_π level to the empty $\sigma^*(z)$, where it helps to labilize ligands along the z axis of the molecule. Irradiation at ν_2 labilizes ligands in the xy plane.

Increased pressure accelerates an associative process because the volume of the transition state $L_nM \cdots L'$ is smaller than that of the separated L_nM and L' molecules; the reverse is true for a dissociative process because $L_{n-1}M \cdots L$ is larger than L_nM. Several hundred atmospheres are required to see substantial effects, however. Van Eldik[55] has shown that pressure accelerates the MLCT photosubstitution of $W(CO)_4(phen)$ but decelerates the LF photosubstitution. As the MLCT excited state is effectively a 17e W species, an A mechanism is reasonable for this process; the LF process is evidently dissociative, probably as a result of populating the M–L σ^* levels.

A complex such as $(\eta^6\text{-}C_7H_8)Cr(CO)_3$ undergoes thermal substitution by loss of cycloheptatriene. Although the triene is a polydentate ligand, this does not make up for the intrinsically much stronger binding of CO. In contrast, photochemical substitution (366 nm) gives $(\eta^6\text{-}C_7H_8)Cr(CO)_2L$. This is probably because monodentate ligands are more affected by occupation of "their" σ^* orbital than a polydentate ligand that binds simultaneously along two or all three axes of the molecule. The arene is lost in photosubstitution of $[CpFe(\eta^6\text{-}PhCH_3)]PF_6$, however, because the Cp is also polydentate.[56]

Hydrides

The second most common photosubstitution is the extrusion of H_2 from a di- or polyhydride[57] discovered in the case of the yellow crystalline complex, Cp_2WH_2 (Eq. 4.54). This is most probably the result of the promotion of an electron into the M–L σ^* orbital corresponding to the MH_2 system. Sometimes the reductive elimination product is stable, or as in Eq. 4.54, it can also be very unstable and can oxidatively add to C—H or other bonds in the solvent or ligands.

$$Cp_2WH_2 \xrightarrow{h\nu,\ benzene} Cp_2WPhH + H_2 \qquad (4.54)$$

In some cases it has been shown that loss of phosphine can occur in preference to reductive elimination of H_2, presumably depending on which σ^* orbital is populated by the wavelength of light used.

$$ReH_5(PR_3)_3 \xrightarrow{h\nu} \{ReH_5(PR_3)_2\} + PR_3 \qquad (4.55)$$

M—M Bonds

Another important photochemical process is the homolysis of M—M bonds. The fragments produced are likely to be odd-electron and therefore substitutionally labile. For example, the photosubstitution of CO in Mn_2CO_{10} by PPh_3 proceeds via the 17e intermediates $\cdot Mn(CO)_5$. Equation 4.57 is an interesting example[58] because the replacement of three COs by the non-π-acceptor NH_3 leads to a buildup of electron density on the metal. This is relieved by an electron transfer from a 19e $Mn(CO)_3(NH_3)_3$ intermediate to a 17e $Mn(CO)_5$ fragment to give the *disproportionation* product **4.20** in a chain mechanism.[59]

$$Mn_2(CO)_{10} + PPh_3 \longrightarrow Mn_2(CO)_9(PPh_3) + CO \qquad (4.56)$$

$$Mn_2^0(CO)_{10} + NH_3 \longrightarrow [Mn^I(CO)_3(NH_3)_3]^+[Mn^{-I}(CO)_5]^- \qquad (4.57)$$

4.20

- Photochemical excitation can promote substitution by labilizing ligands like CO and hydrides or splitting M—M bonds.

4.7   STERIC AND SOLVENT EFFECTS IN SUBSTITUTION

As we saw in Section 4.4, the substitution rate for an associative reaction changes as we change the incoming ligand L, but what properties of L are important in deciding the rate? At first sight this looks complicated because σ effects, π effects, and steric hindrance might all play a role. A promising approach[60] has been to assume that σ effects are dominant and compare observed rates with the pK_a of L. Since the pK_a measures the tendency for L to bind a proton, it correlates with the σ-donor power and the Tolman parameter of L. For small ligands L, the rates are successfully predicted by Eq. 4.58. Both α and β need to be determined by experiment but are constant for any particular complex.

$$\log k_\sigma = \alpha + \beta(pK_a) \qquad (4.58)$$

The α value measures the intrinsic reactivity of the complex, and β measures how much the rate is affected by the σ-donor strength of the ligand. The result is a Hammett-type linear free energy (LFE) relationship. Very bulky ligands show rates slower than predicted, however.

Solvents and Other Weakly Coordinating Ligands

As we have seen in the last few sections, solvents can act as ligands. Of the common solvents, the ones most likely to bind, and therefore perhaps to divert the reaction from its intended goal are MeCN, pyridine, Me_2SO (dimethylsulfoxide, DMSO), and Me_2NCHO (dimethylformamide, DMF). Several species dissolve only in such solvents, which bind to the metal. DMF binds via the carbonyl because the nitrogen lone pair is tied up by resonance with the CO to give:

$$Me_2N^+{=}CH{-}O^-$$

4.21

DMSO is a particularly interesting ligand because it can bind either via the S or the O. Both steric and hard and soft considerations seem to play a role in the choice. Unhindered, soft Rh(I) gives S-bound $[Rh(SOMe_2)_3Cl]$, for example.[61] CS_2 is another solvent that finds restricted use in organometallic chemistry because it reacts with most complexes; SO_2 has been used successfully, especially as a low-temperature NMR solvent.

Tetrahydrofuran (THF), acetone, water, and ethanol are much less strongly ligating and are widely used. Early transition metal complexes can be very sensitive to solvents containing labile protons, but this depends on the case. All of these solvents can act as weak ligands, and their complexes can be synthetically useful. Ketones usually bind in the η^1 mode via O, as in **4.22**, but, as ambidentate ligands, can also bind in the η^2 mode via both C and O, as in **4.23** (Eq. 4.59). The latter is favored by low steric hindrance and by a strongly back-donating metal fragment. Equation 4.59 shows how the η^1-to-η^2 rearrangement of acetone can occur on changing the oxidation state of the metal. The strong π-donor Os(II) favors the η^2 form.[62]

$$Me_2C\overset{O}{\diagup}\!\!\diagdown\, Os^{III}(NH_3)_5{}^{3+} \;\underset{-e^-}{\overset{+e^-}{\rightleftarrows}}\; \underset{Me_2C}{\overset{\displaystyle O\diagdown}{\underset{\diagup}{|}}}\!\!Os^{II}(NH_3)_5{}^{2+} \qquad (4.59)$$

4.22   **4.23**

Halocarbon solvents tend to be oxidizing and can destroy sensitive compounds. Dichloromethane is probably one of the least reactive but $PhCF_3$ is a useful completely nonoxidizing alternative for CH_2Cl_2. Halocarbons can form stable complexes, some of which have been crystallographically characterized, such as $[IrH_2(IMe)_2(PPh_3)_2]^+$.[63a] Although the binding is relatively weak, the presence of the metal greatly increases the rate of attack of nucleophiles at the halocarbon and, by changing the steric bulk, can strongly affect the selectivity of the reaction. For example, halocarbon binding favors the useful C alkylation rather than the usual N alkylation of enamines:[63b]

Arenes can in principle bind to metals, but the reaction is usually either sufficiently slow or thermodynamically unfavorable to permit the satisfactory use of arenes as solvents without significant interference. Alkanes are normally reliably noncoordinating (but see Section 12.4). Many complexes do not have sufficient solubility in the usual alkanes, but solvents such as ethylcyclohexane—the molecules of which pack poorly, leaving gaps in the liquid structure—are significantly better. IR spectra are best recorded in alkanes because they interact least with the solute and give the sharpest absorbtion peaks.

Even xenon is able to act as a ligand, as in Seidel and Seppelt's $[Au(Xe)_4][Sb_2F_{11}]_2$, which is even stable enough for an X-ray structural study.[64]

"Noncoordinating" Anions

In the case of ionic complexes, the choice of counterion may be important because they may bind to the metal. Several anions in common use are optimistically termed "noncoordinating." BF_4^- is useful, but F^- abstraction from BF_4^- by the metal to give the metal fluoro complex is a recognized problem, especially for the early metals, and many complexes are now known in which BF_4^- acts as a ligand via a $B-F-M$ bridge; SbF_6^- appears to be less strongly coordinating. BPh_4^- can form η^6 arene complexes. The $[B(3,5-\{CF_3\}_2C_6H_3)_4]^-$ anion, denoted $[BAr_4^F]^-$, or "barf" anion is one of the very best noncoordinating anions we have today.[65] It has allowed isolation of many extremely electrophilic and low-coordinate cations, for example[66] $[IrH_2(PR_3)_2]^+$. Even so, undesired Ar^F transfer to the metal has been seen in some cases.[67] Among cations, $Ph_3P=N=PPh_3^+$ is one of the most widely used. In each case the counterions of choice are large, so as to stabilize the ionic lattice of the large organometallic ion.

REFERENCES

1. (a) S. H. Strauss et al., *J. Am. Chem. Soc.* **116**, 10003, 1994; (b) W. A. Howard, G. Parkin, and A. L. Rheingold, *Polyhedron* **14**, 25, 1995; (c) H. Willner, F. Aubke et al., *Angew. Chem. Int. Ed.* **35**, 1974, 1996 and **39**, 168, 2000.

2. (a) W. Hieber, *Z. Anorg. Allgem. Chem.* **245**, 295, 1940; (b) M. W. Burk and R. H. Crabtree, *Inorg. Chem.* **25**, 931, 1986.

3. W. Hieber, E. O. Fischer et al., *Z. Anorg. Allgem. Chem.* **269**, 308, 1952.

4. J. E. Ellis, *Adv. Organomet. Chem.* **31**, 1, 1990.

5. M. C. Baird, J. A. Osborn, and G. Wilkinson, *Chem. Commun.* 129, 1966.

6. (a) M. J. Winter, *Adv. Organomet. Chem.* **29**, 101, 1989; (b) F. A. Cotton, *Prog. Inorg. Chem.* **21**, 1, 1976.

7. H. Schnöckel, *Angew. Chem., Int. Ed.* **31**, 638, 1992.

8. L. G. Hubert-Pfalzgraf et al., *Inorg. Chem.* **30**, 3105, 1991; S. J. Lippard et al., *J. Am. Chem. Soc.* **112**, 3230, 1990.

9. R. H. Crabtree, *J. Inorg. Nucl. Chem.* **40**, 1453, 1978; J. Y. Yang, Z. A. Luthey-Schulten, and K. S. Suslick, *Proc. Nat. Acad. Sci.* **100**, 3035, 2003.

10. G. Wilkinson et al., *J. Chem. Soc. A* 2037, 1967.

11. I. S. Butler, *Pure Appl. Chem.* **60**, 1241, 1988.

12. T. W. Hayton, P. Legzdins, and W. B. Sharp, *Coord. Chem. Rev.* **102**, 935, 2002.

13. J. A. McCleverty, *Chem. Rev.* **104**, 403, 2004.

14. J. W. Faller et al., *J. Organomet. Chem.* **383**, 161, 1990.

15. R. S. Armstrong et al., *J. Am. Chem. Soc.* **120**, 10827, 1998.

16. (a) J. P. Collman, P. Farnham, and G. Dolcetti, *J. Am. Chem. Soc.* **93**, 1788, 1971; (b) G. Dolcetti et al., *Inorg. Chem.* **17**, 257, 1978.

17. D. E. Koshland et al., *Science* **258**, 186, 1992; S. Snyder et al., *Science* **259**, 381, 1993.

18. B. I. Swanson and S. K. Satija, *Chem. Commun.* 40, 1973.

19. M. Manassero et al., *Chem. Commun.* 789, 1973.

20. L. F. Dahl et al., *Chem. Commun.* 880, 1970.

21. R. G. Bergman et al., *J. Am. Chem. Soc.* **105**, 3922, 1983.

22. M. Verdaguer et al., *Coord. Chem. Rev.* **190-2**, 1023, 1999.

23. B. A. MacKay and D. Fryzuk, *Chem. Rev.* **104**, 385, 2004.

24. (a) A. G. Orpen, *Chem. Commun.* 1310, 1985; (b) P. B. Dias, M. E. M. Depiedade, and J. A. M. Simoes, *Coord. Chem. Rev.* **135**, 737, 2004.

25. R. J. Morris and G. S. Girolami, *Inorg. Chem.* **29**, 4167, 1990; H. Taube et al., *Inorg. Chem.* **28**, 1310, 1989.

26. C. A. Tolman, *Chem. Rev.* **77**, 313, 1977.

27. D. R. Anton and R. H. Crabtree, *Organometallics* **2**, 621, 1983.

28. (a) P. W. N. M. van Leeuwen, P. C. J. Kamer, J. N. H. Reek, and P. Dierkes, *Chem. Rev.* **100**, 2741, 2000; (b) P. W. N. M. van Leeuwen, *J. Am. Chem. Soc.* **120**, 11616, 1998.

29. H. Werner *J. C. S. Dalton* 3829, 2003.

30. (a) L. Perrin, E. Clot, O. Eisenstein, J. Loch, and R. H. Crabtree, *Inorg. Chem.* **40**, 5806, 2001; (b) T. R. Cundari et al., *J. Am. Chem. Soc.* **125**, 4312, 2003.

31. (a) F. Basolo, *Polyhedron* **9**, 1503, 1990; J. A. S. Howell and P. M. Burkinshaw, *Chem. Rev.* **83**, 557, 1983; (b) D. J. Darensbourg, *Adv. Organomet. Chem.* **21**, 113, 1982.

32. O. Eisenstein et al., *New J. Chem.* **14**, 671, 1990; F. Maseras, O. Eisenstein, et al., *New J. Chem.* **22**, 1493, 1998.

33. R. Poli, *Acc. Chem. Res.* **30**, 494, 1997; S. Shaik, H. Schwartz, et al., *Chem. Eur J.* **4**, 193, 1998.

34. R. R. Schrock et al., *Inorg. Chem.* **36**, 171, 1997.

35. R. J. Cross, *Chem. Soc. Rev.* **14**, 197, 1985; F. Basolo et al., *J. Am. Chem. Soc.* **85**, 3929, 1966; *J. Am. Chem. Soc.* **89**, 4626, 1967.

36. R. G. Pearson, H. B. Gray, and F. Basolo, *J. Am. Chem. Soc.* **82**, 787, 1960.

37. H. Wawersik and F. Basolo, *J. Am. Chem. Soc.* **89**, 4626, 1969.

38. M. J. Calhorda and L. F. Veiros, *Coord. Chem. Rev.* **187**, 183, 1999.

39. K. H. Theopold et al., *Curr. Sci.* **78**, 1345, 2000.

40. D. Astruc et al., *J. Am. Chem. Soc.* **120**, 11693, 1998.

41. D. R. Tyler et al., *Organometallics* **11**, 3856, 1992; D. Astruc, *Chem. Rev.* **88**, 1189, 1988; *Acc. Chem. Res.* **24**, 36, 1991; C. M. Older and J. M. Stryker, *Organometallics*, **19**, 3266, 2000.

42. P. H. Rieger et al., *Chem. Commun.* 265, 1981.

43. D. A. Sweigert et al., *Chem. Commun.* 916, 1993.

44. M. C. Baird, *Chem. Rev.* **88**, 1217, 1988.

45. D. R. Kidd and T. L. Brown, *J. Am. Chem. Soc.* **100**, 4095, 1978.

46. M. Absi-Halabi and T. L. Brown, *J. Am. Chem. Soc.* **99**, 2982, 1977.

47. G. A. Ozin et al., *J. Am. Chem. Soc.* **97**, 7054, 1975.

48. T. L. Brown et al., *J. Am. Chem. Soc.* **104**, 4007, 1982; W. C. Trogler and F. Basolo et al., *J. Am. Chem. Soc.* 4032.

49. J. K. Kochi et al., *Chem. Commun.* 212, 1982.

50. D. R. Tyler, *Coord. Chem. Rev.* **97**, 119, 1990.

51. D. Astruc: (a) *New J. Chem.* **16**, 305, 1992; (b) *J. Am. Chem. Soc.* **114**, 8310, 1992.

52. H. Taube, *Comments Inorg. Chem.* **1**, 17, 1981; A. Poë and M. V. Twigg, *J. Chem. Soc. Dalton* 1860, 1974; M. L. Tobe, *Inorg. Chem.* **7**, 1260, 1968.

53. V. Montiel-Palma, R. N. Perutz, M. W. George, O. S. Jina, and S. Sabo-Etienne, *Chem. Comm.* 1175, 2000.

54. M. Wrighton, *Chem. Rev.* **74**, 401, 1974.

55. R. van Eldik, ed., *Inorganic High Pressure Chemistry: Kinetics and Mechanism*, Elsevier, Amsterdam, 1986.

56. D. Astruc, *J. Organomet. Chem.* **377**, 309, 1989.

57. M. L. H. Green, *Pure Appl. Chem.* **50**, 27, 1978; J. D. Feldman, J. C. Peters, and T. D. Tiltey, *Organometallics* **21**, 4050, 2002.

58. M. Herberhold et al., *J. Organomet. Chem.* **152**, 329, 1978.

59. D. R. Tyler et al., *Coord. Chem. Rev.* **63**, 217, 1985.

60. W. P. Giering et al., *Organometallics* **4**, 1981, 1985; A. J. Poë, *Pure Appl. Chem.* **60**, 1209, 1988; J. K. Kochi et al., *J. Am. Chem. Soc.* **106**, 3771, 1984.

61. D. Milstein et al., *Chem. Comm.* 710, 2002.

62. W. D. Harman, H. Taube et al., *J. Am. Chem. Soc.* **110**, 2439, 1988.

63. (a) R. H. Crabtree, J. W. Faller, et al., *Organometallics* **1**, 1361, 1982; (b) R. J. Kulawiec and R. H. Crabtree, *Coord. Chem. Rev.* **99**, 89, 1990.

64. S. Seidel and K. Seppelt, *Science* **290**, 117, 2000.

65. M. Brookhart et al., *Organometallics* **11**, 3920, 1992.

66. A. C. Cooper, W. E. Streib, O. Eisenstein, and K. G. Caulton, *J. Am. Chem. Soc.* **119**, 9069, 1997.

67. G. J. Kubas, *Chem. Commun.* 1807, 1999.

PROBLEMS

1. (a) Would you expect 18e metal carbonyl halides $M(CO)_nX$, $X =$ halide, to dissociate into halide anions and the metal carbonyl cation as easily as the

hydrides, $X = H$, dissociate into H^+ and the metal carbonyl anion? (b) Given that we have a case where both of the above processes occur, contrast the role of the solvent in the two cases.

2. $Ni(CO)_4$ and $CO(lin\text{-}NO)(CO)_3$ are both tetrahedral. Why does the Ni compound undergo dissociative substitution and the Co compound undergo associative substitution?[55]

3. List the following in the order of decreasing reactivity you would predict for the attack of trimethylamine oxide on their CO groups: $Mo(CO)_6$, $Mn(CO)_6{}^+$, $Mo(CO)_2(dpe)_2$, $Mo(CO)_5{}^{2-}$, $Mo(CO)_4(dpe)$, $Mo(CO)_3(NO)_2$.

4. What single piece of physical data would you choose to measure as an aid to establishing the reactivity order of the carbonyl complexes above?

5. What are the electron counts, oxidation states, and coordination numbers of the metals in Eqs. 4.50–4.53.

6. Amines, NR_3, are usually only weakly coordinating toward low valent metals. Why is this so? Do you think that NF_3 would be a better ligand for these metals? Discuss the factors involved.

7. Ligand dissociation from NiL_4 is only very slight for $L = P(OMe)_3$, yet for $L = PMe_3$ it is almost complete. Given that the two ligands have essentially the same cone angle, discuss the factors that might be responsible.

8. Determine whether associative or dissociative substitution is more likely for the following species (not all of which are stable): $CpFe(CO)_2L^+$, $Mn(CO)_5$, $Pt(PPh_3)_4$, $ReH_7(PPh_3)_2$, $PtCl_2(PPh_3)_2$, $IrCl(CO)(PPh_3)_2$.

9. Propose plausible structures for complexes with the following empirical formulas: $Rh(cod)(BPh_4)$, $(indenyl)_2W(CO)_2$, $PtMe_3I$, $(cot)(PtCl_2)_2$, $(CO)_2RhCl$.

10. Given a complex $M(CO)_6$ undergoing substitution with an entering ligand L', what isomer(s) of the product would you expect to find in the products if L' were (a) monodentate and a higher-trans-effect ligand than CO, or (b) L' were bidentate and had a lower trans effect than did CO.

11. NO^+ is isoelectronic with CO and often replaces CO in a substitution reaction, so it might seem that Eq. 4.60 should be a favorable reaction. Comment on whether the process shown is likely.

$$Mo(CO)_6 + NOBF_4 \longrightarrow Mo(NO)_6(BF_4)_6 + 6CO \qquad (4.60)$$

12. $Fe(CO)_5$ loses CO very slowly, but in the presence of an acid, substitution is greatly accelerated. Suggest possible explanations. For dissociative CO substitutions, the rate tends to be higher as the $v(CO)$ stretching frequency of the carbonyl increases. Suggest a reason.

13. Use the data of Table 2.8 to predict the position of the highest frequency $\nu(CO)$ band in $[Co(CO)_6]^{3+}$ and comment on the result in connection with deciding whether this hypothetical species would be worth trying to synthesize.

14. Tertiary amines, such as NEt_3, tend to form many fewer complexes with low-valent metals [e.g., W(0)] than PEt_3. What factors make two cases so different? In spite of this trend, $(Et_3N)W(CO)_5$ is isolable. What factors are at work to make this species stable?

15. Count electrons for the species in Eq. 4.31 and consider how π bonding might affect the outcome.

5

COMPLEXES OF π-BOUND LIGANDS

In this chapter we continue our survey of the different types of ligand by looking at cases in which the π electrons of an unsaturated organic fragment, rather than a lone pair, are donated to the metal to form the M−L bond.

5.1 ALKENE AND ALKYNE COMPLEXES

In 1827, the Danish chemist Zeise obtained a new compound he took to be KCl·PtCl$_2$·EtOH from the reaction of K$_2$PtCl$_4$ with EtOH. Only in the 1950s was it established that Zeise's salt, **5.1**, is really K[PtCl$_3$(C$_2$H$_4$)]·H$_2$O, containing a coordinated ethylene, formed by dehydration of the ethanol, and a water of crystallization. The metal is bonded to both carbons of the ethylene, but the four C−H bonds bend slightly away from the metal, as shown in **5.4**; this allows the metal to bind efficiently to the π electrons of the alkene. For Zeise's salt, the best bonding picture[1] is given by the *Dewar–Chatt* model. This involves donation of the C=C π electrons to an empty d_σ orbital on the metal, so this electron pair is now delocalized over three centers: M, C, and C′. This is accompanied by back donation from a metal d_π orbital into the ligand LUMO, the C=C π^* level, as shown in **5.2** (occupied orbitals shaded). By analogy with the bonding in CO, we will refer to the former as the "σ bond" and the latter as the "π bond." As is the case for CO, a σ bond is insufficient for tight binding, and so only metals capable of back donation, and not d^0 metals such as Ti(IV), bind alkenes well.

The Organometallic Chemistry of the Transition Metals, Fourth Edition, by Robert H. Crabtree
Copyright © 2005 John Wiley & Sons, Inc.

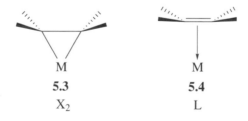

5.1   **5.2**

The C=C bond of the alkene lengthens on binding. The M—alkene σ bond depletes the C=C π bond by partial transfer of these electrons to the metal and so slightly weakens and, therefore, lengthens it. The major factor in lengthening the C=C bond, however, is the strength of back donation from the metal. By filling the π^* orbital of the C=C group, this back donation can sharply lower the C—C bond order of the coordinated alkene. For a weakly π-basic metal this reduction is slight, but for a good π base it can reduce it almost to a single bond. For Zeise's salt itself, M—L σ bonding predominates because the Pt(II) is weakly π basic, and the ligand (C—C: 1.375 Å) more nearly resembles the free alkene (1.337 Å). The substituents are only slightly bent back away from the metal, and the C—C distance is not greatly lengthened compared to free ethylene. Pt(0), in contrast, is much more strongly π basic, and in Pt(PPh$_3$)$_2$C$_2$H$_4$, the C—C distance becomes much longer (1.43 Å). In such a case the metal alkene system approaches the *metalacyclopropane* extreme, **5.3**, as contrasted with the Dewar–Chatt model, **5.4**, involving minimal π back donation; both are considered η^2 structures.

$$
\begin{array}{cc}
\text{M} & \text{M} \\
\textbf{5.3} & \textbf{5.4} \\
\text{X}_2 & \text{L}
\end{array}
$$

In the metalacyclopropane extreme (i.e., cyclopropane with M replacing one CH$_2$ group), the substituents on carbon are strongly folded back away from the metal as the carbons rehybridize from sp^2 to something more closely approaching sp^3. The presence of electron-withdrawing groups on the alkene also encourages back donation and makes the alkene bind more strongly to the metal; for example, Pt(PPh$_3$)$_2$(C$_2$CN$_4$) has an even longer C—C distance (1.49 Å) than the C$_2$H$_4$ complex. In the Dewar–Chatt extreme, we can think of the ligand acting largely as a simple L ligand such as PPh$_3$, but in the metalacyclopropane extreme, we have what is effectively a cyclic dialkyl, and so we can think of it as an X$_2$ (diyl or σ_2) ligand. In both cases we have a 2e ligand on the covalent model, but while the L (ene or π) formulation, **5.4**, leaves the oxidation state unchanged, the X$_2$ picture, **5.3**, makes the oxidation state more positive by two units. By convention, the L model is usually adopted for the assignment of the formal oxidation state.

Structural studies are best for determining where any given alkene complex lies on the structural continuum between **5.3** and **5.4**. The position of any vinyl protons, or of the vinyl carbons in the ^{1}H and ^{13}C NMR, also shows a correlation with the structure. For example, at the metalacyclopropane X_2 extreme, the vinyl protons can resonate 5 ppm, and the vinyl carbons 100 ppm to high field of their position in the free ligand, as is appropriate for a change of hybridization from sp^2 to about sp^3 at carbon. Coordination shifts are usually much lower in the case of the L extreme.

The same factors that lead to lowering of $v(CO)$ in metal carbonyls also lead to greater metalacyclopropane character in alkene complexes: strong donor coligands, a net negative charge on the complex ion, and a particularly low oxidation state for the metal. This means that Pd(II), Hg(II), Ag(I), and Cu(I) alkene complexes tend to be L-type, or Dewar–Chatt, in character, while those of Ni(0), Pd(0), and Pt(0), tend to be X_2, or metalacyclopropane-like.

One chemically significant difference between the two extremes is that **5.4** tends to have a ∂^+ charge on carbon and therefore some of the character of a masked carbonium ion. This is because the ligand to metal σ donation depletes the charge on the ligand, and in the L-type extreme this is not recouped by back dona-tion. These alkene complexes are therefore subject to nucleophilic attack and are resistant to electrophilic attack at the vinyl carbons, Pd(II), being the classic case. Alkenes are more commonly actor, rather than spectator, ligands. Since simple alkenes in the free state are subject to electrophilic but not nucleophilic attack, the effect of binding is very significant. It means that the appropriate metal fragment inverts the chemical character of the alkene, a phenomenon known as *umpolung*. The metal can either promote nucleophilic attack or inhibit electrophilic attack at the ethylene carbons—that is to say, it can act as either an activating group or a protecting group, depending on the reagents involved.

Strained alkenes, such as cyclopropene or norbornene (**5.5**), bind unusually strongly to metals because the rehybridization on binding leads to relief of strain. Much of the strain in a small ring compound arises because the real

5.5

C−C−C angles are constrained to be smaller than the ideal ones. Such an alkene is therefore less strained when complexed because the ideal angles at the vinylic carbons drop from the value of 120°, appropriate for sp^2 hybridization, to close to 109°, appropriate for sp^3 hybridization. In some cases very strained alkenes are only stable in the complexed form. Nonconjugated dienes such as 1,5-cyclooctadiene (cod), and norbornadiene (nbd), can chelate to the metal and so bind more strongly than the corresponding monoenes, but conjugated dienes behave somewhat differently (Section 5.3). Ketenes ($RCH=C=O$) can bind in several ways, including η^2 via the $C=C$ bond.[2]

Synthesis

Alkene complexes are usually synthesized by the methods shown in Eqs. 5.1–5.7:

1. Substitution in a low-valent metal:

$$AgOSO_2CF_3 + C_2H_4 \longrightarrow (C_2H_4)AgOSO_2CF_3 \qquad (5.1)$$

$$PtCl_4{}^{2-} + C_2H_4 \longrightarrow [PtCl_3(C_2H_4)]^- + Cl^- \qquad (5.2)$$

$$Cp(CO)_2Fe(Me_2C=CH_2)^+ \xrightarrow{\text{1-hexene}} Cp(CO)_2Fe(\text{1-hexene})^+ \qquad (5.3)$$

2. Reduction of a higher-valent metal in the presence of an alkene:

$$(cod)PtCl_2 + C_8H_8{}^{2-} + cod \longrightarrow Pt(cod)_2 \qquad (5.4)$$

$$RhCl_3 + nbd + CH_3CH_2OH \longrightarrow [(nbd)Rh(\mu\text{-Cl})]_2$$
$$+ CH_3CHO + HCl \qquad (5.5)$$

3. From alkyls and related species:

$$(5.6)$$

$$Cp_2TaCl_3 \xrightarrow{n\text{-BuMgX}} \{Cp_2TaBu_3\} \xrightarrow{\beta \text{ elim., red. elim.}}$$

$$Cp_2TaH(\text{1-butene}) + \text{butene} + \text{butane} \qquad (5.7)$$

(where red. elim. = reductive elimination).

Reversible binding of alkenes to Ag^+ (Eq. 5.1) is used to separate different alkenes chromatographically on silver-doped gas chromatography columns. Less hindered alkenes usually bind more strongly (Eq. 5.3). The reducing agent in Eq. 5.4 is the dianion of cyclooctatetraene, which the authors may have intended to act as a ligand. If so, this is an example of a common event—a reaction with an unintended outcome. The alcohol solvent is the reductant in Eq. 5.5; this happens by the mechanism of Eq. 3.26. Protonation at the terminal methylene in the η^1-allyl manganese complex of Eq. 5.6 creates a carbonium ion having a metal at the β position. Since the carbonium ion is a zero-electron ligand like a proton, it can coordinate to the 18e metal to give the alkene complex. Equation 5.7 shows a common result of trying to make a metal alkyl in which the alkyl contains a β hydrogen.

Reactions

Perhaps the most important reaction of alkene ligands is their insertion[3,4] into M—X bonds to give alkyls (Eqs. 3.19–3.20). This goes very readily for X = H, often at room temperature. On the other hand, insertion into other M—X bonds is rarer. Strained alkenes and alkynes insert most readily; the first case is promoted by relief of strain in the alkyl product and the second because the product M—vinyl bond strength is unusually high. Fluoroalkenes (e.g., Eq. 5.9) also insert readily because the resulting fluoroalkyl has a very high M—C bond strength (Section 3.5).

$$PtHCl(PEt_3)_2 + C_2H_4 \rightleftharpoons PtEtCl(PEt_3)_2 \tag{5.8}$$

$$AuMe(PPh_3) + CF_2{=}CF_2 \longrightarrow \{(CF_2{=}CF_2)AuMe(PPh_3)\} \longrightarrow$$
$$Au(CF_2{-}CF_2Me)(PPh_3) \tag{5.9}$$

When the metal fragment is a poor π base, the L model (**5.4**) applies and the vinylic carbons bound to the metal behave as masked, metal-stabilized carbonium ions. In such a case we often see nucleophilic attack (e.g., Eq. 5.10).[5] This is an example of a more general reaction type—nucleophilic attack on polyenes or polyenyls (Section 8.3).

$$(Me_2NH)Cl_2Pt - \| \quad + \quad :NHMe_2 \rightleftharpoons (Me_2NH)Cl_2\overset{-}{Pt} \diagdown \diagup \overset{+}{N}HMe_2$$
$$\tag{5.10}$$

Finally, alkenes containing allylic hydrogens can undergo oxidative addition of the C—H bond in what is effectively a cyclometallation to give an allyl hydride complex. In the example shown, a base is also present so as to remove HCl from the metal and trap the allyl product.

(5.11)

Alkyne Complexes

Alkynes behave in ways broadly similar to alkenes, but being more electronegative, they tend to encourage back donation and bind more strongly. The substituents tend to fold back away from the metal by $30°-40°$ in the complex, and the M−C distances are slightly shorter than in the corresponding alkene complexes. The metalacyclopropene model (**5.6**) seems often to be the most appropriate description when alkynes act as 2e donors. More interestingly, alkynes can form complexes that appear to be coordinatively unsaturated. For example, **5.7**[6] appears to be 14e, and **5.8**,[7] is a 16e species if we count the alkyne as a conventional 2e donor. In such cases the alkyne also donates its second C=C π-bonding orbital, which lies at right angles to the first. The alkyne is now a 4e donor[8] and **5.8** can be formulated as an 18e complex. Compound **5.7** might seem to be a 20e complex on this model, but in fact one combination of ligand π orbitals, **5.7a**, finds no match among the d orbitals of the metal, and so the true electron count is 18e. An extreme valence bond formulation of the 4e donor form is the bis-carbene (**5.9**), the bonding of which we look at in Section 11.1. Four electron alkyne complexes are rare for d^6 metals because of a 4e repulsion between the filled metal d_π and the second alkyne C=C π-bonding pair.

When the free alkyne has a structure that leads to bending of the C≡C triple bond, this induces strain, which is partially relieved on binding. Cyclohexyne and benzyne are both highly unstable species that bind very strongly to metals, as in $[(Ph_3P)_2Pt(\eta^2\text{-cyclohexyne})]$ or the product shown in Eq. 5.12.[9] Cyclobutyne, normally inaccessible, has been trapped as its triosmium cluster complex.[10]

$$Cp^*TaMe_3Ph \xrightarrow{\text{heat}} Cp^*TaMe_2(\eta^2\text{-benzyne}) + MeH \qquad (5.12)$$

Alkynes readily bridge an M−M bond, in which case they can act as conventional 2e donors to each metal (**5.10**). The alternative tetrahedrane form (**5.11**) is the equivalent of the metalacyclopropane picture for such a system.

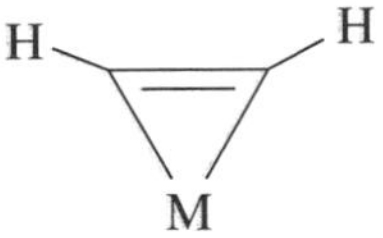

5.6 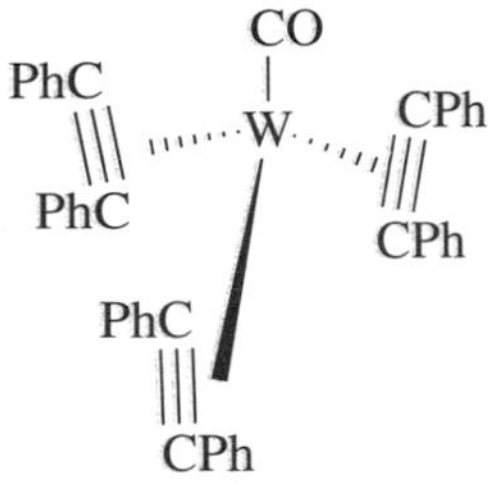   **5.7** 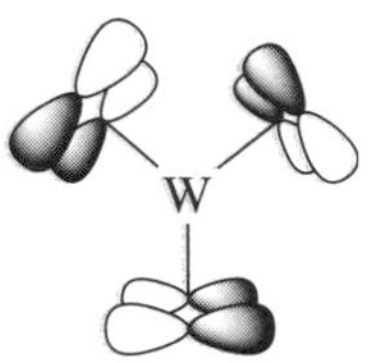   **5.7a**

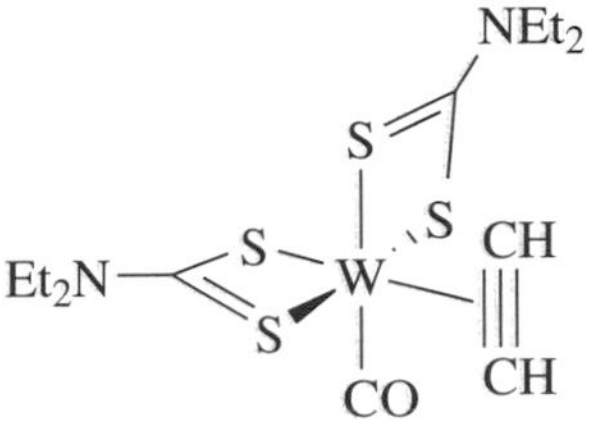

5.8 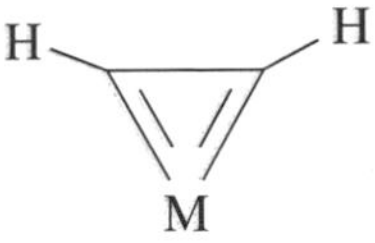   **5.9**

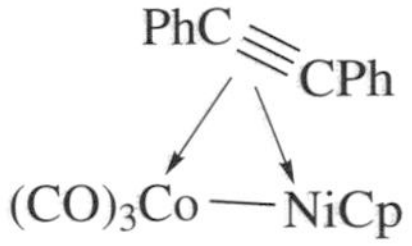

5.10 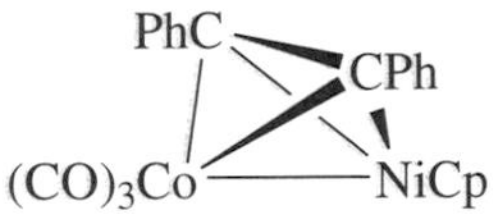   **5.11**

1-Alkynes, RCCH, can easily rearrange to vinylidenes,[11] RHC=C=M, on binding (Section 11.1).

- Alkenes donate via their C=C π bond; back bonding into C=C $\pi*$ weakens the C=C bond.
- Weak back bonding gives a Dewar–Chatt structure with the C=C bond largely retained, but strong back bonding gives a metalacyclobutane structure with the C=C bond largely converted to C–C.
- Dewar–Chatt structures tend to undergo nucleophilic attack (Eq. 5.10).

5.2 ALLYL COMPLEXES

The allyl group,[12] commonly an actor ligand, binds in one of two ways. In the monohapto form, **5.12**, it is a simple 1e X-type ligand like Me, and in the trihapto form, **5.13**, it acts as a 3e LX enyl ligand. It is often useful to think of **5.13** in terms of the resonance forms **5.14a** and **5.14b**. Intermediate cases between **5.12** and **5.13** (η^2-allyls) are also known.[13]

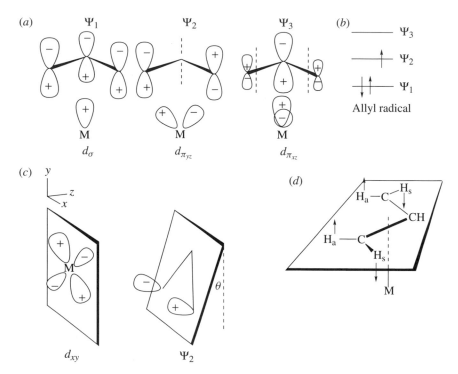

FIGURE 5.1 The electronic structure of the allyl ligand and some features of metal-allyl bonding. Nodes are shown as dotted lines in (a).

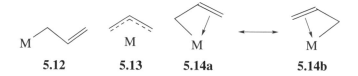

Figure 5.1a shows that of the three molecular orbitals of the allyl fragment, ψ_1 can interact with a suitable metal d_σ orbital, and ψ_2 with an M(d_π) orbital on the metal; ψ_3 is not a frontier orbital and so probably of lesser importance. As the number of nodes increases, the MOs of the free ligand become less stable (Fig. 5.1b). Two peculiarities of the structures of η^3-allyl complexes can be understood on this picture. First, the plane of the allyl is canted at an angle θ with respect to the coordination polyhedron around the metal, as shown in Fig. 5.1c; θ is usually $5°-10°$. The reason is that the interaction between ψ_2 and the d_{xy} orbital on the metal is improved if the allyl group moves in this way, as can be seen in Fig. 5.1c. The structures also show that the terminal CH_2 groups of the allyl are twisted about the C—C vector so as to rotate the anti hydrogens, H_a, away from the metal, and the syn hydrogens, H_s, toward the metal as shown by the arrows in Fig. 5.1d. This allows the p orbital on these carbons to point more directly

toward the metal, thus further improving the overlap.[14] Note the nomenclature of the allyl substituents, which are syn or anti with respect to the central CH.

The η^3-allyl group often shows exchange of the syn and anti substituents. One mechanism goes through an η^1-allyl intermediate, as shown in Eq. 5.13. This kind of exchange can affect the appearance of the ^{1}H NMR spectrum (Section 10.2), and also means that an allyl complex of a given stereochemistry may rearrange with time.

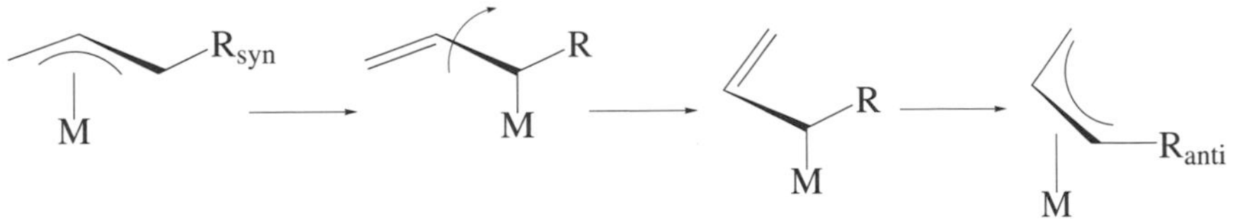

$$(5.13)$$

Synthesis

Typical routes to allyl complexes are shown below.

1. From an alkene (see also Eq. 5.11):

$$Mo(dpe)_2(\eta^2\text{-propene}) \rightleftharpoons Mo(dpe)_2(\eta^3\text{-allyl})H \qquad (5.14)^{15}$$

2. From an allyl compound by nucleophilic attack on the metal:

$$CH_2{=}CHCH_2SnMe_3 + Mn(CO)_5Br \xrightarrow{\text{reflux}}$$

$$(\eta^3\text{-}CH_2CHCH_2)Mn(CO)_4 \qquad (5.15)^{16}$$

3. From an allyl compound by electrophilic attack on the metal:

$$CH_2{=}CHCH_2Cl + Mn(CO)_5{}^- \longrightarrow$$

$$(\eta^1\text{-}CH_2{=}CHCH_2)Mn(CO)_5 \xrightarrow{\text{heat}} (\eta^3\text{-}CH_2CHCH_2)Mn(CO)_4 \quad (5.16)$$

4. From diene complexes:

$$(\eta^2\text{-}CH_2{=}CH{-}CH{=}CH_2)Fe(CO)_3 \xrightarrow{\text{HCl}}$$

$$(\eta^3\text{-}CH_2CHCHMe)Fe(CO)_3Cl \qquad (5.17)$$

$$Cp_2TiCl \xrightarrow[-\text{propene}]{i\text{-PrMgBr}} \{Cp_2TiH\} \xrightarrow{\text{butadiene}}$$

$$Cp_2Ti(\eta^3\text{-MeCHCHCHMe}) \qquad (5.18)$$

$$[PtH(acetone)(PR_3)_2]^+ + CH_2{=}C{=}CH_2 \longrightarrow$$

$$[(\eta^3\text{-allyl})Pt(PR_3)_2]^+ \qquad (5.19)$$

The first route we saw in Section 5.1; the second and third resemble the synthetic reactions most commonly used for alkyl complexes. In Eqs. 5.15 and 5.16, the metal often attacks at the least hindered terminal CH_2 group. Equation 5.17 demonstrates an electrophilic attack on a diene complex; we shall see in the next section why attack takes place at the terminal carbon. Equation 5.18 shows that when one C=C group of a diene undergoes insertion into a M−H bond, the hydrogen tends to attach itself to the terminal carbon of the conjugated chain. This leaves a methylallyl group, which can become η^3 if a vacant site is available. Finally, Eq. 5.19 shows that allenes insert into an M−H bond to put the hydride on the central carbon and generate an allyl group.

Reactions

The most important reactions of allyls are illustrated in Eqs. 5.20–5.23:

1. With nucleophiles (Eq. 5.20):

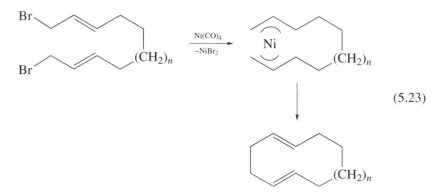

$$(5.20)^{17}$$

2. With electrophiles:[18]

$$Cp(CO)_2FeCH_2CH=CH_2 + E^+ \longrightarrow [Cp(CO)_2Fe(CH_2=CHCH_2E)]^+$$

$$(E^+ = HgCl^+, Me^+, RCO^+, H^+, Br^+) \tag{5.21}$$

3. By insertion:[19]

$$(\eta^3\text{-allyl})_2Ni \xrightarrow{CO_2} (\eta^3\text{-allyl})NiOCOCH_2CH=CH_2 \tag{5.22}$$

4. With reductive elimination (Eq. 5.23):[20]

$$(5.23)$$

Nucleophilic attack at one of the terminal carbons of the allyl group most often takes place from the face of the allyl away from the metal. This happens when the nucleophile attacks directly. On the other hand, cases are known in which the nucleophile first attacks the metal and only then is transferred to the allyl group. The latter route can only take place when a vacant site is made available at the metal. An example of a system that gives products of both stereochemistries is shown in Eq. 5.24.[21]

$$(5.24)$$

Other Ligands

Cyclopropenyl complexes,[22] such as $(\eta^3\text{-Ph}_3\text{C}_3)\text{Co(CO)}_3$, are also known but are less well studied than allyls. Benzyl groups can be persuaded to give η^3-benzyl species, but the aromatic C=C double bond is less available than that of the simple allyl group, so the complexes have a high tendency to go to the η^1 form. One example of such a complex is formed by cocondensing Pd atoms and benzyl chloride in a *metal vapor synthesis*[23] experiment (Eq. 5.25). This technique requires special equipment but allows preparatively useful quantities of metal atoms to be used as reagents. They are formed by firing an electron gun at the metal surface in a vacuum and condensing the atoms with ligand vapor at liquid N_2 temperature. The η^3-benzyl intermediate, shown in Eq. 5.25, has also been invoked in the unusual rearrangement of Eq. 5.26.[24] The η^3-propargyl ligand ($^-\text{CH}_2-\text{C}\equiv\text{CH}$), which sometimes behaves as an η^3-allenyl group ($\text{CH}_2=\text{C}=\text{CH}^-$), as well as η^1 forms of each are also known.[25] The bis-triphenylphosphine Pt(II) η^3-allenyl complex readily undergoes nucleophilic attack at the central carbon.[26]

$$(5.25)$$

$$(s = \text{solvent})$$

$$(5.26)$$

5.3 DIENE COMPLEXES

This ligand usually acts as a 4e donor in its cisoid conformation, as shown in **5.15**. This L_2 (diene or π_2) form is analogous to the Chatt–Dewar extreme for alkenes, while the LX_2 (enediyl or $\sigma_2\pi$) form **5.16** is related to the metalacyclopropane extreme. The first is rarely seen in pure form but (butadiene)$Fe(CO)_3$ has an intermediate character, with the C_1C_2, C_2C_3, and C_3C_4 distances about equal (1.46 Å) and C_1 and C_4 further from the metal than C_2 and C_3. Form **5.16** becomes more important as the back donation increases. Bound to the strongly back-donating $Hf(PMe_3)_2Cl_2$ d^2 system, 1,2-dimethylbutadiene shows an extreme LX_2 bonding pattern.[27] The substituents at C_1 and C_4 twist approximately $20°$–$30°$ out of the plane of the ligand and bend back strongly so that the corresponding p orbitals can overlap better with the metal (**5.17**). The C_1C_2, and C_3C_4 distances [1.46 Å(average)] are much longer than C_2C_3 (1.40 Å), and C_1 and C_4 are closer to the metal than C_2 and C_3 by 0.18 Å.

L_2	LX_2	M
5.15	**5.16**	**5.17**

We expect the frontier orbitals of the butadiene, ψ_2 (HOMO) and ψ_3 (LUMO), to be the most important in bonding to the metal. The MO diagram (Fig. 5.2) shows that both the depletion of electron density in ψ_2 by σ donation to the metal and population of ψ_3 by back donation from the metal lengthens C_1C_2 and shortens C_2C_3 because ψ_2 is C_1C_2 antibonding and ψ_3 is C_2C_3 bonding. Protonation occurs at C_1 (Eq. 5.17) because the HOMO, ψ_2, has its highest coefficient there.

This is quite general—binding to a metal usually depletes the ligand HOMO and fills the ligand LUMO. This is the main reason why binding has such a profound effect on the chemical character of a ligand (see Section 2.6). The structure of the bound form of a ligand is often similar to that of the first excited state of the free ligand because to reach this state we promote an electron from the HOMO to the LUMO, thus partially depleting the former and filling the latter.

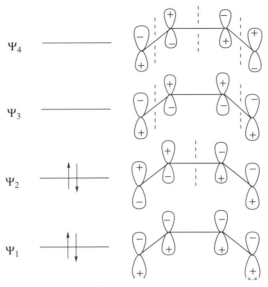

FIGURE 5.2 Electronic structure of butadiene. An electron-rich metal will tend to populate Ψ_3; an electron-poor metal will tend to depopulate Ψ_2.

Butadiene complexes are usually prepared in ways very similar to those used for alkenes, but some methods specific to diene complexes are shown below (Eqs. 5.27–5.29):[28,29]

$$(5.27)$$

$$(5.28)$$

$$(5.29)$$

The binding of butadiene in the transoid form is much rarer. It is found in $Os_3(CO)_{10}(C_4H_6)$, **5.18**, in which the diene is η^2 bound to two different Os centers[30] and in $Cp_2Zr(C_4H_6)$, **5.19**, in which the diene is bound to a single Zr.[31] In the zirconium case, the cisoid isomer also exists, but it rearranges to give a 1:1 thermodynamic mixture of the two forms on standing; photolysis leads to the trans form.

5.18   **5.19**

The "envelope shift" (shown in Eq. 5.30), is sometimes seen.[32] It exchanges the anti and syn substituents on the diene via an X_2-type metalacyclic intermediate, in which the central C=C group must be uncomplexed (**5.20**) because the metal lies in the plane of this C=C group and orthogonal to the C=C π electrons.

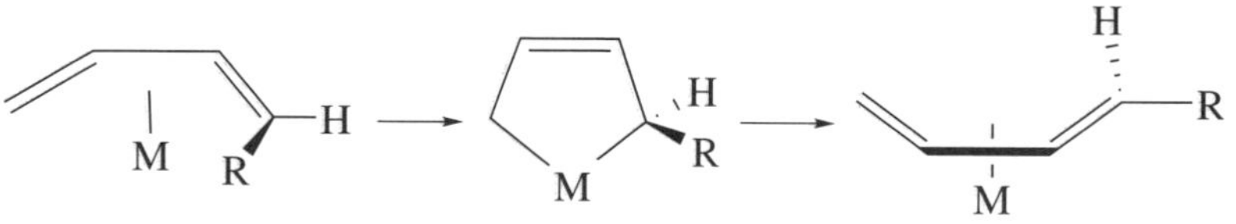

(5.30)

5.20

Cyclobutadiene Complexes

Most the neutral ligands we have studied have been stable in the free state. With cyclobutadiene, the complexes are very stable and have been known for many years, but the free dienes are so highly reactive that stable examples were reported only much later. The free molecule, with four π electrons, is antiaromatic and rectangular, but the ligand is square and seems to be aromatic. The metal must stabilize the diene by populating the LUMO of the free diene by back donation; by gaining partial control of two more π electrons, this gives the diene an electronic structure resembling that of the aromatic six π-electron dianion $R_4C_4^{2-}$; ligand-to-metal σ donation prevents the ligand from accumulating excessive negative charge. This is a good example of the free and bound forms of the ligand being substantially different from one another (Section 2.6).

Some synthetic routes are:[33]

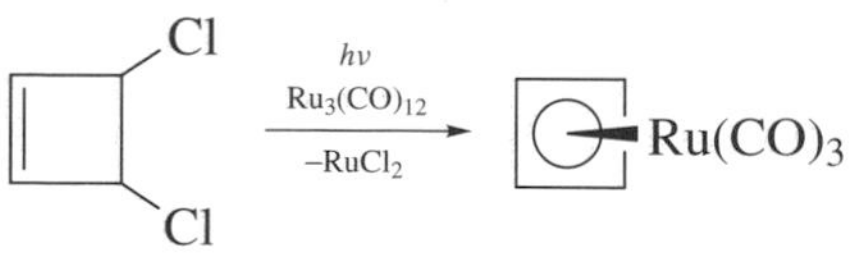

(5.31)

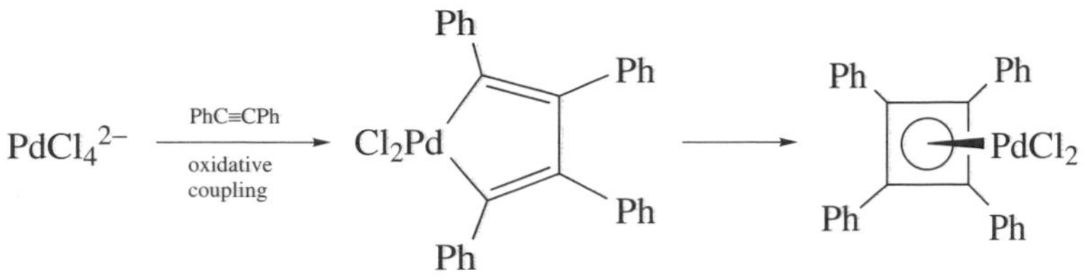

(5.32)

The ruthenium example probably involves oxidative addition of the dihalide to two $Ru(CO)_3$ fragments derived from the photolysis of the cluster; then the metals probably disproportionate, so that one becomes the observed product and the other carries away the halides in the form of undefined Ru(II) halo complexes. The reaction of Eq. 5.32 probably goes by an important general class of reaction, *oxidative coupling* (Section 6.7), to give a metalacycle, followed by a reductive elimination of the cyclobutadiene ligand.

Other Ligands

Another significant tetrahapto ligand that is very unstable in the free state is trimethylenemethane (**5.21**). It can be considered as an LX_2 ligand; one of the resonance forms is shown as **5.22**. The ligand shows an umbrella distortion from the ideal planar conformation, which means that the central carbon lies out of the plane away from the metal. Maintaining good delocalization within the ligand favors the planar form, but distorting allows the p orbitals on the terminal carbons to point more directly toward the metal and improve M−L overlap. In spite of the distortion, the central carbon is still closest to the metal.[34] Two synthetic routes[35] are illustrated in Eq. 5.33.

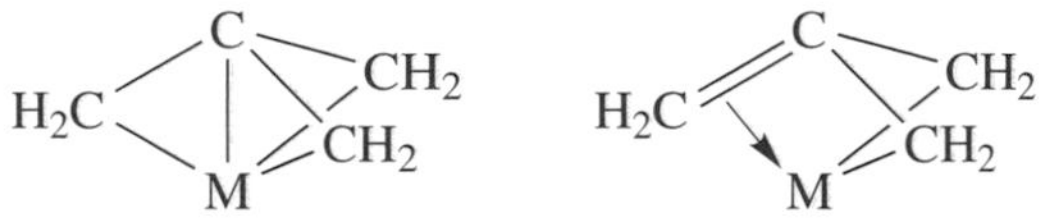

5.21   **5.22**

Nonconjugated diolefins behave much as simple olefin complexes, except that the chelation introduces rigidity and increases the binding constant to the metal. 1,5-Cyclooctadiene (**5.23**), 1,5-heptadiene (**5.24**), and norbornadiene (**5.25**) are typical examples that all form $[(diene)Rh(\mu\text{-}Cl)]_2$ complexes.

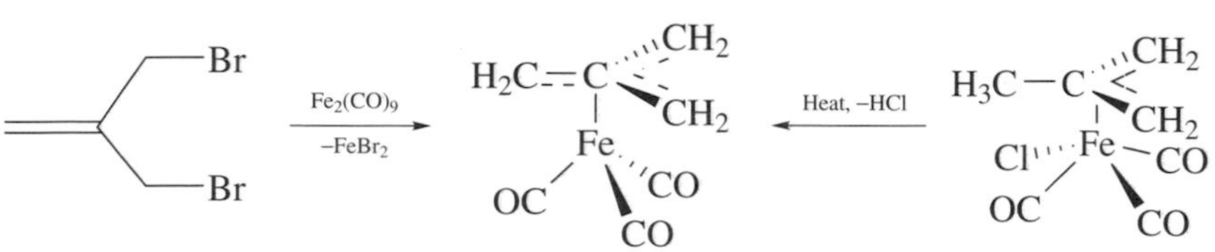

(5.33)

5.23   5.24   5.25

5.4 CYCLOPENTADIENYL COMPLEXES

The cyclopentadienyl group is perhaps the most important of the polyenyls because it is the most firmly bound and the most inert to nucleophilic or electrophilic reagents. This makes it a reliable spectator ligand for a whole series of complexes $CpML_n$ ($n = 2$, 3, or 4) where we want chemistry to occur at the ML_n group. $CpML_n$ are often referred to as "two-, three-, or four-legged piano stools," with the Cp being regarded as the "seat" and the other ligands as the "legs." The *metallocenes*, Cp_2M (see Fig. 5.5) are also important in the historical development of organometallic chemistry, but their chemistry is somewhat less rich than that of the piano stools because fewer ligands can bind to the metallocenes without overstepping the 18e limit. Their most important application is alkene polymerization (Chapter 11).

The sandwich structure of the orange crystalline Cp_2Fe was deduced by Wilkinson and Woodward and by Fischer in 1954.[36] This is usually counted as one of the most significant discoveries during the early development of organotransition metal chemistry and helped to launch it as an independent field in its own right.

The η^1 structure is also found where the coligands are sufficiently firmly bound so that the Cp cannot rearrange to η^5 (e.g., **5.26**). Trihapto-Cp groups are rather rare (e.g., **5.27**); the Cp folds so the uncomplexed C=C group can bend away from the metal. The tendency of an η^5 Cp group to "slip" to an η^3 or η^1 structure is small. Nevertheless, there are cases in which 18e piano stool complexes have been found to undergo substitution by an associative mechanism, and it is therefore assumed that the Cp can slip in the transition state.

H$_{,,}$ FeCp(CO)$_2$ WCp(CO)$_2$

5.26   5.27

$$CpRh(CO)_2 \xrightarrow{\text{L}} \{\eta^3\text{-}CpRh(CO)_2L\} \xrightarrow{\text{--CO}} CpRhL(CO) \qquad (5.34)$$

Rearrangement of an η^5 Cp to a stable η^1 structure on the addition of a ligand is also known:

$$\text{CpReMe(NO)(CO)} \xrightarrow{\text{PMe}_3} \eta^1\text{-CpReMe(NO)(CO)(PMe}_3)_2 \qquad (5.35)^{37}$$

In this case the slip takes place in preference to two other possible rearrangements that might have relieved the electron count on the metal: bending of the NO or methyl migration to CO. It is likely that one of these two processes may be important in the initial attack of the phosphine, but that slip of the Cp gives the stable product shown. η^1-Cp groups tend to show both long and short C—C distances, as appropriate for an uncomplexed diene. The η^5 form has essentially equal C=C distances, and the substituents bend very slightly toward the metal.

In diamagnetic complexes, η^5-Cp groups usually show a resonance in the ^{1}H NMR spectrum at 3.5–5.5δ, as appropriate for an aromatic group. This aromaticity was one of the first properties of the Cp group to attract the attention of Robert Woodward, the celebrated organic chemist, who showed that ferrocene, like benzene, undergoes electrophilic acylation.[38] η^1-Cp groups can show a more complex ^{1}H NMR pattern: The α hydrogen appears at about 3.5δ and the β and γ hydrogens at 5–7δ. As we see in Chapter 10, the η^1-Cp group can be fluxional, in which case the metal rapidly moves around the ring so as to make all the protons equivalent.

The MO scheme for the C_5H_5 group is shown in Fig. 5.3. The five p orbitals on carbon give rise to five MOs for the C_5H_5 group. In Fig. 5.3a, only the nodes are shown for simplicity, but Fig. 5.3b shows the orbitals in full in one case. The most important overlaps are ψ_1 with the metal d_{z^2}, and ψ_2 and ψ_3 with the d_{xz} and d_{yz} orbitals (an example is shown explicitly in Fig. 5.3b); ψ_4 and ψ_5 do not interact very strongly with metal orbitals, and the Cp group is therefore not a particularly good π acceptor. This and the anionic charge means that Cp complexes are generally basic, and that the presence of the Cp encourages back donation from the metal to the other ligands present.

If we put two Cp groups and one metal together, we obtain the MO diagram for a metallocene (Fig. 5.4). We now have to look at the symmetry of *pairs* of Cp orbitals and ask how they will interact with the metal orbitals. As an example, if we take the combination of the ψ_1's of both rings shown in Fig. 5.4b, which has the symmetry label a_{1g}, we find it can interact with the d_{z^2} orbital on the metal, also a_{1g}. Taking the opposite combination of ψ_1's (shown in Fig. 5.4c, and labeled a_{2u}) we find that the interaction now takes place with p_z. Similarly, ψ_2 and ψ_3 combinations are strongly stabilized by interactions with the d_{xz}, d_{yz}, p_x, and p_y orbitals. Although the details of the interactions are more complex in this case, the picture retains L $\rightarrow$ M direct-donation and M $\rightarrow$ L back-donation components as we saw for CO or C_2H_4.

As expected for what is essentially an octahedral complex, the d-orbital splitting pattern for an octahedral crystal field, highlighted in a box in Fig. 5.4a, appears in the final pattern. Because of the different choice of axes in this case (Fig. 5.3c) than previously, it turns out that the labels of the orbitals (d_{xy}, d_{yz}, etc.)

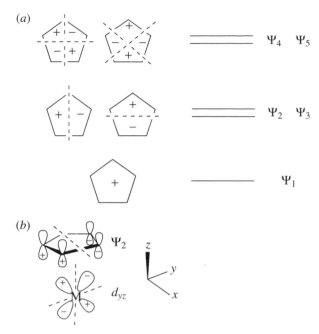

FIGURE 5.3 Electronic structure of the cyclopentadienyl ligand and one of the possible metal–Cp bonding combinations.

are different in this diagram from what they were for the crystal field diagrams we saw before. This does not matter; labels are our convention, not Nature's.

In the case of ferrocene itself, all the bonding and nonbonding orbitals are exactly filled, so it is not surprising that the group 8 metallocenes are the stablest members of the series. Metallocenes from groups 9 and 10 have one or two electrons in antibonding orbitals; this is why $CoCp_2$ and $NiCp_2$ are paramagnetic and much more reactive than ferrocene. Cobalticene also has an 18e cationic form, Cp_2Co^+. Chromocene and vanadocene have fewer than 18e and are also paramagnetic, as the electron occupation diagram (Fig. 5.5) predicts. Because d^5 ions have no crystal field stabilization in their high-spin form, high-spin $MnCp_2$ is very reactive and strongly ionic in character. The higher-field ligand C_5Me_5, on the other hand, gives a low-spin manganocene.

Bent Metallocenes

Metallocenes of group 4, and of the heavier elements of groups 5–7 are capable of binding up to three ligands in addition to the two Cp groups. In doing so, the Cp's bend back away from the ligands as shown in Fig. 5.6. This bending causes mixing of the d, s, and p orbitals so that the three hybrid orbitals shown in **5.28** point out of the open side of the metallocene away from the rings and toward the additional ligands (**5.28**). In ferrocene itself, these are all filled, but one may

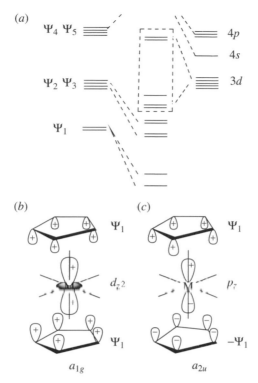

FIGURE 5.4 Qualitative MO diagram for a first-row metallocene. (a) The box shows the crystal field splitting pattern, only slightly distorted from its arrangement in an octahedral field. Because we now have two Cp groups, the sum and difference of each MO has to be considered. For example, Ψ_1 gives $\Psi_1 + \Psi_1'$, of symmetry a_{1g}, which interacts with d_{z^2}, as shown in (b), and $\Psi_1 - \Psi_1'$, of symmetry a_{2u}, which interacts with p_z, as shown in (c). For clarity, only one lobe of the Cp p orbitals is shown.

5.28

still be protonated to give bent Cp_2FeH^+. "Cp_2Re" has one fewer electron and so requires one 1e ligand to give a stable complex (e.g., Cp_2ReCl). "Cp_2Mo" and "Cp_2W" have two fewer electrons than ferrocene and so can bind two 1e ligands or one 2e ligand to reach 18e [e.g., Cp_2MH_2 or $Cp_2M(CO)$]. Only two of the three available orbitals are used in the metallocene dihydrides. One is a lone pair that points between the two substituents and can be protonated to give the water-soluble trihydride cations $Cp_2MH_3^+$. This lone pair can also take part in back donation to stabilize any unsaturated ligands present (e.g., $[Cp_2M(C_2H_4)Me]^+$). The group 5 metals can bind three X ligands (e.g., Cp_2NbCl_3). The group 4 metals

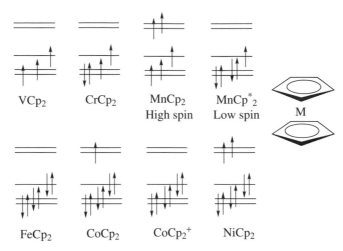

FIGURE 5.5 The d orbital occupation patterns for some first-row metallocenes. The splitting pattern of the d orbitals is approximately the same as for octahedral ML_6.

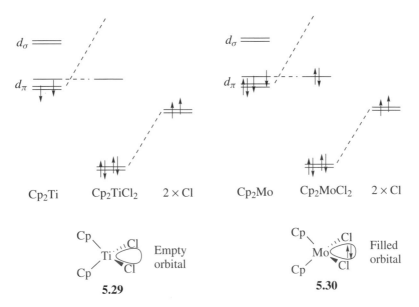

FIGURE 5.6 Bent metallocenes. The d^2 Cp_2Ti fragment can bind two Cl atoms to give the metallocene dichloride Cp_2TiCl_2 in which the single nonbonding orbital is empty and located as shown between the two Cl ligands; this empty orbital makes the final complex a hard 16e species. The d^4 Cp_2Mo fragment can also bind two Cl atoms to give the metallocene dichloride Cp_2MoCl_2 in which the single nonbonding orbital is now full and located as before; this filled orbital, capable of back donation, makes the final complex a soft 18e species.

bind only two X ligands (e.g., Cp_2TiCl_2); having only 4 valence electrons, their maximum oxidation state is M(IV). This leaves the 16e titanocene dihalide with an empty orbital (**5.29**), rather than a filled one as in the molybdocene dihalides (**5.30**). This accounts for many of the striking differences in the chemistry of the group 4 and group 6 metallocene complexes. The former act as hard Lewis acids and tend to bind π-basic ligands such as $-OR$, but the latter act as soft π bases and tend to bind π-acceptor ligands such as ethylene.

The orbital pattern shown in Fig. 5.6 is consistent with the expectation based on the discussion of Fig. 2.2. Since the virtual CN (coordination number) of Cp_2MX_2 is 8 (Cp_2MX_2 is an MX_4L_4 system), we expect $(9 - 8)$ or one non-bonding orbital, as shown in Fig. 5.6.

The η^5-C_5Me_5 ligand, often designated Cp^*, is a popular and important variant of Cp itself. It is not only higher field but also more electron releasing and more bulky. It stabilizes a wider range of organometallic complexes than Cp itself. This is an example of a general strategy for producing more stable versions of interesting compounds—introducing steric hindrance. The Cp^* derivatives are often also more soluble than the Cp compounds. Examples of Cp^* compounds showing properties not shared by their Cp analogs are discussed in Sections 7.1, 11.1, and 15.3.

Synthesis

The synthesis of cyclopentadienyls follows the general pattern shown in Eqs. 5.36–5.41. TlCp is an air-stable reagent that is often useful for making Cp complexes from halides. Some of the syntheses go in rather low yield (e.g., Eq. 5.38 typically gives 30%).

1. From a source of Cp^-:

$$NaCp + FeCl_2 = Cp_2Fe \qquad (5.36)^{39}$$

$$TlCp + Mn(CO)_5Cl \longrightarrow CpMn(CO)_3 \qquad (5.37)^{40}$$

$$MoCl_5 \xrightarrow{NaCp,\ NaBH_4,\ -100°C} Cp_2MoH_2 \qquad (5.38)^{41}$$

2. From a source of Cp^+:

$$CpFe(CO)_2^- + C_5H_5Br \longrightarrow CpFe(CO)_2(\eta^1\text{-}Cp) \xrightarrow{heat,\ -CO} FeCp_2 \quad (5.39)$$

3. From the diene or a related hydrocarbon:

$$C_5Me_5H + MeRe(CO)_5 \longrightarrow Cp^*Re(CO)_3 \qquad (5.40)$$

$$Cyclopentene \xrightarrow{IrH_2S_2L_2^+} CpIrHL_2^+ \qquad (5.41)^{42}$$

$$(S = Me_2CO;\ L = PPh_3)$$

The paramagnetic metallocenes, such as $NiCp_2$, are very reactive (see Fig. 5.7[43]). Compound **5.31** (Fig. 5.7) is an example of a triple-decker sandwich in which the electrons of the center ring are delocalized over the two metal centers. It is rare for a π-bonding carbocyclic ligand to bond to two metals on opposite faces. The reason this happens here is probably that $NiCp_2$ is a 20e compound and so formation of the triple-decker sandwich allows two metals to share the excess electrons.

Two pentahapto ligands that are closely analogous to Cp are cyclohexadienyl **5.32** and pentadienyl **5.33**. In the first, the uncomplexed methylene unit of the ring is bent 30°–40° out of the plane of the rest of the ligand, but the ligand is otherwise much like Cp itself. The pentadienyl group on the other hand, is easily able to shuttle back and forth between the η^1, η^3, and η^5 structures.[44] The η^3 form, being a substituted allyl, can have syn and anti isomers (**5.34** and **5.35**), and the η^1 form can have the metal substituted at the 1 or 3 positions along the chain.

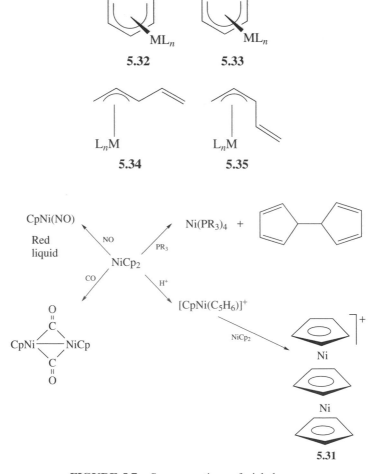

FIGURE 5.7   Some reactions of nickelocene.

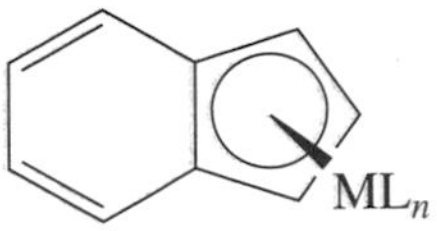

5.36

Ligands Analogous to Cp

As we saw in Section 4.4, the tendency of the indenyl **5.36** to slip from η^5 to η^3 is higher than in Cp because the full aromatic stabilization of the benzo ring is restored in the slipped form. Indenyl is also a better π acceptor than Cp. For example, $[(\eta^5\text{-Ind})IrHL_2]^+$ is deprotonated by NEt_3, but the Cp analog is not deprotonated even by t-BuLi. This is probably nothing to do with slip because the $\eta^5\text{-PhC}_5H_4$ analog is also readily deprotonated.

Tris-pyrazolyl borate (**5.37**), often denoted Tp,[45] is a useful tridentate facial N-donor spectator ligand. Tp complexes have some analogy with Cp, although this is not as close as once thought. Tp has a lower field strength, for example, and Tp_2Fe, unlike Cp_2Fe, is paramagnetic. Tp is an L_2X ligand because it has a negative charge (i.e., $L_3^- = L_2X$). Tp and related ligands can be extensively modified, for example, as shown in **5.38–5.40**.[46] In **5.38** a BH bond can also bind to the metal in an agostic fashion to complete the tridentate coordination.

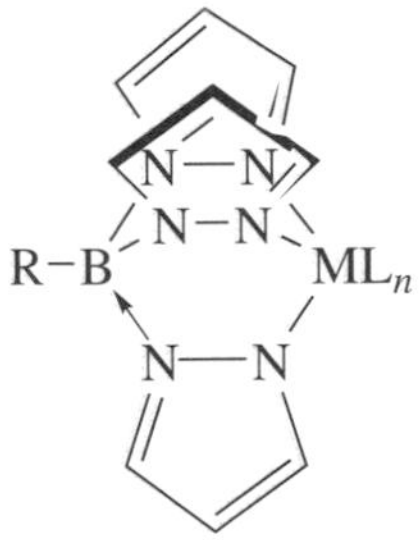

5.37

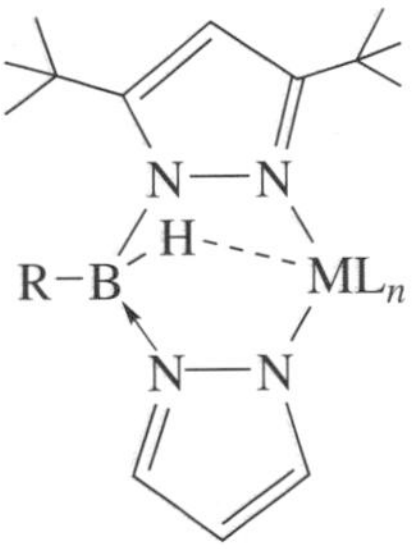

5.38

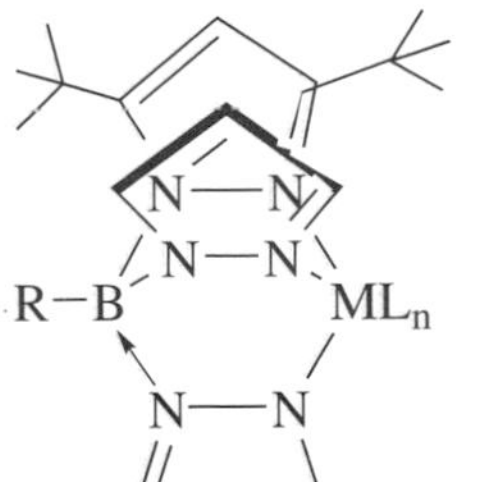

5.39

(Trp = trypticyl)

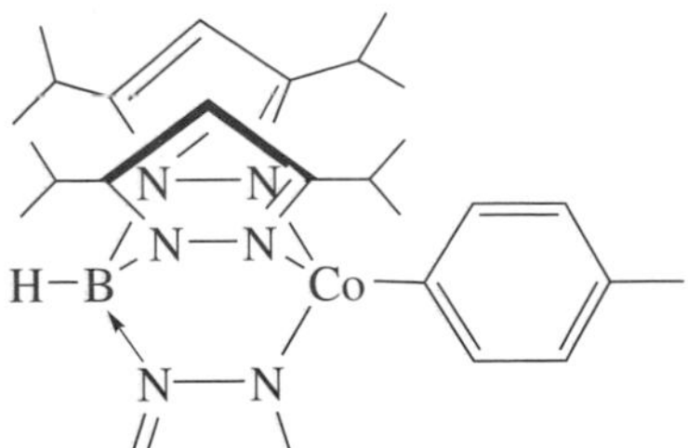

5.40

Bulky Tp ligands, sometimes considered *tetrahedral enforcers*, strongly promote a tetrahedral geometry even when this results in a highly unusual structure, such as 15e paramagnetic Co(II) complex **5.40**.

5.5 ARENES AND OTHER ALICYCLIC LIGANDS

Arenes usually bind to transition metals in the 6e, η^6-form **5.41**, but η^4 (**5.42**) and η^2 (**5.43**) structures are also known.[47] In the η^4 form the ring is usually strongly folded, while an η^6 arene tends to be flat. The C—C distances are usually essentially equal, but slightly longer than in the free arene. Arenes are much more reactive than Cp groups, and they are also more easily lost from the metal so arenes are normally actor, rather than spectator, ligands.

5.41   **5.42**   **5.43**

^{13}C NMR spectroscopy is perhaps the most useful method of characterization, the metal-bound carbons showing a ~25-ppm shift to high field on coordination, due to the increased shielding from the nearby metal.

Synthesis

Typical synthetic routes differ little from those used for alkene complexes:

1. From the arene and a complex of a reduced metal:

$$Cr(CO)_6 + C_6H_6 \xrightarrow{\text{n-Bu}_2\text{O}} (\eta^6\text{-}C_6H_6)Cr(CO)_3 \qquad (5.42)$$

$$Ti(atoms) + PhMe \longrightarrow (\eta^6\text{-PhMe})_2Ti \qquad (5.43)^{48}$$

$$FeCp_2 + C_6H_6 + AlCl_3 \longrightarrow [CpFe(\eta^6\text{-}C_6H_6)]^+[AlCl_4] \quad (5.44)^{49}$$

2. From the arene, a metal salt and a reducing agent:

$$3CrCl_3 + 2Al + AlCl_3 + 6C_6H_6 \longrightarrow$$

$$3[Cr(\eta^6\text{-}C_6H_6)_2]^+ \xrightarrow{\text{reduction}} 3[Cr(\eta^6\text{-}C_6H_6)_2] \qquad (5.45)$$

3. From the diene:

$$1,3\text{-cyclohexadiene} + RuCl_3 \longrightarrow [(\eta^6\text{-}C_6H_6)RuCl(\mu\text{-Cl})]_2 \quad (5.46)^{50}$$

The route of Eq. 5.42 is interesting in that the ether solvent may help stabilize the unsaturated Cr complexes that are probably intermediates. Metal vapor synthesis is used to make $[Cr(\eta^6\text{-}2,4,6\text{-trimethylpyridine})_2]$ **5.44**, which is not accessible by the usual routes. Note how the steric hindrance of the methyl groups on the pyridine discourages the normally more favored η^1 binding via nitrogen. Arenes bind only to low-valent metals, so metal salts of higher oxidation state are often reduced in the presence of the ligand (method 2 above). In the third route, the diene reduces the metal and in so doing provides the arene ligand by an as yet undefined mechanism.

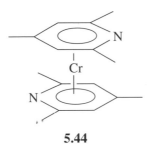

5.44

The MO picture is similar to that for Cp, but the arene ligand is a weaker net donor to the metal. The shift in $\nu(CO)$ of only 50 cm^{-1} to lower energy on going from $Cr(CO)_6$ to $(C_6H_6)Cr(CO)_3$ confirms this picture. Binding depletes the electron density on the ring, which becomes subject to nucleophilic attack. Apart from nucleophilic attack, the metal encourages deprotonation both at the ring protons, because of the increased positive charge on the ring, and α to the ring (e.g., at the benzylic protons of toluene), because the negative charge of the resulting carbanion can be delocalized on to the metal, where it is stabilized by back bonding to the CO groups.

Other Arene Ligands

Polycyclic arenes such as naphthalene also bind to low-valent metals. In this case η^6 binding is still common but the tendency to bind η^4 is enhanced because, as we saw for indenyl, this allows the uncomplexed ring to be fully aromatic. If one ring is different from the other, different isomers, called *haptomers*, can exist in which the metal is bound to one or the other ring. The metal can migrate from one ring to the other in a haptomeric equilibrium.

Moving to the fullerene series,[51] Fig. 5.8 shows how the ellipsoidal molecule C_{70} binds[52] to Vaska's complex. Free C_{70} itself does not give crystallographically useful crystals, and so this structure confirmed the ellipsoidal structure previously deduced from its NMR spectrum. The junctions between six-membered rings seem to be the most reactive in the fullerenes, and this is where the metal binds. It is almost always the Cl and CO groups in the planar Vaska complex that bend back to become cis when an alkene or alkyne binds; here the PPh$_3$ groups bend back, presumably because of steric repulsion by the bulky C_{70} group. Figure 5.8

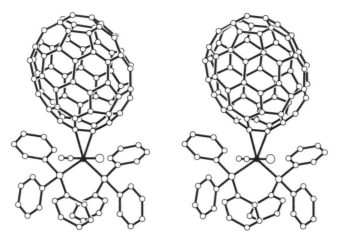

FIGURE 5.8 Stereoscopic drawing of $(\eta^2\text{-}C_{70})Ir(CO)Cl(PPh_3)_2$.[52]

is a stereoscopic diagram of a type commonly seen in research papers. With practice, it is possible to relax the eyes so that the two images formed by each eye are fused to give a three-dimensional representation of the molecule. The metal can also be inside the fullerene cavity, in which case the symbol @ is used, as in $Ca@C_{60}$.[53]

Fullerenes contain both five- and six-membered rings and so could in principle act as η^5-cyclopentadienyl or η^6-arene ligands, but η^2 binding is strongly favored.[54] The reason may be that the carbon p orbitals of fullerenes radiate outward from the center of the fullerene. This is not the situation in conventional Cp or arene ligands where the ring carbon p orbitals are parallel or even point toward the metal. Fullerenes containing several metal atoms within the cage are also known, as in $Sc_3N@C_{80}$, where the larger C_{80} cage allows more room for the triangular metal nitride cluster.[55] Even helium has been incorporated as in $He@C_{60}$, and in the large C_{86} fullerene, a ScC diatomic.[56]

η^7 Ligands

η^7-Cycloheptatrienyl ligands are well known. The ring is planar, and the C−C distances are essentially the same; η^5, η^3, and η^1-bonding modes are also known. The tropylium cation $C_7H_7^+$ is stable, and isolable salts, such as the fluoroborate, are often used in the synthesis of the complexes. Although the aromatic $C_7H_7^+$ and not the antiaromatic $C_7H_7^-$ is the stable form of the free ligand, it is still considered as L_3X (or $C_7H_7^-$) for electron counting and oxidation state assignments.

$$CpCr(C_6H_6) + [C_7H_7]BF_4 \longrightarrow [CpCr(\eta^7\text{-}C_7H_7)]^+ \xrightarrow{\text{reduction}} CpCr(\eta^7\text{-}C_7H_7)$$

$$\textbf{17e} \qquad\qquad\qquad\qquad \textbf{17e} \qquad\qquad\qquad \textbf{18e}$$

$$(5.47)$$

The commonest synthetic method is abstraction of H^- from an η^6 cycloheptatriene complex with Ph_3C^+ (Eq. 5.48) or Et_3O^+; the oxonium cation is the reagent of choice because the by-products, Et_2O and EtH, are both volatile.

$$(5.48)$$

η^8 Ligands

η^8-Cyclooctatetraene (cot) complexes are usually made from the aromatic cot^{2-} dianion. The classic example is $U(cot)_2$.

$$UCl_4 + 2cot^{2-} \longrightarrow U(cot)_2 \qquad (5.49)$$

Early metals that need many electrons to achieve an 18e structure can also give η^8 C_8H_8 complexes, such as $[(\eta^8-C_8H_8)Ti(NtBu)]$.[57]

Fluorocarbons[58a]

Perfluorinated polyenes and polyenyls have a chemistry significantly different from that of their hydrocarbon analogs. Octafluorocyclooctatetraene (ofcot), one of the more extensively studied, has been found to undergo unusual rearrangements and adopt bonding modes unknown for cot (see Eq. 5.50).[58b] Some of the synthetic difficulties are illustrated by the fact that such an apparently simple ligand as $(\eta^5-C_5F_5)$ was first reported only in 1992 in $Cp^*Ru(\eta^5-C_5F_5)$.[58c] As a result of the electron-withdrawing F substituents, these ligands are worse σ donors and better π acceptors than their hydrocarbon analogs.

$$(5.50)$$

5.6 METALACYCLES AND ISOELECTRONIC AND ISOLOBAL REPLACEMENT

We looked at some metalacycles with saturated rings in Eqs. 3.22 and 3.23, and we have seen several metalacyclic descriptions of complexes in this chapter (e.g., **5.3**, **5.6**, **5.9**). *Isoelectronic replacement* is a general strategy for finding new ligand types based on known ones or of drawing comparisons between known types. For example, if one CH_2 in an η^2-alkene complex **5.45** is replaced by O, the result is an η^2-formaldehyde complex **5.46** [e.g., $Cp_2Zr(\eta^2\text{-}CH_2O)$]. In the Zr example, the strong π-donor character of the d^2 metal encourages the η^2-bonding mode. Thioformaldehyde is not isolable in the free state, but η^2 complexes are known, for example, $Os(CO)_2(PPh_3)_2(\eta^2\text{-}CH_2S)$.[59] Replacing both CH_2 groups by O gives an η^2-dioxygen complex (**5.47**), such as $IrCl(\eta^2\text{-}O_2)(CO)(PPh_3)_2$. The presence of the heteroatom also introduces a lone pair and therefore an alternative mode of binding via that lone pair (e.g., **5.48**, **5.49**).

In the unusual heteroatom-substituted species **5.50**, the lone pair of the P is tied up by the $W(CO)_5$ group, leaving the $W(CO)_4$ group to bind the "butadiene" fragment.

$$C_2H_2 \xrightarrow{\text{Fe(CO)}_5} \quad \longleftrightarrow \quad \tag{5.51}$$

An interesting polyphospha analog of ruthenocene is shown as **5.51**.[60] Among η^5 ligands, a common heteroatom type is a ferrole, or ferracyclopentadiene, shown in Eq. 5.51, where it is not free but bonded to a second $Fe(CO)_3$ group. Note the different resonance forms of the product, one including an Fe→Fe donor metal–metal bond.

An ML_n fragment like $Fe(CO)_4$ is not isoelectronic with C fragments like CH_2, the iron fragment has far more electrons. Hoffmann[61] has pointed out that

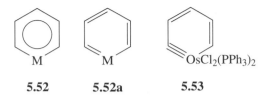

5.52   **5.52a**   **5.53**

particular metal fragments can have the same number, occupation, and shape of their orbitals as, say, CH_2 and can replace CH_2 in organic molecules as if they were isoelectronic; he called these fragments *isolobal* with the organic group. For example, $Fe(CO)_4$ is said to be isolobal with CH_2. This concept, studied in detail in Section 13.2, has been useful in understanding metallabenzenes (**5.52**). These are species in which we replace on CH of benzene by a metal fragment isolobal with CH.[62–64] The X-ray structures show a planar MC_5 ring without the alternating bond lengths that would be expected for the alternative nonaromatic (metalacyclohexatriene) structure **5.52a**. Equation 5.52 shows the sequence used by Haley.[63] If the electron count permits, metallabenzenes readily rearrange to cyclopentadienyls, this no doubt occurred in Eq. 5.52. Structure **5.53** is an example of a metallabenzyne.[65] Nitration and bromination, reactions characteristic of truly aromatic rings, have been seen for an osmabenzene.[66]

$$\text{(5.52)}$$

On a strongly back-donating metal, the normal metallole structure of Eq. 5.51 converts to a bis-carbene. For example,[67] X-ray crystallography shows that **5.54** has the bis-carbene structure **5.54a** and not the usual metallole structure **5.54b**. Note that the metalacycle in **5.54a** is a 4e ligand but in **5.54b** is a 2e ligand, so this conversion of **b** to **a** can happen only if the metal can accept 2e [on the ionic model both ligands are counted as 4e ligands but the metal is counted as d^6 Os(II) in **5.54a** and d^4 Os(IV) in **5.54b**]; on both models. **5.54a** is an 18e complex and **5.54b** is a 16e complex.

5.54a   **5.54b**

5.7 STABILITY OF POLYENE AND POLYENYL COMPLEXES

The stability of the polyene complexes L_n toward dissociation is in general less than that of polyenyl complexes L_nX because the free polyene is usually a stable species, but the polyenyl must dissociate as a less stable anion, cation, or radical. The strongest π-back-bonding and most electron-rich metal fragments generally bind polyenes and polyenyls most tightly. For example, butadiene complexes of strongly π-basic metal fragments have more LX_2 character than those of less basic fragments and so less resemble the free ligand and dissociate less easily. Electron-withdrawing substituents also encourage back donation and can greatly increase complex stability, as we have seen for C_2F_4 in Section 5.1. Conversely, d^0 metals incapable of back donation, such as Ti(IV) and Nb(V), normally bind L_nX ligands like Cp (e.g., Cp_2NbCl_3 or $[Ti(\eta^3-C_3H_5)_4]$) but not L_n ligands like CO, C_2H_4, and C_6H_6. Interesting exceptions exist, however, such as $[Cp^*M(\eta^6-PhCH_3)Me_2][MeB(C_6F_5)_3]$ (M = Ti, Zr, Hf).[68]

Many complexes are known of ligands that are extremely reactive and unstable in the free state. We saw cyclobutadiene in Section 5.3, but alkylidenes [e.g., $Cp_2Ta(=CH_2)Me$, Section 11.1], and benzyne [e.g., $CpTa(\eta^2-C_6H_4)Me_2$] are also good examples. Cyclic polyene and polyenyl ligands tend to be kinetically more stable to dissociation than their open-chain analogs because the latter can more easily dissociate stepwise. The trihapto pentadienyl group is common,[69] but η^3-Cp is very rare. The open-chain ligand merely has to undergo a rotation about a C–C bond to become η^3, while a cyclopentadienyl has to fold out of the plane of the ligand to disengage two carbon atoms from the metal. Just as a cyclic ligand can be kinetically slow to depart, they also tend to be slower to bind to a metal. The synthesis of a Cp or a benzene complex is often found to go in lower yield or more slowly than that of related η^3-allyl or ethylene complexes.

As we go to the right in the periodic table, the ML_n fragments tend to have a higher electron count simply because the contribution from the metal rises. This means that those polyenes that have a large electron count may not be able to bind because the final electron count would exceed 18e. As noted above, uranium, with its 32e rule as a result of the presence of f orbitals, is able to accept 16e from the two cot ligands in $U(\eta^8-C_8H_8)_2$. No d-block element could do this. Titanium manages to take on one $\eta^8-C_8H_8$ ring in **5.55** (Eq. 5.53), chromium one $\eta^6-C_8H_8$ ring in **5.56**, but rhodium does not accept more than 4e from cot in the μ-η^4-C_8H_8 acetylacetonate complex, **5.57**.

$$Ti(Ot\text{-}Bu)_4 + cot + AlEt_3 \longrightarrow \text{LTi} \quad \text{TiL} \tag{5.53}$$

5.55

(L = η^8-cot)

5.56   **5.57**

Although the problem is less severe for η^5-Cp and (η^6-C_6H_6) complexes, these are notably less stable on the right-hand side of the periodic table, for example, for Pd and Pt. The η^4-butadiene and η^3-allyl groups do not seem to bind less strongly until we reach group 11.

Stability of polyene complexes also increases in lower oxidation states. In Eq. 5.54, Co(-I) back-donates so strongly that it gives the η^4-anthracene ligands significant enediyl (LX$_2$) character.[70]

$$ (5.54) $$

- Polyenyl ligands such as cyclopentadienyl tend to be held more strongly than polyene ligands such as benzene.
- A number of ligands have alternative binding modes (e.g., σ-allyl vs. π-allyl) with very different properties.

REFERENCES

1. G. Frenking and N. Frohlich *Chem. Rev.* **100**, 717, 2000.

2. G. L. Geoffroy et al., *Adv. Organometal. Chem.* **28**, 1, 1988.

3. M. S. Kang, A. Sen, L. Zakharov, and A. L. Rheingold, *ACS Symp. Ser.* **857**, 143, 2003.

4. C. P. Casey, T. Y. Lee et al., *J. Am. Chem. Soc.* **125**, 2641, 2003.

5. J. M. Takacs and X. T. Jiang, *Curr. Org. Chem.* **7**, 369, 2003.

6. R. M. Laine, R. E. Moriarty, and R. Bau, *J. Am. Chem. Soc.* **94**, 1402, 1972.

7. R. Weiss et al., *J. Am. Chem. Soc.* **100**, 1318, 1978.

8. J. L. Templeton, *Adv. Organomet. Chem.* **29**, 1, 1989; M. A. Esteruelas et al., *Organometallics* **21**, 305, 2002.

9. R. R. Schrock et al., *J. Am. Chem. Soc.* **101**, 263, 1979.

10. R. D. Adams et al., *J. Am. Chem. Soc.* **114**, 10977, 1992.

11. X. Li, C. D. Incarvito, and R. H. Crabtree, *J. Am. Chem. Soc.* **125**, 3698, 2003.

12. B. M. Trost, *Acc. Chem. Res.* **35,** 695, 2002.

13. J. W. Faller and N. Sarantopoulos, *Organometallics* **23**, 2008, 2004.

14. C. Kruger et al., *Organometallics* **4**, 285, 1985.

15. J. A. Osborn, *J. Am. Chem. Soc.* **97**, 3871, 1975.

16. E. E. Abel et al., *J. Chem. Soc., Dalton* 1706, 1973.

17. R. Hoffmann, J. W. Faller et al., *J. Am. Chem. Soc.* **101**, 592, 2570, 1979.

18. M. Rosenblum, *Acct. Chem. Res.* **7**, **125**, 1974.

19. T. Saegusa et al., *Synth. Commun.* **9**, 427, 1979.

20. E. J. Corey, M. F. Semmelhack et al., *J. Am. Chem. Soc.* **94**, 667, 1972.

21. B. M. Trost et al., *J. Org. Chem.* **44**, 3448, 1979.

22. C. W. Chang, Y. C. Lin, G. H. Lee, and Y. Wang, *Organometallics* **19**, 3211, 2000.

23. P. L. Arnold, F. G. N. Cloke, T. Geldbach, and P. B. Hitchcock, *Organometallics* **18**, 3228, 1999.

24. R. H. Crabtree, M. F. Mellea, and J. M. Quirk, *J. Am. Chem. Soc.* **106**, 2913, 1984.

25. S. Ogoshi, H. Kurosawa et al, *J. Am. Chem. Soc.* **120**, 1938, 1998; A Wojcicki, *Inorg. Chem. Commun.* **5**, 82, 2002.

26. T. T. Chen et al., *J. Am. Chem. Soc.* **115**, 1170, 1993.

27. M. L. H. Green, J. A. K. Howard et al., *J. Chem. Soc., Dalton* 2641, 1992.

28. M. L. H. Green et al., *J. Chem. Soc., Dalton* 1325, 1974.

29. L. A. P. Kane-Maguire et al., *J. Chem. Soc., Dalton* 873, 1979.

30. C. G. Pierpoint et al., *Inorg. Chem.* **17**, 78, 1976.

31. G. Erker, C. Kruger et al., *J. Am. Chem. Soc.* **102**, 6344, 1980.

32. J. W. Faller and A. M. Rosan, *J. Am. Chem. Soc.* **99**, 4858, 1977.

33. (a) R. Pettit et al., *Chem. Commun.* 1208, 1967; (b) P. M. Maitlis et al., *Can. J. Chem.* **42**, 183, 1964.

34. M. R. Churchill et al., *Inorg. Chem.* **8**, **401**, 1969.

35. G. G. Emerson and K. Ehrlich, *J. Am. Chem. Soc.* **94**, 2464, 1972.

36. G. Wilkinson, *J. Organomet. Chem.* **100**, 273, 1975.

37. C. P. Casey and W. D. Jones, *J. Am. Chem. Soc.* **102**, 6156, 1980.

38. P. Laszlo and R. Hoffmann, *Angew. Chem. Int. Ed.*, **39**, 123, 2000.

39. G. Wilkinson, *J. Organomet. Chem.* **100**, 273, 1975; E. O. Fischer et al., *Z. Naturforsch.* **106**, 665, 1965; E. O. Fischer and R. Jira, *J. Organometal. Chem.*, **637-9**, 7, 2001.

40. P. L. Pauson, *Inorg. Synth.* **19**, 154, 1979.

41. M. L. H. Green, J. A. McCleverty, J. Pratt, and G. Wilkinson, *J. Chem. Soc.* 4854, 1961.

42. R. H. Crabtree, M. F. Mellea, J. M. Mihelcic, and J. M. Quirk, *J. Am. Chem. Soc.* **104**, 107, 1982.

43. P. W. Jolly and G. Wilke, *The Organic Chemistry of Nickel*, Academic, New York, 1974.

44. R. D. Ernst, *Acc. Chem. Res.* **18**, 56, 1985; *Chem. Rev.* **88**, 1255, 1988; *Organometallics*, **22**, 1923, 2003; J. R. Bleeke and W.-J. Peng, *Organometallics* **5**, 635, 1986.

45. S. Trofimenko, *Chem. Rev.* **93**, 943, 1993.

46. G. Parkin et al., *New J. Chem.* **23**, 961, 1999; M. Akita et al, *Organometallics*, **21**, 3762, 2002.

47. H. Le Bozec, D. Touchard, and P. H. Dixneuf, *Adv. Organomet. Chem.* **29**, 163, 1989.

48. M. L. H. Green et al., *Chem. Commun.* 866, 1973.

49. A. N. Nesmeyanov et al., *Tetrahedron Lett.* 1625, 1963.

50. M. A. Bennett and A. K. Smith, *J. Chem. Soc., Dalton* 233, 1974.

51. H. W. Kroto, *Angew. Chem., Int. Ed.* **31**, 111, 1991.

52. A. L. Balch et al., *J. Am. Chem. Soc.* **113**, 8953, 1991.

53. K. J. Fisher et al., *Chem. Commun.* 941, 1993.

54. R. C. Haddon *J. Comput. Chem.* **19**, 139, 1998.

55. S. Stevenson et al., *Nature,* **401**, 55, 1999.

56. Y. Rubin, T. Jarrosson, G. W. Wang, M. D. Bartberger, K. N. Houk, G. Schick, M. Saunders, and R. J. Cross, *Angew. Chem. Int. Ed.* **40**, 1543, 2001; H. Shinohara et al., *Angew. Chem. Int. Ed.* **40**, 397, 2001.

57. J. C. Green, P. Mountford et al., *Chem. Commun.* 1235, 1998.

58. R. P. Hughes et al.: (a) *Adv. Organomet. Chem.* **31**, 183, 1990; (b) *Chem. Commun.* 306, 1986, and references cited therein; (c) *J. Am. Chem. Soc.* **114**, 5895, 1992.

59. W. D. Roper et al., *J. Organomet. Chem.* **159**, 73, 1978.

60. O. J. Scherer, *Angew. Chem., Int. Ed.* **26**, 59, 1987.

61. R. Hoffmann, *Angew. Chem., Int. Ed.* **21**, 711, 1982.

62. H. Masui, *Coord. Chem. Rev.* **219**, 957, 2001.

63. M. M. Haley et al., *J. Am. Chem. Soc.,* **121**, 2597, 1999.

64. J. R. Bleeke, *Acc. Chem. Res.* **24**, 271, 1991; *Chem. Rev.* **101**, 1205, 2001.

65. G. Jia et al., *Angew. Chem. Int. Ed.* **40** 1951, 2001.

66. W. R. Roper and L. J. Wright, *Angew. Chem., Int. Ed.* **39**, 750, 2000.

67. H. Taube et al., *J. Am. Chem. Soc.* **114**, 7609, 1992.

68. M. C. Baird et al., *Organometallics* **15**, 3600, 1996; C. Floriani et al., *Inorg. Chem.* **33**, 2018, 1994.

69. J. R. Bleeke, *Organometallics* **4**, **194**, 1985.

70. J. E. Ellis et al., *Angew. Chem. Int. Ed.* **41**, 1211, 2002.

PROBLEMS

1. Rank the following pairs of metal fragments in order of increasing tendency for an attached alkene to undergo nucleophilic attack: (a) $PdCl_2(H_2O)$, $PtCl_2(H_2O)$; (b) $Pd(PPh_3)_2$, $Pd(PPh_3)_2Cl^+$; (c) $CpMo(NO)P(OMe)_3^+$, $CpMo(NO)PMe_3^+$.

2. Although $L_nMCH_2CH_2ML'_n$ can be thought of as a bridging ethylene complex, examples of this type of structure are rarely made from ethylene itself. Propose a general route that does not involve ethylene and explain how you would know that the complex had the bridging structure, without using crystallography. What might go wrong with the synthesis?

3. Among the products formed from $PhC{\equiv}CPh$ and $Fe_2(CO)_9$, is 2,3,4,5,-tetraphenylcyclopentadienone. Propose a mechanism for the formation of

this product. Do you think the dienone would be likely to form metal complexes? Suggest a specific example and how you might try to make such a complex.

4. Suggest a synthesis of $Cp_2Mo(C_2H_4)Me^+$ from Cp_2MoCl_2. What orientation would you expect for the ethylene ligand? Given that there is no free rotation of the alkene, how would you show what orientation is adopted?

5. What structural distortions would you expect to occur in the complex $L_nM(\eta^4\text{-butadiene})$ if the ligands L were made more electron releasing?

6. 1,3-cod (= cyclooctadiene) can be converted into free 1,5-cod by treatment with $[(C_2H_4)IrCl]_2$, followed by $P(OMe)_3$. What do you think is the mechanism? Since 1,5-cod is thermodynamically unstable with respect to 1,3-cod (why is this so?), what provides the driving force for the rearrangement?

7. How many isomers would you expect for $[PtCl_3(propene)]^-$?

8. [TpCoCp] is high spin. Write its d-orbital occupation pattern following Fig. 5.5 and predict how many unpaired electrons it has (see *Chem. Comm.* 2052, 2001).

9. $IrH_2(H_2O)_2(PPh_3)_2^+$ reacts with indene, C_9H_8, to give $(C_9H_{10})Ir(PPh_3)_2^+$. On heating, this species rearranges with loss of H_2 to give $(C_9H_7)IrH(PPh_3)_2^+$. Only the first of the two species mentioned reacts with ligands such as CO to displace C_9H_7. What do you think are the structures of these complexes?

indene

10. $Cp_2Ti(X)_2(CO)$ is not known for $X = Cl$ but exists for $X = H$. Comment.

6

OXIDATIVE ADDITION AND REDUCTIVE ELIMINATION

We have seen how neutral ligands such as C_2H_4 or CO can enter the coordination sphere of a metal by substitution. We now look at a general method for simultaneously introducing pairs of anionic ligands, A and B, by the *oxidative addition* of an A—B molecule such as H_2 or CH_3-I (Eq. 6.1), a reaction of great importance in both synthesis and catalysis (Chapter 9). The reverse reaction, *reductive elimination*, leads to the extrusion of A—B from an M(A)(B) complex and is often the product-forming step in a catalytic reaction. In the oxidative direction, we break the A—B bond and form an M—A and an M—B bond. Since A and B are X-type ligands, the oxidation state, electron count, and coordination number all increase by two units during the reaction. It is the change in formal oxidation state (OS) that gives rise to the oxidative and reductive part of the reaction names.

$$L_nM \ + \ A\!-\!B \ \underset{\text{reductive elimination}}{\overset{\text{oxidative addition}}{\rightleftarrows}} \ L_nM\!\!\begin{array}{c} \nearrow A \\ \searrow B \end{array}$$

$$16e \qquad\qquad\qquad\qquad 18e$$

$$\Delta OS = +2$$
$$\Delta CN = +2$$

(6.1)

Oxidative additions proceed by a great variety of mechanisms, but the fact that the electron count increases by two units in Eq. 6.1 means that a vacant 2e site is always required on the metal. We can either start with a 16e complex or a

159

2e site must be opened up in an 18e complex by the loss of a ligand. The change in oxidation state means that a metal complex of a given oxidation state must also have a stable oxidation state two units higher to undergo oxidative addition (and vice versa for reductive elimination).

Equation 6.2 shows *binuclear oxidative addition*, in which each of two metals changes its oxidation state, electron count, and coordination number by one unit instead of two. This typically occurs in the case of a 17e complex or a binuclear 18e complex with an M−M bond where the metal has a stable oxidation state more positive by one unit. Table 6.1 systematizes the more common types of oxidative addition reactions by d^n configuration and position in the periodic table. Whatever the mechanism, there is a net transfer of a pair of electrons from the metal into the σ^* orbital of the A−B bond, and of the A−B σ electrons to the

TABLE 6.1   Common Types of Oxidative Addition Reaction[a]

Change in d^n Configuration	Change in Coordination Geometry	Examples	Group	Remarks
$d^{10} \rightarrow d^8$	Lin. $\xrightarrow{X_2}$ Sq. Pl.	Au(I) → (III)	11	
	Tet. $\xrightarrow{-2L, X_2}$ Sq. Pl.	Pt, Pd(0) → (II)	10	
$d^8 \rightarrow d^6$	Sq. Pl. $\xrightarrow{X_2}$ Oct.	M(II) → (IV)	10	M = Pd, Pt
		Rh, Ir(I) → (III)	9	Very common
		M(0) → (II)	8	Rare
	TBP. $\xrightarrow{-L, X_2}$ Oct.	M(I) → (III)	9	
		M(0) → (II)	8	
$d^7 \rightarrow d^6$	2Sq. Pyr. $\xrightarrow{X_2}$ 2Oct.	2Co(II) → (III)	8	Binuclear
	2Oct. $\xrightarrow{-L, X_2}$ 2Oct.	2Co(II) → (III)	8	Binuclear
$d^6 \rightarrow d^4$	Oct. $\xrightarrow{X_2}$ 7-c	Re(I) → (III)	7	
		M(0) → (II)	6	
		V(−I) → (I)	5	
$d^4 \rightarrow d^3$	2Sq. Pyr. $\xrightarrow{X_2}$ 2Oct.	2Cr(II) → (III)	6	Binuclear
	2Oct. $\xrightarrow{-L, X_2}$ 2Oct.	2Cr(II) → (III)	6	Binuclear
$d^4 \rightarrow d^2$	Oct. $\xrightarrow{X_2}$ 8-c	Mo, W(II) → (IV)	6	
$d^2 \rightarrow d^0$	Various	M(III) → (V)	5	
		M(II) → (IV)	4	

[a]Abbreviations: Lin. = linear, Tet. = tetrahedral, Oct. = octahedral, Sq. Pl. = square planar, TBP = trigonal bipyramidal, Sq. Pyr. = square pyramidal; 7-c, 8-c = 7- and 8-coordinate.

metal. This cleaves the A−B bond and makes an M−A and an M−B bond. The reaction is promoted by starting with a metal in a reduced state; only rarely do metals in an oxidation state higher than +2 retain sufficient reducing character to undergo oxidative addition, except with powerful oxidants, like Cl_2. Conversely, a highly oxidized metal is more likely to undergo reductive elimination.

$$2L_nM \quad (\text{or } L_nM\text{-}ML_n) \xrightarrow{\text{A}-\text{B}} L_nM\text{-}A + L_nM\text{-}B$$

$$\begin{array}{cccc} 17e & 18e & 18e & 18e \\ & & \Delta O\,S\ =+1 & \Delta O\,S\ =+1 \\ & & \Delta C\,N\ =+1 & \Delta C\,N\ =+1 \end{array}$$

$$(6.2)$$

As we have seen, oxidative addition is the inverse of reductive elimination and vice versa. In principle, these reactions are reversible, but in practice they tend to go in the oxidative or reductive direction only. The position of equilibrium in any particular case is governed by the overall thermodynamics; this in turn depends on the relative stabilities of the two oxidation states and the balance of the A−B versus the M−A and M−B bond strengths. Alkyl hydride complexes commonly eliminate alkane, but only rarely do alkanes oxidatively add to a metal. Conversely, alkyl halides commonly add to metal complexes, but the adducts rarely reductively eliminate the alkyl halide. Third-row elements, which tend to have stronger metal−ligand bonds, tend to give more stable adducts. Occasionally, an equilibrium is established in which both the forward and back reactions are observed.

$$[Ir(cod)_2]^+ \underset{\text{warm}}{\overset{H_2,\ \text{fast},\ -80°}{\rightleftharpoons}} cis\text{-}[IrH_2(cod)_2]^+ \qquad (6.3)^1$$

Reaction in the oxidizing direction is usually favored by strongly donor ligands because these stabilize the oxidized state. While the formal oxidation state change is always +2 for Eq. 6.1, the real charge on the metal changes much less than that because A and B do not end up with pure −1 charges in $L_nM(A)(B)$. The change in real charge depends mostly on the electronegativity of A and B in Eq. 6.1, so that the following reagents are more oxidizing in the order: $H_2 < HCl < Cl_2$. We can estimate the oxidizing power of different reagents experimentally by measuring $\Delta v(CO)$ on going from $IrCl(CO)L_2$ to $Ir(A)(B)Cl(CO)L_2$ (Table 6.2) because a more oxidizing reagent will reduce M−CO back bonding and make $\Delta v(CO)$ more negative.

These reactions are not limited to transition metals; perhaps the most familiar oxidative addition is the formation of Grignard reagents (Eq. 6.4), but it can occur whenever an element has two accessible oxidation states two units apart. Equation 6.5 illustrates oxidative addition to P(III).

$$Me-Br + Mg \longrightarrow Me-Mg-Br \qquad (6.4)$$

$$Cl-Cl + PCl_3 \longrightarrow PCl_5 \qquad (6.5)$$

TABLE 6.2   Carbonyl Stretching Frequencies of the Oxidative Addition Products from Vaska's Complex

Reagent	$\nu(CO)$ (cm^{-1})	$\Delta\nu(CO)$ (cm^{-1})
None	1967	0
O_2	2015	48
$D_2{}^a$	2034	67
HCl	2046	79
MeI	2047	80
C_2F_4	2052	85
I_2	2067	100
Cl_2	2075	108

[a] The D isotope is used because the Ir—H stretching vibrations have a similar frequency to $\nu(CO)$ and so couple with CO stretching and cause $\nu(CO)$ to shift for reasons that have nothing to do with the electronic character of the metal (see Chapter 10).

The unusual feature of oxidative addition reactions of transition metals is the unusually wide range of reagents A—B that can be involved, including such normally relatively unreactive molecules as silanes, H_2, and even alkanes. Oxidative additions are very diverse mechanistically, and we therefore consider each type separately.

6.1   CONCERTED ADDITIONS

Concerted, or three-center, oxidative addition is really an associative reaction in which the incoming ligand, for example, H_2, first binds as a σ complex and then undergoes H—H bond breaking as a result of strong back donation from the metal into the H—H σ^* orbital. Nonpolar reagents, such as H_2, or compounds containing C—H and Si—H bonds all tend to react via a transition state—or more probably an intermediate—of this type (**6.1**; A = H; B = H, C, or Si). Step *a* in Eq. 6.6, the associative step, involves formation of a σ complex; sometimes this is stable and the reaction stops here. Step *b* is the oxidative part of the reaction in which metal electrons are formally transferred to the σ^* orbital of A—B. The best-studied case is the addition of H_2 to 16e square planar d^8 species, such as $IrCl(CO)(PPh_3)_2$, *Vaska's complex*,[2] to give 18e d^6 octahedral dihydrides (Eq. 6.6). Normally two ligands that are trans in the Ir(I) complex fold back to give the cis dihydride isomer, but subsequent rearrangement can occur. Conversely, in a reductive elimination such as the loss of H_2 from the dihydride, the two ligands to be eliminated normally have to be cis to one another.

$$L_nM \ + \ A\!-\!B \ \xrightarrow{\ a\ } \ L_nM\!-\!\!\begin{matrix}A\\|\\B\end{matrix} \ \xrightarrow{\ b\ } \ L_nM\!\!\begin{matrix}{\nearrow}A\\{\searrow}B\end{matrix} \quad\quad (6.6)$$

$$\begin{matrix} \text{16e} & & \textbf{6.1} & & \textbf{6.2} \\ \text{M(0)} & & \text{18e, M(0)} & & \text{18e, M(II)} \end{matrix}$$

square planar	trig. bipyramid	octahedral
16e, Ir(I)	18e, Ir(I)	18e, Ir(III)

In oxidative addition of H_2 to Vaska's complex, the *trans*-Cl(CO) set of ligands usually folds back to become cis in the product (Eq. 6.7); the alternate product with the phosphines folded back is rarely seen.[3,4] As a powerful π acceptor, the CO prefers to be in the equatorial plane of the resulting TBP transition state, which may resemble the intermediate H_2 complex in Eq. 6.7. This tendency for a pair of trans π-acceptor ligands to fold back can be so strong that the square planar d^8 ML_4 starting material already strongly distorts even in the absence of an incoming ligand. Such is the case for *trans*-[Ru(CO)₂(P{t-Bu}₂Me)₂], where the two trans COs fold back strongly (OC—Ru—CO = 133°). As we saw in Section 4.4, high-trans-effect ligands such as CO prefer to be in the equator of a TBP, or in this case of a TBP-like geometry. One might suspect that the steric bulk of the phosphines could play a role, but theoretical work shows that *trans*-[Ru(CO)₂(PH₃)₂] has essentially the same geometry.[5]

6.3

A 2e site must be present on the metal for the reaction to occur, so that in 18e complexes, such as $[Ir(CO)_2L_3]^+$, ligand dissociation must take place first.

The reactions are usually second order and show negative entropies of activation (ca. −20 eu) consistent with an ordered transition state such as **6.1**. They are little affected by the polarity of the solvent but may be accelerated to some extent by electron-releasing ligands. The C—H and Si—H bonds of various hydrocarbons and silanes can also oxidatively add to metals. Among different types of C—H bonds, those of arenes are particularly prone to do this because of the high thermodynamic stability of the aryl hydride adduct.

Agostic complexes, σ complexes of C—H bonds, can be thought of as lying along the pathway for oxidative addition but arrested at different points. A study[6] of the structures of a series of these complexes allowed the kinetic pathway for Eq. 6.8 to be mapped out. This is a general strategy[7] for studying reaction trajectories. The C—H bond seems to approach with the H atom pointing toward the metal and then the C–H bond pivots around the hydrogen to bring the carbon closer to the metal in a side-on arrangement, followed by C–H bond cleavage.[6] The addition goes with retention of stereochemistry at carbon, as expected on this mechanism.

$$C-H + M \rightarrow C-M-H \qquad (6.8)$$

Carbon–carbon bonds do not normally oxidatively add, but a classic early case, the reaction of cyclopropane with Pt(II) to give a Pt(IV) metalacyclobutane shown below, illustrates how the reaction can be driven by ring strain. Biphenylene[8] readily reacts in the same way; in the example shown, the high trans effect of the resulting biphenyl ligand makes this a 16e product with distorted TBP geometry in spite of the normal preference for 18e octahedral geometry in Ir(III).

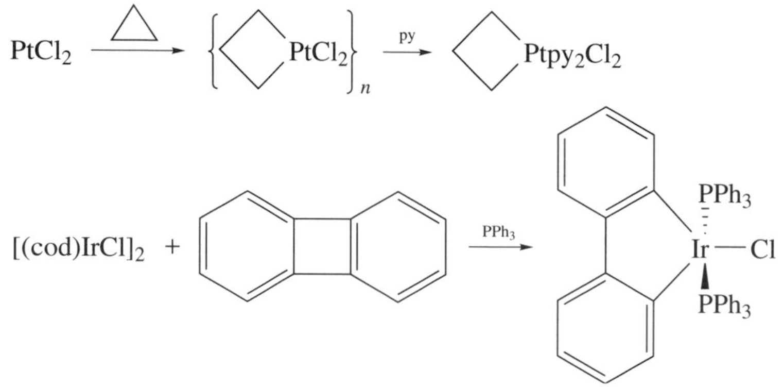

$$(6.9)$$

When Vaska's complex adds O_2, the metal reduces the O_2 to O_2^{2-}, the peroxide ion, which coordinates to the Ir(III) to give **6.4**. Why not envisage the reaction as a ligand addition by regarding O_2 as a 2e donor such as ethylene? This is the same problem we looked at in Section 2.7, and is a result of the different formal oxidation states assigned to the L and X_2 extreme pictures of binding. In fact, ethylene is much closer to the L extreme, as shown in **6.5**, while O_2 is very close to the X_2 extreme (**6.6**). This means that the conventional descriptions of ethylene binding as a ligand addition or simple substitution and of O_2 binding as an oxidative addition are the most appropriate. For ligands, such as $CF_2=CH_2$, which bind in a fashion that is approximately equidistant between the two extremes, there is clearly a gray area in which the choice between the two descriptions is arbitrary. This emphasizes that categories such as "oxidative addition" are mental constructs and have their limitations.

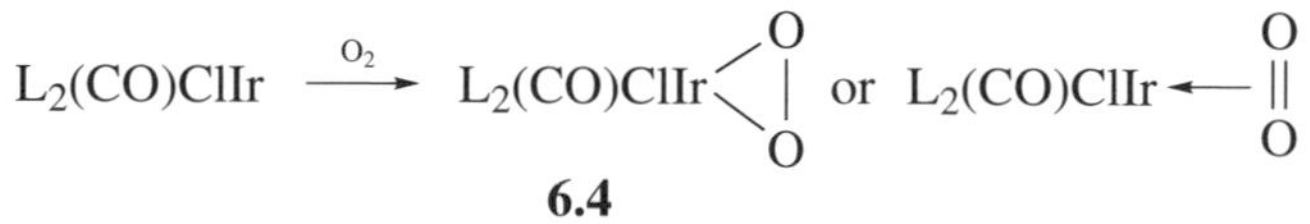

$$(6.10)$$

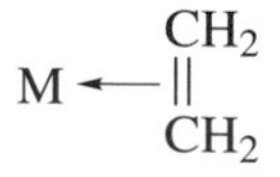

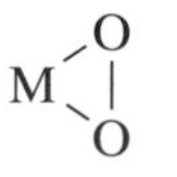

Aryl halides can also react via a concerted mechanism. For example, [Pd(P{Ar}$_3$)$_2$] reacts with Ar'Br in this way (Ar = o-tolyl; Ar' = t-BuC$_6$H$_4$). Prior loss of PAr$_3$ is required to give the reactive 1-coordinate intermediate, Pd(PAr$_3$), that reacts with the aryl halide to give [(PAr$_3$)(Ar')Pd(μ-Br)]$_2$ as the final product.[9]

6.2 S$_N$2 REACTIONS

In all oxidative additions, a pair of electrons from the metal is used to break the A−B bond in the reagent. In the S$_N$2 pathway (Eq. 6.11), adopted for polarized AB substrates such as alkyl halides, the metal electron pair of L$_n$M directly attacks the A−B σ^* orbital by an in-line attack at the least electronegative atom (where σ^* is largest; compare Fig. 4.3) formally to give L$_n$M^{2+}, A$^-$, and B$^-$ fragments on the ionic model.

The S$_N$2 mechanism is often found in the addition of methyl, allyl, acyl, and benzyl halides to species such as Vaska's complex. Like the concerted type, they are second-order reactions, but they are accelerated in polar solvents and show negative entropies of activation ($\Delta S^{\ddagger} = -40$ to -50 eu).[10] This is consistent with an ordered, polar transition state, as in organic S$_N$2 reactions. Inversion at carbon has been found in suitably substituted halides. Equation 6.11 shows how the stereochemistry at the carbon of the oxidative addition product was determined by carbonylation to give the metal acyl followed by methanolysis to give the ester. Both of these reactions are known to leave the configuration at carbon unchanged, and the configuration of the ester can be determined unambiguously from the measured optical rotation.[11] R and X may end up cis or trans to one another in the final product, as expected for the recombination of the ion pair formed in the first step. Equation 6.12 shows a case in which the product is trans. This happens because the high-trans-effect Me group prefers to remain trans to the vacancy in the 16e square pyramidal intermediate shown; this is reminiscent of the situation for dissociative substitution (Section 4.3).

(6.11)

(6.12)

Of the two steps in Eq. 6.12, the first involves oxidation by two units but no change in the electron count (Me^+ is a 0e reagent), and the second involves an increase by 2e in the electron count (I^- is a 2e reagent) but no change in the oxidation state. Only the two steps together constitute the full oxidative addition. When an 18e complex is involved, the first step can therefore proceed without the necessity of losing a ligand first. Only the second step requires a vacant 2e site. In some cases the product of the first step is stable and does not lose a ligand to admit the halide anion. This is sometimes loosely called an oxidative addition, but it is better considered as an electrophilic attack at the metal, for example:

$$CpIr(CO)L + MeI \rightarrow [CpIr(CO)LMe]I \qquad (6.13)$$

The more nucleophilic the metal, the greater its reactivity in S_N2 additions, as illustrated by the reactivity order for some Ni(0) complexes: $Ni(PR_3)_4 >$ $Ni(PAr_3)_4 > Ni(PR_3)_2(alkene) > Ni(PAr_3)_2(alkene) > Ni(cod)_2$ (R = alkyl; Ar = aryl).[12] Steric hindrance at carbon slows the reaction, so we find the reactivity order: MeI > EtI > i-PrI. A better leaving group, X at carbon, accelerates the reaction, which gives rise to the reactivity order $ROSO_2(C_6H_4Me) > RI >$ RBr > RCl for this mechanism.

Halide ions can increase the nucleophilicity of the metal and hence exert a powerful catalytic effect on S_N2 oxidative additions. Such an effect is seen, for example, for iodide ions in the oxidative addition of MeI to $RhI(CO)(PPh_3)_2$ to give $Rh(Me)I_2(CO)(PPh_3)_2$. Iodide ion initially replaces PPh_3 at the metal center to give an intermediate $[RhI_2(CO)(PPh_3)]^-$ that reacts very rapidly with MeI.[13]

R_3Sn-X, another reagent with a strong tendency to give S_N2 additions (X = Cl, Br, I), gives the following rapid, reversible addition/elimination equilibrium:[14]

$(N\widehat{}N = 4,4'\text{-di-}t\text{-Bu-}2,2'\text{-dipyridyl})$

6.3 RADICAL MECHANISMS

Radical mechanisms[15] in oxidative additions were recognized later than the S_N2 and the concerted processes. A troublesome feature of these reactions is that minor changes in the structure of the substrate, the complex, or in impurities present in the reagents of solvents can sometimes be enough to change the rate, and even the predominant mechanism of a given reaction. They can also be photoinitiated.[15c] Sharp disagreements have turned on questions of repeatability and on what types of experimental evidence should be considered as valid mechanistic criteria. For example, the use of radical traps, such as RNO•, has been criticized on the grounds that these may initiate a radical pathway for a reaction that otherwise would have followed a nonradical mechanism in the absence of trap.

Two subtypes of radical process are now distinguished: the nonchain and the chain. The nonchain variant is believed to operate in the additions of certain alkyl halides, RX, to $Pt(PPh_3)_3$ ($R = Me$, Et; $X = I$; $R = PhCH_2$; $X = Br$).[16]

$$PtL_3 \xrightarrow{\text{fast}} PtL_2 \qquad (6.14)$$

$$PtL_2 + RX \xrightarrow{\text{slow}} \cdot PtL_2 + \cdot RX^- \rightarrow \cdot PtXL_2 + R\cdot \qquad (6.15)$$

$$\cdot PtXL_2 + R\cdot \xrightarrow{\text{fast}} RPtXL_2 \qquad (6.16)$$

The key feature is one electron transfer from M to the RX σ^* to form M^+ and RX^-. After X^- transfer to M^+, the R$\cdot$ radical is liberated. This produces the pair of radicals shown in Eq. 6.15, which rapidly recombine to give the product before either can escape from the solvent cage. Like the S_N2 process, the radical mechanism is faster the more basic the metal, and the more readily electron transfer takes place, which gives the reactivity order $RI > RBr > RCl$. Unlike the S_N2 process, the reaction is very slow for alkyl tosylates [e.g., $ROSO_2(C_6H_4Me)$], and it goes faster as the alkyl radical, R, becomes more stable and so easier to form, giving rise to the order of R group reactivity: $3° > 2° > 1° > Me$. In the reaction of NiL_3 with aryl halides, the Ni(I) complex, $NiXL_3$, formed in the first step is sufficiently stable to survive as an observable product of the reaction; the Ar$\cdot$ radical abstracts a hydrogen atom from the solvent to give ArH. There are also cases where the organic radical R$\cdot$ is sufficiently stable to survive and become a product of the reaction, for example, in the reaction of certain quinones with NiL_3.

The second general kind of reaction is the *radical chain*.[17] This has been identified in the case of the reaction of EtBr and $PhCH_2Br$ with the PMe_3 analog of Vaska's complex. Equations 6.14 and 6.15 can lead to a chain process if the radicals formed can escape from the solvent cage without recombination. Otherwise, a radical *initiator*, Q$\cdot$, (e.g., a trace of air) may be required to set the process going (Eq. 6.17 with Q$\cdot$ replacing R$\cdot$). This can lead to an *induction period* (a period of dead time before the reaction starts). In either case, a metal-centered radical abstracts X$\cdot$ from the halide (Eq. 6.18), to leave the *chain carrier,* R$\cdot$. The chain consists of Eqs. 6.17 and 6.18. Chain termination steps such as Eq. 6.19 limit the number of cycles possible per R$\cdot$. The alkyl group always loses any stereochemistry at the α carbon (because $RR'R''C\cdot$ is planar at the central carbon). Unlike the nonchain case, the reactions slow down or stop in the presence of radical inhibitors, such as the hindered phenol, 2,6-di-t-butylphenol; these quench the radical R$\cdot$ to give $R-H$ and the stable, and therefore unreactive, aryloxy radical, ArO$\cdot$.

$$R\cdot + Ir^I Cl(CO)L_2 \longrightarrow RIr^{II}\cdot Cl(CO)L_2 \qquad (6.17)$$

$$RIr^{II}\cdot Cl(CO)L_2 + RX \longrightarrow RXIr^{III}Cl(CO)L_2 + R\cdot \qquad (6.18)$$

$$2R\cdot \longrightarrow R_2 \qquad (6.19)$$

Certain substrates are particularly useful for determining what happens to the stereochemistry at the α carbon during oxidative addition or in other reactions. For example, **6.7** can be obtained with a defined relative stereochemistry at the α and β carbons. This has the advantage that we do not need to resolve anything, we have both enantiomers of **6.8** present, but we assume that the reaction will do nothing to the stereochemistry at the β carbon, so we can look at the configuration at the α position relative to that at the β. This is easily done by ^{1}H NMR spectroscopy. The conformation adopted by these substituted ethanes puts the two bulky groups t-Bu and ML_n or t-Bu and X anti to one another. This in turn puts the α and β protons gauche or anti to one another according to whether the stereochemistry at the α position has been retained or inverted. By the Karplus relationship, which tells us the HCCH$'$ dihedral angle from the observed ^3J(H, H$'$); the very different coupling constant in the two cases serves to identify the stereochemistry of the product. For example, **6.9** would be the product of an S_N2 reaction.

$$\begin{array}{cc} \beta & \alpha \end{array}$$
$$t\text{-BuCHDCHDI}$$
6.7

6.8   **6.9**

Other useful tests for radicals rely on the fact that some free radicals are known to rearrange very rapidly (radical clock, Eq. 6.20). For example, if hexenyl bromide gives a cyclopentylmethyl metal complex, then a radical intermediate is strongly indicated. Cyclopropylmethyl radicals ($C_3H_5CH_2\bullet$), on the other hand, rearrange by ring opening to give $CH_2{=}CHCH_2CH_2\bullet$. Other common reactions of radicals are Cl$\bullet$ abstraction from a chlorinated solvent to give RCl, and dimerization to give R$-$R. An NMR method, called *chemically induced dynamic nuclear polarization* (CIDNP),[18] can be useful in certain cases. The method relies on the fact that the product of a radical recombination can have very unusual distributions of α and β spins. This implies that the ^{1}H NMR *may* show large positive (if α spins are in excess) or negative peaks (if β), if the conditions are right. It is not easy to tell how much of the reaction is going by a radical route because the intensity of the effect is variable and difficult to predict.

(6.20)

Binuclear oxidative additions, because they involve 1e rather than 2e changes at the metals, often go via radicals. One of the best known examples is shown in Eq. 6.21.

$$2[Co^{II}(CN)_5]^{3-} + RX \longrightarrow [RCo^{III}(CN)_5]^{3-} + [XCo^{III}(CN)_5]^{3-} \quad (6.21)$$

The rate-determining step is net abstraction of a halogen atom from RX by the odd-electron d^7 Co(II) (Eq. 6.22); the resulting R• combines with a second Co(II) center:[19]

$$\bullet[Co(CN)_5]^{3-} + RX \xrightarrow{\text{slow}} [XCo(CN)_5]^{3-} + R\bullet \quad (6.22)$$

$$R\bullet + \bullet[Co(CN)_5]^{3-} \xrightarrow{\text{fast}} [RCo(CN)_5]^{3-} \quad (6.23)$$

In reactions involving radicals, it is important to use solvents that do not react fast with R•; alkane, C_6H_6, AcOH, CH_3CN, and water are usually satisfactory.

6.4 IONIC MECHANISMS

Hydrogen halides are often largely dissociated in solution, and the anion and proton tend to add to metal complexes in separate steps. Two variants have been recognized. In the more common one, the complex is basic enough to protonate, after which the anion binds to give the final product. Rarer is the opposite case in which the halide ion attacks first, followed by protonation of the intermediate. The first route is favored by basic ligands and a low-oxidation-state metal, the second by electron-acceptor ligands and by a net positive charge on the complex. Polar solvents encourage both types; examples are given in Eqs. 6.24 and 6.25:[20]

$$Pt(PPh_3)_4 + H^+ + Cl^- \xrightarrow{-PPh_3} [HPt(PPh_3)_3]^+ + Cl^- \xrightarrow{-PPh_3} [HPtCl(PPh_3)_2]$$
18e d^{10} tetrahedral   16e d^8 square planar   16e d^8 square planar

$$(6.24)$$

$$[Ir(cod)L_2]^+ + Cl^- + H^+ \longrightarrow [IrCl(cod)L_2] + H^+ \longrightarrow [IrHCl(cod)L_2]^+$$
16e d^8 square planar   18e d^8 TBP   18e d^6 octahedral

$$(6.25)$$

The rate of the first type generally follows Eq. 6.26, suggesting that protonation is the slow step. This can be carried out independently by using an acid with a noncoordinating anion: HBF_4 and HPF_6 are the most often used. The anion has insufficient nucleophilicity to carry out the second step, and so the intermediate can be isolated. This is an example of a general strategy in which (e.g., $[PtH(PPh_3)_3]$) a "noncoordinating" anion is used to isolate reactive cations as stable salts.

$$\text{Rate} = k[\text{complex}][H^+] \quad (6.26)$$

The rate of the second type (Eq. 6.25) usually follows the rate equation shown in Eq. 6.27, suggesting that Cl^- addition is the slow step. This step can be carried out independently with LiCl alone, but no reaction is observed with HBF_4 alone because the cationic iridium complex is not basic enough to protonate and BF_4^- is a noncoordinating anion.

$$\text{Rate} = k[\text{complex}][Cl^-] \tag{6.27}$$

Other acids (or Lewis acids), which are ionized to some extent in solution, such as RCO_2H and $HgCl_2$ (Eqs. 6.28 and 6.29), may well react by the same mechanism, but this has not yet been studied in detail.

$$IrCl(CO)L_2 \xrightarrow{RCO_2H} IrH(\kappa^1\text{-OCOR})Cl(CO)L_2 \tag{6.28}$$

$$IrCl(CO)L_2 \xrightarrow{HgCl_2} Ir(HgCl)Cl_2(CO)L_2 \tag{6.29}$$

As we saw in Chapter 3, alkyls $L_nM(R)$ can often be cleaved with acid to give the alkane. In some cases simple protonation of the metal to give $L_nM(R)H^+$ or of the M–R bond to give the σ complex $L_nM(H-R)^+$ is the likely mechanism, but in others (e.g., Eq. 6.30) there is a dependence of the rate, and in some cases even of the products,[21] on the counterion; in such cases, an oxidative addition–reductive elimination mechanism seems more likely:

$$PtR_2(PR_3)_2 \xrightarrow{HCl} PtHClR_2(PR_3)_2 \xrightarrow{-RH} PtRCl(PR_3)_2 \tag{6.30}^{21}$$

> • Oxidative addition increases both the oxidation state and the coordination number by two.
> • Oxidative addition can go by many different mechanisms.

6.5   REDUCTIVE ELIMINATION

Reductive elimination, the reverse of oxidative addition, is most often seen in higher oxidation states because the formal oxidation state of the metal is reduced by two units in the reaction. The reaction is especially efficient for intermediate oxidation states, such as the d^8 metals Ni(II), Pd(II), and Au(III) and the d^6 metals Pt(IV), Pd(IV), Ir(III), and Rh(III). Reductive elimination can be stimulated by oxidation[22] or photolysis: The case of photoextrusion of H_2 from dihydrides is the best known (Section 12.4).

Certain groups are more easily eliminated than others, for example, Eqs. 6.31–6.35 often proceed to the right for thermodynamic reasons. Reactions that involve H, such as Eqs. 6.31 and 6.33, are particularly fast, probably because the transition state energy is lowered by the formation of a relatively stable σ-bond complex $L_nM(H-X)$ along the pathway; such complexes are known to be stable only where at least one H is eliminated.

$$L_nMRH \longrightarrow L_nM + R-H \tag{6.31}$$

$$L_nMR_2 \longrightarrow L_nM + R-R \tag{6.32}$$

$$L_nMH(COR) \longrightarrow L_nM + RCHO \tag{6.33}$$

$$L_nMR(COR) \longrightarrow L_nM + R_2CO \tag{6.34}$$

$$L_nMR(SiR_3) \longrightarrow L_nM + R-SiR_3 \tag{6.35}$$

In catalysis reactions (Chapter 9), a reductive elimination is often the last step in a catalytic cycle, and the resulting L_nM fragment must be able to survive long enough to react with the substrates for the organic reaction and so reenter the catalytic cycle. The eliminations of Eqs. 6.31–6.35 are analogous to the concerted oxidative additions in that they are believed to go by a nonpolar, nonradical three-center transition state, such as **6.10**. Retention of stereochemistry at carbon is a characteristic feature of this group of reactions.

6.10

Since there are several mechanisms for oxidative addition (Section 6.4) the principle of microscopic reversibility (which holds that a reversible reaction proceeds by the same mechanism in both forward and reverse directions) suggests that reductive eliminations should show the same variety. We only discuss the concerted pathway here.

Octahedral Complexes

Octahedral d^6 complexes of Pt(IV), Pd(IV), Ir(III), and Rh(III) tend to undergo reductive elimination readily but often with initial loss of a ligand to generate a 5-coordinate intermediate, a much more reactive species than the starting 6-coordinate complex. When ligand dissociation does not occur, reductive elimination can be slow, even when it would otherwise be expected to be very favorable. For example, complexes with an alkyl group cis to a hydride are rare because reductive elimination of an alkane (Eq. 6.31) is usually very thermodynamically favorable. A stable example of this type is *mer*-[IrH(Me)Cl(PMe$_3$)$_3$], with H and Me cis, which survives heating to 100°C. The Rh analog, **6.11**, with its weaker M−PMe$_3$ bonds, gives reductive elimination even at 30°C.[23] It is the PMe$_3$ trans to the high-trans-effect hydride ligand that is lost because this site is labeled by treating **6.11** with P(CD$_3$)$_3$ at 30°C. The 5-coordinate intermediate may be more reactive because it can more readily distort to reach the transition state for reductive elimination. It might be expected to be able to gain access to the Y-type distorted trigonal bipyramidal structure, **6.12**, (Section 4.3) shown in Fig. 6.1.[24] A Y structure is favored where one π-donor ligand, Cl in this case, is located at the basal position of the Y, as shown in **6.12**. This structure brings the two groups to be eliminated, R and H, very close together. The typical small R−M−H angle for these groups, 70°, may facilitate achievement of the proposed transition state (**6.13**) for reductive elimination. After reductive elimination, a T-shaped 3-coordinate species is formed.

If reductive elimination of 6-coordinate d^6 species goes by **6.13**, then the reverse reaction, oxidative addition to 4-coordinate d^8 species, is also expected to go by **6.13** by reversal of the pathway of Fig. 6.1. Indeed, Halpern[25] showed that RhCl(PPh$_3$)$_2$, formed by loss of a PPh$_3$ group from RhCl(PPh$_3$)$_3$, gives oxidative addition with hydrogen at a rate at least 10^4 times faster than the 4-coordinate complex.

The reversibility argument also applies to reductive elimination of alkyl halides for which an S$_N$2 pathway (Fig. 6.2) applies for the oxidative addition direction. Iodide attacks the coordinated methyl trans to the open site and nucleophilically displaces the Pt complex, which is a good leaving group. The reactive 5-coordinate intermediate, which can even be isolated in some cases, can also undergo concerted reductive elimination of ethane if the I$^-$ concentration is low.[26]

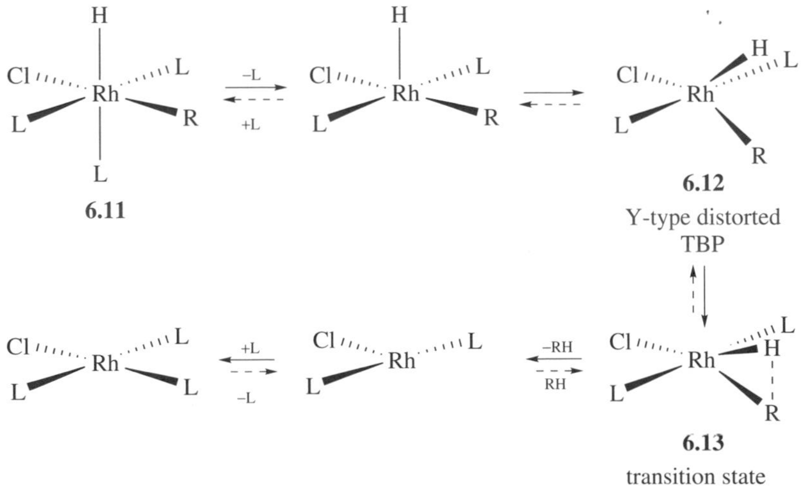

FIGURE 6.1 Example of a common general mechanism for reductive elimination in Milstein's octahedral d^6 species (L = PMe$_3$; R = CH$_2$COMe). The reverse mechanism (dotted arrows) often holds for oxidative addition to square planar d^8 species (e.g., R = H).

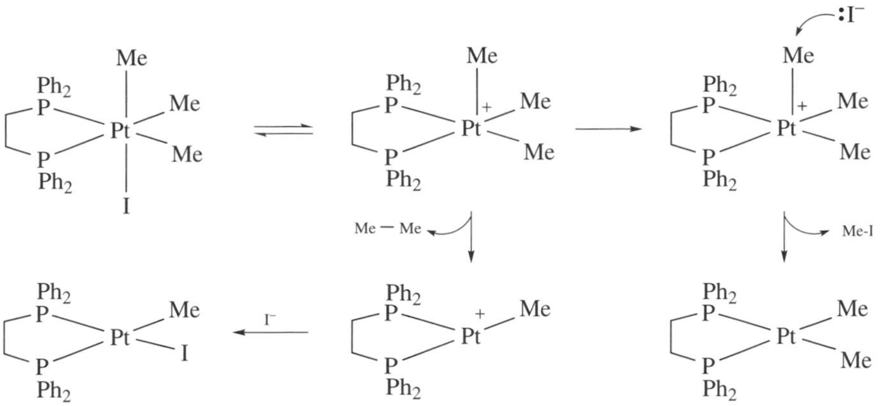

FIGURE 6.2 The mechanism for reductive elimination to form C−C and C−Hal bonds in octahedral d^6 species in Goldberg's complex. The reverse mechanism holds for oxidative addition to square planar d^8 species.

Other Complexes

Square planar d^8 complexes show a variety of reductive elimination mechanisms: dissociative, nondissociative, and associative.[27–31] Sometimes a ligand dissociates from MRXL$_2$, and the elimination occurs from the 3-coordinate MRXL intermediate, resulting in initial formation of a 1-coordinate ML metal fragment; this is the case[28,29] for PdR$_2$L$_2$ and several Au(III) species. In some cases, the 4-coordinate *trans*-MRXL$_2$ species can reductively eliminate but usually only after initial isomerization from trans to cis to put the two groups to be eliminated next to one another. Occasionally, a fifth ligand associates, and elimination occurs from a 5-coordinate TBP intermediate; this has been found for Ni(II).[31] Driver and Hartwig[32] have analyzed the kinetics for the case of *trans*-[PdAr(N{tolyl}$_2$)(PPh$_3$)$_2$] (**6.14**) where reductive elimination of Ar−N{tolyl}$_2$ takes place via competing dissociative and nondissociative pathways. Pt(II) is often slow to eliminate, perhaps because ligand dissociation is harder, but oxidative addition of RX to give a Pt(IV) intermediate can promote reductive elimination.[31] Some progress has been made in understanding these mechanistic differences in MO terms.[27]

Mechanisms are probed via the kinetics; for example, in the dissociative reductive elimination of Me−Me from *trans*-[PdMe$_2$(PPh$_3$)$_2$] (**6.15**), added PPh$_3$ retards the reaction in an inverse first-order way (the rate is proportional to 1/[PPh$_3$]), suggesting that loss of phosphine takes place to give the 3-coordinate intermediate PdR$_2$L. The retardation might alternatively have been due to formation of PdR$_2$L$_3$, which would have to be less reactive than PdR$_2$L$_2$ itself; it can be shown by NMR that this does not happen, however.

The chelating diphosphine complex **6.16** loses phosphine much less easily than do the analogs containing monodentate phosphines and undergoes elimination 100 times more slowly.[28] The "transphos" complex **6.17** does not eliminate ethane under conditions where the corresponding cis derivative **6.16** does so very readily.[28] The groups to be eliminated therefore need to be cis; transphos locks this system in a trans geometry.

6.14   **6.15**   **6.16**

In an important general mechanistic experiment that is useful for this problem, the *crossover* experiment, a mixture of *cis*-Pd(CH$_3$)$_2$L$_2$ and *cis*-Pd(CD$_3$)$_2$L$_2$, is thermolyzed. We find that only C$_2$H$_6$ and C$_2$D$_6$ are formed, showing that the reaction is *intramolecular*; that is, R groups can couple only within the same molecule of starting complex. This experiment rules out coupling between R groups originating in different molecules of the complex (the *intermolecular* route). The crossover product, CH$_3$CD$_3$, would have been formed if alkyl groups eliminated in a binuclear way, or free methyl radicals had been involved because they are sufficiently long-lived to migrate through the solution from one molecule of palladium complex to the next. We always need to do proper control experiments; for example, even if CH$_3$CD$_3$ is formed, we need to check whether scrambling happens in the reaction or whether the CH$_3$ and CD$_3$ groups exchange between the starting materials before reductive elimination takes place or in the analytical method used to detect crossover. This can be done by isolating the starting materials after partial conversion to products to make sure that no Pd(CH$_3$)(CD$_3$)L$_2$ is present.

As we saw in Table 6.1, Pd(IV) is not a very stable oxidation state, but it often acts as a transient intermediate in reactions; the transphos complex **6.17** reacts with CD$_3$I to give CD$_3$CH$_3$. This probably goes via the unstable Pd(IV) intermediate **6.18**.

6.17   **6.18**

Dialkyls containing β hydrogens often β-eliminate to give an alkyl hydride and alkene before they reductively eliminate R—H. In PdEt$_2$(PR$_3$)$_2$, the cis isomer reductively eliminates butane, but in the trans isomer, in which the two R groups are not properly oriented for reductive elimination, the β-elimination–reductive elimination path is followed to give ethylene and ethane.[28]

The catalytic decarbonylation of aldehydes by $[Rh(triphos)CO]BF_4$ illustrates the reductive elimination of an alkane from an octahedral Rh(III) intermediate (Eq. 6.36, L = triphos); note some similarities with Fig. 6.1.

$$RCHO \xrightarrow{\text{LIr(CO)}} LIr(CO)(H)(COR) \xrightarrow{\text{-CO}} LIr(CO)(H)(R) \xrightarrow{\text{-RH}} LIr(CO) \quad (6.36)$$

Reductive elimination involving acyl groups is easier than for alkyls. For example, the cobalt dimethyl shown in Eq. 6.37 does not lose ethane but undergoes migratory insertion with added CO to give an acyl alkyl complex, which subsequently loses acetone; a crossover experiment with the protonated d_0 and deuterated d_6 dialkyls showed that this reaction is also intramolecular:

$$CpCoMe_2L \xrightarrow{\text{CO}} CpCo(COMe)MeL \longrightarrow CpCo(CO)L + Me_2CO \quad (6.37)^{[34]}$$

Formation of new carbon–heteroatom (O,N,S) bonds is also possible by reductive elimination.[35] Goldberg et al. found a methyl platinum(IV) acetate that forms methyl acetate in this way,[36] and Hartwig discovered a series of carbon heteroatom reductive eliminations as well as catalytic reactions that involve these steps.[35,37,38] These are of great use in organic synthesis (e.g., Buchwald–Hartwig reaction, Section 9.6).

Binuclear Reductive Elimination

We saw earlier that a binuclear version of oxidative addition is important for those metals that prefer to change their oxidation state by one, rather than two units. The same is true of reductive elimination:

$$2MeCH=CHCu(PBu_3) \xrightarrow{\text{heat}} MeCH=CHCH=CHMe \quad (6.38)^{[39]}$$

$$ArCOMn(CO)_5 + HMn(CO)_5 \longrightarrow ArCHO + Mn_2(CO)_{10} \quad (6.39)^{[40]}$$

Unexpectedly, the binuclear variant can even occur when an intramolecular path ought to be available as in Norton's[41] $Os(CO)_4RH$ (R = Me or Et). Alkyl hydrides normally eliminate rapidly to give the alkane, but here the usual intramolecular process (Eq. 6.40) is not observed. CO loss to give a 5-coordinate intermediate does not readily occur because this would give "$Os(CO)_4$," a highly unstable species. As a group 8 element, Os(0) greatly prefers the 5-coordinate geometry over 4-coordinate (d^8 metals of group 9 tend to prefer, and group 10 strongly prefer the 4-coordinate geometry). In addition, carbonyls strongly prefer the 18e over the 16e configuration:

$$cis\text{-}Os(CO)_4RH \xrightarrow{\text{X}} \text{"}Os(CO)_4\text{"} + R-H \quad (6.40)$$

Norton's mechanism provides a bimolecular way to eliminate alkane that does not involve 4-coordinate Os(0). The slow step is migratory insertion to give a coordinatively unsaturated 16e acyl. The resulting vacant site is filled by an Os—H bond from a second molecule of the alkyl hydride. Crossover products are seen (e.g., d_1 and d_3 methane from the d_0 and d_4 methyl hydrides), and so the R group of the acyl now eliminates with the hydride from the metal to give a binuclear complex containing an Os—Os bond. A hydride seems to be required for this mechanism to operate, probably because hydrides bridge so easily. The analogous complex cis-Os(CO)$_4$Me$_2$, lacking a hydride, decomposes only slowly at 160°C and even then does not give an elimination, binuclear or otherwise:

$$cis\text{-Os(CO)}_4\text{RH} \longrightarrow \text{(RCO)Os(CO)}_3\text{H} \xrightarrow{\text{Os(CO)}_4\text{RH}}$$

$$\text{(RCO)H(CO)}_3\text{Os}-\text{H}-\text{OsR(CO)}_4 \xrightarrow{-\text{RH}} \text{H(CO)}_4\text{Os}-\text{OsR(CO)}_4 \quad (6.41)$$

6.6 σ-BOND METATHESIS

Apparent oxidative addition–reductive elimination sequences can in fact be σ-*bond metathesis* reactions.[42] These are best recognized for d^0 early metal complexes such as Cp$_2$ZrRCl or WMe$_6$ because oxidative addition is forbidden in these cases. (The oxidative addition product would unambiguously exceed the maximum permitted oxidation state (Section 2.4).) In a reaction of such a complex with H$_2$ (Eq. 6.42), the metal therefore cannot follow mechanism a of Eq. 6.43. Instead a concerted process (path b of Eq. 6.43) is believed to operate. Path b probably goes via formation of an intermediate H$_2$ complex that is permitted even for d^0 species. The strong proton donor character of M(H$_2$) species may encourage proton transfer to the R group:

$$\text{Cp}_2\text{ZrRCl} + \text{H}_2 \longrightarrow \text{Cp}_2\text{ZrHCl} + \text{RH} \quad (6.42)$$

$$(6.43)$$

For the same reason, reaction of d^0 alkyls with acids cannot go via initial protonation at the metal (step a in Eq. 6.44) because as a d^0 system, the metal has no M(d_π) lone pairs. Instead, protonation of the M—R bond must take place.

Formation of an alkane σ-bond complex then would lead to loss of alkane. For late metals, where both pathways *a* and *b* are formally allowed in Eq. 6.43 and 6.44, it is hard to tell which pathway is followed; pathway *a* is normally assumed to operate in the absence of specific evidence to the contrary.

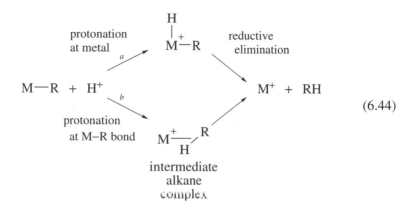

(6.44)

- Reductive elimination, the reverse of oxidative addition, decreases both the oxidation state and the coordination number by two.
- σ-Bond metathesis gives the same outcome as oxidative addition followed by reductive elimination; the two situations are hard to tell apart.

6.7   OXIDATIVE COUPLING AND REDUCTIVE CLEAVAGE

In oxidative coupling, Eq. 6.45, the metal induces a coupling reaction between two alkene ligands to give a metalacycle. The formal oxidation state of the metal increases by two units; hence the "oxidative" part of the name. The electron count decreases by two, but the coordination number stays the same. The reverse reaction, which is perhaps best called "reductive fragmentation" is more rarely seen. It cleaves a relatively unactivated $C-C$ bond to give back the two unsaturated ligands.

$$\text{M} \xrightleftharpoons[\text{reductive fragmentation}]{\text{oxidative coupling}} \text{M} \qquad (6.45)$$

Alkynes undergo the reaction more easily than do alkenes. Alkenes can be activated by electron-withdrawing substituents or by strain. Simple alkenes will still undergo the reaction if the metal is sufficiently π basic. Some examples follow:[43,44]

$$Fe(CO)_5 \xrightarrow{C_2F_4} (CO)_4Fe\begin{array}{c} CF_2-CF_2 \\ | \quad\quad | \\ CF_2-CF_2 \end{array} \tag{6.46}$$

$$\tag{6.47}$$

$$Cp^*Cl_2Ta\triangleleft \; + \; \diagup\diagdown\diagup\diagdown\diagup\diagdown \longrightarrow Cp^*Cl_2Ta\langle\text{bicyclo} \tag{6.48}$$

$$L_2Ni\triangleleft \xrightarrow{C_2H_4} L_2Ni\langle \tag{6.49}$$

$$Cp^*_2Zr-N_2-ZrCp^*_2 \xrightarrow{=C=} Cp^*_2Zr\langle \tag{6.50}$$
$$\quad\quad | \quad\quad\quad | $$
$$\quad N_2 \quad\quad N_2$$

$$FeL_4 \xrightarrow{MeC\equiv CMe} L_4Fe\langle \tag{6.51}$$

Intermediates with one coordinated alkene are often seen (e.g., Eq. 6.49), but the bis-alkene species is probably the immediate precursor of the coupled product.[45] The products from alkynes are often stable and are known as *metalloles* (Eq. 6.51) but they can also reductively eliminate to give cyclobutadiene complexes (Eq. 6.47).

Coupling is not limited to alkenes and alkynes. A particularly important, case involves carbenes and alkenes going to metalacyclobutanes (Eq. 6.52), the key step of the alkene metathesis reaction (Section 11.3). The reverse reaction constitutes another example of a C–C bond cleavage reaction, as we also saw in Eq. 6.45.

$$
\text{M} \diagup\!\!\!\diagdown \quad \rightleftharpoons \quad \text{M} \diamond \tag{6.52}
$$

Carbenes, $M{=}CR_2$, can couple to give the alkenes $R_2C{=}CR_2$, (Chapter 11), and the coupling of isonitriles and carbonyls[46] has been effected by the reduction of a 7-coordinate Mo complex. Note how the alkyne in Eq. 6.53 behaves as a 4e donor in the product.

$$
\text{Br(RNC)}_4\text{Mo}^+ \underset{C\equiv NR}{\overset{C\equiv NR}{<}} \quad \xrightarrow{\text{Zn}} \quad \text{Br(RNC)}_4\text{Mo}^+ - \underset{C\diagdown NHR}{\overset{C\diagup NHR}{\|||}} \tag{6.53}
$$

The same oxidation state ambiguity that we have seen several times before also operates here. Equation 6.54 shows that if the alkenes are considered to be in the metalacyclopropane (X_2 or σ_2 form), the coupling reaction proceeds with formal reduction at the metal and resembles a reductive elimination of two alkyl groups. Parkin et al.[47] have a case of a reductive coupling in Eq. 6.55, which shows how two terminal telluride ligands in a W(IV) precursor can be coupled to give a W(II)η^2Te$_2$ complex by addition of an isonitrile, t-BuNC.

$$
\text{M} \diagup\!\!\!\diagdown\!\!\!\diagup \quad \rightleftharpoons \quad \text{M} \pentagon \tag{6.54}
$$

6.19

$$
\begin{array}{c} \text{Te} \\ \text{Me}_3\text{P} \diagdown \| \diagup \text{PMe}_3 \\ \text{W} \\ \text{Me}_3\text{P} \diagup \| \diagdown \text{PMe}_3 \\ \text{Te} \end{array} \xrightarrow{\text{RNC}} \begin{array}{c} \text{CNR} \\ \text{Me}_3\text{P} \diagdown | \diagup \text{CNR} \\ \text{Te} \diagdown \text{W} \\ \diagup | \diagdown \text{CNR} \\ \text{Te} \quad \text{CNR} \end{array} \tag{6.55}
$$

- Oxidative coupling, like oxidative addition, has oxidation state and coordination number both increasing by two, but two new ligands are involved and a new bond is formed between them (Eq. 6.45).

REFERENCES

1. R. H. Crabtree, *Acc. Chem. Res.* **12**, 331, 1979.
2. L. Vaska, *Acc. Chem. Res.* **1**, 335, 1968; A. L. Sargent and M. B. Hall, *Inorg. Chem.* **31**, 317, 1992.
3. R. Eisenberg et al., *J. Am. Chem. Soc.* **107**, 3148, 1985.
4. R. H. Crabtree et al., *J. Am. Chem., Soc.* **110**, 5034, 1988.
5. O. Eisenstein and K. G. Caulton et al., *J. Am. Chem. Soc.* **117**, 8869, 1995; **121**, 3242, 1999.
6. R. H. Crabtree et al., *Inorg. Chem.* **24**, 1986, 1985.
7. H. B. Bürgi and J. D. Dunitz, *Acc. Chem. Res.* **16**, 153, 1983.
8. W. D. Jones et al., *J. Mol. Catal. A* **189**, 157, 2002.
9. J. F. Hartwig and F. Paul, *J. Am. Chem. Soc.* **117**, 5373, 1995.
10. P. B. Chock and J. Halpern., *J. Am. Chem. Soc.* **88**, 3511, 1966.
11. J. K. Stille, *Acc. Chem. Res.* **10**, 434, 1977; *J. Am. Chem. Soc.* **100**, 838, 845, 1975.
12. E. Uhlig and D. Walton, *Coord. Chem. Rev.* **33**, 3, 1980.
13. C. M. Thomas and G. Suss-Fink, *Coord. Chem. Revs.* **243**, 125, 2003.
14. C. J. Levy and R. J. Puddephatt, *J. Am. Chem. Soc.* **119**, 10127, 1999.
15. (a) J. K. Kochi *Adv. Phys. Org. Chem.*, **28**, 185, 1994; (b) W. C. Trogler *Organometallic Radical Chemistry*, Elsevier, New York, 1990; (c) E. Hoggan *Coord. Chem. Rev.* **159**, 2, 1997.
16. M. F. Lappert and P. W. Lednor, *J. Chem. Soc., Dalton* 1448, 1980; *Adv. Organomet. Chem.* **14**, 345, 1976.
17. J. A. Osborn et al., *Inorg. Chem.* **19**, 3230, 3236, 1980; J. Halpern et al., *J. Am. Chem. Soc.* **107**, 4333, 1985; W. Kaim, *Top. Curr. Chem.* **169**, 231, 1994.
18. J. R. Woodward, *Prog. React. Kinet.* **27**, 165, 2002.
19. J. Halpern, *Pure Appl. Chem.* **51**, 2171, 1979; *J. Am. Chem. Soc.* **106**, 8317, 1984.
20. W. J. Luow et al., *J. Chem. Soc., Dalton* **340**, 1978; R. H. Crabtree et al., *J. Organomet. Chem.* **181**, 203, 1979.
21. P. Uguagliati, *J. Organomet. Chem.* **169**, 115, 1979; R. Romeo et al., *Inorg. Chim. Acta* **19**, L55, 1976; *Inorg. Chem.* **17**, 2813, 1978.
22. G. L. Hillhouse, *Organometallics* **14**, 4421, 1995.
23. D. Milstein, *Acc. Chem. Res.* **17**, 221, 1984.
24. F. Maseras, O. Eisenstein, A. L. Rheingold, and R. H. Crabtree, *New J. Chem.* **23**, 1493, 1998, and references cited therein; G. S. Hill and R. J. Puddephatt, *Organometallics* **17**, 1478, 1998.
25. J. Halpern and C. S. Wong, *Chem. Commun.* 629, 1973.
26. K. I. Goldberg et al., *J. Am. Chem. Soc.* **122**, 962, 2000; **123**, 6423, 2001; *Adv. Inorg. Chem.* **54**, 259, 2003.
27. R. Hoffmann, in *Frontiers in Chemistry*, J. K. Laidler, ed., Pergamom, Oxford, 1982, p. 247.
28. J. K. Stille et al., *J. Am. Chem. Soc.* **102**, 4933, 1980; *J. Am. Chem. Soc.* **103**, 2143, 1981.
29. R. Hoffmann and J. K. Kochi et al., *J. Am. Chem. Soc.* **98**, 7255, 1976.

30. G. M. Whitesides, *Pure Appl. Chem.* **53**, 287, 1981; R. J. Puddephatt et al., *J. Chem. Soc., Dalton* 2457, 1974.

31. A. Yamamoto et al., *J. Am. Chem. Soc.* **93**, 3360, 1971.

32. M. S. Driver and J. F. Hartwig, *J. Am. Chem. Soc.* **119**, 8232, 1997.

33. C. M. Beck et al., *Organometallics*, **18**, 5311, 1999.

34. R. Bergman, *Acc. Chem. Res.* **13**, 113, 1980.

35. J. F. Hartwig, *Acc. Chem. Res.* **31**, 852, 1998; **36**, 234, 2003.

36. K. I. Goldberg et al., *J. Am. Chem. Soc.* **121**, 252, 1999.

37. J. F. Hartwig, et al., *J. Am. Chem. Soc.* **117**, 4708, 1995; **120**, 9205, 1998; **125**, 16347, 2003; **120**, 5344, 2004.

38. S. L. Buchwald et al., *J. Am. Chem. Soc.* **120**, 6504, 1998; **125**, 11818, 2003.

39. G. M. Whitesides, C. P. Casey et al., *J. Am. Chem. Soc.* **93**, 1379, 1971.

40. J. A. Gladysz et al., *J. Am. Chem. Soc.* **101**, 1589, 1979.

41. J. Norton, *Acc. Chem. Res.* **12**, 139, 1979.

42. S. Q. Niu and M. B. Hall, *Chem. Rev.* **100**, 353, 2000.

43. F. G. A. Stone et al., *J. Organomet. Chem.* **100**, 257, 1975.

44. Y. Yamamoto, T. Arakawa, and K. Itoh, *Organometallics* **23**, 3610, 2004.

45. R. D. W. Kemmit et al., *J. Organomet. Chem.* **187**, C1, 1980.

46. S. J. Lippard et al., *J. Am. Chem. Soc.* **114**, 4166, 1992.

47. G. Parkin et al., *Inorg. Chem.* **34**, 6341, 1995; *Chem. Commun.* 1099, 1995.

PROBLEMS

1. An oxidative addition to a metal complex **A** is found to take place with $MeOSO_2Me$ but not with i-PrI. A second complex, **B**, reacts with i-PrI but not with $MeOSO_2Me$. What mechanism(s) do you think is (are) operating in the two cases? Which of the two complexes, **A** or **B**, would be more likely to react with MeI? What further tests could you apply to confirm the mechanism(s)?

2. Suppose we are able to discover that the equilibrium constants for Eq. 6.1 are in the order $CH_3-H < Ph-H < H-H < Et_3Si-H$ for a given square planar Ir(I) complex. Can we say anything about the relative metal–ligand bond strengths in the adducts? Justify any assumptions that you make.

3. A given complex ML_n forms only a dihydrogen complex $(\eta^2-H_2)ML_n$, not the true oxidative addition product H_2ML_n with H_2. Would the true oxidative addition product be more or less likely to form as we move to (a) more electron-releasing ligands L, (b) from a third- to a first-row metal, M, or (c) to the 1e oxidation product $H_2ML_n^+$? Would you expect the same metal fragment to form an ethylene complex, $(C_2H_4)ML_n$, with predominant Dewar–Chatt or metalacyclopropane character? Explain.

4. Complexes of the type $Pt(PR_3)_4$ can form $PtCl_2(PR_3)_2$ with HCl. How do you explain this result? The same product can also be formed from t-BuCl

and Pt(PR$_3$)$_4$. What do you think is happening here? In each case a different non-metal-containing product is also formed; what do you think they are?

5. A 16e metal complex L$_n$M is found to react with ethylene to give 1-butene and L$_n$M. Provide a reasonable mechanism involving oxidative coupling.

6. Predict the order of reactivity of the following in oxidative addition of HCl: **A**, IrCl(CO)(PPh$_3$)$_2$; **B**, IrCl(CO)(PMe$_3$)$_2$; **C**, IrMe(CO)(PMe$_3$)$_2$; **D**, IrPh(CO)(PMe$_3$)$_2$. Would you expect the v(CO) frequencies of **A–D** to (i) be different from one another or (ii) to change in going to the oxidative addition products? Explain and justify any assumptions you make.

7. The products from HCl addition to **C** and **D** in Problem 6 are unstable, but the addition products to **A** and **B** are stable. Explain, and state how **C** and **D** will decompose.

8. WMe$_6$ reacts with H$_2$ and PMe$_3$ to give WH$_2$(PMe$_3$)$_5$. Propose a reasonable mechanism.

9. H$_2$ adds to Ir(dppe)(CO)Br to give a kinetic product **A**, in which the cis H ligands are trans to P and CO, and a thermodynamic product **B**, in which the cis H ligands are trans to P and Br. Write the structures of **A** and **B**. How would you tell whether the rearrangement of **A** to **B** occurs by initial loss of H$_2$ or by a simple intramolecular rearrangement of **A**?

10. Pt(PEt$_3$)$_2$, generated electrochemically, reacts with the PhCN solvent to give PhPt(CN)(PEt$_3$)$_2$. Oxidative addition of a C–C bond is very rare. Discuss the factors that make it possible in this case.

11. Complexes **6.20** are formed by the route of Eq. 6.56. Suggest a plausible pathway for this reaction if epoxide **6.21** gives complex **6.22**.

                                                                (6.56)

6.20

6.21   **6.22**

12. In light of the discussion in Section 6.1 about the preference for COs to lie in the equatorial plane of a TBP, comment on the structure shown in Fig. 5.8 and suggest reasons for any unexpected features.

7

INSERTION AND ELIMINATION

Oxidative addition and substitution allow us to assemble 1e and 2e ligands on the metal. With insertion, and its reverse reaction, elimination, we can now combine and transform these ligands within the coordination sphere, and ultimately expel these transformed ligands to form free organic compounds. In insertion, a coordinated 2e ligand, A=B, can insert itself into an M—X bond to give M—(AB)—X, where ABX is a new 1e ligand in which a bond has been formed between AB and both M and X.

There are two main types of insertion—1,1 and 1,2—as shown in Eqs. 7.1 and 7.2, in which the metal and the X ligand end up bound to the same (1,1) or adjacent (1,2) atoms of an L-type ligand. The type of insertion observed in any given case depends on the nature of the 2e inserting ligand. For example, CO gives only 1,1 insertion: that is, both the M and the X group end up attached to the CO carbon. On the other hand, ethylene gives only 1,2 insertion, in which the M and the X end up on adjacent atoms of what was the 2e ligand. In general, η^1 ligands tend to give 1,1 insertion and η^2 ligands give 1,2 insertion. SO_2 is the only common ligand that can give both types of insertion; as a ligand, SO_2 can be η^1 (S) or η^2 (S, O).

$$
\begin{array}{ccc}
\overset{\displaystyle X}{\underset{\displaystyle M-C=O}{|}} & \xrightarrow{\text{1,1-migratory insertion}} & \overset{\displaystyle \square}{\underset{\displaystyle M-C}{|}}\overset{\displaystyle X}{\underset{\displaystyle O}{\diagup}} \\
\text{18e} & & \text{16e}
\end{array} \tag{7.1}
$$

The Organometallic Chemistry of the Transition Metals, Fourth Edition, by Robert H. Crabtree
Copyright © 2005 John Wiley & Sons, Inc.

183

$$
\begin{array}{c}
X \\
| \quad C \\
M{-}\underset{C}{\overset{||}{}} \\
\end{array}
\quad \xrightarrow{\text{1,2-migratory insertion}} \quad
\begin{array}{c}
\square \quad X \\
| \qquad C \\
M{-}C \\
\end{array}
\tag{7.2}
$$

$$18e \qquad\qquad\qquad\qquad 16e$$

In principle, insertion reactions are reversible, but just as we saw for oxidative addition and reductive elimination in Chapter 6, for many ligands only one of the two possible directions is observed in practice, probably because this direction is strongly favored thermodynamically. For example, SO_2 commonly inserts into $M{-}R$ bonds to give alkyl sulfinate complexes, but these rarely eliminate SO_2. Conversely, diazoarene complexes readily eliminate N_2, but N_2 has not yet been observed to insert into a metal–aryl bond.

$$M{-}R + SO_2 \longrightarrow M{-}SO_2R \tag{7.3}$$

$$M{-}N{=}N{-}Ar \longrightarrow M{-}Ar + N_2 \tag{7.4}$$

The immediate precursor to the final insertion product usually has both the 1e and 2e ligands coordinated. This means that a net 3e set of ligands is converted to a 1e insertion product (ionic model: $4e \rightarrow 2e$), so that a 2e vacant site is generated by the insertion. This site can be occupied by an external 2e ligand and the insertion product trapped. Conversely, the elimination requires a vacant site, so that an 18e complex will not undergo the reaction unless a ligand first dissociates. The insertion also requires a cis arrangement of the 1e and 2e ligands, while the elimination generates a cis arrangement of these ligands. The formal oxidation state does not change during the reaction; Eq 7.5 shows the typical case of CO insertion.

$$
\begin{array}{c}
X \\
| \\
M{-}C{=}O \\
\end{array}
\quad \longrightarrow \quad
\begin{array}{c}
\square \\
| \quad X \\
M{-}C \\
\qquad\ \ O \\
\end{array}
\tag{7.5}
$$

$$\Delta OS = 0$$
$$\Delta(e\ count) = -2$$

One way to picture insertion reactions is to consider that the X ligand migrates with its $M{-}X$ bonding electrons (e.g., H^- or Me^-) to attack the π^* orbital of the $A{=}B$ ligand.

- 1,1-Insertion (Eq. 7.1) occurs with η^1-bound ligands such as CO.
- 1,2-Insertion (Eq. 7.2) occurs with η^2-bound ligands such as $H_2C{=}CH_2$.

7.1 REACTIONS INVOLVING CO

CO shows a strong tendency to insert into metal–alkyl bonds to give metal acyls. The reaction has been carefully studied for a number of systems. Although the details may differ, most follow the pattern set by the best-known case:[1]

$$\underset{\text{(CO)}_4\text{Mn}-\text{C}=\text{O}}{\overset{\text{Me}}{|}} \quad \underset{-\text{CO}}{\overset{\text{CO}}{\rightleftharpoons}} \quad \text{(CO)}_4\overset{\overset{\text{CO}}{|}}{\text{Mn}}-\text{C}\overset{\text{Me}}{\underset{\text{O}}{\diagdown}} \tag{7.6}$$

The mechanism[2] of migratory insertion shown in Eq. 7.7 applies in many cases. The alkyl group in the reagent (Rgt) undergoes a migration to the CO to give an acyl intermediate (Int.) that is trapped by added ligand, L, to give the final product (Pdct).

$$\underset{\text{Rgt}}{\underset{\text{Me}-\text{Mn(CO)}_4}{\overset{\overset{\overset{\text{O}}{\|}}{\text{C}}}{|}}} \quad \underset{k_{-1}}{\overset{k_1}{\rightleftharpoons}} \quad \underset{\text{Int.}}{\underset{\square-\text{Mn(CO)}_4}{\overset{\overset{\text{Me}\diagdown\diagup^{\text{O}}}{\text{C}}}{|}}} \quad \underset{\text{slow}}{\overset{\text{L}, k_2}{\longrightarrow}} \quad \underset{\text{Pdct}}{\underset{\text{L}-\text{Mn(CO)}_4}{\overset{\overset{\text{Me}\diagdown\diagup^{\text{O}}}{\text{C}}}{|}}} \tag{7.7}$$

The kinetics in this situation are reminiscent of dissociative substitution (Section 4.3) except that the 2e site is formed at the metal in the migratory step, not by loss of a ligand. Using the usual[2] steady-state method, the rate is given by

$$\text{Rate} = \frac{-d[\text{Rgt}]}{dt} = \frac{k_1 k_2 [\text{L}][\text{Rgt}]}{k_{-1} + k_2 [\text{L}]}$$

There are three possible regimes,[2] all of which are found in real cases:

1. If k_{-1} is very small relative to $k_2[\text{L}]$, [L] cancels and the equation reduces to

$$\text{Rate} = \frac{-d[\text{Rgt}]}{dt} = k_1 [\text{Rgt}]$$

Because k_{-1} is small, L always traps the intermediate; this means the rate of the overall reaction is governed by k_1 and we have a first-order reaction.

2. If k_{-1} is very large relative to $k_2[\text{L}]$, then the equation becomes

$$\text{Rate} = \frac{-d[\text{Rgt}]}{dt} = \frac{k_1 k_2 [\text{L}][\text{Rgt}]}{k_{-1}}$$

In this case the intermediate almost always goes back to the starting reagent and the second step, attack by L, governs the overall rate, so we have second-order kinetics.

3. If k_{-1} is comparable to $k_2[L]$, then the situation is more complicated and the equation is usually rewritten as

$$\text{Rate} = \frac{-d[\text{Rgt}]}{dt} = k_{\text{obs}}[\text{Rgt}]$$

$$k_{\text{obs}} = \frac{k_1 k_2 [L]}{k_{-1} + k_2[L]}$$

In this case, the intermediate is trapped by L at a rate that is comparable to the reverse migration. This is handled by plotting $1/k_{\text{obs}}$ versus $1/[L]$. In such a case, the intercept is $1/k_1$ and the slope is $k_{-1}/(k_1 k_2)$. Dividing the slope by the intercept gives k_{-1}/k_2; this tells us how the intermediate partitions between the forward (k_2) and back (k_{-1}) reactions.

$$\frac{1}{k_{\text{obs}}} = \underbrace{\frac{k_{-1}}{k_1 k_2}}_{\text{slope}} \frac{1}{[L]} + \underbrace{\frac{1}{k_1}}_{\text{intercept}}$$

When the incoming ligand is ^{13}CO, the product contains only one labeled CO, which is cis to the newly formed acetyl group. This shows that the methyl group migrates to a coordinated CO, rather than free CO attacking the Mn—Me bond. Any stereochemistry at the alkyl carbon is retained in the insertion as is consistent with the mechanism of Eq. 7.7.[1,3] We can tell where the labeled CO is located in the product because there is a characteristic shift of the $\nu(\text{CO})$ stretching frequency to lower energy in the IR spectrum of the complex as a result of the greater mass of ^{13}C over normal carbon (see Section 10.9).

By studying the reverse reaction (Eq. 7.8), elimination of CO from $\text{Me}^{13}\text{COMn(CO)}_5$, where we can easily label the acyl carbon with ^{13}C (by reaction of $\text{Mn(CO)}_5{}^-$ with $\text{Me}^{13}\text{COCl}$), we find that the label ends up in a CO cis to the methyl. This is an example of a general strategy in which we examine the reverse of a reaction to learn something about the forward process.

(7.8)

(where $\text{C}^* = {}^{13}\text{C}$).

This relies on *microscopic reversibility*, according to which the forward and reverse reactions of a thermal process must follow the same path. In this case, if the labeled CO ends up cis to Me, the CO to which a methyl group migrates in the forward reaction, must also be cis to methyl. We are fortunate in seeing the kinetic products of these reactions. If a subsequent scrambling of the COs had been fast, we could have deduced nothing.

It is also possible to use reversibility arguments to show that the acetyl ligand in the product is bound at a site cis to the original methyl, rather than anywhere else. To do this we look at CO elimination in *cis*-(MeCO)Mn(CO)$_4$(^{13}CO), in which the label is cis to the acetyl group. If the acetyl CO migrates during the elimination, then the methyl in the product will stay where it is and so remain cis to the label. If the methyl migrates, then it will end up both cis and trans to the label, as is in fact observed:

$$(7.9)$$

This observation implies that the methyl also migrates when the reaction is carried out in the direction of insertion. The *cis*-(MeCO)Mn(CO)$_4$(^{13}CO) required for this experiment can be prepared by the photolytic method discussed in Section 4.7, and we again use the IR spectrum to tell where the label has gone in the products. This is the only feature of migratory insertion in MeMn(CO)$_5$ that does not reliably carry over to other systems, where the product acyl is occasionally found at the site originally occupied by the alkyl.[3c]

Enhancing Insertion Rates

Steric bulk in the ligand set accelerates insertion, no doubt because the acetyl in the L$_n$M(COMe) product, occupying one coordination site, is far less bulky than the alkyl and carbonyl, occupying two sites, in the starting complex, L$_n$M(Me)(CO).[4] Lewis acids such as AlCl$_3$ or H$^+$ can increase the rate of migratory insertion by as much as 10^8-fold. Metal acyls (**7.1**) are more basic at oxygen than are the corresponding carbonyls by virtue of the resonance form **7.2**. By binding to the oxygen, the Lewis acid would be expected to stabilize the transition state and speed up trapping by L and therefore speed up the reaction.[5] Polar solvents such as acetone also significantly enhance the rate.

Another important way of promoting the reaction is oxidation of the metal. $Cp(CO)_2FeMe$ is normally very slow to insert, but 1e oxidation at $-78°C$ in MeCN using Ce(IV) salts (or electrochemically) gives the acyl $[CpFe(MeCN)(CO)(COMe)]^+$, in which the solvent has played the role of incoming ligand.[6] As we saw in Chapter 4, 17e complexes can be very labile, but another factor here may be the increased electrophilicity (decreased π basicity) of the metal, leading to a larger partial positive charge on the CO carbon. The migration of Me^- to an electron-deficient CO carbon seems to be a good description of the CO insertion, and so the rate of the reaction may increase in response to the increase in the ∂^+ charge on the CO carbon. Oxidation would also speed up trapping by phosphine, but this may no longer be the rate-determining step.

Under CO, trityl cation, Ph_3C^+, can catalyze migratory insertion in complexes such as $Cp(CO)_2FeMe$, by oxidation to $[Cp(CO)_2FeMe]^{\cdot+}$. This 17e radical cation then undergoes migratory insertion with CO as the incoming ligand. The trityl radical, formed in the first step, then reduces the 17e insertion product to the 18e $Cp(CO)_2FeCOMe$ and the starting trityl cation.[7] The rates of insertion are also increased to some extent by using more nucleophilic solvents, suggesting that the solvent may act as a temporary ligand to stabilize an initial, solvated insertion product.[8]

Early metals are Lewis acids in their own right and tend to bind oxygen ligands (see the discussion of oxophilicity in Section 3.2); they can therefore act as their own Lewis acid catalysts for insertion. The product is an η^2-acyl as shown in Eq. 7.10:[9]

$$(7.10)$$

By altering the thermodynamics in favor of the adduct, this effect is even sufficient to promote the normally unfavorable CO insertion into an M−H bond, as shown in Eq. 7.11:[10]

$$(7.11)$$

In each of these reactions, the formation of an intermediate carbonyl complex is proposed. Zr(IV) and Th(IV) are both poor π bases, and so these intermediates must be very unstable;[11] limited back bonding should make the CO much more reactive toward insertion, however.

Apparent Insertions

These can in fact go by an entirely different route as shown in Eq. 7.14. The late metal alkoxide is unstable (since MeO is a good π donor bound to a π-donor metal) and the MeO group dissociates as MeO$^-$ to give an ion pair with a 2e vacancy at the metal. The free CO present then binds to this 2e site and is strongly activated toward nucleophilic attack at the CO carbon owing to the positive charge on the metal. The product is the interesting metala–ester complex shown in Eq. 7.12:[12a]

$$L_2(CO)Ir\text{---}OMe \quad \xrightarrow{\text{CO}} \quad L_2(CO)Ir^+\text{---}CO \; + \; OMe^-$$

nucleophilic attack
of OMe$^-$ on CO $\qquad$ (7.12)

$$L_2(CO)Ir\text{---}C\overset{\displaystyle OMe}{\underset{\displaystyle O}{\diagdown}}$$

Genuine migratory insertions into M—O bonds are also possible. For *trans*-[Pt(Me)(OMe)(dppe)], CO inserts into the Pt—OMe bond, while for [Ni(Me)(O-p-C$_6$H$_4$CN)(bipy)], CO inserts into Ni—Me. This follows from theoretical work. For Ni, the M—Me bond is significantly stronger than M—OMe, but migratory insertion with M—Me is marginally preferred owing to the weaker C—O bond of the aryloxy-carbonyl. For Pt, M—Me and M—OMe bonds are equally strong, so the stronger methoxycarbonyl C—O bond results in reaction with the M—OMe bond.[12b]

Double Insertion

Given that the methyl group migrates to the CO, why stop there? Why does the resulting acyl group not migrate to another CO to give an MeCOCO ligand? If migration happened repeatedly, we might even get the unknown R(CO)$_m$ML$_n$ polymer, a material that is believed to be thermodynamically unstable with respect to CO itself. The complex that would have been formed in a double insertion can be made by an independent route from MeCOCOCl and Mn(CO)$_5$$^-$. It easily eliminates CO to give MeCOMn(CO)$_5$, which suggests that the double-insertion product is thermodynamically unstable with respect to MeCOMn(CO)$_5$. The —CHO and CF$_3$CO— groups share with MeCOCO— the property of eliminating CO irreversibly to give hydride and trifluoromethyl complexes, respectively. The reason is again probably thermodynamic because the M—COMe, M—H, and M—CF$_3$ bonds are all distinctly stronger than M—CH$_3$, the bond formed in CO elimination from the acetyl (Chapter 3). In contrast, isonitriles do undergo repeated migratory insertion to give polymers with as many as 100 isocyanide units inserted:

$$Cl(R_3P)_2Pt-C\equiv C-Pt(PR_3)_2Cl + nRNC \longrightarrow$$

$$Cl(R_3P)_2Pt-(C=NR)_nC\equiv C-Pt(PR_3)_2Cl \quad (7.13)$$

Products that *appear* to arise from double migratory insertion of CO have been found by Yamamoto[13] in the following catalytic reaction:

$$R_2NH + CO + ArI \xrightarrow{\text{Pd catalyst}} R_2NCOCOAr \quad (7.14)$$

In fact, the reaction goes via the cycle shown in Fig. 7.1, in which a reductive elimination reaction, not a migratory insertion, forms the $R_2N(CO)-(CO)Ar$ carbon–carbon bond. In the first step, oxidative addition of ArI forms a Pd–Ar bond. The Ar(CO) ligand is then formed by a conventional migratory insertion, and the $R_2N(CO)$ group arises by a nucleophilic attack of R_2NH on a second CO.

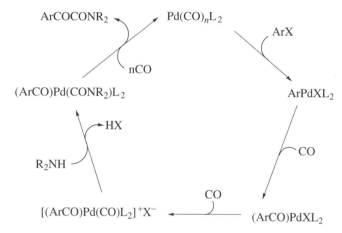

FIGURE 7.1   The catalytic cycle proposed by Yamamoto for the formation of an apparent "double insertion" product.[13]

Isonitriles

Isonitriles insert very readily into M–R and even M–H bonds to give iminoacyls, which can be η^2 bound for the early metals (Eq. 7.15)[14] or in metal clusters (Eq. 7.16):[15]

$$Cp_3U-R \xrightarrow{R'NC} Cp_3U\begin{subarray}{l} CR \\ \| \\ NR' \end{subarray} \quad (7.15)$$

$$\text{(CO)}_3\text{Os} \underset{\text{H}}{\overset{\text{Os(CO)}_4}{\diagdown}} \text{OsH(CO)}_3 \quad \longrightarrow \quad \text{(CO)}_3\text{Os} \overset{\text{Os(CO)}_4}{\underset{}{\diagup}} \text{Os(CO)}_3$$

(7.16)

We look at insertions involving carbenes in Chapter 11.

7.2 INSERTIONS INVOLVING ALKENES

The insertion of coordinated alkenes into M—H bonds is a very important reaction because it gives alkyls and constitutes a key step in a variety of catalytic reactions (see Chapter 9) including alkene polymerization (Chapter 12), perhaps the most commercially important organometallic reaction. As η^2-ligands, alkenes give 1,2 insertion. This is the reverse of the familiar β-elimination reaction (Eq. 7.17). Some insertion reactions are known to give agostic (**7.3**) rather than classical alkyls, and species of type **7.3** probably lie on the pathway for insertion into M—H bonds.[16a] The position of equilibrium is decided by the thermodynamics of the particular system, and depends strongly on the alkene. For simple alkenes, such as ethylene (Eq. 7.18),[16b] the equilibrium tends to lie to the left (i.e., the alkyl β-eliminates), but for alkenes with electron-withdrawing ligands (e.g., C_2F_4, Eq. 7.18), the alkyl is particularly stable, and the equilibrium lies entirely to the right; alkyls such as $L_nMCF_2CF_2H$ with electron-withdrawing substituents have M—C bonds that are particularly strong (Section 3.1), and the alkyls are stable even though they have a β hydrogen.

$$\text{M—H} + \text{C}{=}\text{C} \underset{\beta \text{ elimination}}{\overset{1,2\text{-insertion}}{\rightleftharpoons}} \text{M} \overset{\diagup}{\underset{\text{H}}{\diagdown}} \text{C} \rightleftharpoons \text{M} \overset{\text{C—C}}{\underset{\text{H}}{\diagdown}}$$

7.3
Proposed
agostic
intermediate

(7.17)

$$\text{Cl} \overset{\text{PR}_3}{\underset{\text{R}_3\text{P}}{\diagup}} \text{Pt—H} \underset{}{\overset{CX_2{=}CX_2}{\rightleftharpoons}} \text{Cl} \overset{\text{PR}_3}{\underset{\text{R}_3\text{P}}{\diagup}} \text{Pt—CX}_2$$

$$\text{CX}_2\text{-H}$$

(X = H or F)

(7.18)[16]

$$(7.19)^{17}$$

$$(7.20)$$

Equation 7.20 shows a strategy that Spencer and co-workers[16c] have used to determine the mode of Pd—H addition to an unsymmetrical cis alkene. On isomerization to the trans form, the deuterium initially incorporated in the insertion step is retained by the substrate and labels the β carbon of the product. The mode of insertion seen in Eq. 7.20 is consistent with the PdH hydride acting as a hydridic δ^- group.

The usual stereochemistry of the insertion is syn, and so the stereochemistry at both carbons is retained, as shown by the alkyne example in Eq. 7.19, but the initially formed *cis*-vinyl complex, if 16e, can sometimes rearrange to the trans isomer, via an η^2-vinyl (Eq. 7.21). This can lead to a net anti addition of a variety of X—H groups (Eq. 7.22, where X = R$_3$Si) to alkynes:[18]

$$(7.21)$$

$$\text{(7.22)}$$

As we saw for CO insertions and eliminations, a 2e vacant site is generated by the insertion (and required for the elimination). Reversible insertion/elimination equilibria are also known.[19] The vacant site may be filled by any suitable ligand, such as the solvent, excess alkene, an agostic CH bond or a phosphine.

$$\text{(7.23)}$$

The transition state for insertion, **7.4**, has an essentially coplanar M−C−C−H arrangement, and this implies that both insertion and elimination also require the M−C−C−H system to be capable of becoming coplanar. We have seen in Section 3.1 how we can stabilize alkyls against β elimination by having a noncoplanar M−C−C−H system. The same principles apply to stabilizing alkene hydride complexes. Compound **7.5** undergoes insertion at least 40 times more rapidly than **7.6**, although the alkene and M−H groups are cis in both cases, only in **7.6** is there a noncoplanar M−C−C−H arrangement.[20]

7.4   **7.5**   **7.6**

One application of alkene insertion is hydrozirconation of alkenes by Cp_2ZrHCl.[21] Terminal alkenes insert in the anti-Markownikov direction to give a stable 1° alkyl (Eq. 7.24). Internal alkenes, such as 2-butene, insert to give an unstable 2° alkyl, which is not observed. This intermediate β eliminates to give 1- and 2-butene. The 1-butene can now give the stable 1° alkyl, the observed product (Eq. 7.25). This is a particularly noteworthy reaction because the terminal alkene is less stable than the internal alkene. The insertion goes in the way it does because the 1° alkyl is thermodynamically more stable than any 2° alkyl, probably for steric reasons. The 1° alkyl can now be functionalized in a number of ways as discussed in Chapter 14.

$$Cp_2ClZr—H \xrightarrow{\hspace{2cm}} Cp_2ClZr \diagup\diagdown\diagup\diagdown \qquad (7.24)^{21}$$

$$Cp_2ClZr—H \xrightarrow{\hspace{2cm}} Cp_2ClZr \xrightarrow{\hspace{2cm}} Cp_2ClZr—H$$

unstable

$$Cp_2ClZr \diagdown\diagup\diagdown \longleftarrow \diagup\diagdown\diagup \qquad +$$

observed product

$$(7.25)^{21}$$

Insertion into M—H Versus M—R

We saw in the last section that for thermodynamic reasons, CO insertion generally takes place into M—R but not into M—H bonds. Alkene insertion, in contrast, is common for M—H, but much less common for M—R. Alkene polymerization is a reaction that involves repeated alkene insertion into an M—R bond (Section 12.2). The thermodynamics still favor the reaction with M—R, so its comparative rarity must be due to kinetic factors. Brookhart and co-workers[22a] have compared the barriers for insertion of ethylene into the M—R bond in $[Cp^*\{(MeO)_3P\}MR(C_2H_4)]^+$, where R is H or Et and M is Rh or Co. The reaction involving M—H has a 6–10-kcal/mol lower barrier (Table 7.1). This corresponds to a migratory aptitude ratio k_H/k_{Et} of 10^6–10^8. As we have seen

TABLE 7.1   Comparison of Barriers (kcal/mol) for Insertion in $[Cp^*\{(MeO)_3P\}MR(C_2H_4)]^+$ for R = H and R = Et[22a]

M	R = H[a]	R = Et[b]	Difference
Rh	12.2	22.4	10.2
Co	6–8 (est.)	14.3	6–8 (est.)

[a] ±0.1 kcal/mol.
[b] ±0.2 kcal/mol.

before, reactions involving M−H are almost always kinetically more facile than reactions of M−R. This means that an alkene probably has less intrinsic kinetic facility for insertion than does CO. Looking at the reverse reaction (Eq. 7.26), elimination, we see that this implies that β-H elimination in an alkyl will be kinetically very much easier than β-alkyl elimination, and it will also give a thermodynamically more stable product, so it is not surprising that β-alkyl elimination is extremely rare. In those cases where it is observed, there is always some special factor that modifies the thermodynamics or the kinetics or both. For example, for f-block metals M−R bonds appear to be comparable in strength, or stronger than M−H bonds and both β-H and β-alkyl elimination can be observed:[22b]

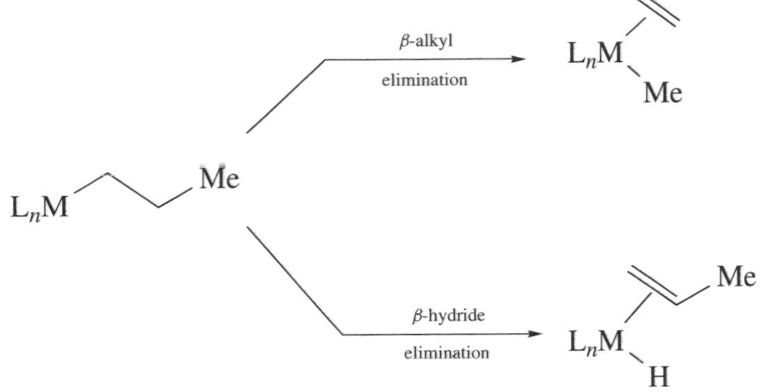

(7.26)

Strain, or the presence of electronegative substituents on the alkene, or moving to an alkyne are some of the other factors that can bias both the thermodynamics and the kinetics in favor of insertion, as shown in Eqs. 7.27 and 7.28:[23,24]

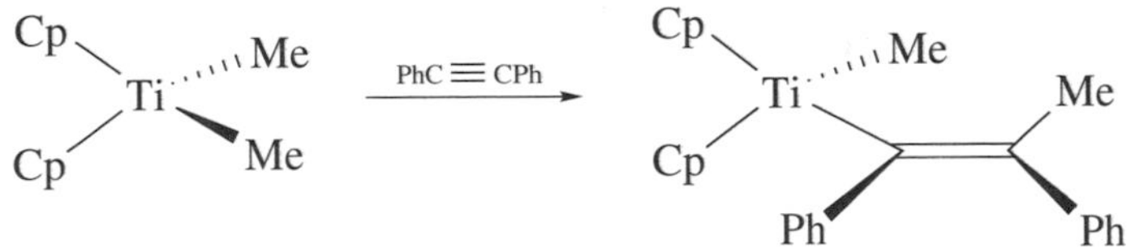

(7.27)

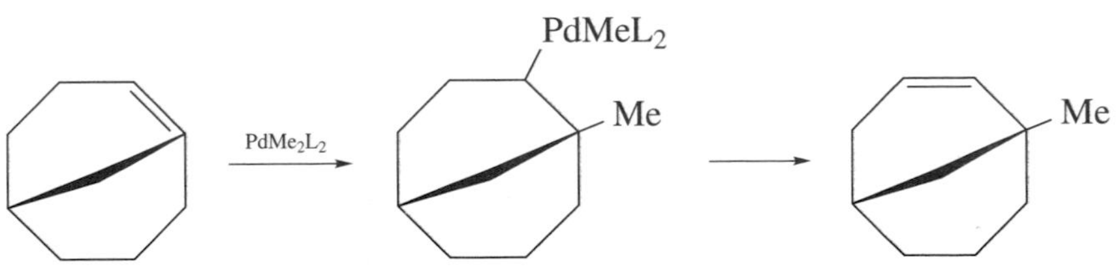

(7.28)

A case in which ethylene inserts into an M−R bond was described by Bergman et al.[25] The insertion mechanism was confirmed by the labeling scheme shown in

Eq. 7.29 (L = PPh$_3$, □ = vacant site). Insertion into M–M bonds is also known, as shown in Eq. 7.30:[26a]

$$(7.29)$$

$$(7.30)$$

Styrene can insert into the M–M bond of [Rh(OEP)]$_2$ (OEP = octaethylpor-phyrin). Initial homolysis gives the 15e metalloradical •Rh(OEP), which adds to the alkene to give PhCH•–CH$_2$Rh(OEP) (the Ph group stabilizes the adjacent C radical) and then (OEP)RhCHPh–CH$_2$Rh(OEP). [Rh(OEP)]$_2$ also initiates rad-ical photopolymerization of CH$_2$=CHCOOR, in which case the intermediate C radicals add repetitively to acrylate rather than recombine with the metalloradical as is the case for styrene.[26b]

Dienes

As we saw in Sections 5.2 and 5.3, butadiene and allene react with a variety of hydrides by 1,2 insertion, but butadienes also react with HMn(CO)$_5$ to give an apparent 1,4 insertion. Since this 18e hydride has no vacant site and CO dissociation is slow, an indirect mechanism must be operating; this is thought to be H atom transfer to give a 1,1-dimethylallyl radical that is subsequently trapped by the metal (Eq. 7.31).[26c] Only substrates that form especially stable radicals can react (e.g., 1,3-diene → allyl radical); CIDNP effects can be seen in such cases.[26d]

$$HMn(CO)_5 \longrightarrow \dot{M}n(CO)_5 \;+\; \longrightarrow \;-Mn(CO)_5$$

$$(7.31)$$

Alternating CO/Alkene Insertion

A series of cationic Pd catalysts such as $[(phen)PdMe(CO)]^+$ is known[27] to copolymerize CO with an alkene such as ethylene to give a strictly alternating copolymer, $(CH_2CH_2CO)_n$. The polymer is interesting from a practical standpoint because it has carbonyl functionality and so lends itself to useful chemical modification. The polymerization reaction is also of mechanistic interest because of the essentially perfect alternation of alkene and CO insertions that is involved.

$$LPdMe(CO)^+ \xrightarrow{\;C_2H_4\;} LPd(C_2H_4)(COMe)^+ \xrightarrow{\;C_2H_4\;}$$

$$LPd(CO)(CH_2CH_2COMe)^+$$

Of the possible erroneous insertions, double carbonyl insertion is forbidden for the thermodynamic reasons discussed in Section 7.1, and double alkene insertion is very rare because of the high affinity of the catalyst for CO together with the much slower intrinsic rate—by a factor of 2000 in a typical case—of alkene insertion versus CO insertion.

> - The alkyl normally migrates to the CO position as the first step in CO insertion.
> - Alkene insertion is kinetically favored into M−H versus M−R.

7.3   OTHER INSERTIONS

Sulfur dioxide is a strongly electrophilic species with a vacant orbital on sulfur, which it can use to attack even 18e metal complexes. Wojcicki and co-workers[28a] have studied these reactions in detail and find that the SO_2 can give electrophilic attack at the α carbon of the alkyl from the side opposite the metal, which leads to the formation of an alkyl sulfinate ion (RSO_2^-) with inversion at carbon. Since the anion has much of its negative charge on the oxygens, it is not surprising that the kinetic product of ion recombination is the O-bound sulfinato complex. On the other hand, the thermodynamic product is usually the S-bound sulfinate, as is appropriate for a soft metal (since S is softer than O). This sequence constitutes a 1,2- (if the sulfinate is O bound in the product) or a 1,1-insertion of SO_2 (if S bound). A notable feature of this mechanism is that the SO_2 does not need to bind to the metal, and so a 2e vacant site is not needed, and SO_2 can attack 18e complexes:

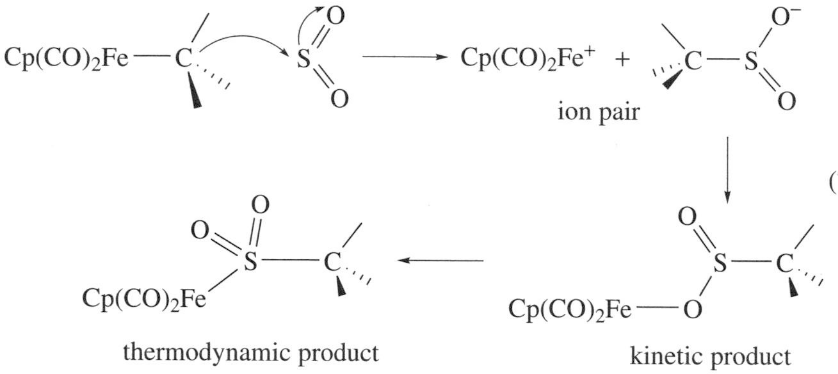

$$(7.32)$$

As expected for this mechanism, the reactivity falls off as the alkyl group becomes more bulky and as the substituents become more electron attracting. By carrying out a crossover reaction on a mixture of RS and SR isomers of $[CpFe^*(CO)L\{CH_2C^*H(Me)Ph\}]$, which is chiral at both Fe and the carbon shown, very little of the crossover products, the R,R and S,S isomers of the sulfinate complex, are seen. This shows that the postulated ion pairs must stay together, and that the intermediate iron cation also has stereochemical stability. Ion pairing is very common in organic solvents, and the ion pairs seem to have a well-defined solution structure.[28b] Ion pairing is probably quite general for organometallic compounds and can even change the reaction products formed.

An unusual insertion of SO_2 has been reported in a Pd methyl complex where the resulting $MeSO_2$ ligand bridges between the two metals.[29a]

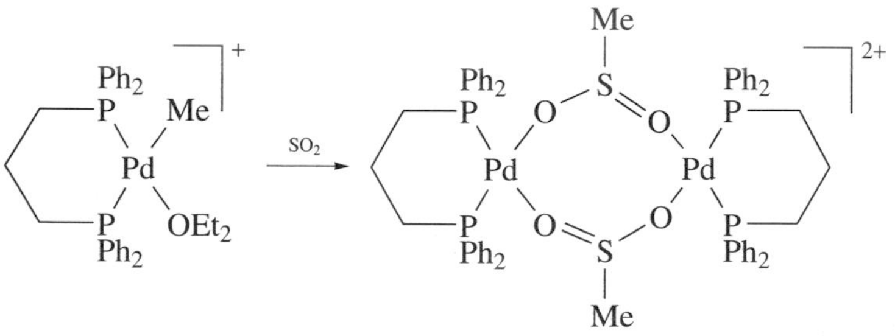

Goldberg and co-workers have described a rare example of O_2 insertion into a M—X bond; their case, involving Pt(IV)-H, is shown below:[29b]

$$Tp^*PtMe_2H + O_2 \rightarrow Tp^*PtMe_2(-O-O-H)$$

Insertions involving CO_2 are discussed in Section 12.3.

7.4   α, β, γ, AND δ ELIMINATION

β Elimination

As we saw in Chapter 3, β elimination is the chief decomposition pathway for alkyls that have β-H substituents. A 2e vacant site is required at the metal, and there has to be a roughly coplanar M−C−C−H arrangement that brings the β-H close to the metal. A complicating feature of this process is that the alkene often reinserts into the metal hydride, and this can give rise to isomerization of the alkene or of the starting alkyl, as we saw for hydrozirconation in Section 7.2. The alkene is rarely coordinated in the final products of a β elimination because it is usually displaced by the ligand that originally dissociated to open up a 2e vacant site at the metal, or by some other ligand in the reaction mixture. Rare cases are known in which both the alkyl and the alkene hydride can be observed directly:[30]

An 18e complex has to lose a ligand to open up a site for elimination (e.g., Eq. 7.34), but this process may[31] or may not[32] be rate limiting. In each case the addition of excess ligand inhibits the reaction by quenching the open site. Only if the elimination itself is rate limiting will we see a kinetic isotope effect for elimination of H over D (e.g., by comparing the rate of elimination of $L_nMC_2H_5$ vs. $L_nMC_2D_5$). The appearance of an isotope effect ($k_H > k_D$) implies that C−H(D) bond breaking is important in the slow step.

$$Cp(CO)LFe(n\text{-}Bu) \xrightarrow{-L} Cp(CO)Fe(n\text{-}Bu)(\square) \longrightarrow$$

$$Cp(CO)FeH(1\text{-}butene) \xrightarrow{+L} Cp(CO)LFeH + 1\text{-}butene \quad (7.34)^{32}$$

In 16e complexes, a 2e site is usually available, except for Pd(II), and especially for Pt(II), which tend to avoid the 18e configuration. Yamamoto and co-workers[33] found that *trans*-[PdL$_2$Et$_2$] complexes (L = 3° phosphine), tend to decompose by beta elimination directly from the 16e starting complex via an 18e transition state, but phosphine dissociation is often required for β elimination in d^8 alkyls such as [PtL$_2$Bu$_2$] (**7.7**).[34] The related metalacycle **7.8** β-eliminates 10^4-fold more slowly than **7.7**, presumably because a coplanar M−C−C−H arrangement is harder to achieve:[34]

$$n\text{-Bu} \diagdown \!\!\!\diagup\!\!\!\diagup\!\!\! \diagup L \\ \qquad Pt \\ L \diagup \quad \diagdown n\text{-Bu}$$

7.7

$$L_2Pt \bigcirc \longrightarrow L_2Pt \diagup\!\!\!\searrow_H \diagdown\!\!\!\diagdown \longrightarrow L_2Pt \; + \; \diagup\!\!\diagdown\!\!\diagup\!\!\diagdown$$ (7.35)

7.8

Grubbs and co-workers[35] have studied the analogous nickel complexes in the presence and absence of excess phosphine and have found that there are three decomposition pathways, one for each of the different intermediates, 14e, 16e, and 18e, that can be formed (Eq. 7.36).

$$L_3Ni \bigcirc \; \underset{}{\overset{-L}{\rightleftharpoons}} \; L_2Ni \bigcirc \; \underset{}{\overset{-L}{\rightleftharpoons}} \; LNi \bigcirc$$

$$\big\downarrow \text{red. clvg.} \qquad\qquad \big\downarrow \text{red. elim.} \qquad\qquad \big\downarrow \beta \text{ elim./ red. elim.} \qquad (7.36)$$

$$2C_2H_4 \qquad\qquad\qquad \square \qquad\qquad\qquad \text{butenes}$$

Alkoxide complexes readily undergo β elimination to give ketones or aldehydes, accounting for the ability of basic isopropanol to reduce many metal halides to hydrides with formation of acetone. β Elimination of amides and amines to imines also occurs but tends to be slow.[36]

α Elimination

If an alkyl has no β hydrogens, it may break a C–H bond in the α, γ, or δ position. The simplest case is a methyl group, which has no β hydrogens and can undergo only α elimination to give the methylene hydride. While the β process gives an alkene, a stable species that can dissociate from the metal, the methylene ligand formed from the α elimination is very unstable in the free state and so does not dissociate. Methylene hydride complexes are unstable with respect to the starting methyl complex, and so the products of α elimination can be intermediates in a reaction but are seldom seen as isolable species. For this reason, the α-elimination process is less well characterized than β elimination. Studies of both molybdenum and tantalum alkyls suggest that α elimination can be up to 10^6 times faster than β elimination even in cases in which both α- and β-H substituents are present.[37,38] In some cases, a coordinatively unsaturated methyl complex seems to be in equilibrium with a methylene hydride[39] species, which can sometimes be trapped, either by nucleophilic attack at the carbene carbon (Eq. 7.37) or by removing the hydride by reductive elimination with a second alkyl present on the metal (Eq. 7.38):[40]

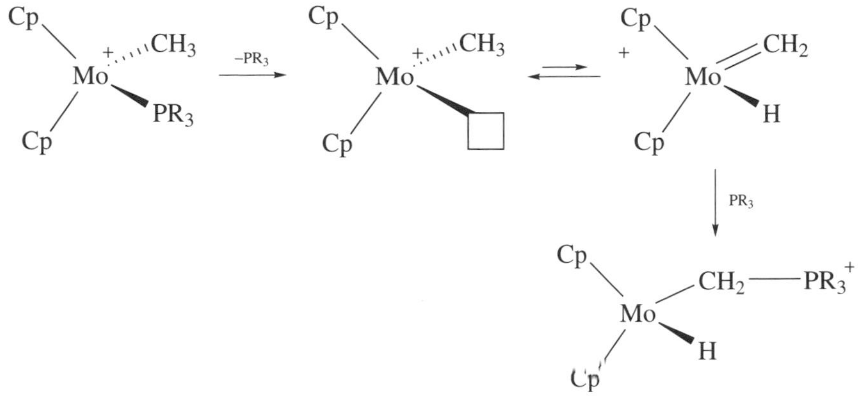

(7.37)

$$\text{TaCl}_2(\text{CH}_2\text{Ph})_3 \xrightarrow{\text{LiCp}^*} \text{Cp}^*\text{TaCl}(\text{CH}_2\text{Ph})_3 \xrightarrow{-\text{PhCH}_3} \text{Cp}^*\text{Ta}(=\text{CHPh})\text{Cl}(\text{CH}_2\text{Ph})$$

(7.38)

Schrock and co-workers[41a] have found an interesting case of α and β elimination taking place competitively in a tantalum complex, the two tautomers of which can be observed in solution by ^1H NMR.

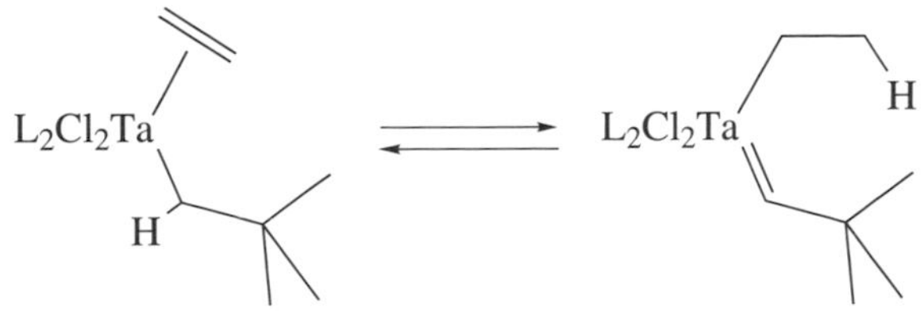

(7.39)

Because a heteroatom can strongly stabilize a carbene (Section 11.1), α elimination is strongly preferred to β elimination in Eq. 3.4. A photolytic α elimination via transfer of a hydrogen atom from carbon to an oxo group in CH_3-ReO_3 has been seen in a low-temperature matrix.[41b]

$$\text{CH}_3-\text{ReO}_3 \xrightarrow{h\nu} \text{CH}_2=\text{ReO}_2(\text{OH})$$

Other Eliminations

In addition to alkyls, a great variety of other ligands have no β-H but do have γ- or δ-Hs and can undergo γ or δ elimination to give cyclic products; some examples of these cyclometallation reactions are shown in Eqs. 7.40–7.42:

$$Ir(PMe_2Ph)_4^+ \xrightarrow{80°C} (Me_2PhP)_3Ir \underset{H}{\overset{Me_2P}{\diagdown}} \qquad (7.40)$$

$$MeMn(CO)_5 \xrightarrow{PhCH_2SMe} (CO)_4Mn \tag{7.41}$$

$$Ir(PMe_3)_4^+ \xrightarrow{LiCH_2(COMe)} Me_3P-Ir \underset{Me_3P}{\overset{PMe_3}{\diagup}} \xrightarrow{-PMe_3} \underset{Me_3P}{\overset{Me_3P}{\diagup}} Ir \diagdown H$$

$$(7.42)$$

All these elimination reactions can be thought of as being related to oxidative additions of a C−H bond to the metal. This is seen more clearly for β elimination if we write the metalacyclopropane (X_2) form of the alkene hydride product (Eq. 7.43), and for α elimination if we consider the X_2 form for the product carbene hydride (Eq. 7.44). Both γ and δ elimination are more obvious examples of oxidative addition.

$$\underset{M}{\diagup} \qquad \underset{H}{\diagdown} \longrightarrow \underset{M-H}{\triangleleft} \qquad (7.43)$$

$$M-C \overset{H_2}{\underset{H}{\diagdown}} \longrightarrow \underset{H}{\overset{M=CH_2}{|}} \qquad (7.44)$$

Neopentyl platinum compounds tend to decompose by γ elimination (Eq. 7.45), in contrast to the α elimination found for the Ta complexes shown in Eq. 7.39. This may imply that the mechanism in the two cases is different; for example, in the Ta case, a σ-bond metathesis is possible in which one alkyl might be deprotonated at the activated α-H by a second alkyl group, rather than undergo an oxidative addition of a C−H bond, which is more favorable for low-valent Pt.[42] Related examples of γ, δ, and ε elimination are shown in Eq. 7.46.

$$L_2Pt \xrightarrow{-CMe_4} L_2Pt \qquad (7.45)$$

$$(7.46)$$

23%   68%   9%

REFERENCES

1. F. Calderazzo, *Angew. Chem., Int. Ed.* **16**, 299, 1977.

2. A. D. Kovacs and D. S. Marynick, *J. Am. Chem. Soc.* **122**, 2078, 2000.

3. (a) F. Calderazzo, *Coord. Chem. Rev.* **1**, 118, 1966; (b) G. M. Whitesides et al., *J. Am. Chem. Soc.* **96**, 2814, 1974; (c) H. H. Brunner, I. Bernal, et al., *Organometallics* **2**, 1595, 1983.

4. A. Haynes et al., *J. Am. Chem. Soc.*, **121**, 11233, 1999.

5. A. Cutler et al., *Organometallics* **4**, 1247, 1985.

6. W. P. Giering et al., *J. Am. Chem. Soc.* **102**, 6887, 1980.

7. R. S. Bly, *Organometallics* **4**, 1247, 1985.

8. R. G. Bergman, *J. Am. Chem. Soc.* **103**, 7028, 1981.

9. G. Erker et al., *Angew. Chem. Int. Ed.* **17**, 605, 1978.

10. T. J. Marks et al., *J. Am. Chem. Soc.* **103**, 6959, 1981.

11. L. Marko et al., *J. Organomet. Chem.* **199**, C31, 1980.

12. (a) J. D. Atwood et al., *Organometallics* **4**, 402, 1985; (b) S. A. Macgregor and G. W. Neave, *Organometallics* **23**, 891, 2004 and refs. cited.

13. A. Yamamoto, *J. Am. Chem. Soc.* **107**, 3235, 1985.

14. A. Dormond, *Chem. Commun.* 749, 1984.

15. R. D. Adams, *Acct. Chem. Res.* **16**, 67, 1983.

16. (a) T. Ziegler et al., *J. Am. Chem. Soc.* **118**, 4434, 1996; (b) J. Chatt and B. L. Shaw, *J. Chem. Soc.* 5075, 1962; (c) J. B. Spencer et al., *J. Am. Chem. Soc.* **119**, 5257, 1997.

17. H. C. Clark, *J. Organomet. Chem.* **200**, 63, 1980.

18. R. S. Tanke and R. H. Crabtree, *J. Am. Chem. Soc.* **112**, 7984, 1990; *New J. Chem.* **27**, 771, 2003.

19. T. R. Lee et al., *Organometallics*, **23**, 1448, 2004.

20. R. H. Crabtree, *Acct. Chem. Res.* **12**, 331, 1979.

21. P. Wipf et al., *Pure Appl. Chem.* **70**, 1077, 1998.

22. (a) M. Brookhart et al., *J. Am. Chem. Soc.* **112**, 5634, 1990; **114**, 10349, 1992; (b) P. L. Watson and D. C. Roe, *J. Am. Chem. Soc.* **104**, 6471, 1982.

23. W. H. Knoth, *Inorg. Chem.* **14**, 1566, 1975.

24. S. A. Godleski et al., *Organometallics* **4**, 296, 1985; R. F. Jordan et al., *J. Am. Chem. Soc.* **115**, 4902, 1993.

25. R. G. Bergman et al., *J. Am. Chem. Soc.* **101**, 3973, 1979.

26. (a) J. L. Davidson et al., *J. Organometal. Chem.* **46**, C47, 1972; (b) B. B. Wayland et al., *Organometallics* **11**, 3534, 1992, and references cited therein; (c) V. A. Kormer et al., *J. Organometal. Chem.* **162**, 343, 1978; *Dokl. Acad. Nauk SSSR* **246**, 1372, 1979; (d) J. Kocher and M. Lehnig, *Organometallics*, **3**, 937, 1984.

27. A. Sen, *Acc. Chem. Res.* **26**, 303, 1993; M. Brookhart et al., *J. Am. Chem. Soc.* **118**, 4746, 1996; *Organometallics*, **20**, 5266, 2001.

28. (a) A. Wojcicki et al., *Inorg. Chem.* **12**, 717, 1973; *J. Am. Chem. Soc.* **95**, 6962, 1973; *Inorg. Chim. Acta* **10**, 229, 1974; (b) A. Macchioni, C. Zuccaccia, E. Clot, K. Gruet, and R. H. Crabtree, *Organometallics*, **20**, 2367, 2001; (c) A. Kovacevic, S. Gründemann, J. R. Miecznikowski, E. Clot, O. Eisenstein, and R. H. Crabtree, *Chem. Commun.*, 2550, 2002.

29. (a) M. Brookhart et al., *Chem. Commun.* 47, 2000; (b) K. I. Goldberg et al., *J. Am. Chem. Soc.* **121**, 11900, 1999.

30. F. N. Tebbe and G. W. Parshall, *J. Am. Chem. Soc.* **93**, 3793, 1971; T. R. Lee et al., *Organometallics*, **23**, 1448, 2004.

31. A. Yamamoto et al., *J. Organomet. Chem.* **120**, 257, 1976.

32. D. L. Reger and E. C. Culbertson, *J. Am. Chem. Soc.* **98**, 2789, 1976; M. R. Churchill, R. R. Schrock et al., *J. Am. Chem. Soc.* **100**, 647, 1978.

33. A. Yamamoto et al., *J. Am. Chem. Soc.* **102**, 6457, 1980.

34. G. M. Whitesides et al., *J. Am. Chem. Soc.* **94**, 5258, 1972; J. F. Hartwig et al., *J. Am. Chem. Soc.* **123**, 7220, 2001.

35. R. H. Grubbs et al., *Chem. Commun.* 864, 1977; *J. Am. Chem. Soc.* **99**, 3663, 1977; **100**, 1300, 2418, 7416, 7418, 1978.

36. J. F. Hartwig et al., *J. Am. Chem. Soc.* **118**, 3626, 1996; D. E. Fogg and B. R. James, *Inorg. Chem.*, **34**, 2557, 1995.

37. R. R. Schrock et al., *J. Am. Chem. Soc.* **119**, 11876, 1997.

38. G. Parkin, J. E. Bercaw, et al., *J. Mol. Catal.* **41**, 21, 1987.

39. J. R. Shapley et al., *Organometallics*, **19**, 761, 2000.

40. M. L. H. Green, *Pure Appl. Chem.* **50**, 27, 1978; R. R. Schrock, *Acc. Chem. Res.* **12**, 98, 1979.

41. (a) R. R. Schrock et al., *Organometallics*, **1**, 481, 1982; (b) L. J. Morris et al., *Chem. Commun.* 67, 2000.

42. G. M. Whitesides, *J. Am. Chem. Soc.* **103**, 948, 1981.

PROBLEMS

1. Predict the structures of the products (if any would be expected) from the following: (a) $CpRu(CO)_2Me + PPh_3$, (b) $Cp_2ZrHCl +$ butadiene, (c) $CpFe(CO)_2Me + SO_2$, (d) $Mn(CO)_5CF_3 + CO$.

2. Me$_2$NCH$_2$Ph reacts with PdCl$_2$ to give **A**; then **A** reacts with 2,2-dimethyl-cyclopropene and pyridine to give a mixture of **C** and **D**. Identify **A** and explain what is happening. Why is it that Me$_2$NPh does not give a product of type **A**, and that **A** does not insert ethylene.

C   **D**

3. In the pyrolysis of TiMe$_4$, both ethylene and methane are observed; explain.

4. Suggest mechanisms for the following:

5. The reaction of *trans*-PdAr$_2$L$_2$ (**A**, Ar = *m*-tolyl, L = PEt$_2$Ph) with MeI gives 75% of *o*-xylene (**B**) and 25% of 3,3′-bitolyl (**C**). Explain how these

products might be formed and list the possible Pd-containing products of the reactions. When the reaction was carried out with CD_3I in the presence of d_0-PdMeIL$_2$ (**D**), both d_0- and d_3-xylene (**B**) were formed. **A** also reacts with **D** to give **B** and **C**. How does this modify your view of the mechanism?

6. $[Cp^*Co\{P(OMe)_3\}Et]^+$ has an agostic interaction involving the β-H of the ethyl group. Draw the structure. It reacts with ethylene to form polyethylene. How might this reaction proceed? $RhCl_3$/EtOH and other late metal systems usually only dimerize ethylene to a mixture of butenes. Given that a Rh(I) hydride is the active catalyst in the dimerization, what mechanism would you propose? Try to identify and explain the key difference(s) between the two systems.

7. Design an alkyl ligand that will be resistant to β elimination (but not the ones mentioned in the text; try to be as original as possible). Design a second ligand, which may be an alkyl or an aryl-substituted alkyl, that you would expect to be resistant to β elimination but have a high tendency to undergo β-C—C bond cleavage. What products are expected?

8. Given the existence of the equilibrium

$$L_nM(Me)(CO) \rightleftharpoons L_nM(COMe)(solv)$$

how would you change L, M, and the solvent to favor (a) the right-hand side and (b) the left-hand side of the equation?

9. *trans*-PtCl(CH$_2$CMe$_3$)$\{P(C_5H_9)_3\}_2$ gives 1,1-dimethylcyclopropane on heating. What mechanism is most likely, and what Pt-containing product would you expect to be formed? If the neopentyl group is replaced by $-CH_2Nb$ (Nb = 1-norbornyl), then CH_3Nb is formed instead. What metal complex would you expect to find as the other product?

10. In mononuclear metal complexes, β elimination of ethyl groups is almost always observed, rather than α elimination to the ethylidene hydride $L_nM(=CHCH_3)H$. In cluster compounds, such as $HOs_3(CO)_{10}(Et)$, on the other hand, α elimination to give the bridging ethylidene $H_2Os_3(CO)_{10}(\eta^1, \mu_2$-CHCH$_3$) is observed in preference to β elimination. Suggest reasons for this difference.

8

NUCLEOPHILIC AND ELECTROPHILIC ADDITION AND ABSTRACTION

For a metal to bring about the reaction of two organic fragments, both of them generally have to be coordinated. In contrast, we see in this chapter how a metal can activate an unsaturated ligand so that direct attack of an external reagent can take place on the ligand without prior binding of the reagent to the metal.

Types of Reaction

The attacking reagent is normally either an electrophile or a nucleophile. Nucleophilic attack is favored when the metal fragment L_nM is a poor π base but a good σ acid, for example, if the complex bears a net positive charge or has electron-withdrawing ligands. In such a case, one of the ligands L' is depleted of electron density to such an extent that the nucleophile, Nu^- (e.g., LiMe, OH^-, etc.), can attack L'.

Electrophilic attack is favored when the metal is a weak σ acid but a strong π base, for example, if the complex has a net anionic charge, a low oxidation state, and ligands L that are good donors. The electron density of one of the ligands is enhanced by back donation so that it now becomes susceptible to attack by electrophiles, E^+ (H^+, MeI, etc.).

Two possible modes of nucleophilic or electrophilic attack are found. The reagent can become covalently attached to the ligand L' so that a bond is formed between the reagent and L'. In this case, the newly modified ligand stays on the metal and we have an *addition*. When the added electrophile is a proton, the reaction is normally considered as a protonation by an acid.

The Organometallic Chemistry of the Transition Metals, Fourth Edition, by Robert H. Crabtree
Copyright © 2005 John Wiley & Sons, Inc.

Alternatively, the reagent can detach a fragment from the ligand L' or even detach the entire ligand, in which case the modified reagent leaves the coordination sphere of the metal and we have an *abstraction*. A nucleophile abstracts a cationic fragment, such as H^+ or Me^+, while an electrophile abstracts an anionic fragment, such as H^- or Cl^-. When a nucleophile abstracts H^+, we normally consider the reaction as deprotonation by a base. Often, reaction with an electrophile generates a positive charge on the complex and prepares it for subsequent attack by a nucleophile. We will see examples of alternating attack in Eqs. 8.17, 8.20, and 8.31; the reverse order of addition is seen in Eq. 8.10.

Some examples are shown in Eqs. 8.1–8.9. You can see that the nucleophiles tend to reduce the hapticity of the ligands to which they add because they displace the metal from the carbon to which the addition takes place. In Eq. 8.2, we convert an η^5-L_2X into an η^4-L_2 ligand and make the net ionic charge on the complex one unit more negative, for a net change in the electron count of zero. In general, an L_nX ligand is converted to an L_n ligand, and an L_n ligand is converted to an $L_{n-1}X$ ligand. Electrophilic reagents, in contrast, tend to increase the hapticity of the ligand to which they add. Electrophilic attack on a ligand gives rise to a deficiency of electron density on that ligand, which is compensated by the attack of a metal lone pair on the ligand. For instance, in Eq. 8.7, an η^4-L_2 diene ligand becomes an η^5-L_2X pentadienyl ligand. At the same time, a net positive charge is added to the complex, which leaves the overall electron count unchanged. In general, an L_nX ligand is converted to an L_{n+1} ligand and an L_n ligand is converted to an L_nX ligand. Equations 8.3 and 8.4 show nucleophilic abstraction of H^+, which is simply ligand deprotonation. Nucleophilic abstraction of a methyl cation from Pt(IV) by iodide was the key step in the reductive elimination mechanism of Fig. 6.2.

1. Nucleophilic addition:[1,2]

$$Cl_2(py)Pt \longleftarrow \| \quad \xrightarrow{py} \quad Cl_2(py)\overset{-}{Pt} \diagup \diagdown \diagup \overset{+}{N} \quad \bigcirc \tag{8.1}$$

$$\tag{8.2}$$

2. Nucleophilic abstraction:[3–5]

$$Cp_2TaMe_2{}^+ + Me_3PCH_2 \longrightarrow Cp_2Ta(=CH_2)Me + Me_4P^+ \tag{8.3}$$

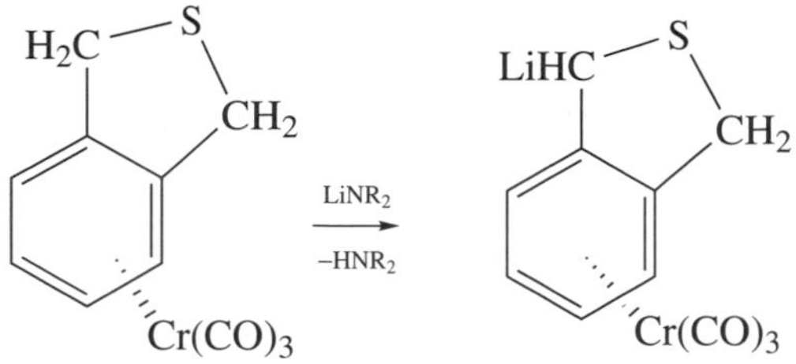

(8.4)

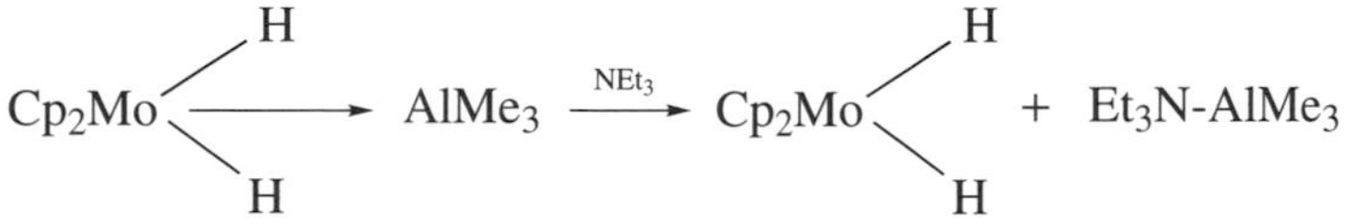

(8.5)

3. Electrophilic addition:[6,7]

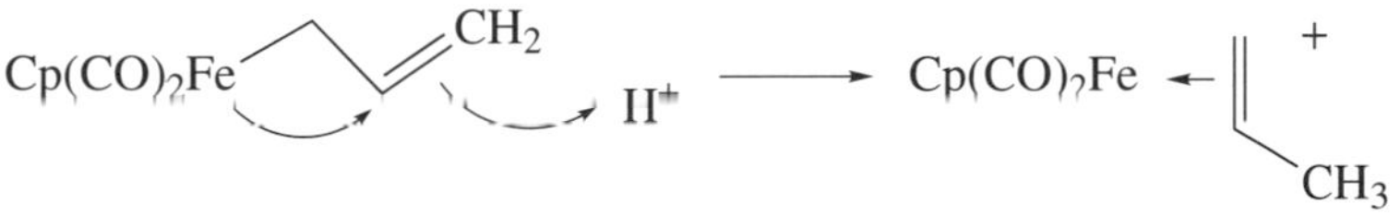

(8.6)

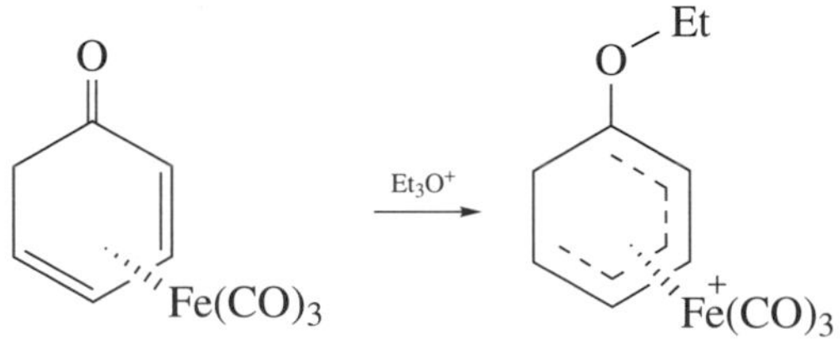

(8.7)

4. Electrophilic abstraction:[8,9]

$$\text{Cp}^*(\text{CO})_2\text{FeH} + [\text{Ph}_3\text{C}]\text{BF}_4 \longrightarrow [\text{Cp}^*(\text{CO})_2\text{FeFBF}_3] + \text{Ph}_3\text{CH} \quad (8.8)$$

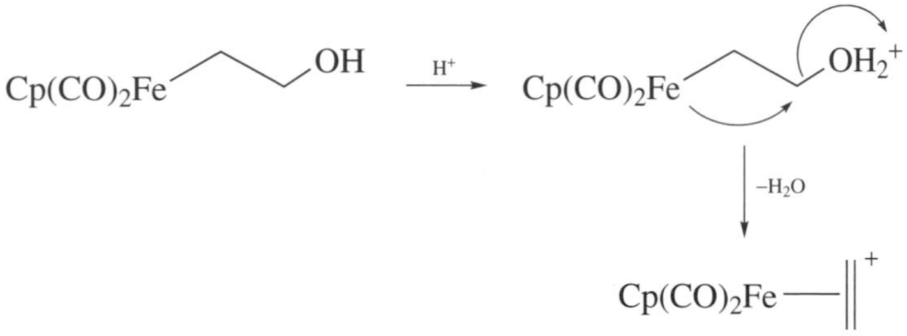

(8.9)

Attack at the metal, rather than at the ligands, is often observed. In the case of a nucleophile, this is simply associative substitution (Section 4.4) and can lead to the displacement of the polyene. If the original metal complex is 16e, attack may take place directly on the metal, if 18e, a ligand must usually dissociate first. A nucleophile is therefore more likely to attack a ligand, rather than the metal, if the complex is 18e. The pyridine in Eq. 8.1 is a potential 2e ligand, but it does not attack the metal because an 18e configuration is not a favorable situation for Pt(II). As we have seen, by attacking the ligand, the nucleophile does not increase the metal electron count.

For electrophilic attack, the situation is different. As a 0e reagent, an electrophile does not increase the electron count of the metal whether it attacks at the metal or at the ligand, and so attack at the metal is always a possible alternative pathway even for an 18e complex (except for d^0 complexes that have no metal-based lone pairs). Of course, large electrophiles, such as Ph_3C^+, may have steric problems in attacking the metal directly.

As 1e reagents, organic free radicals can also give addition and abstraction reactions, but these reactions are less well understood. Radical addition and abstraction also tends to occur only as part of a larger reaction scheme in which radicals are formed and quickly react (e.g., Section 16.2).

The attack of nucleophiles at the metal has been discussed under substitution in Chapter 4; we also looked at the attack of electrophiles and of radicals at the metal in connection with oxidative addition in Chapter 6.

8.1 NUCLEOPHILIC ADDITION TO CO

CO is very sensitive to nucleophilic attack when coordinated to metal sites of low π basicity. On such a site, the CO carbon is positively charged because L-to-M σ donation is not compensated by M-to-L back donation, and the CO π^* orbitals are open to attack by the nucleophile. Nucleophilic lithium reagents convert a number of metal carbonyls to the corresponding anionic acyls. The net negative charge now makes the acyl liable to electrophilic addition to the acyl oxygen to give the Fischer (heteroatom-stabilized) carbene complex, **8.1**.[10]

$$Fe(CO)_5 \xrightarrow{\text{LiNEt}_2} (CO)_4Fe=C\begin{smallmatrix} \nearrow NEt_2 \\ \searrow OLi \end{smallmatrix} \xrightarrow{\text{Me}_3O^+} (CO)_4Fe=C\begin{smallmatrix} \nearrow NEt_2 \\ \searrow OMe \end{smallmatrix} \quad (8.10)$$

8.1

The cationic charge on $[Mn(CO)_6]^+$ makes it much more sensitive to nucleophilic attack than is $[Mo(CO)_6]$. In this case, hydroxide, or even water, can attack coordinated CO to give an unstable metalacarboxylic acid intermediate. These decompose to CO_2 and the metal hydride by β elimination. This can be synthetically useful as a way of removing a CO from the metal, something that is difficult to do in other ways because CO can be one of the most tightly bound ligands.

$$(8.11)$$

The nucleophilic attack of methanol instead of water can give a metalaester, $L_nM(COOR)$, which is stable because it has no β-H.

Note how the displacement of Cl^- is favored in the first step of Eq. 8.12 over displacement of PPh_3. This is a consequence of the polar solvent used and sets the stage for the subsequent nucleophilic attack by putting a positive charge on the complex ion, which activates the CO. Acid can reverse the addition reaction by protonating the methoxy group, which leads to loss of methanol. This is, of course, a methoxide abstraction reaction and is an example of a nucleophilic addition being reversed by a subsequent electrophilic abstraction. This is common and means that the product of an addition reaction may even decompose via its inverse reaction if unsuitable workup conditions are used. For example, the product of a nucleophilic addition may revert to the starting material if excess acid is added to the reaction mixture with the object of neutralizing the excess nucleophile:[11]

$$(PR_3)_2PtCl_2 \xrightarrow{\text{CO}} [(PR_3)_2PtCl(CO)]^+Cl^-$$

$$\xrightarrow{\text{MeOH, Et}_3\text{N}} [(PR_3)_2PtCl(COOMe)] + (Et_3NH)Cl \quad (8.12)$$

We saw in Chapter 4 that Et_3NO is an excellent reagent for removing coordinated COs from 18e metal complexes.[12] Very nucleophilic oxygen ($Et_3N^+\text{-}O^-$) is capable of attacking the CO carbon to give a species that can break down to Et_3N, CO_2, and the corresponding 16e metal fragment (Eq. 8.13). Note how the cis-disubstituted product is obtained selectively in Eq. 8.13 because a CO trans to another CO has less back donation from the metal and hence is more activated toward nucleophilic attack at carbon than is the CO trans to the weak π-acid PR_3. Unfortunately, the amine formed can sometimes coordinate to the metal if no better ligand is available. A second problem with the method is that successive carbonyls become harder and harder to remove as the back bonding to the remaining CO groups increases because their sensitivity to nucleophilic attack decreases, and so we are usually unable to remove more than one CO in this way.

$$(8.13)$$

Isonitrile complexes are more easily attacked by nucleophiles than are CO complexes, but isonitriles tend to bind to higher oxidation state metals where back donation is less effective; the final product is a carbene.[13]

$$(8.14)$$

- Nucleophilic attack at a ligand is favored when the metal is a weak back bonder, and electrophilic attack when the metal is a strong back bonder.
- In an addition, the reagent stays in the coordination sphere.
- In an abstraction, the reagent abstracts a ligand (or ligand fragment) from the coordination sphere.

8.2 NUCLEOPHILIC ADDITION TO POLYENE AND POLYENYL LIGANDS

Simple polyenes in the free state, such as benzene and ethylene, normally undergo electrophilic, not nucleophilic, attack. It is a measure of the power of complexation to alter the chemical character of a group that both of these polyenes, as ligands, become sensitive to nucleophilic, and inert to electrophilic, attack. This reversal of the chemical character of a compound is known as *umpolung*. If we are interested in inhibiting electrophilic attack, we would regard the metal as a protecting group. On the other hand, if we are interested in promoting nucleophilic attack, we would regard the same metal fragment as an activating group.

In the vast majority of cases, the nucleophile adds to the face of the polyene opposite to the metal. Since the metal is likely to have bound to the least hindered face of the free polyene, we may therefore see a selective attack of the nucleophile on what was the more hindered face in the free polyene; this is often useful in organic synthetic applications.

Green–Davies–Mingos Rules

It is not unusual for a single complex to have several polyene or polyenyl ligands, in which case we often see selective attack at one site of one ligand only. Green, Davies, and Mingos[14] noticed certain patterns in these reactions and from them devised a set of rules that usually allow us to predict the site of addition:

Rule 1 Polyenes (even or L_n ligands) react before polyenyls (odd or L_nX ligands).

Rule 2 Open ligands react before closed.

Rule 3 Open polyenes: terminal addition in all cases. Open polyenyls: usually terminal attack, but nonterminal if L_nM is electron donating.

Rule 1 takes precedence over rule 2 whenever they conflict. Polyenes or even ligands are simply ones having an even electron count on the covalent model (e.g., η^2-C_2H_4, η^6-C_6H_6); odd ligands have an odd electron count (e.g., η^3-C_3H_5, η^5-C_5H_5). Closed ligands are ones like Cp in which the coordinated π system of the polyene or -enyl is conjugated in a ring; in open ligands like allyl, the conjugation is interrupted. Some ligands and their classification according to these rules are illustrated in **8.2–8.5**:

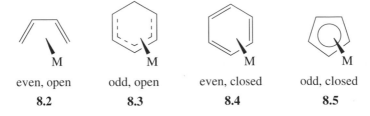

| even, open | odd, open | even, closed | odd, closed |
| **8.2** | **8.3** | **8.4** | **8.5** |

Diagrams **8.6–8.13** show these rules in action. In **8.6**, addition of a variety of nucleophiles takes place at the arene ring (indicated by the arrow in the diagram), as predicted by rule 1. A second nucleophile can also add, but to the other ring, as predicted by rule 1. Diagram **8.7** shows that addition takes place to the even, open butadiene ligand, rather than to the even, closed arene (rule 2) and at the terminal position (rule 3). In **8.8**, we see that the even closed arene is attacked rather than the odd open allyl; we must be careful in a case such as this to apply rule 1 before rule 2. Diagram **8.9** shows a rare example of attack at a Cp ring; as an odd closed system, this only happens if there is no other π-bonded ligand present. Cp is a stabilizing ligand in studies on nucleophilic attack because Cp is usually very resistant to attack and therefore directs addition to other ligands present on the metal.

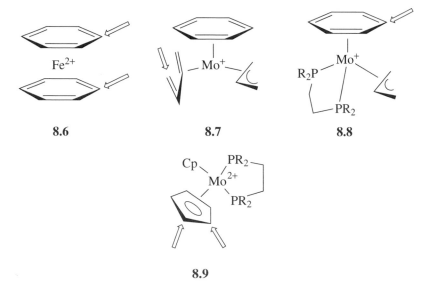

8.6	**8.7**	**8.8**

8.9

In **8.10**, we can treat the alkene and the allyl parts of the bicyclooctadienyl as independent entities; the even alkene part is attacked. CO is an even ligand but is among the least reactive of these, as shown in **8.11** and **8.13**. The examples also illustrate what might be called the "zeroth rule" of nucleophilic addition: A nucleophile usually adds once to a monocation (e.g., **8.7** and **8.8**) twice to a dication (e.g., **8.6** and **8.9**), and so on.

8.10	**8.11**	**8.12**	**8.13**

Although the rules were first developed empirically, an MO study has shown that they often successfully predict the location of the atom having the highest coefficient of the LUMO. Under kinetic control, we would expect addition at the point where this empty acceptor orbital is largest. Qualitatively, we can understand the rules as follows. Ligands having a higher X character will tend to be more negatively charged and therefore will tend to resist nucleophilic attack relative to L ligands. The coordinated allyl group, as an L_2X ligand, has more anionic character than ethylene, an L ligand. This picture even predicts the relative reactivity of different ligands in the same class, a point not covered by the rules. For example, it is found that pentadienyl (L_2X) reacts before allyl (LX); we can understand this because the former has the lower X character. Ethylene reacts before butadiene; as we saw in Section 5.3, the LX_2 form is always a significant contributor to the structure of butadiene complexes.

The reason the terminal carbons of even open ligands are the sites of addition is that the coefficients of the LUMO are larger there. As an example, look at ψ_3 in butadiene as depicted in Fig. 5.2. An odd, open polyenyl gives terminal addition only if the metal is sufficiently electron withdrawing. Reference to the MO picture for the allyl group (Fig. 5.1) will show that the usual LUMO, ψ_2, has a large coefficient at the terminus, but ψ_3 has a large coefficient at the central carbon. As we go to a less electron-withdrawing metal, we tend to fill ψ_2 and to the extent that ψ_3 becomes the new LUMO, and so we may no longer see terminal attack. An example of nonterminal attack in an allyl is shown by $[Cp_2W(\eta^3-C_3H_5)]^+$ (Eq. 8.15)—as a d^2 fragment, Cp_2W is strongly electron donating in character.

$$(8.15)$$

It is surprising that these simple rules do so well in most cases. The situation can sometimes be much more complicated, however, as shown by Eq. 8.16.[15] Here, the methoxide ion attacks at every possible site, as the mixture is warmed from $-80°C$ to room temperature. Initially, addition takes place at the metal (which must be preceded by a decrease in the hapticity of the cycloheptatrienyl to generate an open site) and later at the CO and C_7H_7 sites. Had the reaction been carried out above $0°C$, the normal product would have been observed, and the complications would have escaped detection.

Substituents on an arene tend to direct addition in the way one might expect. Electron-releasing substituents usually direct attack meta, and electron-attracting ones, Q, ortho rather than para, perhaps because that puts Q at the terminus of the conjugated system of the resulting open polyenyl.

The arene chromium tricarbonyls have been studied intensively[2,4] with regard to their reactions with nucleophiles.

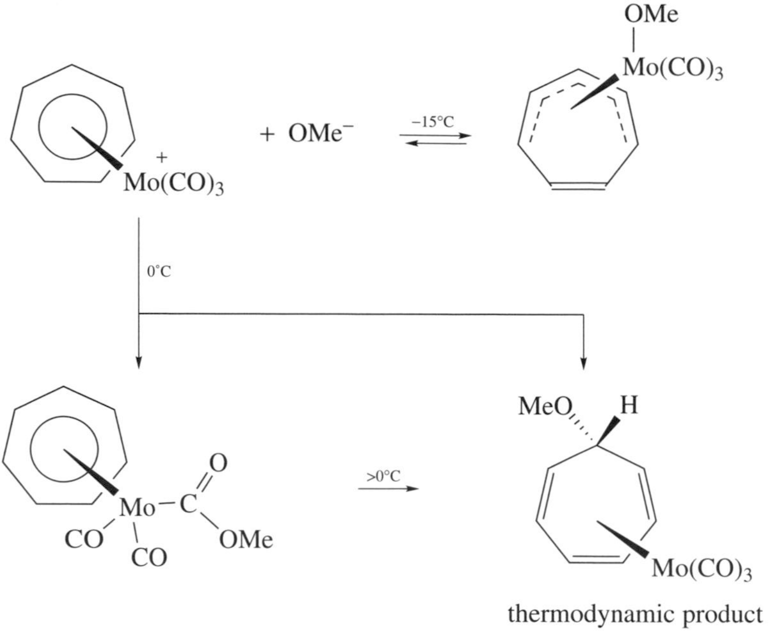

(8.16)

thermodynamic product

Cyclohexadienyl complexes react with nucleophiles to give 1,3-diene complexes.[16] An example is shown in Eq. 8.17; the arrow refers to the point of attachment of the nucleophile to the polyene ligand. The synthesis of the starting complex by an electrophilic abstraction is also shown; this activates the ligand for nucleophilic attack. Once again, directing effects can be used to advantage: A 2-OMe substituent directs attack to the C-5 position of the cyclohexadienyl, for example.[17]

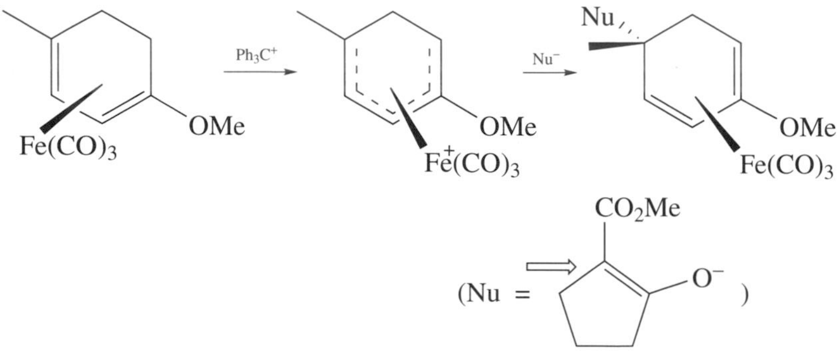

(8.17)

Diene complexes give allyls on nucleophilic attack. Note how the cisoid conformation of the butadiene in Eq. 8.18 gives rise to an anti methylallyl (in the nomenclature of allyl complexes, a substituent is considered as syn or anti with respect to the central CH proton).[18] Equation 8.19 is interesting in that the amine acts in this case as a carbon, not as a nitrogen nucleophile.[19]

$$(Ind)(CO)_2Mo^+ \longrightarrow \quad \xrightarrow{\text{H}^-} \quad (Ind)(CO)_2Mo \longrightarrow \bigg\rangle \quad \text{CH}_3 \qquad (8.18)$$

$$(8.19)$$

η^3-Allyls are also readily attacked. Note how **8.14** in Eq. 8.20 is activated toward nucleophilic attack by substituting the bromide ion with CO,[20] which gives the complex a net positive charge. The nucleophile adds selectively on the end of the allyl cis to NO.

8.14   (Nu = OH⁻, D⁻)

$$(8.20)$$

This gives control over the stereochemistry of the product because **8.14** can be resolved, thanks to the presence of the optically active group (R*) on the Cp ring, in which case carrying out the addition with one enantiomer of the metal complex means that the new asymmetric center on the ligand is formed with very high asymmetric induction. This reaction therefore constitutes a chiral synthesis of the alkenes shown.

Wacker Process

Alkene complexes undergo nucleophilic attack to give metal alkyls, which can often rearrange to give other products. This is the basis of an important industrial process, the Wacker process, now used to make about 4 million tons a year of aldehydes from alkenes. The fact that aqueous $PdCl_2$ oxidizes ethylene to acetaldehyde had been known[21] (although not understood) since the nineteenth century; the reaction consumes the $PdCl_2$ and deposits metallic Pd(0). It took considerable imagination to see that such a reaction might be useful on an industrial scale because $PdCl_2$ is far too expensive to use as a stoichiometric reagent in the synthesis. The key is catalysis, which allows the Pd to be recycled almost indefinitely. J. Smidt[22] of Wacker Chemie realized in the late 1950s that it is possible to intercept the Pd(0) before it has a chance to precipitate by using $CuCl_2$, which reoxidizes the palladium and is itself reduced to cuprous chloride. This is air sensitive and is reoxidized back to Cu(II). The resulting set of reactions (Eq. 8.21) are an elegantly simple solution to the problem and resemble the coupled reactions of biochemical catalysis.

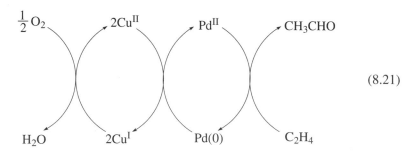

$$(8.21)$$

Later mechanistic work revealed the following rate equation:

$$\text{Rate} = \frac{k[PdCl_4^{2-}][C_2H_4]}{[Cl^-]^2[H^+]} \tag{8.22}$$

Equation 8.22 implies that the complex, in going from its normal state in solution, $PdCl_4^{2-}$, to the transition state of the slow step of the reaction has to gain a C_2H_4 and lose two Cl^- ions and a proton. It was originally argued that the proton must be lost from a coordinated water, and so $[Pd(OH)(C_2H_4)Cl_2]^-$ was invoked as the key intermediate; it was assumed that this might undergo olefin insertion into the Pd$-$OH bond, or the OH might attack the coordinated ethylene as a nucleophile. The resulting hydroxyethyl palladium complex might β-eliminate to give vinyl alcohol, CH_2=CHOH, which is known to tautomerize to acetaldehyde.

In fact, this mechanism is wrong, something that was only discovered 20 years later as a result of stereochemical work by Bäckvall[23] and by Stille.[24a] According to the original intramolecular mechanism, whether the reaction goes by insertion or by nucleophilic addition from a coordinated OH, the stereochemistry at each carbon of the ethylene should remain unchanged. This can be tested if we use cis- or trans-CHD=CHD as the alkene and trap the intermediate alkyl. We have to trap the alkyl because the rearrangement to acetaldehyde destroys the stereo-chemical information. Equation 8.23 shows one way of trapping the alkyl, using CO. You can see that if the hydroxyethyl is carbonylated, the OH group can curl back and effect a nucleophilic abstraction on the acyl to give a free lactone, the stereochemistry of which can be determined by a number of methods, including NMR and microwave spectroscopy. In fact, the stereochemistry of the two car-bons in the product is not the same as that of the starting material, which rules out the older mechanisms.

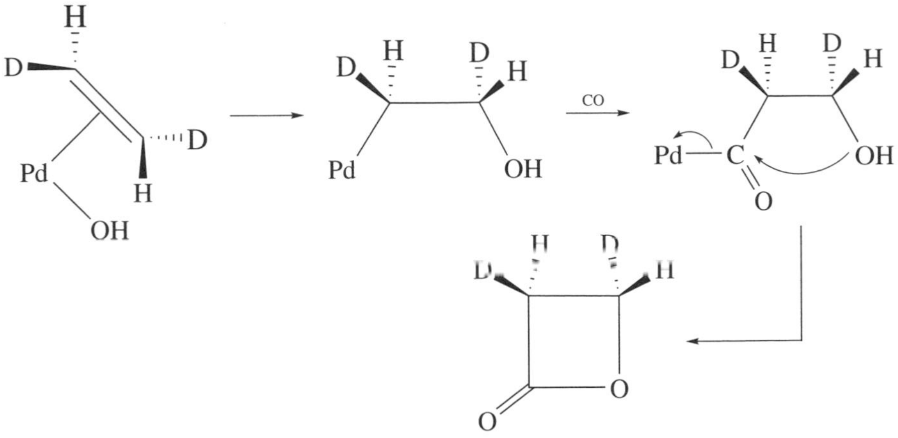

(8.23)

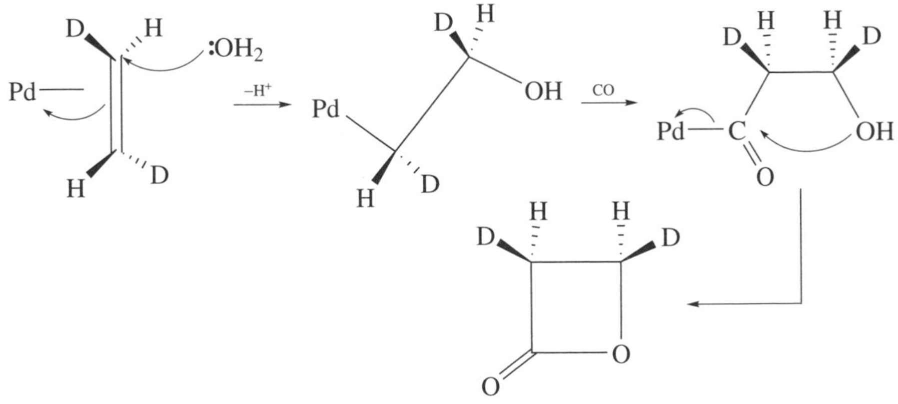

(8.24)

The currently accepted mechanism involves attack of a free water molecule from the solvent on the coordinated ethylene. Equation 8.24 shows how this inverts the stereochemistry at one of the carbons, as opposed to the old insertion mechanism (Eq. 8.23).

The loss of two Cl^- ions removes the anionic charge from the metal, which would otherwise prevent nucleophilic attack. Equations 8.25–8.28 show the sequence of events as now understood. This mechanism implies that an $[H_2O]^2$ term should be present in the rate equation, and if it could have been seen, the mechanistic problem would have been solved earlier, but one cannot normally alter the concentration of a solvent and get meaningful rate data because changing the solvent composition leads to unpredictable solvent effects on the rate.

$$\begin{array}{ccc}
\text{Cl} & & \\
\text{Cl}\!\!-\!\!\overset{\text{Cl}}{\underset{\text{Cl}}{\text{Pd}}}\!\!-\!\!\text{Cl} & \rightleftharpoons & \text{Cl}\!\!-\!\!\overset{\text{Cl}}{\underset{\text{Cl}}{\text{Pd}}}\!\!-\!\!\|
\end{array} \qquad (8.25)$$

$$\text{Cl}\!\!-\!\!\overset{\text{Cl}}{\underset{\text{Cl}}{\text{Pd}}}\!\!-\!\!\| \quad \overset{H_2O}{\rightleftharpoons} \quad H_2O\!\!-\!\!\overset{\text{Cl}}{\underset{\text{Cl}}{\text{Pd}}}\!\!-\!\!\| \qquad (8.26)$$

$$H_2O\!\!-\!\!\overset{\text{Cl}}{\underset{\text{Cl}}{\text{Pd}}}\!\!-\!\!\| \quad \overset{H_2O}{\underset{-H^+}{\longrightarrow}} \quad H_2O\!\!-\!\!\overset{\text{Cl}}{\underset{\text{Cl}}{\text{Pd}}}\!\!-\!\!\text{OH} \qquad (8.27)$$

$$H_2O\!\!-\!\!\overset{\text{Cl}}{\underset{\text{Cl}}{\text{Pd}}}\!\!-\!\!\text{OH} \quad \overset{slow}{\longrightarrow} \quad CH_3CHO + Pd(0) + 2Cl^- + H^+ \qquad (8.28)$$

No deuterium is incorporated into the acetaldehyde when the reaction is carried out in D_2O, which would happen if vinyl alcohol were released. This must be another case in which the β-elimination product never leaves the coordination sphere of the metal until it has had time to rearrange on the metal by multiple insertion–elimination steps:

$$Pd\diagup\!\!\!\diagdown\text{OH} \longrightarrow Pd\diagup\overset{H}{\diagdown}\text{OH} \longrightarrow Pd\diagdown\overset{}{\underset{H}{\diagup}}O \qquad (8.29)$$

$$\longrightarrow \diagup\!\!\diagdown\overset{O}{\|} + Pd\!\!-\!\!H$$

Nucleophilic addition to alkyne complexes gives vinylmetal species. A particularly interesting variant of this reaction is addition to an 18e complex containing a 4e alkyne (see Section 5.1), written in its bis-carbene form in Eq. 8.30. If the product were a η^1-vinyl, then the complex would be 16e, and so a 3e η^2-vinyl is usually found:[24b]

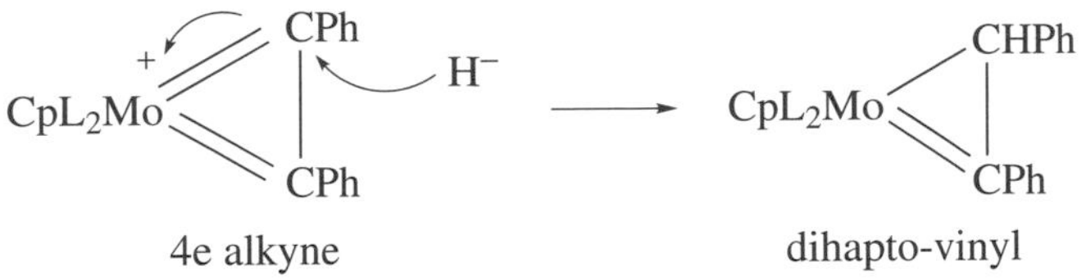

$$\begin{array}{ccc}
\text{CpL}_2\text{Mo} \overset{\text{CPh}}{\underset{\text{CPh}}{}} + \text{H}^- & \longrightarrow & \text{CpL}_2\text{Mo} \overset{\text{CHPh}}{\underset{\text{CPh}}{}}
\end{array} \qquad (8.30)$$

$$\qquad\qquad \text{4e alkyne} \qquad\qquad\qquad \text{dihapto-vinyl}$$

- The position of nucleophilic attack is usually that predicted by Green–Davies–Mingos rules.
- The Wacker process (Eq. 8.21) involves nucleophilic attack at coordinated C_2H_4.

8.3 NUCLEOPHILIC ABSTRACTION IN HYDRIDES, ALKYLS, AND ACYLS

Hydrides

Deprotonation of a metal hydride can produce a nucleophilic metal anion. For example, $ReH_7L_2(L_2 = dppe)$ does not lose H_2 easily as does the $L = PPh_3$ complex. To generate the ReH_5L_2 fragment for the dppe case, Ephritikhine and Felkin[24c] first formed the anion with BuLi and made the intermediate methyl hydride with MeI (Eq. 8.31). The driving force for methane loss is higher than for H_2 loss, and the required fragment was formed and intercepted with cyclopentadiene to give the unusual polyhydride $CpReH_2L_2$.

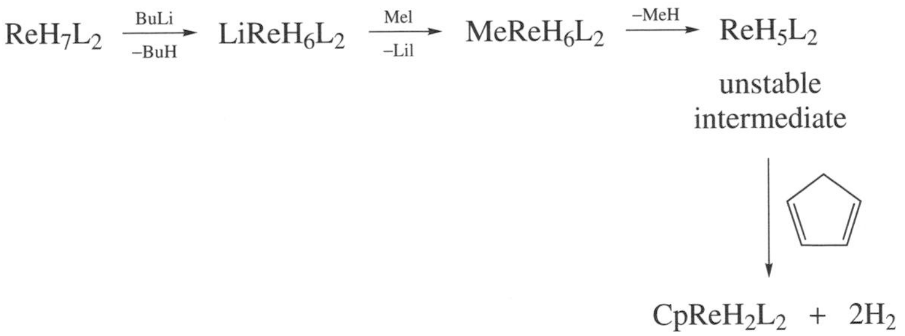

$$ReH_7L_2 \xrightarrow[-BuH]{BuLi} LiReH_6L_2 \xrightarrow[-LiI]{MeI} MeReH_6L_2 \xrightarrow{-MeH} ReH_5L_2$$

$$\text{unstable} \\ \text{intermediate}$$

$$(8.31)$$

$$CpReH_2L_2 \; + \; 2H_2$$

Alkyls and Acyls

Alkyl groups can be exchanged between metals with inversion at carbon. This transmetalation reaction provides a route for the racemization of a metal alkyl during the early stages of an oxidative addition reaction, while there is still some of the low-valent metal left in the reaction mixture. In the case shown in Eq. 8.32, exchange of a $(CR_3)^+$ fragment between the metals oxidizes the Pd(0) nucleophile to Pd(II), and reduces the Pd(II) complex to Pd(0). Mechanistic interpretation of the stereochemical outcome of an oxidative addition can be clouded by exchange reactions such as the one shown. Nucleophilic abstraction of acyls is particularly useful in organic synthetic applications, as we shall see in Chapter 14.

$$(8.32)$$

Acyls are very readily abstracted by nucleophiles, as in the last step of Eqs. 8.23 and 8.24. As in the abstraction of Eqs. 8.32, the reaction goes with reduction of the metal by two units, so a Pd(II) acyl is ideal because the Pd(0) state is easily accessible.

The recurrence of Pd(II) in this chapter is no accident. It has a very high tendency to encourage nucleophilic attack at the ligands in its complexes. As an element on the far-right-hand side of the transition metal block, it is very electronegative (Pauling electronegativity: 2.2), and its d orbitals are very stable. This means that polyene to metal electron donation is more important than metal d_π to polyene π^* back donation, and so the polyene is left with a net positive charge.

8.4 ELECTROPHILIC ADDITION

As a zero-electron reagent, an electrophile such as H^+ or Me^+ can attack the ligand, or the M−L bonds, or the metal—even in an 18e complex. Particularly in the case of the proton, initial attack may occur at one site, followed by rearrangement with transfer to a second site, so the location of the electrophile in the final product may be misleading. Electrophilic addition to metal complexes can therefore be mechanistically complex;[25] it is also less easily controllable and less often used than nucleophilic addition.

Addition to the Metal

Oxidative addition by the S_N2 or by the ionic mechanisms involves electrophilic addition to the metal (Eq. 8.33 and Sections 6.2 and 6.4) in the first step:

$$L_nM + MeI \xrightarrow{\text{electrophilic addition}} [L_nMMe]^+I^- \xrightarrow{-L} [L_{(n-1)}MMeI] \qquad (8.33)$$

In some cases the second step does not take place, and the counterion never binds to the metal. This makes the reaction an electrophilic addition, rather than an oxidative addition to the metal, although the latter term is sometimes seen in the literature to describe this type of reaction. An example is the reaction of the highly nucleophilic Co(I) anion, $[Co(dmg)_2py]^-$, with an alkyl triflate, a reaction known to go with inversion. Protonation of metal complexes to give metal hydrides is also very common (Eqs. 3.30–3.31).

The addition of any zero-electron ligand to the metal can be regarded as an electrophilic addition: $AlMe_3$, BF_3, $HgCl_2$, Cu^+, and even CO_2, when it binds in an η^1 fashion via carbon, can all act in this way. Each of these reagents has an empty orbital by which it can accept a d-type lone pair from the metal. Since the acceptor atoms of these ligands are generally more electronegative than the metal, the metal is conventionally regarded as being oxidized by two units for each 0e ligand that binds. So, for example, $Cp_2H_2W \rightarrow AlMe_3$ is conventionally a W(VI) complex because $AlMe_3$ is formally removed as the closed-shell $AlMe_3{}^{2-}$. Complex formation of this type is more likely the more basic the starting complex, and the more powerful the Lewis acid.

Chung and co-workers[26] result of Eq 8.34 shows how a conventional deprotonation, followed by a nucleophilic attack, leaves the metal open to electrophilic attack by the carbonyl carbon in the last step.

$$(8.34)$$

Addition to a Metal–Ligand Bond

Protonation reactions are the most common. For example, a hydride can give a dihydrogen complex.[27]

$$[Pt(diphos)_2] \xrightarrow{H^+} [HPt(diphos)_2]^+ \xrightarrow{H^+} [(H_2)Pt(diphos)_2]^{2+}$$

Early metal alkyls such as Cp_2TiMe_2 are readily cleaved by acid to liberate the alkane; a transient alkane complex is the probable intermediate in such cases. Protonation of the alkene complex shown below can occur by two simultaneous paths: (1) direct protonation at the metal and (2) initial protonation at the alkene followed by β elimination. Path 2 leads to incorporation of label from DCl into the alkene ligands.[25]

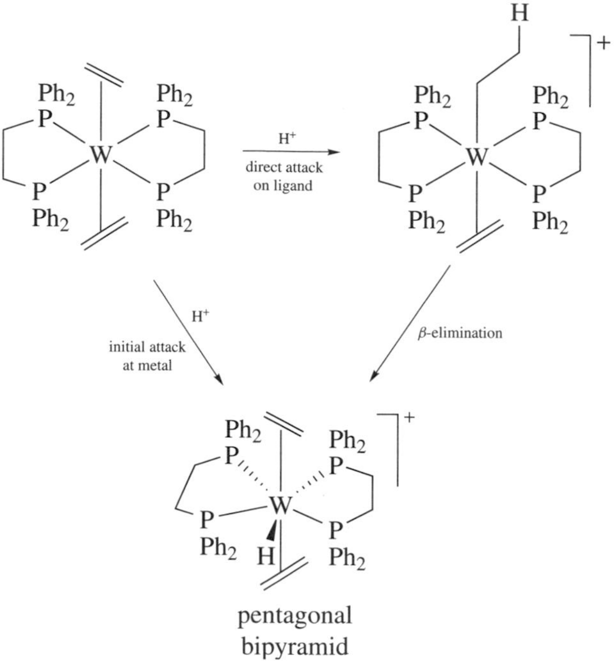

pentagonal
bipyramid

Addition to Ligand

Simple addition to the ligand occurs in protonation of Cp_2Ni, as shown by the exo attack and lack of scrambling of the deuterium label.

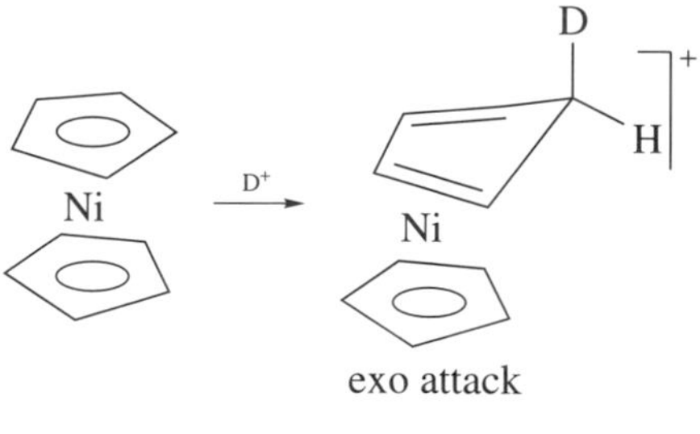

exo attack

Unlike the nucleophilic case, where exo attack is the normal rule, for electrophiles an endo addition is also possible as a result of initial attack of the electrophile at the metal, followed by transfer to the endo face of the ligand. This tends to occur where the electrophile is soft like $Hg(OAc)_2$. Exo-proton abstraction by OAc^- completes the sequence.

For a hard electrophile such as CH_3CO^+, direct exo attack at the ligand tends to be seen, as in acetylation of ferrocene. The preference for an exo-proton abstraction means that in an exo attack of the electrophile, the endo-deuterium has to be transferred to the endo position of the other ring. This leads to loss

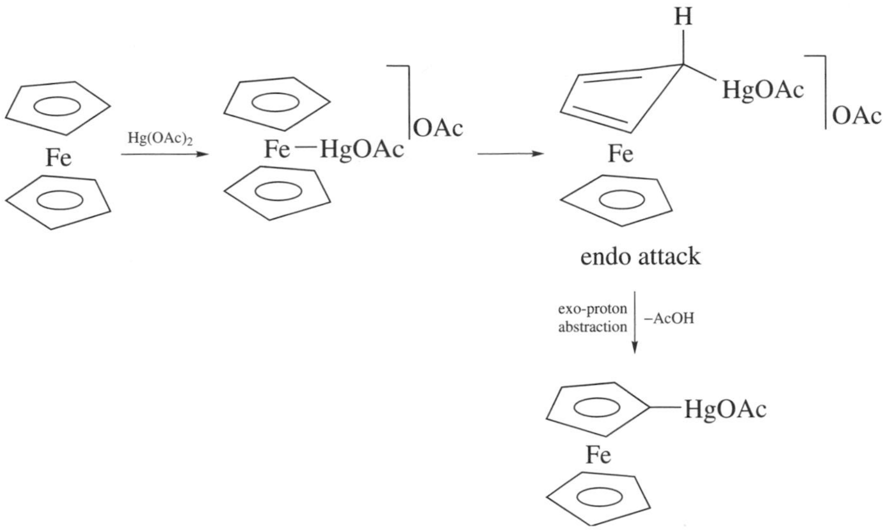
Fe
Hg(OAc)2
Fe—HgOAc
OAc
H
HgOAc
OAc
Fe
endo attack
exo-proton abstraction
−AcOH
HgOAc
Fe

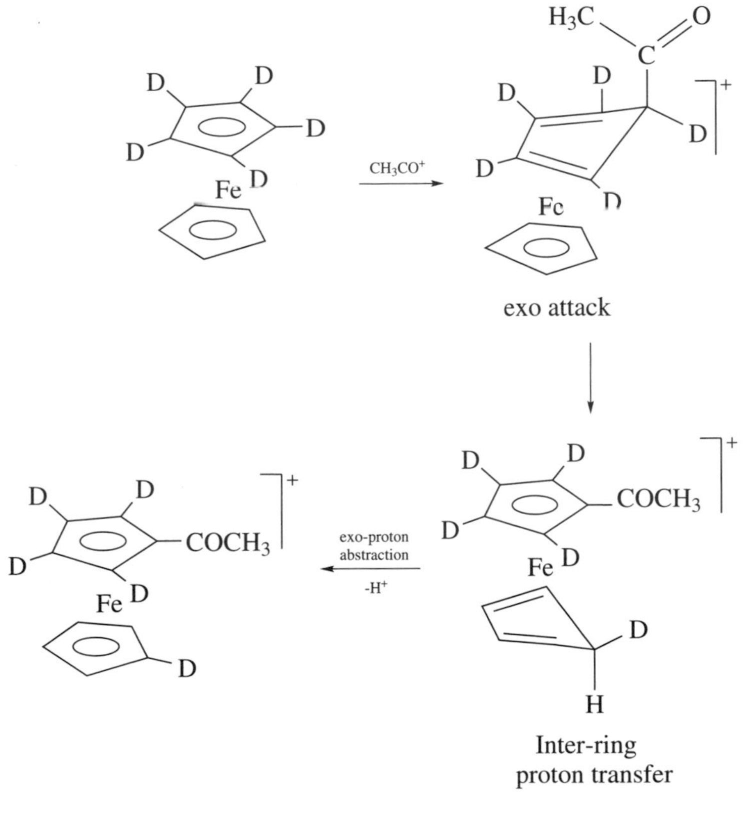
D
Fe
CH3CO+
H3C
O
C
D
+
Fc
exo attack
COCH3
+
Fe
H
Inter-ring proton transfer
exo-proton abstraction
-H+
COCH3
+
Fe

of the resulting exo-hydrogen; for this reason, all five starting D atoms tend to be retained in the complex. The acetyl attacks both rings, but only the result of attack at the C_5D_5 ring is shown.[28]

8.5 ELECTROPHILIC ABSTRACTION OF ALKYL GROUPS

Electrophilic metal ions, notably Hg^{2+}, can cleave transition metal alkyl bonds relatively easily. Two main pathways have been identified, one of which is attack at the α carbon of the alkyl, which can lead to inversion of configuration at that carbon (Eq. 8.35). In the other, attack occurs at the metal or at the M—C bond and retention of configuration is found (Eq. 8.36). The difference has been ascribed to the greater basicity of the metal in the iron case.[29,30]

$$(8.35)$$

$$(8.36)$$

As a 0e ligand, $HgCl_2$, or more likely $HgCl^+$, can bind to an 18e metal exactly in the same way as can a proton. It is not yet clear whether the electrophilic attack takes place at the M—C bond or at the metal. The first pathway can give RHgCl directly (Eq. 8.37), the second gives an alkylmetal mercuric chloride, which can reductively eliminate to give the same product (Eq. 8.38). In the absence of an isolable intermediate, it is very difficult to tell the two paths apart. This is an important process: As we will see in Chapter 16, electrophilic attack by Hg(II) on the methyl derivative of coenzyme B_{12} is the route by which mercuric ion from various sources is converted into the toxic methylmercury cation in natural waters.

$$(8.37)$$

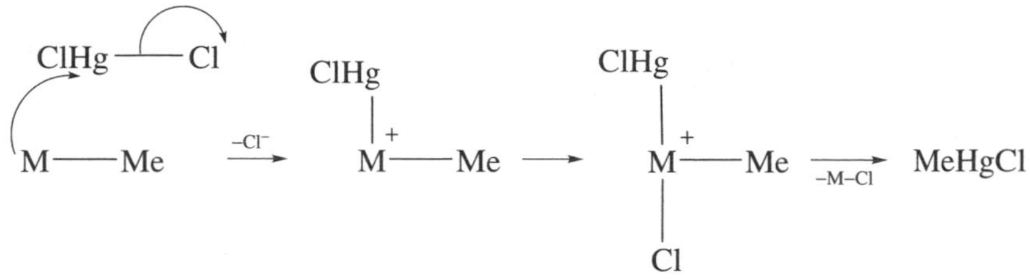

$$(8.38)$$

The proton is often able to cleave metal alkyls. This happens most readily for the electropositive metals, where the alkyl has a higher negative charge. Even water is a good enough source of protons for RLi, RMgBr, and many of the early metal alkyls. The later metals need stronger acids and more vigorous conditions.[31]

$$CpL_2FeMe + CF_3COOH \longrightarrow [CpL_2FeMeH]^+ (CF_3COO)^- \xrightarrow{-CH_4}$$

$$[CpL_2Fe(OCOCF_3)] \quad (8.39)$$

Other electrophiles are known to abstract transition metal alkyls, as shown below:

$$Cp_2TaMe_3 + Ph_3C^+ \longrightarrow [Cp_2TaMe_2]^+ + Ph_3CMe \quad (8.40)^3$$

$$[Co(CN)_5(CH_2Ph)]^{3-} + NO^+ \longrightarrow [Co(CN)_5(H_2O)]^{2-} + PhCH_2NO\bullet \longrightarrow$$

$$PhCH=NOH + other\ products \quad (8.41)^{32}$$

Retention of configuration is not always observed in the electrophilic abstraction of vinyl groups because the electrophile sometimes gives an initial reversible addition to the β carbon (Eq. 8.43). Free rotation about a C−C single bond in the carbene intermediate then leads to loss of stereochemistry.

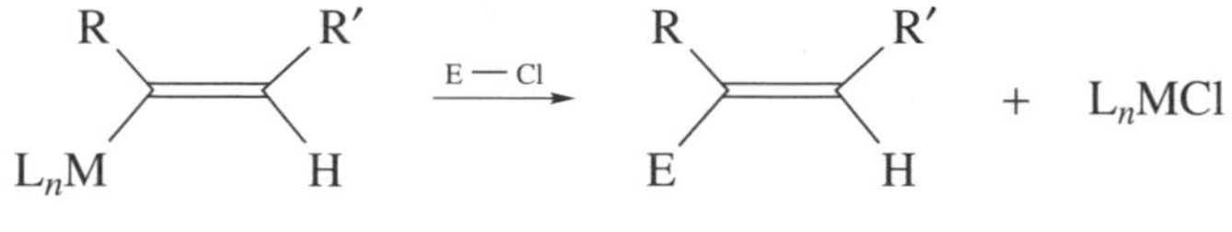

$$(8.42)$$

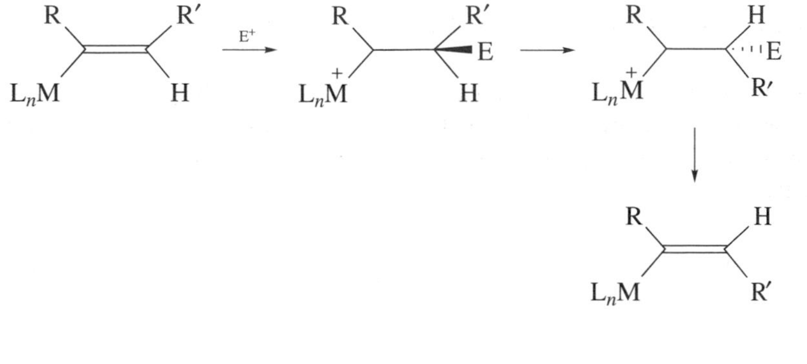

$$(8.43)$$

Halogens are electrophilic reagents and can readily cleave many metal alkyls to give the free alkyl halides. One common mechanism (Eq. 8.44: L = dmg; R = n-hexyl) involves oxidation of the metal. This increases the electrophilic character of the alkyl group and generates halide ion, so that, paradoxically, it is *nucleophilic* abstraction of the alkyl group by halide ion that leads to the final products. Co(III) alkyls are known to behave in the same way, and the intermediate Co(IV) species are stable enough to be detected by EPR at $-50°C$.[33]

$$(8.44)$$

As we saw in Section 7.3, some reactions that lead to overall insertion into an M—R bond go by the electrophilic abstraction of an alkyl as the first step. SO_2 insertion is the best known, but it is thought that SO_3, $(CN)_2C=C(CN)_2$, and $CF_3C\equiv CCF_3$ may be able to react in the same way.

An alternative pathway for the reaction of a metal alkyl with an electrophile is the abstraction of a substituent at the α carbon to form a carbene.

$$Tp(CO)LWCH_3 + Ph_3C^+ \longrightarrow [Tp(CO)LW=CH_2]^+ \qquad (8.45)^{[34]}$$

$$Cp(CO)_2Fe-CF_3 + BF_3 \longrightarrow [Cp(CO)_2Fe=CF_2]^+BF_4^- \qquad (8.46)^{[35]}$$

8.6 SINGLE-ELECTRON TRANSFER PATHWAYS

It is sometimes difficult to differentiate between a true electrophilic abstraction or addition, a one-step process in which a pair of electrons is implicated, and a two-step process involving a single-electron transfer (SET) step to give radical intermediates.

$$L_nM-R \xrightarrow[\text{electrophilic}\atop\text{abstraction}]{E^+} L_n\overset{+}{M}-\square \ + \ E-R \qquad (8.47)$$

$$L_nM-R + E^+ \xrightarrow{\text{Set}} L_n\overset{\bullet+}{M}-R + E\bullet \xrightarrow[\text{radical abstraction}]{} L_n\overset{+}{M}-\square + E-R \quad (8.48)$$

An analogous ambiguity holds for nucleophilic reactions. We have already seen one facet of this problem in the oxidative addition of alkyl halides to metals (Section 6.3), which can go either by an electrophilic addition to the metal, the S_N2 process, or by SET and the intermediacy of radicals. The two processes can often give the same products. Other related cases we have seen are the promotion of migratory insertion and nucleophilic abstraction by SET oxidation of the metal (Section 7.1) and electrophilic abstraction of alkyl groups by halogen (Section 8.5).

Hayes and Cooper[36a] have described abstraction reactions from a metal alkyl by an electrophilic reagent that goes by an SET route. Instead of the normal β abstraction of hydride from an ethyl group, which occurs in the usual electrophilic abstraction, he finds a preference for α abstraction from a methyl group. Since H atom abstraction usually takes place at the weakest C$-$H bond, the metal substituent presumably weakens the α- more than the β-C$-$H bonds of the alkyl.

observed product

Nucleophiles can also give SET reactions. Lapinte and co-workers[36b] have shown that $[Cp^*Mo(CO)_3(PMe_3)]^+$ reacts with $LiAlH_4$ to give paramagnetic $[Cp^*Mo(CO)_3(PMe_3)]$, observed by EPR. Loss of CO, easy in this 19e species, leads to $Cp^*Mo(CO)_2(PMe_3)$, which abstracts H$\bullet$, probably from the THF solvent, to give the final product, $Cp^*MoH(CO)_2(PMe_3)$.

8.7 REACTIONS OF ORGANIC FREE RADICALS WITH METAL COMPLEXES

The reactions of organic free radicals with metal complexes is much less well understood than the attack of electrophiles and nucleophiles. If the starting material is an 18e complex, the product will be a 17e or 19e species and therefore reactive, so the nature of this initial reaction product may have to be inferred from the final products. Addition to the metal is well recognized and is easiest to detect when the starting complex is 17e, so that the product becomes 18e. For example, organic radicals are known to react very rapidly with $[Co^{II}(dmg)_2py]$ as follows:

$$[Co^{II}(dmg)_2py] + R\cdot \longrightarrow [RCo^{III}(dmg)_2py] \qquad (8.50)$$

We saw an example of this process as part of larger mechanistic schemes in the radical-based oxidative additions of Section 6.3. We also saw typical radical rearrangements used to detect the presence of radical intermediates (e.g., Eq. 6.20).

Since organic radicals react rapidly by the pathways shown in Eqs. 8.51 and 8.52, only a rapid reaction with a metal complex can successfully compete.

$$RCH\cdot CH_3 \longrightarrow \underset{\substack{\text{recombination}\\\text{product}}}{R(CH_3)HC-CH(CH_3)R} + \underset{\substack{\text{disproportionation}\\\text{products}}}{RCH=CH_2 + RCH_2CH_3} \quad (8.51)$$

$$R\cdot + solvH \longrightarrow RH + solv\cdot \qquad (8.52)$$

The reaction of Eq. 8.52 means that the solvent has to be chosen with care or solvent-derived radicals may attack the metal complex. Solvents with strong $X-H$ bonds, such as water, t-BuOH, n-alkane, benzene, and acetic acid, are resistant to loss of an H atom via Eq. 8.52 but THF, Et_2O, $CH_3C_6H_5$, and $(CH_3)_2CO$ are much less resistant.

Radical abstraction is still uncommon, but it constitutes one step of Eq. 8.49, and it has been proposed to explain the acceleration in the rate of substitution of $Cp_2Fe_2(CO)_4$ caused by O_2/BEt_3. In this case Et$\cdot$ radicals, formed from BEt_3 and O_2, are thought to abstract CO from the complex to give EtCO$\cdot$ and a coordinatively unsaturated 17e Fe species, which undergoes substitution.[37] A related radical addition to a CO group of $IrCl(CO)_2(PMe_3)_2$ has been proposed by Boese and Goldman,[38] who generated $C_6H_{11}\cdot$ (cyclohexyl = Cy$\cdot$) by photolysis of benzophenone in cyclohexane and saw CyCHO as the organic product:

$$Ph_2CO \xrightarrow{\;h\nu\;} Ph_2CO^* \xrightarrow{\;CyH\;} Ph_2C\cdot -OH + Cy\cdot \qquad (8.53)$$

$$Cy\cdot \xrightarrow{\;M(CO)_n\;} (CyCO)M\cdot(CO)_{n-1} \xrightarrow{\;Ph_2C\cdot-OH\;CO\;} CyCHO + M(CO)_n + Ph_2CO \quad (8.54)$$

Baird and co-workers[39] have looked at phenylazotriphenylmethane, $PhN=NCPh_3$, as a thermal source of phenyl and triphenylmethyl (trityl) radicals that react with compounds of the types $(\eta^3$-allyl)$PdCl(PPh_3)$ and $[(\eta^3$-allyl)$Pd(PPh_3)_2]Cl$ to give palladium phenyl compounds, $[PdPhCl(PPh_3)]_2$ and trans-$PdPhCl(PPh_3)_2$, respectively, and $Ph_3CCH_2CH=CH_2$ formed from trityl abstraction of the allyl. Cyclohexyl (Cy) radicals, formed from photolysis of $[Cy(dmg)_2(pyridine)cobalt(III)]$, react with $(\eta^3$-allyl)$PdCl(PPh_3)$ to give cyclohexene and propene, presumably via Eq. 8.55; no Pd-containing products were characterizable.

$$(8.55)$$

Radical traps like galvinoxyl, TEMPO, and DPPH (Q•) are sometimes used as a test for the presence of radicals, R•, in solution; in such a case the adduct Q-R is expected as product. Unfortunately, this procedure can be misleading in organometallic chemistry becase typical Q• abstract H from some palladium hydrides at rates competitive with those of typical organometallic reactions; [PdHCl(PPh$_3$)$_2$] reacts in this way but [PdH(PEt$_3$)$_3$]BPh$_4$ is stable.[40]

> • Electrophilic and radical reactions have less easily predictable outcomes than nucleophilic reactions.

REFERENCES

1. G. Natile et al., *J. Chem. Soc., Dalton* 651, 1977.

2. F. Rose-Munch, E. Rose et al., *Organometallics* **23**, 184, 2004.

3. R. R. Schrock, *Acc. Chem. Res.* **97**, 6577, 1975.

4. S. E. Gibson et al., *Chem. Comm.* 2465, 2000.

5. C. P. Casey, R. L. Anderson et al., *J. Am. Chem. Soc.* **94**, 8947, 1972; **96**, 1230, 1974.

6. J. W. Faller et al., *J. Organomet. Chem.* **88**, 101, 1975.

7. A. J. Birch et al., *Tetrahedron Lett.* 115, 1975.

8. R. M. Bullock et al., *Organometallics* **21**, 2325, 2002.

9. M. Rosenblum, W. P. Giering et al., *J. Am. Chem. Soc.* **94**, 7170, 1972.

10. E. O. Fischer, *Adv. Organomet. Chem.* **14**, 1, 1976.

11. H. C. Clark et al., *Synth. React. Inorg. Met.-Org. Chem.* **4**, 355, 1974.

12. Y. Shvo, *Chem. Commun.* 336, 1974; T. L. Brown et al., *J. Organomet. Chem.* **71**, 173, 1975.

13. J. Vicente, M. C. Ramirez de Arellano, P. G. Jones, M. G. Humphrey, M. Samoc et al., *Organometallics* **19**, 2698, 2000.

14. S. G. Davies, M. L. H. Green, and D. M. P. Mingos, *Tetrahedron* **34**, 3047, 1978; R. D. Pike and D. A. Sweigert, *Coord. Chem. Rev.* **187**, 183, 1999.

15. D. A. Brown et al., *Organometallics* **5**, 158, 1986.

16. A. J. Birch et al., *J. Chem. Soc., Perkin I* 1882, 1900, 1973; *Tetrahedron Lett.* 871, 979, 2455, 1980.

17. A. J. Pearson, *J. Chem. Soc., Perkin I* 1980, 395.

18. J. W. Faller and A. M. Rosan, *J. Am. Chem. Soc.* **99**, 4858, 1977.

19. J. C. Calabrese, S. D. Ittel, S. G. Davies, and D. A. Sweigert, *Organometallics* **2**, 226, 1983.

20. J. W. Faller et al., *Organometallics* **3**, 927, 1231, 1984.

21. F. C. Phillips, *Am. Chem. J.* **16**, 255, 1894.

22. J. Smidt et al., *Angew. Chem.* **71**, 176, 1959; **74**, 93, 1962; J. M. Takacs, *Curr. Org. Chem.* **7**, 369, 2003.

23. J. E. Bäckvall, B. Åkermark et al., *J. Am. Chem. Soc.* **101**, 2411, 1979.

24. (a) J. K. Stille et al., *J. Organomet. Chem.* **169**, 239, 1979; (b) M. Mori et al., *J. Am. Chem. Soc.* **93**, 1529, 1971; (c) N. Aktoglu, D. Baudry, D. Cox, M. Ephritikhine, H. Felkin, R. Holmes–Smith, J. Zakrzewski, *Bull. Ch. Soc. France*, 381, 1985.

25. R. A. Henderson, *Angew. Chem., Int. Ed.* **35**, 946, 1996.

26. Y. K. Chung et al., *Organometallics*, **21**, 3417, 2002.

27. D. L. DuBois et al., *Organometallics* **23**, 2670, 2004.

28. A. F. Cunningham, *Organometallics*, **16**, 1114, 1997; J. Weber et al., *Organometallics* **17**, 4983, 1998.

29. M. C. Baird et al., *Inorg. Chem.* **18**, 188, 1979.

30. G. M. Whitesides et al., *J. Am. Chem. Soc.* **96**, 2814, 2826, 1974.

31. A. Wojcicki et al., *J. Organomet. Chem.* **193**, 359, 1980.

32. M. D. Johnson et al., *J. Chem. Soc. A* 177, 1966.

33. J. Halpern, M. E. Vol'pin et al., *Chem. Commun.* 44, 1978.

34. J. L. Templeton et al., *Organometallics* **16**, 4865, 1997.

35. A. Crespi and D. F. Shriver, *Organometallics* **4**, 1830, 1985.

36. (a) J. C. Hayes and N. J. Cooper, *J. Am. Chem. Soc.* **104**, 5570, 1982; (b) C. Lapinte et al., *Organometallics* **11**, 1419, 1992.

37. S. Nakanishi et al., *Chem. Commun.* 709, 1993.

38. W. T. Boese and A. S. Goldman, *J. Am. Chem. Soc.* **114**, 350, 1992.

39. M. C. Baird et al., *J. Organometal. Chem.* **689**, 1257, 2004.

40. A. C. Albeniz, P. Espinet, R. López-Fernández, and A. Sen, *J. Am. Chem. Soc.* **124**, 11278, 2002.

PROBLEMS

1. Where would a hydride ion attack each of the following?

$$[(\eta^5\text{-cyclohexadienyl})(\eta^5\text{-Cp})(C_2H_4)MoMe]^+$$

$$[(\eta^5\text{-cyclohexadienyl})(CO)_3Fe]^+$$

$$[(\eta^4\text{-}C_4H_4)(\eta^4\text{-butadiene})(\eta^3\text{-allyl})MoMe]^+$$

2. Predict the outcome of the reaction of $CpFe(PPh_3)(CO)Me$ with each of the following: HCl, Cl_2, $HgCl_2$, and HBF_4/THF.

3. Explain the outcome of the reaction shown below:

$$\text{Butadiene} + \text{PhI} + R_2NH \xrightarrow{\text{Pd(OAc)}_2, \text{ PPh}_3}$$

$$PhCH_2CH=CHCH_2NR_2 + PhCH=CHCH=CH_2 \quad (8.55)$$

4. $[CpCo(dppe)(CO)]^{2+}$ (A) reacts with 1° alcohols, ROH, to give $[CpCo(dppe)(COOR)]^+$, a reaction known for very few CO complexes. The $\nu(CO)$ frequency for A is 2100 cm^{-1}, extremely high for a CO complex. Br$^-$ does not usually displace CO from a carbonyl complex, but it does so with A. Why is A so reactive?

5. Nucleophilic addition of MeO$^-$ to free PhCl is negligibly slow under conditions for which the reaction with $(\eta^6\text{-}C_6H_5Cl)Cr(CO)_3$ is fast. What product would you expect, and why is the reaction accelerated by coordination?

6. Given a stereochemically defined starting material (either erythro or threo), what stereochemistry would you expect for the products of the following electrophilic abstraction reaction:

$$CpFe(CO)_2(CHDCHDCMe_3) + Ph_3C^+ \longrightarrow$$

$$CHD=CHCMe_3 + CHD=CDCMe_3 \quad (8.56)$$

Let us say that for a related 16e complex $L_nM(CHDCHDCMe_3)$ gave precisely the same products, but of opposite stereochemistries. What mechanism would you suspect for the reaction?

7. You are trying to make a methane complex $L_nM(\eta^1\text{-H}-CH_3)^+$ (8.15, unknown as a stable species at the time of writing), by protonation of a methyl complex L_nMMe with an acid HA. Identify three things that might go wrong and suggest ways to guard against each. (If you try this and it works, send me a reprint.)

$$L_nM \longleftarrow \overset{H}{\underset{CH_3}{\diagdown}}$$

8.15

8. (cod)PtCl$_2$ reacts with MeOH/NaOAc to give a species $[\{C_8H_{12}(OMe)\}PtCl]_2$. This in turn reacts with PR$_3$ to give 1-methoxy cyclooctadiene (8.16) and PtHCl(PR$_3$)$_2$. How do you think this might go?

MeO

8.16

9. $[CpFe(CO)(PPh_3)(MeC{\equiv}CMe)]^+$ reacts with (i) $LiMe_2Cu$ (a source of Me^-) and (ii) I_2 to give compound **8.17**; explain this reaction. What product do you think might be formed from $LiEt_2Cu$?

8.17

10. Me_3NO is a good reagent for removing CO from a metal, but why does Me_3PO not work? Why does Me_3NO not work in the case of $Mo(dppe)_2(CO)_2$? Can you suggest an O-donor reagent that might be more reactive than Me_3NO?

9

HOMOGENEOUS CATALYSIS

The catalysis of organic reactions[1-4] is one of the most important applications of organometallic chemistry and has been a significant factor in the rapid development of the whole field. Organometallic catalysts have long been used in industrial processes but are now being routinely applied in organic synthetic problems as well. A *catalyst* is an additive used in substoichiometric amount to bring about a reaction at a temperature below that required for the uncatalyzed thermal reaction. It binds the reactants, or *substrates*, for the catalytic reaction, causes the desired reaction, and then liberates the reaction products to regenerate the catalyst. The catalyst reenters the catalytic cycle by binding the reagents once more. A typical catalyst may participate in the catalytic cycle 10^1–10^6 times or more, allowing its use in modest or even trace amount. The catalysts we look at are soluble complexes, or *homogeneous catalysts*, as opposed to catalysts such as palladium on carbon, or *heterogeneous catalysts*.[4] These terms are used because the catalyst and substrates for the reaction are in the same phase in the homogeneous, but not in the heterogeneous, type, where catalysis takes place at the surface of a solid catalyst. Some reactions, such as hydrogenation, are amenable to both types of catalysis, but others are currently limited to one or the other, for example, O_2 oxidation of ethylene to the epoxide over a heterogeneous silver catalyst or Wacker air oxidation of ethylene to acetaldehyde with homogeneous Pd(II) catalysts.

The term *homogeneous catalysis* also covers simple acid catalysts and non-organometallic catalysis, such as the decomposition of H_2O_2 by Fe^{2+}. Catalytic mechanisms are considerably easier to study in homogeneous systems, where

The Organometallic Chemistry of the Transition Metals, Fourth Edition, by Robert H. Crabtree
Copyright © 2005 John Wiley & Sons, Inc.

such powerful methods as NMR can be used to both assign structures and follow reaction kinetics. Homogeneous catalysts have the disadvantage that they can be difficult to separate from the product. Sometimes this requires special separation techniques, but in other cases, such as polymer synthesis, the product can be sold with the catalyst still embedded in it. Homogeneous catalysts can also be chemically grafted on to solid supports for greater ease of separation of the catalyst from the reaction products. Although the catalyst is now technically heterogeneous, it often retains the characteristic reactivity pattern that it showed as a homogeneous catalyst, and its properties are usually distinct from those of any of the classical heterogeneous catalysts—these are sometimes called "heterogenized" homogeneous catalysts. The mechanistic ideas developed in homogeneous catalysis are also becoming more influential in the field of classical heterogeneous catalysis by suggesting structures for intermediates and mechanisms for reaction steps.

By bringing about a reaction at lower temperature, a catalyst can save energy in commercial applications. It often gives higher selectivity for the desired product, minimizing product separation problems and avoiding the need to discard the undesired product as waste. The selectivity can be changed by altering the ligands, allowing synthesis of products not formed in the uncatalyzed process. With growing regulatory pressure to synthesize drugs in enantiopure form, asymmetric catalysis has come to the fore, along with enzyme catalysis, as the only practical way to make such products on a large scale. Older commercial processes tended to give side products that had to be discarded, such as inorganic salts. Environmental concerns have promoted the idea of *atom economy*, which values a process most highly when all the atoms in the reagents are used to form the product, minimizing waste. For example, the Monsanto process of Section 12.1 is atom economic because it converts MeOH and CO to MeCOOH with no atoms left over.

A typical reaction (Eq. 9.1) that is catalyzed by many transition metal complexes is the isomerization of allylbenzene (the substrate) into propenylbenzene (the product). Normally, the substrate for the reaction will coordinate to the metal complex that serves as catalyst. The metal then brings about the rearrangement, and the product dissociates, leaving the metal fragment free to bind a new molecule of substrate and participate in the catalytic cycle once again.

$$\text{Ph}\diagup\diagup\diagdown \xrightarrow{\text{catalyst}} \text{Ph}\diagup\diagdown\sim\sim \qquad (9.1)$$

9.1   **9.2**

Before setting out to find a catalyst for a given reaction, say the one shown in Eq. 9.1, the first consideration is thermodynamic: whether the reaction is favorable. If the desired reaction were thermodynamically strongly disfavored, as is the conversion of H_2O to H_2 and O_2, for example, then no catalyst, however efficient, could *on its own* bring about the reaction. If we wanted to bring about an unfavorable reaction of this sort, we would have to provide the necessary driving force in some way. There are ways of doing this, such as coupling a strongly favorable process to the unfavorable one you want to drive, as Nature commonly does with the hydrolysis of ATP (adenosine triphosphate), or we could use the energy of a light photon, as in photosynthesis or we can selectively remove the products (e.g., by distillation).

Normally, the catalyst only increases the rate of a process but does not alter its position of equilibrium, which is decided by the relative thermodynamic stabilities of substrate and products (we discuss ways of getting around this restriction in Section 12.4). Fig. 9.1*a* illustrates this point: The substrate S is slightly less stable than the product P, so the reaction will eventually reach an equilibrium favoring P. In the case of **9.1** going to **9.2**, the additional conjugation present in **9.2** is sufficient to ensure that the product is thermodynamically more stable than the starting material and so the reaction is indeed favorable. Normally, the substrate binds to the metal before it undergoes the rearrangement. This substrate–catalyst complex is represented as "M.S" in Fig. 9.1. It might be thought that strong binding would be needed. A moment's reflection will show why this is not so. If the binding is too strong, M.S will be too stable, and the activation energy to get to "M.TS" will be just as large as it was in going from S to TS in the uncatalyzed reaction. S cannot bind too weakly because it may otherwise be excluded from the metal and fail to be activated by the metal at all. Similarly, the product P will normally be formed as the complex M.P. Product P must be the least strongly bound of all because if it is not then S will not be able to displace P, and the catalyst will be effectively poisoned by the products of the reaction. Many of these ideas also apply to the chemistry of Nature's homogeneous catalysts, enzymes.[5]

Each time the complete catalyst cycle occurs, we consider one catalytic turnover (one mole of product formed per mole of catalyst) to have been completed. The catalytic rate can be conveniently given in terms of the turnover frequency (TOF) measured in turnovers per unit time (often per hour); the lifetime of the catalyst before deactivation is measured in terms of total turnovers.

For most transition metal catalysts, the catalyzed pathway is completely changed from the pathway of the uncatalyzed reaction, as shown in Fig. 9.1*a*. Instead of passing by way of the high-energy uncatalyzed transition state TS, the catalyzed reaction normally goes by a multistep mechanism in which the metal stabilizes intermediates that are stable only when bound to the metal. One new transition state M.TS′ is shown in Fig. 9.1. The TS′ structure in the absence of the metal would be extremely unstable, but the energy of binding is so high that M.TS′ is now much more favorable than TS and the reaction all passes through the catalyzed route. Different metal species may be able to stabilize other transition states TS″—which may lead to entirely different products—hence different catalysts can give different products from the same starting materials.

In a stoichiometric reaction, the passage through M.TS′ would be the slow, or rate-determining, step. In a catalytic reaction the cyclic nature of the system means that the rates of all steps are identical. On a circular track, on average the same number of trains must pass each point per unit time. The slow step in a catalytic process is called the *turnover limiting step.* Any change that lowers the barrier for this step will increase the turnover frequency (TOF). Changes in

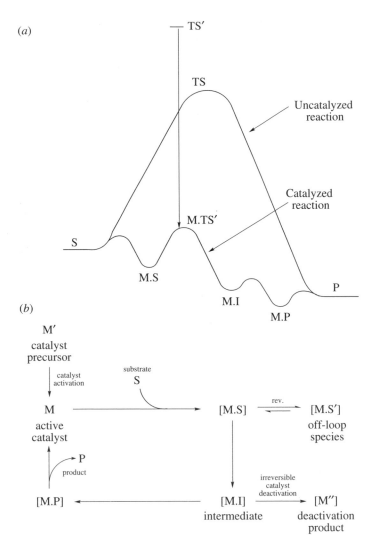

FIGURE 9.1 (*a*) A catalyst lowers the activation energy for a chemical reaction. Here the uncatalyzed conversion of substrate S to product P passes by way of the high-energy transition state TS. In this case the metal-catalyzed version goes via a different transition state TS′, which is very unstable in the free state but becomes viable on binding to the catalyst as M.TS′. The arrow represents the M–TS′ binding energy. The uncatalyzed and catalyzed processes do not necessarily lead to the same product as is the case here. (*b*) Typical catalytic cycle in schematic form.

other barriers will not affect the TOF. For a high TOF, we require that none of the intermediates be bound too strongly (otherwise they may be too stable and not react further) and that none of the transition states be prohibitively high in energy. Indeed, the whole reaction profile must not stray from a rather narrow range of free energies, accessible at the reaction temperature. Even if all this is arranged, a catalyst may undergo a catalytic cycle only a few times and then "die." This happens if undesired deactivation reactions are faster than the productive reactions of the catalytic cycle itself. There are many ways in which a catalyst can fail, and for success it is often necessary to look hard for the right metal, ligand set, solvent, temperature range, and other conditions. Many of the reactions that occupied the attention of the early workers were relatively forgiving in terms of the range of possible catalysts and conditions. Some of the problems that are under study today, notably alkane conversions, constitute more searching tests of the efficiency of homogeneous catalysts.

Figure 9.1*b* shows a schematic catalytic cycle. The *active catalyst* M is often rather unstable and is only formed in situ from the *catalyst precursor* (or precatalyst), M′. If during the reaction we observe the system, for example, by NMR, we normally see only the disappearance of S and the appearance of P. Decreasing the substrate concentration [S] and increasing the metal concentration [M] may allow us to see the complex. We may still see only M′ because only a small fraction of the metal is likely to be on the loop at any given time. Even if an observed species appears to be an intermediate, we still cannot be sure it is not M · S′, an off-loop species. If a species builds up steadily during the reaction, it might be a *catalyst deactivation product* M″, in which case the catalytic rate will fall as [M″] rises. Excellent reviews are available on the determination of mechanism in catalytic reactions.[3]

9.1 ALKENE ISOMERIZATION

Many transition metal complexes are capable of catalyzing the 1,3-migration of hydrogen substituents in alkenes, a reaction that has the net effect of moving the C=C group along the chain of the molecule (e.g., Eq. 9.1). This is often a side reaction in other types of catalytic alkene reaction, desired or not according to circumstances. Two mechanisms are most commonly found: the first goes via alkyl intermediates (Fig. 9.2*a*); the second, by η^3-allyls (Fig. 9.2*b*). Note that in each cycle, all the steps are reversible, so that the substrates and products are in equilibrium, and therefore although a nonthermodynamic ratio of alkenes can be formed at early reaction times, the thermodynamic ratio is eventually formed if the catalyst remains active long enough. In other catalytic reactions, we sometimes find that the last step is irreversible. As we shall see later, this distinction has important practical consequences in allowing the formation of grossly nonthermodynamic ratios (e.g., in asymmetric catalysis).

Alkyl Mechanism

In the alkyl route, we require an M−H bond and a vacant site. The alkene binds and undergoes insertion to give the alkyl. For 1-butene, the alkyl might be the 1° or the 2° one, according to the regiochemistry of the insertion. If the 1° alkyl is

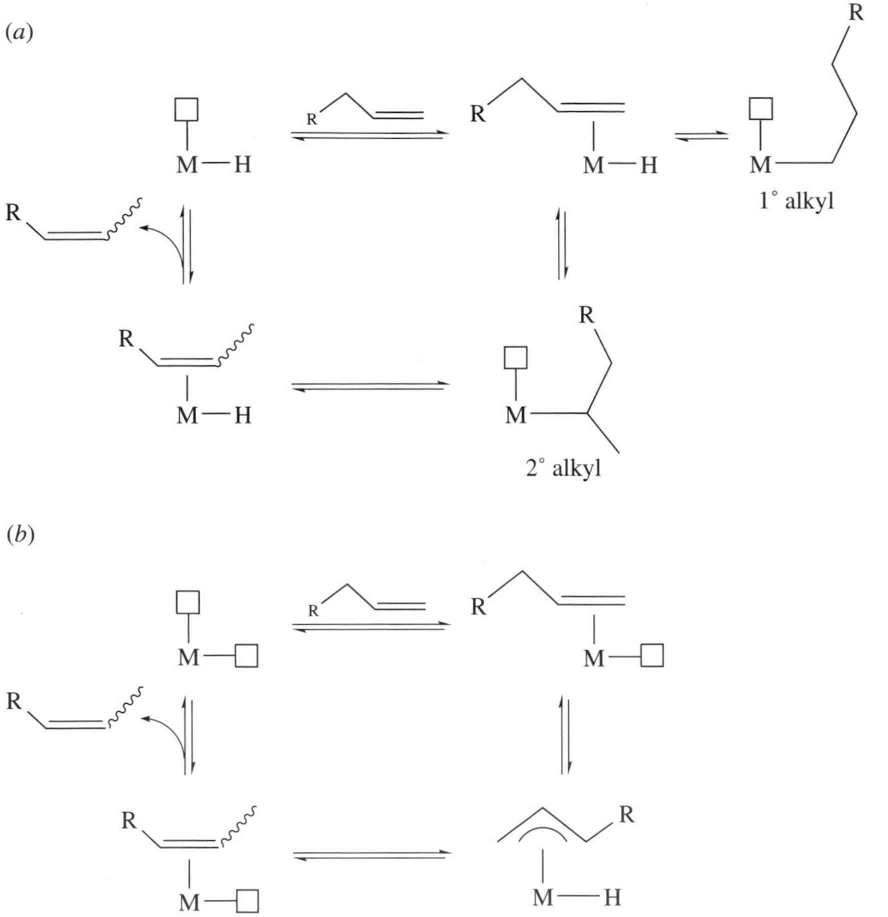

FIGURE 9.2 The (a) alkyl and (b) allyl mechanisms of alkene isomerization. The open box represents a 2e vacancy or potential vacancy in the form of a labile 2e ligand.

formed, β elimination can give back only 1-butene, but β elimination in the 2° alkyl, often faster, can give both 1- and cis- and trans-2-butene. Since insertion to give the 1° alkyl is favored for many catalysts, nonproductive cycling of the 1-butene back to 1-butene is common, and productive isomerization may be slower. The *initial* cis/trans ratio in the 2-butenes formed depends on the catalyst; the cis isomer is often favored. The final ratio depends only on the thermodynamics, and the trans isomer is preferred. A typical isomerization catalyst is RhH(CO)L$_3$ (L = PPh$_3$).[6] As this is a *coordinatively saturated* 18e species it must lose a ligand, PPh$_3$ in this case, to form a *coordinatively unsaturated* intermediate (<18e), able to bind the alkene.

Allyl Mechanism

The second common mechanism involves allyl intermediates and is adopted by those metal fragments that have two 2e vacant sites but no hydrides. It has been established for the case of $Fe_3(CO)_{12}$ as catalyst, a system in which "$Fe(CO)_3$," formed by fragmentation of the cluster on heating, is believed to be the active species.[7] The cluster itself is an example of a catalyst precursor. As a 14e species, $Fe(CO)_3$ may not have an independent existence in solution, but may always be tied up with substrate or product. The open square in Fig. 9.2 represents a vacant site or a labile ligand. In this mechanism the C−H bond at the activated allylic position of the alkene undergoes an oxidative addition to the metal. The product is an η^3-allyl hydride. Now, we only need a reductive elimination to give back the alkene. Again, we can have nonproductive cycling if the H returns to the same site it left, rather than to the opposite end of the allyl group.

An experimental distinction[7] can be made between the two routes with a crossover experiment (Section 6.5) using the mixture of C_5 and C_7 alkenes of Eq. 9.2. For the allyl mechanism, we expect the D in **9.3a** to end up only in the corresponding product **9.3b** having undergone an intramolecular 1,3 shift. For the hydride mechanism, the D will be transferred to the catalyst that can in turn transfer it by crossover to the C_5 product.

$$\text{(9.2)}$$

9.3a **9.3b**

- Catalytic cycles involve a series of reaction steps of the types seen in Chapters 5–8.
- The catalyst precursor is usually transformed in some way before entering the cycle.
- Observable species in the catalytic solution are often off-loop species (Fig 9.1).

9.2 ALKENE HYDROGENATION

Hydrogenation catalysts[3] add molecular hydrogen to the C=C group of an alkene to give an alkane. Three general types have been distinguished, according to the way each type activates H_2. This can happen by (1) oxidative addition, (2) heterolytic activation, and (3) homolytic activation.

Oxidative Addition

Perhaps the most important group employs oxidative addition, of which $RhCl(PPh_3)_3$ (**9.4**, Wilkinson's catalyst) is the best known. A catalytic cycle that is important under certain conditions is shown in Fig. 9.3. Hydrogen addition to give a dihydride leads to labilization of one of the PPh_3 ligands (high trans effect of H) to give a site at which the alkene binds.

The alkene inserts, as in isomerization, but the intermediate alkyl is irreversibly trapped by reductive elimination with the second hydride to give an alkane. This is an idealized mechanism.[3] In fact, **9.4** can also lose PPh_3 to give $RhCl(PPh_3)_2$, and dimerize via halide bridges and each of these species have their own separate catalytic cycles[3c] that can be important under different conditions. Indeed, $RhClL_2$ reacts so much faster with H_2 than does $RhClL_3$ that the vast majority of the catalytic reaction goes through $RhClL_2$ under most conditions. By reversibility arguments, the more rapid oxidative addition of H_2 to the 3-coordinate d^8 $RhClL_2$ to give 5-coordinate d^6 RhH_2ClL_2 relative to the corresponding 4-coordinate → 6-coordinate conversion is consistent with the tendency for faster reductive elimination from 5-coordinate d^6 species that was discussed in Section 6.5. In a key study by Tolman and co-workers[3d], the dihydride was directly seen by ^{31}P NMR under H_2 and the reversible loss of the PPh_3 trans to a hydride detected from a broadening of the appropriate resonance, as discussed in Section 10.5. Figure 9.3 represents the *hydride mechanism* in which H_2 adds before the olefin. Sometimes the olefin adds first (the *olefin mechanism*) as is found for $[Rh(dpe)(MeOH)_2]BF_4$.[3e]

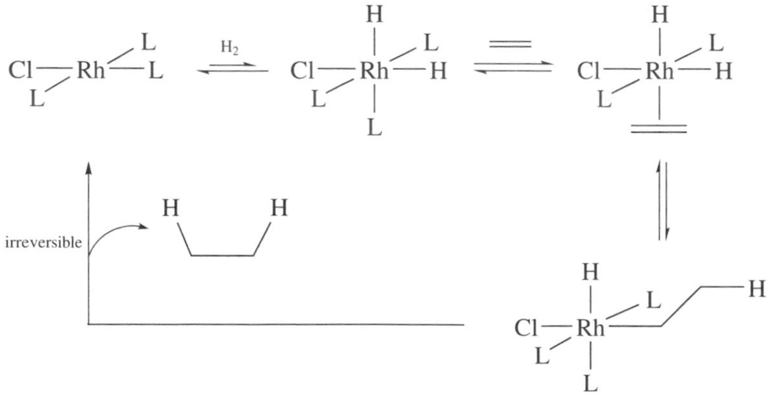

FIGURE 9.3 Mechanism for the hydrogenation of alkenes by Wilkinson's catalyst. Other pathways also operate in this system, however.

$$Cl-\!\!\!-Rh\overset{PPh_3}{\underset{PPh_3}{-\!\!\!-PPh_3}}$$

9.4

Since we need to bind two hydrides and the alkene, for a total electron count of 4e, the 16e catalyst RhCl(PPh₃)₃ needs to dissociate a ligand, PPh₃ in this case, to do this. The PEt₃ analog of **9.4** reacts with H₂ to give a stable and catalytically inactive dihydride RhH₂Cl(PEt₃)₃, **9.5**. The smaller PEt₃ ligand does not dissociate and so **9.5** is not an active catalyst. All we have to do to make the PEt₃ analog active is artificially arrange to generate the desired RhH₂ClL₂ intermediate by forming it in situ by starting with 0.5 equiv of [(nbd)Rh(μ-Cl)]₂ and adding 2 equiv of PEt₃, for a final P/Rh ratio of 2. Under H₂, the norbornadiene (nbd) is removed by hydrogenation, and we get RhH₂Cl(PEt₃)₂, which is an active hydrogenation catalyst under these conditions.[8] A key prerequisite for catalysis in many systems is coordinative unsaturation, that is, an open site at the metal.

$$H-\!\!\!-Rh\overset{PEt_3}{\underset{PEt_3}{-\!\!\!-PEt_3}}$$

9.5

As predicted by the mechanism of Fig. 9.3, the hydrogen gives syn addition to the alkene, although it is possible to tell only this in certain cases. For example,

$$(9.3)$$

Isomerization is often a minor pathway in a hydrogenation catalyst because the intermediate alkyl may β-eliminate before it has a chance to reductively eliminate. The more desirable catalysts, such as **9.4**, tend to give little isomerization. The selectivity for different alkenes (the hydrogenation rates change in the following order: monosubstituted > disubstituted > trisubstituted > tetrasubstituted = 0) is determined by how easily they can bind to the metal; the poorer ligands among them are reduced slowly, if at all. This means that **9.4** reduces the triene **9.6** largely to the octalin **9.7** (Eq. 9.4). Heterogeneous catalysts give none of this product, but only the fully saturated decalin (**9.9**), and the isomerization product, tetralin (**9.8**) (Eq. 9.4). The C=O and C=N double bonds of ketones and imines are successfully reduced only by certain catalysts. Other functional groups that can be reduced by heterogeneous catalysts, such as $-CN$, $-NO_2$, $-Ph$, and $-CO_2Me$ are not reduced by the usual homogeneous catalysts.

9.6

9.7
hydrogenation product
(major)

9.8
isomerization
product (minor)

9.9

$$(9.4)$$

IrCl(PPh$_3$)$_3$, the iridium analog of **9.4**, is inactive because of the failure of the dihydride IrH$_2$Cl(PPh$_3$)$_3$ to lose phosphine; this is a result of the stronger metal–ligand bond strengths usually found for the third-row metals. Using the same general strategy we saw for Rh, [(cod)Ir(μ-Cl)]$_2$ is active if only 2 mol of phosphine are added per metal. A more useful catalyst is obtained by replacing the chloride with a "noncoordinating" anion and changing the ligands to give the precursors [(cod)Ir(PMePh$_2$)$_2$]$^+$PF$_6^-$, **9.10**, and [(cod)Ir(py)(PCy$_3$)]$^+$PF$_6^-$, **9.11**.[8] These catalysts tend to bind a solvent, such as EtOH, much more firmly than do such uncharged catalysts as **9.4**, for example, to give the isolable species [IrH$_2$(solv)$_2$(PMePh$_2$)$_2$]$^+$PF$_6^-$ (solv = acetone, ethanol, or water). This results from the presence of hydrides and the net cationic charge, which tend to make the metal a harder Lewis acid. Unlike many noncationic catalysts, these species are also air stable and even tolerate halocarbons. As a result, the catalyst can be used in CH$_2$Cl$_2$, a much more weakly coordinating solvent than EtOH. Compound **9.11** has the unusual feature that it can reduce even highly hindered alkenes such as Me$_2$C=CMe$_2$. This is probably because these alkenes do not have to compete with dissociated phosphine or a coordinating solvent for a site on the metal, and perhaps also because the {Ir(py)(PCy$_3$)}$^+$ fragment is not very bulky. The high activity of **9.10** at first escaped attention because it was initially tested in EtOH, which at that time was the conventional solvent[6] for hydrogenation. Screening a new catalyst under a variety of conditions is therefore advisable.

9.10

9.11

Directing Effects

The catalyst **9.11** shows strong directing effects, which can be very useful in organic synthetic applications (see Section 14.2).[9] This means that H_2 is added to one face of the substrate, if there is a coordinating group (e.g., $-OH$, $-COMe$, $-OMe$) on that face (Eq. 9.5). The net positive ionic charge makes the metal hard enough to bind to the directing group and, as IrL_2^+ is a 12e fragment, it still has enough vacant sites left to bind both H_2 and the alkene to give the key intermediate **9.12**. Of the four possible geometrical isomers of the saturated ketone, only one is formed, H_2 having been added cis to the directing group.

$$(9.5)$$

Asymmetric Catalysis

The corresponding "RhL_2^+" catalysts were developed by Schrock and Osborn.[10a] Their most important application is asymmetric catalysis.[10b] Equation 9.6 shows how the achiral alkene **9.13** can give two enantiomers **9.14** and **9.15** on hydrogenation.

$$(9.6)$$

Any alkene having this property is called *prochiral*, which implies that the two faces of the molecule are different. In **9.13**, one face has a clockwise arrangement of R, R′, and $=CH_2$ about the central carbon; the other face has an anticlockwise arrangement of these groups. If the H_2 is added from one face, one enantiomer is formed; if from the other face, the other enantiomer is the product. If we were to bias the addition of H_2 to one face, then we would have an asymmetric synthesis. As shown in Eq. 9.7, when a prochiral alkene binds to an achiral metal, two enantiomers are formed; that is, the complex is chiral even though neither the ligand nor the metal were chiral before the complex was formed. One way of thinking about this is to regard the carbon indicated by the asterisk as having four different substituents, one of which is the metal.

$$
\begin{array}{c}
\underset{R'}{\overset{R}{\diagdown}}C{=}CH_2 \xrightarrow{ML_n}
\underset{R'}{\overset{R}{\diagdown}}\overset{*}{C}{=}CH_2 \;+\; \underset{R'}{\overset{R}{\diagdown}}\overset{*}{C}{=}CH_2
\end{array} \tag{9.7}
$$

9.16   **9.17**

enantiomers

$$
\begin{array}{c}
\underset{R'}{\overset{R}{\diagdown}}C{=}CH_2 \xrightarrow{ML_n^{*}}
\underset{R'}{\overset{R}{\diagdown}}\overset{*}{C}{=}CH_2 \;+\; \underset{R'}{\overset{R}{\diagdown}}\overset{*}{C}{=}CH_2
\end{array} \tag{9.8}
$$

9.18   **9.19**

diastereomers

The key point is that if the ML_n catalyst fragment can also be made chiral (say because a ligand L has an asymmetric carbon), then we can use one resolved enantiomer of the chiral complex as catalyst. In Eq. 9.8, instead of forming two enantiomeric complexes such as **9.16** and **9.17**, which react at equal rates to give achiral products, we will have diastereomeric alkene–catalyst complexes, **9.18** and **9.19**, because we now have two asymmetric centers present, C^* in the coordinated alkene and the asymmetric ML_n^* fragment. Since diastereomers generally have different chemical properties, **9.18** and **9.19** normally have different rates of hydrogenation. This bias on the rates of hydrogenation can selectively give us one of the pair of enantiomers **9.14** or **9.15** over the other. In summary, one enantiomer of the catalyst should preferentially give one enantiomer of the hydrogenated alkene, and the other enantiomer give the other product. This is an extremely valuable method because we can obtain a large amount of one enantiomeric product from a small amount of resolved material (the catalyst). This is precisely the method Nature uses to make pure enantiomers; enzymes are such efficient asymmetric catalysts that essentially only one enantiomer is normally formed in most enzymatic processes.

In asymmetric hydrogenation, 95–99% enantiomeric excess [e.e. = $100 \times$ {amount of major isomer − amount of minor isomer}/{total of both isomers}] can be obtained in favorable cases. The first alkenes to be reduced with high asymmetric induction contained a coordinating group, examples of which are shown as **9.20** and **9.21**.

$$
\underset{Ph}{\diagup}\overset{NHCOMe}{\underset{CO_2Me}{\diagdown}} \qquad\qquad \overset{NHCOMe}{{=}\underset{CO_2Me}{\diagdown}}
$$

9.20   **9.21**

These are believed to bind to the metal via the amide carbonyl just as we saw happen in directed hydrogenation. This improves the rigidity of the alkene–catalyst complex, which in turn increases the chiral discrimination of the system. As in directed hydrogenation, a 12e catalyst fragment, such as that formed from the Schrock–Osborn catalyst, is required.

One of the best ways of making the metal chiral is to use the ligand shown as **9.22**, called BINAP. This ligand has a so-called C_2 axis; this simply means that it has the symmetry of a helical bolt, which can, of course, either have a left-handed or a right-handed thread. The chiral centers impose a twist on the conformation of the BINAP-metal complex, which in turn leads to a chiral, propeller-like arrangement of the phenyl groups on phosphorus (**9.23**). These phenyl groups can be thought of as transmitting the chiral information from the asymmetric centers to the binding site for the alkene. The advantage of a C_2 symmetry is that the substrate sees the same chirality however it binds; we can think of the substrate as being analogous to a nut with a left-hand thread that will mate with a left-handed (but not a right-handed) bolt, whichever face of the nut is tried.

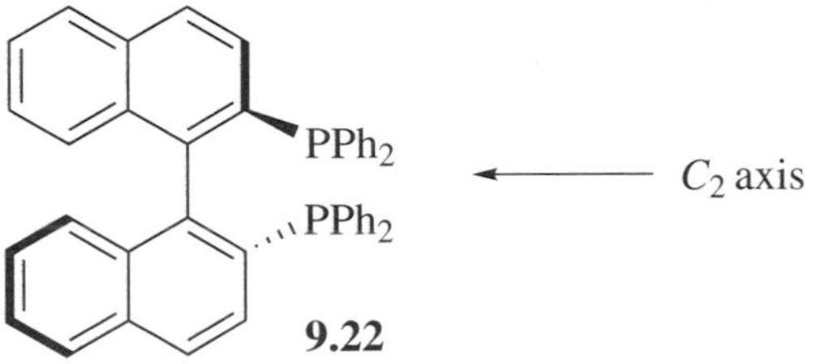

9.22

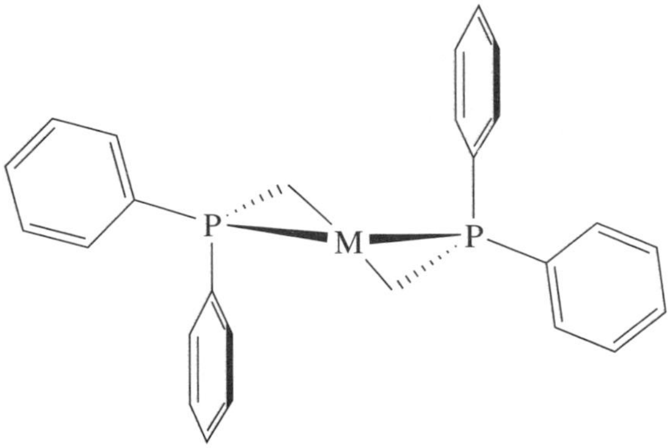

9.23

In the simplest case, one face of the substrate binds better to the catalyst than does the other. Let us say that, if H$_2$ were added to this face, we would get the *S* hydrogenation product. It was once thought that this preferential binding of the substrate always determines the sense of asymmetric induction. Halpern and co-workers[11] showed that in a system that gives the R product in good yield, the metal is bound to the "wrong" face in the major diastereomer (**9.24**), the face that would be expected to give the S product, and so it is the minor isomer of the catalyst–alkene complex that gives rise to most of the product. This in turn means that the minor isomer must react at about 10^3 times the rate of the major isomer (Eq. 9.9). Since **9.24** and **9.25** interconvert rapidly, **9.24** is continually converting into **9.25** because the faster hydrogenation of **9.25** continually depletes

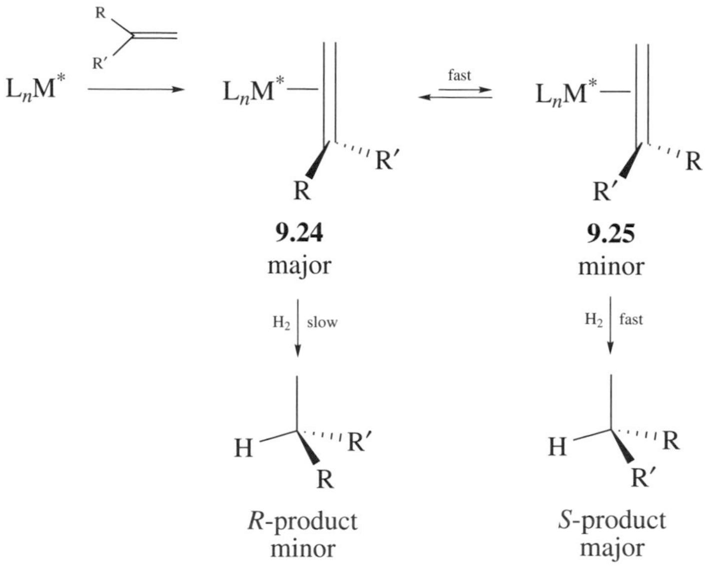

9.24
major

9.25
minor

(9.9)

R-product
minor

S-product
major

the concentration of this minor isomer. Note that Eq. 9.9 is an example of the "olefin mechanism."

Asymmetric alkene hydrogenation was used in the successful commercial production of the Parkinson's disease drug L-DOPA by hydrogenation of the alkene **9.26** and of the pain reliever, naproxen.[12a]

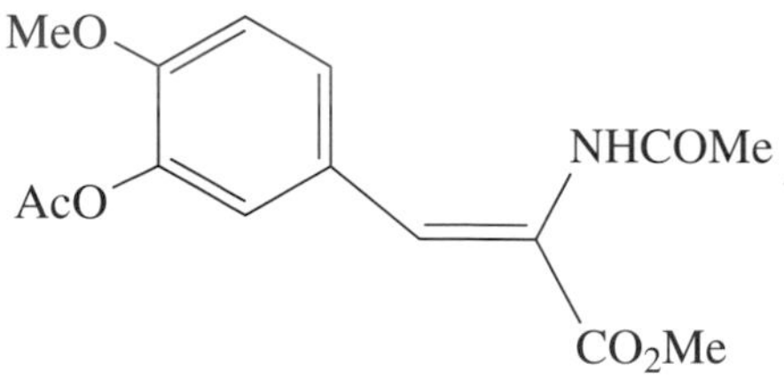

9.26

Another commercial success, this time for Ciba–Geigy, now Novartis, has been the synthesis of the herbicide, (S)-metolachlor, from the imine shown below using an iridium catalyst. The key advantage of iridium is the extremely impressive rate (>200,000 turnovers h^{-1}) and catalyst lifetime ($\sim 10^6$ turnovers) at the expense of some loss in e.e. relative to rhodium.[12b]

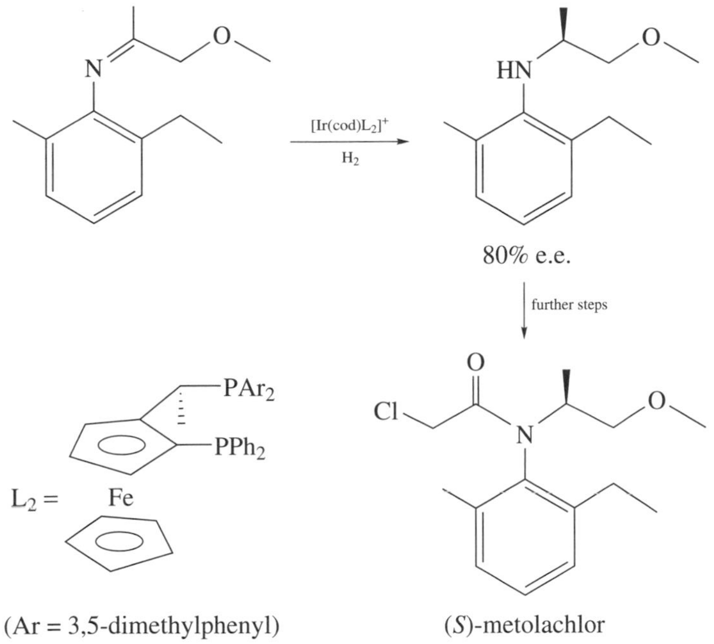

80% e.e.

further steps

$L_2 =$

(Ar = 3,5-dimethylphenyl)

(S)-metolachlor

Kinetic Competence

A useful general point emerges from this work—catalysis is a kinetic phenomenon, and so the activity of a system may rely on a minor, even minuscule, component of a catalyst. This emphasizes the danger on relying too heavily on spectroscopic methods in studying catalysts. The fact that a series of plausible intermediates can all be seen by, say, NMR in the catalytic mixtures does not mean these are the true intermediates. What we need to do is to show that each of the proposed intermediates reacts sufficiently fast to account for the formation of products, that is, that they are *kinetically competent* to do the reaction.

A particularly unpleasant version of this situation is the decomposition of some or all of the complex to give a highly reactive form of the free metal, which now acts as a heterogeneous catalyst. Organometallic chemists like to find examples of homogeneous catalysts that catalyze reactions previously known to be catalyzed heterogeneously only. The Fischer–Tropsch reaction (Section 12.3) and alkane activation (Section 9.6) are examples. It is therefore embarrassing to discover that your unique "homogeneous" catalyst is just a well-known heterogeneous catalyst in disguise. Many of the "homogeneous" hydrogenation catalysts reported in the early days of the development of the field contained a platinum metal halide in a polar solvent under H_2. Viewed with the jaundiced eye of the modern observer, many of these look like preparations of colloidal, and therefore heterogeneous, platinum group metal. (The platinum group metals are Ru, Os, Rh, Ir, Pd, and Pt.) The standard test is the addition of liquid Hg, which selectively poisons any heterogeneous Pt group metal catalyst by absorbtion of Hg onto the active sites.[13a–d] These colloidal particles have been studied in considerable detail.[13g]

Reversibility

In hydrogenation the final step, the reductive elimination of the product, is *irreversible*. This contrasts with the situation in alkene isomerization. In a reversible cycle the products can equilibrate among themselves, and a thermodynamic mixture is always obtained if we wait long enough and if the catalyst retains its activity. This is not the case in hydrogenation, if it were, the *R* and *S* products would eventually come to equilibrium and the e.e. would go to zero with time in an asymmetric hydrogenation. Only an irreversible catalytic cycle (i.e., one in which the last step is irreversible) can give a nonthermodynamic final product ratio. This is very useful because it means we can obtain different (kinetic) product ratios by using different catalysts, and we do not need to be concerned that the products will equilibrate if we leave them in contact with the catalyst. A reversible catalyst can give a nonthermodynamic product ratio initially, but the final ratio will be thermodynamic.

Chiral Poisoning

A new method that can be useful in asymmetric catalysis is *chiral poisoning*, in which an enantiomerically pure compound, P*, selectively binds to and poisons one enantiomer of a racemic catalyst. An e.e. of 49% has been achieved using racemic [(chiraphos)Rh]$_2$(BF$_4$)$_2$ and (*S*)-[Ph$_2$POCH$_2$CH(NMe$_2$)CH$_2$CH$_2$SMe] as poison with a Rh:P* ratio of 0.7. An advantage is that P* can be easily made from methionine, itself easily available optically pure.[13e] A related result is seen with partially resolved [(chiraphos)Rh]$_2$(BF$_4$)$_2$, where the minor enantiomer prefers to form an inactive dimer with the other, leaving the major enantiomer predominating in the pool of catalytically active free monomers. In such a *chiral amplification*,[13f] the product of the catalytic reaction has a higher e.e. than one would expect from the optical purity of the starting catalyst because the major enantiomer of the catalyst acts as a chiral poison for the minor enantiomer. The structure of the dimer is shown below; its 18e configuration makes it catalytically inactive until it dissociates.

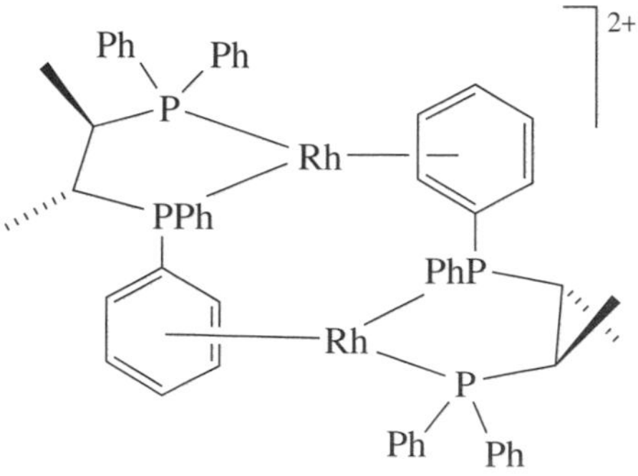

Heterolytic H_2 Activation

We now look at the second mechanistic class of hydrogenation catalyst. $RuCl_2(PPh_3)_3{}^{14a}$ is believed to activate H_2 heterolytically, a reaction accelerated by bases, such as NEt_3. The base abstracts a proton from H_2, leaving an H^- bound to the metal (Eq. 9.10) ultimately giving RuH_2L_3, the true catalyst.[14b]

$$RuCl_2(PPh_3)_3 \xrightarrow{H_2} \underset{\partial+ \; H}{\overset{\partial- \; Cl}{\diagdown RuCl(PPh_3)_3}} \xrightarrow{-HCl} RuHCl(PPh_3)_3$$

$$(9.10)$$

Equation 9.10 is a simple example of a *σ-bond metathesis*,[15] a reaction that has the general form of Eq. 9.11, and in which Y is often a hydrogen atom.

$$X-Y + M-Z \longrightarrow M-X + Y-Z \qquad (9.11)$$

It now seems very likely that the intermediate in the heterolytic activation of H_2 is a dihydrogen complex (Section 3.4). The protons of a dihydrogen ligand are known to be more acidic than those of free H_2, and many H_2 complexes can be deprotonated by NEt_3.[16] In this way the metal gives the same products that would have been obtained by an oxidative addition–reductive elimination pathway, but by avoiding the oxidative addition, the metal avoids becoming Ru(IV), not a very stable state for ruthenium; even $RuH_4(PPh_3)_3$, long thought to be Ru(IV), is now known to have the structure $Ru(H_2)H_2(PPh_3)_3$.[16] Other than in their method of activating H_2, these catalysts act very similarly to the oxidative addition group. As a 16e hydride complex, $RuCl_2(PPh_3)_3$ can coordinate the alkene, undergo insertion to give the alkyl, then liberate the alkyl by a heterolytic activation of H_2, in which the alkyl group takes the proton and the H^- goes to the metal to regenerate the catalyst.

$$(9.12)$$

In the example shown below, a coordinated H_2 is split by a pendant basic amino group. Depending on the size of the phosphine, the H_2 complex or the splitting product can be observed.[16b] The size of the phosphine determines the position of the counterion, which in turn decide the isomer formed.

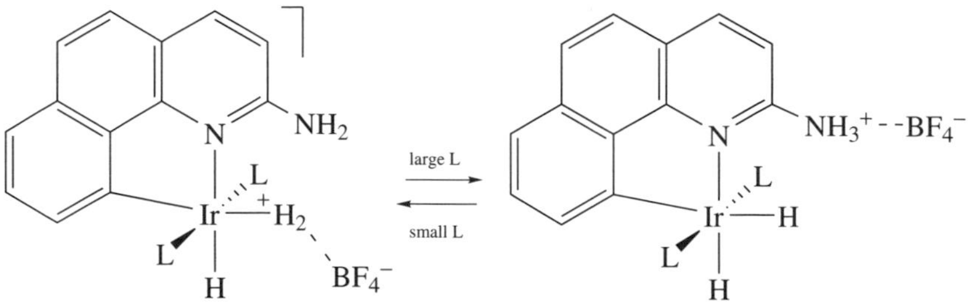

$(L = PR_3)$

Homolytic H_2 Activation

Iguchi's[17] paramagnetic d^7 $Co(CN)_5^{3-}$ system was a very early (1942) example of a homogeneous hydrogenation catalyst. It is an example of the third and rarest group of catalysts, which activate hydrogen homolytically. Another way of looking at this is to say the cobalt system activates H_2 by a binuclear oxidative addition. This is not unreasonable for this Co(II) complex ion, a metal-centered radical that has a very stable oxidation state, Co(III), one unit more positive. Once $CoH(CN)_5^{3-}$ has been formed, a hydrogen atom is transferred to the substrate in the second step, a reaction that does not require a vacant site at the metal, but does require the resulting organic radical to be moderately stable—hence the fact that the Iguchi catalyst will reduce only activated alkenes, such as cinnamate ion, in which the radical is benzylic and therefore stabilized by resonance. Finally, the organic radical abstracts H• from a second molecule of the cobalt hydride to give the final product.

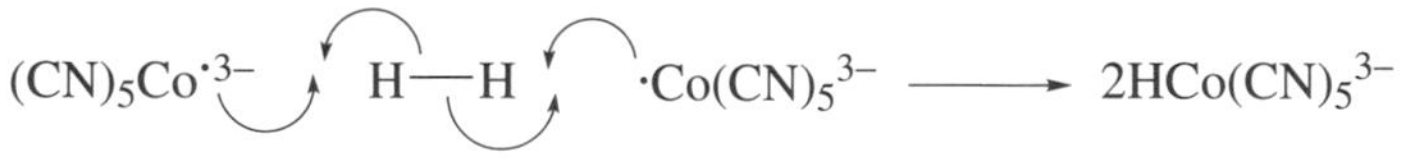

$$(CN)_5Co^{\cdot 3-} \quad H{-}H \quad {}^{\cdot}Co(CN)_5^{3-} \longrightarrow 2HCo(CN)_5^{3-} \quad (9.13)$$

$$HCo(CN)_5^{3-} + Ph\diagup\diagdown CO_2^- \longrightarrow {}^{\cdot}Co(CN)_5^{3-} + Ph\diagup\diagdown CO_2^- \quad (9.14)$$

$$HCo(CN)_5^{3-} + Ph\diagup\diagdown CO_2^- \longrightarrow {}^{\cdot}Co(CN)_5^{3-} + Ph\diagup\diagdown CO_2^- \quad (9.15)$$

Arene Hydrogenation

Although heterogeneous hydrogenation catalysts such as Rh/C readily reduce arenes, none of the homogeneous catalysts discussed up to now are effective for this reaction. A few homogeneous catalysts have been found, however, of which $(\eta^3\text{-allyl})Co\{P(OMe)_3\}_3$ is the best known.[18] When benzene is reduced with this catalyst using D_2, the all-cis isomer of $C_6H_6D_6$ is obtained, and no propane or propene is formed. Heterogeneous catalysts do not give this all-cis product, so we can rule out decomposition of the catalyst to colloidal metal. This suggests that the role of the allyl group may be to open up to the η^1 form to allow the arene to bind in the η^4 form. Phosphite dissociation is still required to allow the H_2 to bind; plausible first steps of the reduction are as follow:

$$(9.16)$$

Transfer Hydrogenation[19a]

In this important variant of hydrogenation, the source of the hydrogen is not free H_2 but an easily oxidizable substrate, such as isopropanol.

$$Me_2CHOH + RCH{=}CH_2 \longrightarrow Me_2C{=}O + RCH_2CH_3 \qquad (9.17)$$

Transfer hydrogenation is particularly good for the reduction of ketones and imines that are somewhat more difficult to reduce with H_2 than are C=C bonds. Bäckvall and co-workers[19a] have shown how $RuCl_2(PPh_3)_3$ is effective at 80°C with added base as catalyst promoter. The role of the base is no doubt to form the isopropoxide ion, which presumably coordinates to Ru and by β elimination forms a hydride and acetone. Noyori and co-workers[19b] have has a remarkable asymmetric catalytic hydrogen transfer that goes without direct coordination of the C=O bond to the metal. Instead, the metal donates a hydride to the C=O carbon while the adjacent Ru-NH_2R group donates a proton to the C=O oxygen.

In ionic hydrogenation, somewhat analogous to the Noyori et al. mechanism, the substrate is protonated and the resulting carbonium ion quenched with a hydride, such as $CpW(CO)_3H$. This is effective for ketones and hindered alkenes but has not yet been made catalytic.[19c]

- Homogeneous catalysts can have a selectivity that is both high and tunable.
- Asymmetric catalysis uses enantiopure ligands to generate products of high enantiomeric excess.

9.3   ALKENE HYDROFORMYLATION

In the late 1930s, Otto Roelen at Ruhrchemie discovered hydroformylation, sometimes called the *oxo* process, one of the first commercially important reactions to use a homogeneous catalyst. He found that an alkene can be converted to the homologous aldehyde by the addition of H_2 and CO, catalyzed by $Co_2(CO)_8$; further reduction to the alcohol is observed under some conditions (Eq. 9.18). More than 4 million tons of aldehydes are made annually in this way.

$$(9.18)$$

A schematic mechanism of this reaction is shown in Fig. 9.4. The $Co_2(CO)_8$ first reacts with H_2 via a binuclear oxidative addition to give $HCo(CO)_4$, which is the active catalyst. The proposed catalytic cycle[20a] is shown in Fig. 9.4: CO dissociation generates the vacant sites required for the alkene and H_2. The first steps resemble hydrogenation in that an alkyl is formed by alkene insertion. Note that at this stage there is no hydride on the metal, so that instead of being trapped by reductive elimination with a hydride, as happens in hydrogenation, the alkyl undergoes a migratory insertion to give the corresponding acyl. H_2 probably binds to give an H_2 complex, followed by a heterolytic H_2 cleavage (e.g., Eq. 9.10) to give the product aldehyde and regenerate the catalyst.[20a] This route avoids oxidative addition of H_2, which has a high activation energy in this system. $HCo(CO)_4$ can also cleave the acyl to give the aldehyde by a binuclear reductive elimination, but this is probably a minor pathway in the catalytic cycle.

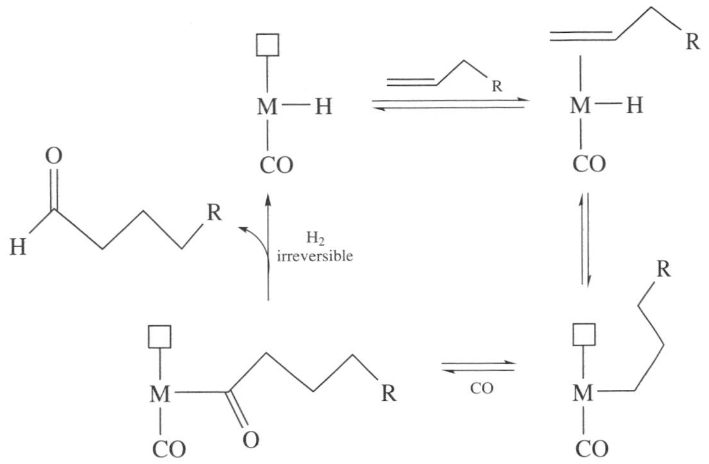

FIGURE 9.4 Catalytic cycle proposed for hydroformylation with HCo(CO)$_4$ as catalyst. Alkene insertion also takes place in the opposite direction to give the 2° alkyl, which goes on to the branched aldehyde RCH(Me)CHO, but this parallel and usually less important cycle is not shown.

Either 1° or 2° aldehydes can be formed from an alkene such as propene; the linear 1° material is much more valuable commercially. Since this is an irreversible cycle, the 1° and 2° products do not come to equilibrium, the kinetic ratio of products being retained. It is not normally the regiochemistry of alkene insertion that decides this ratio but the rate at which the 1° and 2° alkyls are trapped by migration to CO. Slaugh and Mullineaux [20b] made the commercially important discovery that the addition of phosphines, such as P(n-Bu)$_3$, gives a catalyst that is not only much more active (5–10 atm pressure are required versus 100–300 atm for the unmodified catalyst),[1] but which also favors the 1° over the 2° aldehyde to a greater extent (n : iso ratio 8 : 1 vs. 4 : 1). It is believed that the steric bulk of the phosphine both encourages the formation of the less hindered 1° alkyl and speeds up migratory insertion.

With some substrates, HCo(CO)$_4$ is thought to transfer an H atom (H•) to the alkene. This tends to happen when the substrate radical is specially stabilized (e.g., PhCH•–CH$_3$ from PhCH=CH$_2$). The radical may then recombine with the Co to give an alkyl. This accounts for the preferential formation of the 2° aldehyde from styrene.

The more highly phosphine-substituted rhodium species RhH(CO)(PPh$_3$)$_3$ is an even more active catalyst, 1 atm pressure and 25°C being sufficient, and it is even more selective for the 1° product. Rh$_4$(CO)$_{12}$ is also very active but has very poor selectivity, so once again, the presence of phosphine improves the selectivity. The mechanism is broadly similar to the Co-catalyzed process. In practice, excess PPh$_3$ is added to the reaction mixture to prevent the formation of the less selective HRh(CO)$_4$ and HRhL(CO)$_3$ species by phosphine dissociation. The system is also an active isomerization catalyst because much the same mixture of aldehydes is formed from any of the possible isomers of the starting alkene. This is a very useful property of the catalyst because internal isomers of an alkene are easier to obtain than the terminal one. The commercially valuable terminal aldehydes can still be obtained from these internal alkenes. The catalyst first converts the internal alkene, such as 2-butene, to a mixture of isomers including the terminal one. The latter is hydroformylated much more rapidly than the internal ones, accounting for the predominant 1° aldehyde product. Since the terminal alkene can only ever be a minor constituent of the alkene mixture (because it is thermodynamically less stable than the other isomers), this reaction provides another example of a catalytic process in which the major product is formed from a minor intermediate:

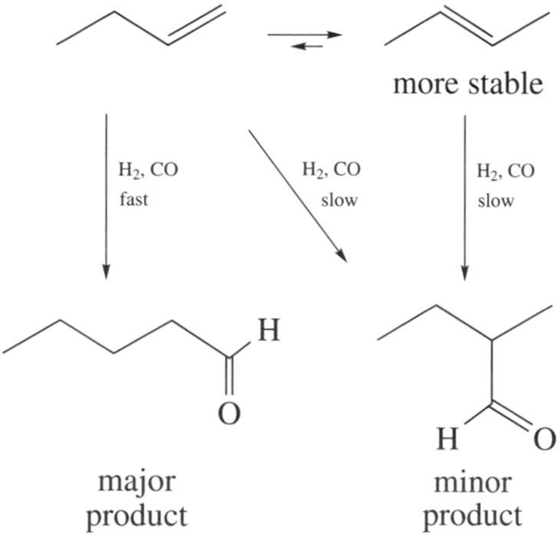

(9.19)

Binuclear Noncluster Catalysts

Stanley et al.[21a] have shown how a rhodium complex that is a poor catalyst in monomeric form becomes very active and selective when connected in a binuclear system with a methylene bridge as shown below. Linear to branched ratios as high as 27 : 1 can be achieved. A rhodium hydride is believed to attack a RhCOR group at the neighboring site in the product forming step. This shows how the proximity of two metals can provide useful chemical effects *without* their being permanently connected by a metal–metal bond.

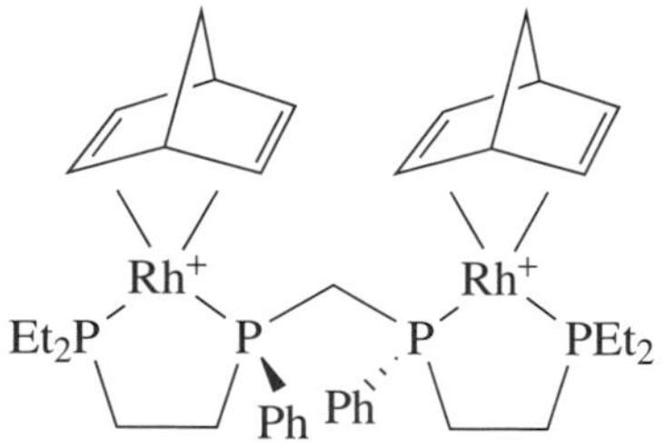

Chelating and Phosphite Ligands

Chelating ligands have proved useful in rhodium-catalyzed hydroformylation, where the n : iso ratio seems to increase with the bite angle (preferred P—M—P angle) of the phosphine. The large cone angle and bite angle[21b] are beneficial in commercial hydroformylation using an interesting chelating phosphite ligand shown. The catalyst is made in situ by adding the phosphite to Rh(CO)$_2$(acac). Cuny and Buchwald[21c] have found that this type of catalyst has a high degree of substrate functional group tolerance (e.g., RCN, RBr, RI, thioether, amide, ketal) and gives very high n : iso ratios.

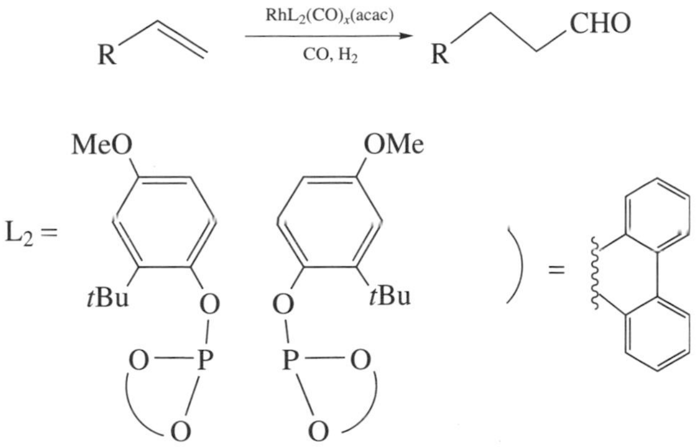

9.4 HYDROCYANATION OF BUTADIENE[22]

The existence of proteins (**9.27**) suggested to Carothers at du Pont that the peptide link, —NHCO—, might be useful for making artificial polymers. Out of this work came nylon-6,6 (**9.28**), one of the first useful petroleum-based polymers.

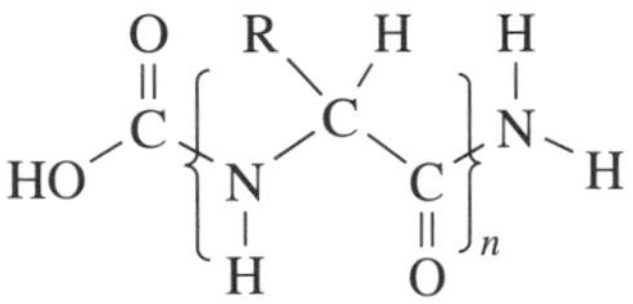

9.27

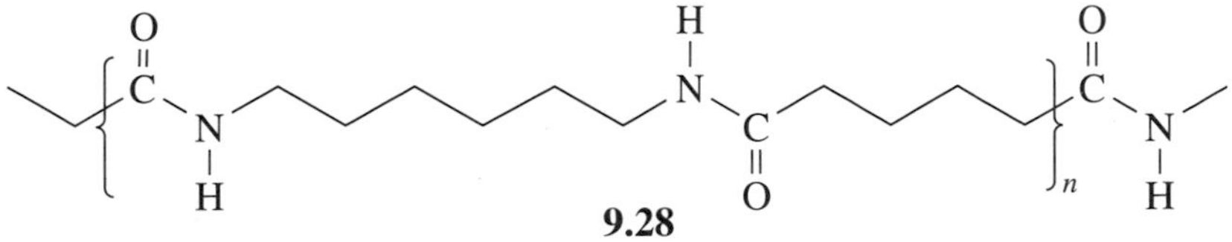

9.28

nylon-6, 6

Now that the original patents have long expired, the key to making this material commercially is having the least expensive source of adiponitrile. The polymer itself is made from adipoyl chloride and hexamethylene diamine, both of which are made from adiponitrile. This nitrile was originally made commercially by the chlorination of butadiene (Eqs. 9.20–9.22). This route involves Cl_2, which leads to corrosion difficulties, only to give NaCl as a by-product, which involves disposal problems. All large commercial concerns defend their key intermediates by trying to find better routes to them before their competitors do. The advent of homogeneous catalysis provided an opportunity to improve the synthesis of adiponitrile very considerably. Fortunately for du Pont it was in its laboratories that the new route was discovered by Drinkard.

$$CH_2=CHCH=CH_2 \xrightarrow{Cl_2} ClCH_2CH=CHCH_2Cl \xrightarrow{NaCN}$$

$$NCCH_2CH=CHCH_2CN \xrightarrow{H_2, \text{ catalyst}} NC(CH_2)_4CN \quad (9.20)$$
$$\text{adiponitrile}$$

$$H_2NCH_2(CH_2)_4CH_2NH_2 \xleftarrow{H_2, \text{ catalyst}} NC(CH_2)_4CN \xrightarrow{\text{(i) } H_2O, \text{ (ii) } PCl_3}$$
$$\text{hexamethylene diamine} \qquad\qquad \text{adiponitrile}$$

$$ClCO(CH_2)_4COCl \quad (9.21)$$
$$\text{adipoyl chloride}$$

$$H_2NCH_2(CH_2)_4CH_2NH_2 + ClCO(CH_2)_4COCl \longrightarrow \text{nylon} \quad (9.22)$$

In the hydrocyanation of butadiene, 2 mol of HCN are added to butadiene with a nickel complex as catalyst to obtain adiponitrile directly.

$$CH_2=CHCH=CH_2 \xrightarrow{2HCN, \; NiL_4} NC(CH_2)_4CN \quad\quad (9.23)$$

For simplicity, we first look at the hydrocyanation of ethylene, for which the cycle shown in Fig. 9.5 is believed to operate. The oxidative addition of HCN to the metal gives a 16e nickel hydride that undergoes ethylene insertion to give an ethyl complex. Reductive elimination of EtCN gives the product. The reaction with butadiene is more complex. In the alkene insertion, the product is an allyl complex (Fig. 9.6); reductive elimination now gives 3-pentene nitrile. This internal alkene cannot be directly hydrocyanated to give adiponitrile but has to be isomerized first. $HNiL_3^+$, present in the reaction mixtures, is a very active isomerization catalyst by the hydride mechanism. The internal alkene is therefore isomerized to the terminal alkene and hydrocyanated to give adiponitrile. One remarkable feature of the isomerization is that the most stable alkene isomer, 2-pentene nitrile, is formed only at a negligible rate. This is fortunate because once formed, it cannot revert to the 3- and 4-isomers, nor is it hydrocyanated, so remains as a contaminating by-product. The terminal alkene,

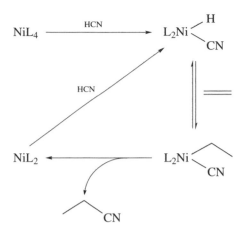

FIGURE 9.5 Hydrocyanation of ethylene by $NiL_4[L = P(O\text{-}o\text{-}tolyl)_3]$.

4-pentene nitrile, once formed, is rapidly hydrocyanated selectively to the linear adiponitrile product; all the other possible dinitriles are formed at a much slower rate.

An important step at several points in the catalytic cycle is loss of L to open up a vacant site at the metal. The rate and equilibrium constant for these dissociative steps are controlled largely by the bulk of the ligand. Electron-withdrawing ligands are required to facilitate the other steps in the cycle, so that one of the best is o-tolyl phosphite, which combines steric bulk with a strongly electron-withdrawing character.

When the first HCN adds to butadiene, some undesired branched 2-methyl-3-butenenitrile, **9.29** in Fig. 9.6, is formed along with the desired linear 3-butenenitrile. Interestingly, the first HCN addition to butadiene is reversible because the branched nitrile can be isomerized to the linear form with NiL_4. This means that **9.29**, which is an activated allylic nitrile, can oxidatively add to the nickel to give back the η^3-allyl nickel cyanide. Labeling studies suggest that this intermediate goes back to HCN and butadiene, before readdition to give the linear nitrile. The formation of the saturated dinitriles is irreversible, however.

The Lewis acid BPh_3 is a useful co-catalyst for the reaction. Such additives are often termed *promoters*. In this case the promoter improves the selectivity of the system for linear product (it is not clear exactly why) and improves the life of the catalyst. A catalyst deactivates when it loses some or all of its activity by going down an irreversible path that leads to an inactive form of the metal complex. In this case, the formation of the inactive $Ni(CN)_2$ is the principal deactivation step. This can happen in several ways; an example is shown here:

$$HNiL_2(CN) \xrightarrow{\text{HCN}} [H_2NiL_2(CN)]^+CN^- \xrightarrow{-H_2}$$

$$[NiL_2(CN)]^+CN^- \xrightarrow{-L} Ni(CN)_2 \quad (9.24)$$

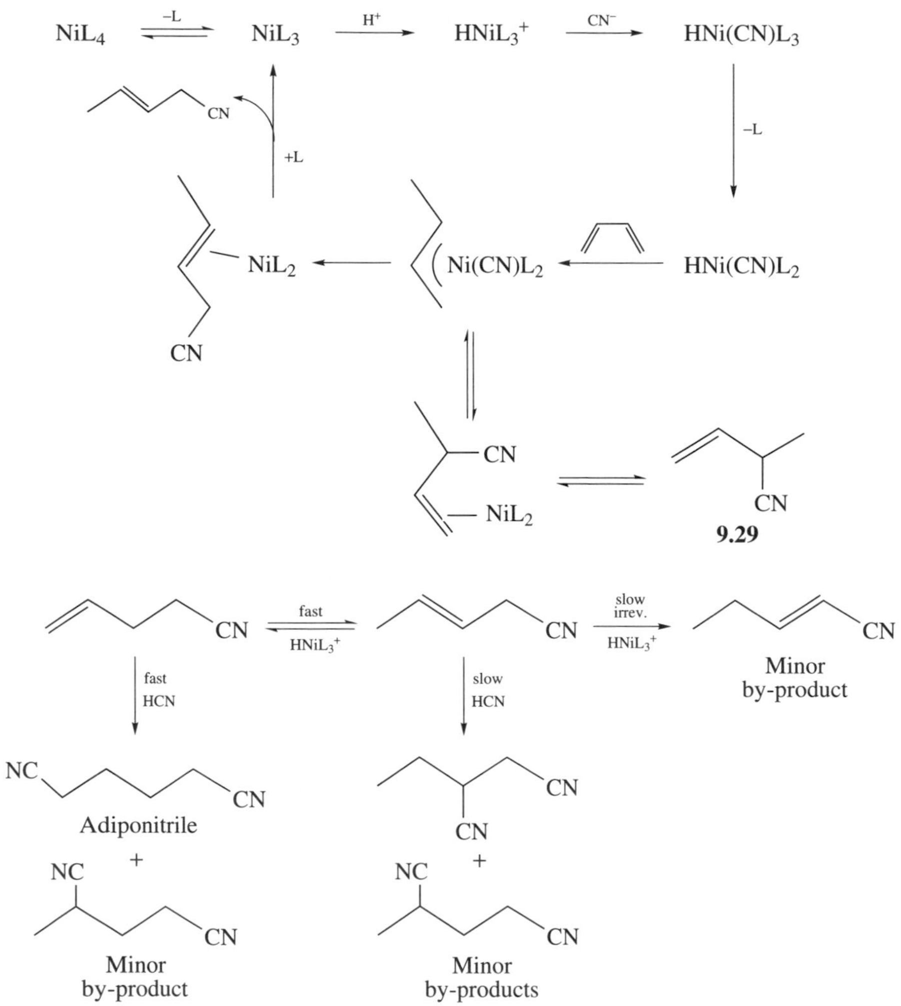

FIGURE 9.6   Hydrocyanation of butadiene by $NiL_4[L = P(O\text{-}o\text{-tolyl})_3]$.

The promoter is believed to inhibit the reaction in Eq. 9.24 by binding to the NiCN groups by the lone pair on nitrogen. This lowers the basicity of the metal and makes it less likely to protonate. Binding of the promoter to the CN group can be detected by IR spectroscopy: On adding BPh_3 to a solution of $HNiL_2(CN)$, the $\nu(CN)$ stretching vibration moves 56 cm^{-1} to higher frequency and the intensity increases. This is because the lone pair on nitrogen has some C–N antibonding character, so depleting the electron density in this orbital by transfer of some of the electron density to boron strengthens the C–N bond and moves the corresponding vibration to higher frequency. The intensity of IR bands is controlled by the change in dipole moment during the vibration $(d\mu/dr)$; by further polarizing the ligand, the Lewis acid increases $d\mu/dr$.

9.5 ALKENE HYDROSILATION AND HYDROBORATION

Hydrosilation

This is the addition of a silane R_3Si-H across a $C=C$ double bond as illustrated in Eq. 9.25. It is a reaction of some commercial importance for the synthesis of silicon-containing monomers, for use in such products as the self-curing silicone rubber formulations sold for domestic use.

$$Cl_3Si-H + H_2C=CH_2 \longrightarrow Cl_3Si-CH_2-CH_3 \tag{9.25}$$

One of the earliest catalysts (1957), H_2PtCl_6, or Speier's catalyst,[23] is extremely active; 0.1 ppm of catalyst is effective. Commercially, the catalyst is normally not even recovered from the product, even though Pt is a precious metal. There is an induction period before hydrosilation begins, which is attributed to reduction of H_2PtCl_6 to the active catalyst, which was taken to be a Pt(II) species. The Chalk–Harrod mechanism,[24] shown in Fig. 9.7a, was accepted for many years. Only recently has it been suggested[13d,25] that the true catalyst is colloidal platinum metal. A colloid of this type is a suspension of very fine particles ($\sim$10–1000 Å radius) of metal in a liquid, which will not settle out of the liquid even on prolonged standing. This implies that in its active form Speier's catalyst is a heterogeneous catalyst. In spite of this new development, other hydrosilation catalysts, such as $C_2(CO)_8$, $Ni(cod)_2$, $NiCl_2(PPh_3)_2$, and $RhCl(PPh_3)_3$, do seem likely to be authentically homogeneous and may well operate by the Chalk–Harrod mechanism.

As in hydroformylation, both linear and branched products can be obtained from a substituted alkene such as $RCH=CH_2$:

$$RCH=CH_2 + R'_3Si-H \xrightarrow{\text{catal.}} RCH_2CH_2SiR'_3 + RCH(Me)SiR'_3 \tag{9.26}$$

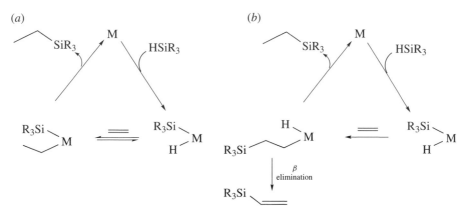

FIGURE 9.7 (a) The Chalk-Harrod mechanism for alkene hydrosilation. (b) An alternative mechanism in which insertion takes place into the M−Si bond. This accounts for the formation of vinylsilane, sometimes seen as a by-product in hydrosilation.

The amount of each product obtained depends on the catalyst and the nature of R and R', but the linear form generally tends to predominate. The unsaturated vinylsilane, RCH=CHSiR'$_3$, is also a product. Although minor in most cases, conditions can be found in which it predominates. The Chalk–Harrod mechanism cannot explain the formation of this *dehydrogenative silation* product, but the alternate mechanism of Fig. 9.7*b*, in which the alkene inserts into the M–Si bond first, does explain it because β elimination of the intermediate alkyl leads directly to the vinylsilane. As in hydrogenation, syn addition is generally observed. Apparent anti addition is due to isomerization of the intermediate metal vinyl, as we saw in Eqs. 7.21 and 7.22, a reaction in which initial insertion of alkyne into the M–Si bond must predominate (>99%).[26a] Co$_2$(CO)$_8$ also catalyzes a number of other reactions of silanes, as shown in Fig. 9.8. Hydrostannylation, Sn–H addition to an alkene, is also attracting attention.[26b]

Hydroboration

RhCl(PPh$_3$)$_3$ catalyzes the addition of the B–H bond in catecholborane (**9.30**) to alkenes (Eq. 9.27). This reaction also goes without catalyst, but the catalytic reaction has usefully different chemo-, regio-, and stereoselectivities.[27] Oxidative workup of the alkylboron product normally gives the corresponding alcohol. The catalytic cycle may be complex, with more than one species contributing to activity, and the results depend on whether aged or freshly prepared catalyst is used. For example, fresh catalyst (or aged catalyst with excess PPh$_3$) gives >99% branched product, PhCHOHMe, from styrene, while aged catalyst gives approximately 1 : 4 branched : linear alcohol. The uncatalyzed reaction gives linear alcohol. In certain cases, dehydrogenative hydroboration is seen, and the vinylboron product appears as an aldehyde or ketone on oxidative workup. As in hydrosilation this may be the result of C=C insertion into Rh–X (X = B or Si) bonds, followed by β elimination. In the stoichiometric reaction of catecholborane with RhCl(PPh$_3$)$_3$, one product is the B–H oxidation

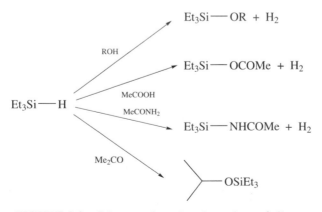

FIGURE 9.8   Other metal-catalyzed reactions of silanes.

product, $RhHCl(BR_2)(PPh_3)$.

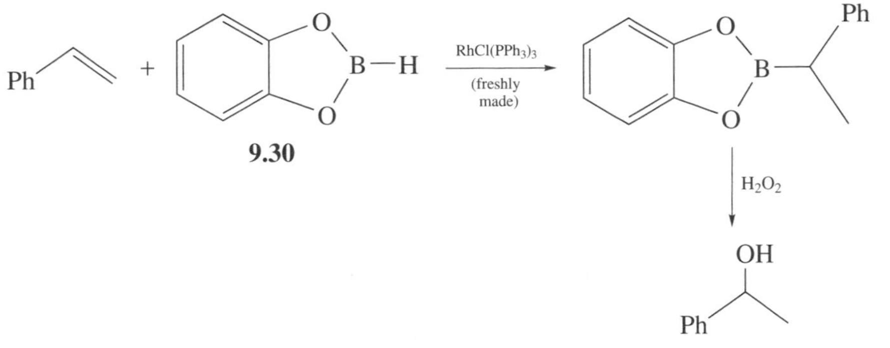

$$(9.27)$$

9.6 COUPLING REACTIONS

Organometallic chemistry has provided several important new methods to carry out carbon–carbon or carbon–heteroatom bond formation. Such processes, termed *coupling reactions*,[28] developed from early work by Tsuji and by Trost, now have a central place in organic chemistry and in the pharmaceutical industry. Among the most useful are shown below. The reaction names are given because research papers often refer to them in this way.

All of these reactions are catalyzed by a number of palladium complexes or simply by a mixture of $Pd(OAc)_2$ and PR_3. They all probably involve initial reduction of Pd(II) to Pd(0), followed by oxidative addition of RX to generate an R–Pd(II) intermediate. R normally has to be an aryl or vinyl group; otherwise β elimination can cause decomposition of the required R–Pd intermediate. Basic, bulky phosphines such as $P(t\text{-}Bu)_3$ can facilitate the oxidative addition step; this may take place to PdL as intermediate, in line with the idea that the microscopic reverse process, reductive elimination from Pd(II), often takes place from a three-coordinate LPd(R)(X) intermediate (Section 6.5).

In the Tsuji–Trost reaction, an allylic acetate first oxidatively adds to the Pd(0) catalyst to give a π-allyl complex, which undergoes nucleophilic attack by the carbanion derived from the deprotonated active methylene compound to give the coupled product; allyl alcohols and aldehydes can be coupled by a related procedure.[29]

In the Mizoroki–Heck reaction,[30a] the Pd–R species undergoes an insertion with the alkene co-substrate, followed by β elimination to give the product. Base, such as NaOAc, is necessary to recycle the palladium by removing HX in the last step. The role of the electron-withdrawing group (EWG), R′, on the alkene cosubstrate, is to ensure that the insertion step takes place in the direction shown, to give R′CH=CHR, not CH_2=CRR′. The Pd–R bond seems to be polarized in the direction Pd^+–R^-, and so the R group attacks the positive end of the C=C double bond, which is the one remote from an EWG.

Tsuji–Trost reaction:

$$R{-}OAc + CH_2E_2 \xrightarrow[\text{base (B)}]{\text{Pd catalyst}} R\text{-}CHE_2 + [BH]OAc$$

allyl   active

acetate methylene   (E = COOMe)

Mizoroki–Heck reaction:

$$R{-}X + \underset{\text{alkene}}{=\!\!\!\diagdown_{R'}} \xrightarrow[\text{base (B)}]{\text{Pd catalyst}} \overset{R}{\diagdown}\!\!=\!\!\diagdown_{R'} + [BH]X$$

aryl or vinyl
halide

Miyaura–Suzuki coupling:

$$R{-}X + ArB(OH)_2 \xrightarrow{\text{Pd catalyst}} R{-}Ar + XB(OH)_2$$

aryl or vinyl
halide

Stille coupling:

$$R{-}X + ArSnR'_3 \xrightarrow{\text{Pd catalyst}} R{-}Ar + XSnR'_3$$

aryl or vinyl
halide

Buchwald–Hartwig amination:

$$Ar{-}X + R'_2NH \xrightarrow{\text{Pd catalyst}} Ar{-}NR'_2 + [H_2NR'_2]X$$

aryl halide

In the other three coupling reactions,[31] RX oxidative addition again occurs, but the anionic group, X, is then replaced by a nucleophilic group from the co-substrate, either an aryl group or an NR_2 group. In the final step, reductive elimination gives the product. Other nucleophiles also work, for example, C−O coupling to form aryl ethers is possible with aryl halides and phenolates.[32]

The mechanisms shown in Figs. 9.9 and 9.10 indicate only the main steps proposed for these reactions because the details may vary with the exact conditions.[33] The Pd(0)/(II) oxidation state couple is usually proposed but Pd(II)/(IV) may be involved in some cases.

Asymmetric versions of some of these reactions are known; an example is the Mizoroki–Heck reaction of a dihydrofuran shown on p. 266. Note that in this case the alkene insertion takes place with the opposite regiochemistry than normal—that is, the R group becomes attached to C-1, next to the ring ether substituent of the C=C double bond. This is because the ether substituent at C-1 is an electron donor, by virtue of the oxygen π-lone-pair electrons, and the polarization of the C=C double bond is inverted from the classic Mizoroki–Heck situation with an EWG substituent. The syn addition of the Pd−R bond means

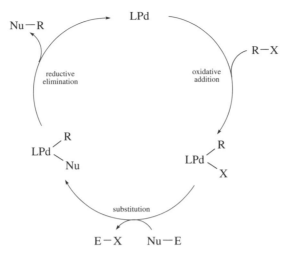

FIGURE 9.9 Schematic mechanism for the Miyaura–Suzuki (R = aryl or vinyl; Nu = aryl; E = B(OH)$_2$), Stille (R = aryl or vinyl; Nu = aryl; E = SnR$_3'$), and Buchwald–Hartwig (R = aryl or vinyl; Nu = NR$_2'$; E = H) coupling reactions.

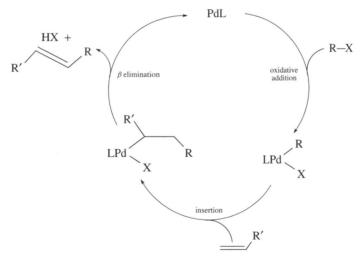

FIGURE 9.10 Schematic mechanism for the Mizoroki–Heck (R' = electron-withdrawing group) coupling reaction.

that β elimination has to occur at the 3 position with the only available syn-β hydrogen. This leaves intact the stereochemistry at the 1 position, now a chiral center. Using chiral phosphines such as the one shown can lead to asymmetric induction with very high enantiomeric excesses.[34]

97% e.e.
87% yield

9.7   SURFACE AND SUPPORTED ORGANOMETALLIC CATALYSIS

Organometallic complexes can be supported in a variety of ways to give catalysts that are more readily separable from the soluble products of the reaction than are soluble homogeneous catalysts.

Surface Organometallic Chemistry

Classical heterogeneous catalysts, consisting of metal particles supported on a solid surface such as silica or carbon, are of great commercial importance. It has also proved possible to support a variety of organometallic species on silica so as to obtain mononuclear complexes covalently anchored to the silica surface. Silica has surface SiOH groups, often denoted $\equiv$Si—O—H, which can form $\equiv$Si—O—M links to the attached metal. The oxophilic early metals are particularly well suited to this approach.

   Once the organometallic species is bound to a surface, many of the usual solution characterization methods no longer apply. A combination of EXAFS (extended X-ray absorption fine structure: see Chapter 16), solid-state NMR, and IR spectroscopy, however, can often give sufficient information to characterize the surface-bound species.

   Unusual reactivity can be seen, probably as a result of *site isolation*, which prevents the formation of inactive $M(\mu\text{-}OR)_n M$ dimers. For example, $ZrNp_4$ (Np = neopentyl) reacts with silica to give what is believed to be a surface-bound $\equiv$Si—O—$ZrNp_3$.[35] Only one Si—O—Zr bond is formed because only one equivalent of NpD is liberated. Reaction of this species with H_2 at 150°C gives CH_4 and EtH (not NpH) and gives a monohydride believed to be $(\equiv$Si—O$)_3$ZrH (**9.31**) with $\nu(Zr\text{-}H)$ at 1625 cm^{-1} together with SiH_2 groups with $\nu(SiH)$ at 2196 and 2263 cm^{-1}. The transformation is thought to go as shown in Eq. 9.28, where an initially formed zirconium hydride reacts with the SiOSi bridges of the silica surface. The evolution of the CH_4 and EtH, not NpH, in the hydrogenolysis reaction is a result of the catalytic conversion of the initially formed NpH to CH_4 and EtH by the $(\equiv SiO)_3ZrH$ species, **9.31**, in the presence of H_2.

$$(9.28)$$

This remarkable C—C cleavage reaction also takes place on treatment of the hydride **9.31** with external NpH. The initial products in this case are MeH and Me_3CH. The reaction probably goes by a preferential β-alkyl elimination of a neopentyl zirconium species as outlined in Eq. 9.29. The resulting isobutene is hydrogenated to Me_3CH and the zirconium methyl hydrogenated to methane.

9.31

$$(9.29)$$

A number of commercially important catalysts consist of organometallic compounds covalently attached to surfaces. In the Phillips alkene polymerization catalyst,[36] for example, $CrCp_2$ is supported on silica. While there is not full agreement on the nature of the species formed, the Si—OH groups of the silica surface are believed to bind the metal via Si—O—Cr linkages in a similar way to **9.31**.

Late metals can also be bound to silica. For example, $[MeRhL_2(CO)]$ (L = PMe_3) reacts with silica surface hydroxyl groups to give a surface-bound Rh(I) species, **9.32**, with release of methane.[37]

$$\equiv Si-OH + MeRhL_2(CO) \longrightarrow \equiv Si-O-RhL_2(CO) + MeH \qquad (9.30)$$
$$\textbf{9.32}$$

$$\equiv Si-O-RhL_2(CO) + L' \longrightarrow \equiv Si-O-RhL_2(L') + CO \qquad (9.31)$$

$$[L' = {}^{13}CO \text{ or } P(i-Pr)_3]$$

$$\equiv Si-O-RhL_2(CO) + RX \longrightarrow \equiv Si-O-RhL_2(H)X(CO) \qquad (9.32)$$

$$(R = H, X = Cl; \text{or } R = Me, X = I)$$

Species **9.32** was shown to undergo substitution with ^{13}CO (Eq. 9.31) and P(i-Pr)$_3$ as well as give oxidative addition with HCl and MeI (Eq. 9.32).

Polymer-Bound Organometallics

Cross-linked polystyrene forms small beads that swell in organic solvents. This produces a "Swiss cheese" structure in which a large internal surface area becomes accessible to external reagents. Merrifield showed that this surface could be chloromethylated so as to allow a variety of species to be grafted to the bead. Equation 9.33 shows how the CH$_2$Cl groups of the Merrifield resin can be converted into −PPh$_2$ groups; the sphere represents the polymer bead.

$$(9.33)$$

Species **9.33** is the bead equivalent of PR$_3$ and can thus readily bind metals that react with PR$_3$.[37] The resulting complexes are active catalysts and have the advantage of being readily removed from the reaction medium by filtration. Supported catalysis in principle allows multiple catalysts to operate independently, without mutual interference, to bring about multistep syntheses.

Homogeneous/Heterogeneous Catalysis Ambiguity

Since metal surfaces are heterogeneous catalysts, some or all of any homogeneous catalyst may decompose to give nanoparticles (metal particles of ca. 10–1000 Å in size) of the metal that can mimic the appearance of a homogeneous solution yet, in fact, be a heterogeneous catalyst. In many early "homogeneous" catalysts, metal salts in polar solvents were subjected to reducing agents—ideal conditions for synthesis of nanoparticles. The active species from a number of pincer precatalysts for the Heck coupling have been identified as nanoparticles[38] using the Hg(0) test. Metallic Hg deactivates heterogeneous catalysts but normally leaves homogeneous catalysts unaffected.[39] In one recent case, an impressive level of asymmetric catalysis (90% e.e.) proved possible by modification of the nanoparticle surface by an asymmetric "ligand,"[40] so asymmetric catalysis is not proof of homogeneity. Palladium is particularly prone to metal formation and the occurrence of two catalytic reactions that are not normally seen for true homogeneous catalysts can be considered a "red flag": nitrobenzene reduction and arene hydrogenation. Careful work in homogeneous catalysis includes tests for heterogeneity.[41]

Future Prospects

An area in which we may expect future developments is in the imaginative application of currently known catalytic reactions to the commercial and laboratory synthesis of new classes of compounds.[2] Our understanding of the catalysis of oxidation is still in a much more primitive state than is the case for the reductive reactions discussed in this chapter, and this remains a great challenge. The current interest in developing environmentally sound synthetic routes in the chemical industry will also provide important new goals for homogeneous catalysis for the future.

- A wide variety of different organic reactions, many with very great industrial importance, can be catalyzed.
- A disadvantage of homogeneous catalysis, separation of the catalyst from the product, can be mitigated by supporting the catalyst on a filterable solid.

REFERENCES

1. B. Cornils and W. A. Herrmann, *Applied Homogeneous Catalysis*, Wiley VCH, Weinheim, 2002.

2. H. Kurosawa and A. Yamamoto, *Fundamentals of Molecular Catalysis*, Elsevier, Amsterdam, 2003.

3. (a) R. B. Jordan, *Reaction Mechanisms of Inorganic and Organometallic Systems*, Oxford University Press, Oxford, 1991; (b) B. R. James, *Adv. Organometal. Chem.* **17**, 319, 1979; (c) J. Halpern and S. Wong, *J. Chem. Soc., Chem Commun.* 629, 1973; (d) P. Meakin, J. P. Jesson, and C. A. Tolman, *J. Am. Chem. Soc.* **94**, 3240, 1972; (e) J. Halpern et al., *J. Am. Chem. Soc.* **99**, 8055, 1977; (f) R. A. Sánchez–Delgado and M. Rosales, *Coord. Chem. Rev.* **196**, 249, 2000.

4. M. Bowker, *Basis and Applications of Heterogeneous Catalysis*, OUP, New York, 1998.

5. A. Fersht, *Enzyme Structure and Mechanism*, WH Freeman, San Francisco, 1984; D. L. Purich, *Enzyme Kinetics and Mechanism*, Academic, San Diego, 2002.

6. D. Evans, J. Osborn, and G. Wilkinson, *J. Chem. Soc. A* 3133, 1968.

7. C. P. Casey and C. R. Cyr, *J. Am. Chem. Soc.* **95**, 2248, 1973.

8. R. H. Crabtree, *Acc. Chem. Res.* **12**, 331, 1979.

9. J. M. Brown et al., *Chem. Commun.* 348, 1982; *Tetrahedron Lett.* **25**, 1393, 1984; R. H. Crabtree and M. W. Davis, *Organometallics* **2**, 681, 1983; G. Stork et al., *J. Am. Chem. Soc.* **105**, 1072, 1983; D. A. Evans et al., *J. Am. Chem. Soc.* **106**, 3866, 1984.

10. (a) R. R. Schrock and J. A. Osborn, *J. Am. Chem. Soc.* **98**, 2134, 2143, 4450, 1976; (b) M. J. Burk, *Accts. Chem. Res.* **33**, 363, 2000; R. Noyori, *Angew. Chem. Int. Ed.* **41**, 2008, 2002.

11. J. Halpern et al., *J. Am. Chem. Soc.* **102**, 5954, 1980.

12. (a) R. D. Larsen et al., *J. Am. Chem. Soc.* **111**, 7650, 1989; (b) H. U. Blaser, *Adv. Synth. Catal.* **344**, 17, 2002; *Chem. Commun.*, 293, 2003.

13. (a) P. Maitlis et al., *J. Mol. Catal.* **7**, 543, 1980; (b) D. R. Anton and R. H. Crabtree, *Organometallics* **2**, 855, 1983; (c) J. P. Collman et al., *J. Am. Chem. Soc.* **106**, 2569, 1984; (d) L. N. Lewis and N. Lewis, *J. Am. Chem. Soc.* **108**, 7228, 1986; (e) J. W. Faller and J. Parr, *J. Am. Chem. Soc.* **115**, 804, 1993; (f) D. G. Blackmond, *Accts. Chem. Res.* **33**, 402, 2000; (g) J. S. Bradley and B. Chaudret, *New J. Chem.* **22**, 1177, 1998 (special issue); B. Chaudret, *Adv. Synth. Catal.* **346**, 72, 2004.

14. (a) D. Evans, J. Osborn, J. A. Jardine, and G. Wilkinson, *Nature* **208**, 1203, 1965; (b) J. E. Backvall et al., *Chem. Comm.* 351, 1999.

15. J. E. Bercaw et al., *Tet. Lett.* **41**, 7609, 2000.

16. (a) G. J. Kubas, *Metal Dihydrogen and Sigma Bond Complexes*, Kluwer, New York, 2001. (b) E. Clot, O. Eisenstein, D. H. Lee, A. Macchioni, R. H. Crabtree et al., *New J. Chem.* **27**, 80, 2003.

17. J. Iguchi, *J. Chem. Soc. Jpn.* **60**, 1287, 1939; R. Mason and D. W. Meek, *Angew. Chem., Int. Ed.* **17**, 183, 1978.

18. L. S. Stuhl, M. Rakowski-Dubois, F. J. Hirsekorn, J. R. Bleeke, A. Z. Stevens, and E. L. Muetterties, *J. Am. Chem. Soc.* **97**, 237, 1975; E. L. Muetterties and J. R. Bleeke, *Acc. Chem. Res.* **12**, 324, 1979; see also K. Jonas, *J. Organometal. Chem.* **400**, 165, 1990.

19. (a) J. E. Bäckvall et al., *Chem. Commun.* 1063, 1991; 337, 980, 1992; (b) M. Yamakawa, I. Yamada, and R. Noyori, *Angew. Chem. Int. Ed.* **40**, 2818, 2000; (c) R. M. Bullock, J. R. Norton et al., *Angew Chem., Int. Ed.* **31**, 1233, 1992.

20. (a) T. Ziegler and L. Versluis, *Adv. Chem. Ser.* **230**, 75, 1992; (b) L. H. Slaugh and R. D. Mullineaux, *J. Organometal. Chem.* **13**, 469, 1968.

21. (a) G. B. Stanley et al., *Science* **260**, 1784, 1993; (b) J. R. Briggs, and G. T. Whiteker, *Chem. Comm.* 2174, 2001; (c) G. D. Cuny and S. L. Buchwald, *J. Am. Chem. Soc.* **115**, 2066, 1993.

22. W. C. Siedel and C. A. Tolman, *Ann. N. Y. Acad. Sci.* **415**, 201, 1983; S. Sabo-Etienne, B. Chaudret et al *Organometallics* **23**, 3363, 2004.

23. J. S. Speier, *Adv. Organometal. Chem.* **17**, 407, 1979.

24. J. F. Harrod and A. J. Chalk, *J. Am. Chem. Soc.* **88**, 3491, 1966.

25. (a) In 1961,[25b] it was suggested that the Speier catalyst was heterogeneous, but that view was later reversed;[25c] (b) R. A. Benkeser et al., *J. Am. Chem. Soc.* **83**, 4385, 1961; (c) R. A. Benkeser et al., *J. Am. Chem. Soc.* **90**, 1871, 1968.

26. (a) R. S. Tanke and R. H. Crabtree, *J. Am. Chem. Soc.* **112**, 7984, 1990. (b) N. D. Smith, J. Mancuso, and M. Lautens, *Chem. Rev.* **100**, 3257, 2000.

27. D. A. Evans and G. C. Fu, *J. Org. Chem.* **55**, 2280, 1990; K. Burgess, R. T. Baker, et al., *J. Am. Chem. Soc.* **114**, 9350, 1992; K. Burgess, *Chem. Rev.* **91**, 1179, 1991.

28. Y. Yamamoto and E. Negishi, eds., *J. Organomet. Chem.* **576**, 1–317, 1999 (special issue); J. Tsuji, *Palladium Reagents and Catalysts*, Wiley, New York, 1995; F. Diederich and P. J. Stang, *Metal-catalyzed Cross Coupling*, Wiley, New York, 1998.

29. Y. Tamaru et al., *J. Am. Chem. Soc.* **123**, 10401, 2001.

30. G. T. Crisp, *Chem. Soc. Rev.* **26**, 427, 1998; W. A. Herrmann, V. P. W. Boehm and C. P. Reisinger, *J. Organomet. Chem.* **576**, 23, 1999.

31. A. Suzuki, *J. Organomet. Chem.* 576, 147, 1999; B. H. Yang and S. L. Buchwald, *J. Organomet. Chem.* **576**, 125, 1999; J. F. Hartwig, *Acc. Chem. Res.* **31**, 852, 1998.

32. J. F. Hartwig et al., *J. Am. Chem. Soc.* **121**, 3224, 1999.

33. C. Amatore and A. Jutand, *J. Organomet. Chem.* **576**, 254, 1999.

34. A. Pfaltz et al., *J. Organomet. Chem.* **576**, 16, 1999.

35. F. Lefebvre and J. M. Basset, *J. Mol. Catal.* **146A**, 3, 1999; M. Tafaoulk, E. Schwab, M. Schultz, D. Vanoppen, M Walter, J. Thivolle, and J. M. Basset, *Chem. Comm.* 1434, 2004.

36. F. H. Karol et al., *J. Pol. Sci. A* **10**, 2621, 1972; O. -M. Bade et al., *Organometallics*, **17**, 2524, 1998.

37. S. L. Scott et al., *J. Am. Chem. Soc.* **120**, 1883, 1998; R. H. Fish et al., *React. Polym.* **6**, 337, 1987.

38. M. R. Eberhard, *Org. Lett.* **6**, 2125, 2004.

39. D. R. Anton and R. H. Crabtree, *Organometallics* **2**, 855 1983.

40. S. Jansat, M. Gomez, K. Philippot, G. Muller, E. Guiu, C. Claver, S. Castillo, and B. Chaudret, *J. Am. Chem. Soc.* **126**, 1592, 2004.

41. M. Poyatos, F. Marquez, E. Peris, C. Claver, and E. Fernandez, *New J. Chem.* **27**, 425, 2003.

PROBLEMS

1. Compound **9.34** is hydrogenated with a number of homogeneous catalysts. The major product in all cases is a ketone, $C_{10}H_{16}O$, but small amounts of an acidic compound $C_{10}H_{12}O$, **9.35**, are also formed. What is the most reasonable structure for **9.35**, and how is it formed?

9.34

2. Would you expect Rh(triphos)Cl to be a hydrogenation catalyst for alkenes (triphos = $Ph_2PCH_2CH_2CH_2PPhCH_2CH_2CH_2PPh_2$)? How might the addition of BF_3 or $TlPF_6$ affect the result?

3. Predict what you would expect to happen in the hydrocyanation of 1,3-pentadiene with HCN and Ni{P(OR)$_3$}$_4$?

4. Write out a mechanism for arene hydrogenation with (η^3-allyl)Co{P(OMe)$_3$}$_3$, invoking the first steps shown in Eq. 9.16. Why do you think arene hydrogenation is so rare for homogeneous catalysts? Do you think that diphenyl or naphthalene would be more or less easy to reduce than benzene? Explain your answer.

5. Suggest plausible mechanisms for the reactions shown below, which are catalyzed by a Rh(I) complex, such as RhCl(PPh$_3$)$_3$.

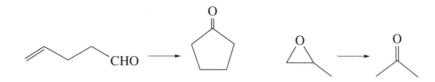

6. Comment on the possibility of finding catalysts for each of the following:

$$Propane \longrightarrow cyclopropane + H_2$$
$$CO_2 \longrightarrow CO + \tfrac{1}{2}O_2$$
$$CH_4 + \tfrac{1}{2}O_2 \longrightarrow CH_3OH$$

7. What do you think is the proper structural formulation for H$_2$PtCl$_6$? Why do you think the compound is commonly called chloroplatinic *acid*? Make sure that your formulation gives a reasonable electron count and oxidation state.

8. In some homogeneous alkyne hydrosilations, a second product (B) is sometimes found in addition to the usual one (A). How do you think B is formed? Try to write a balanced equation for the reaction, assuming an A/B ratio of 1 : 1 and you will see that A and B cannot be the only products. Suggest the most likely identity for a third *organic* product C, which is always formed in equimolar amounts with B.

$$RC{\equiv}CH + R_3SiH \longrightarrow RCH{=}CHSiR_3 + RC{\equiv}CSiR_3$$
$$\qquad\qquad\qquad\qquad\qquad\qquad A \qquad\qquad B$$

9. The reaction

$$2CH_2{=}CHCO_2Et \longrightarrow EtOOCCH{=}CHCH_2CH_2COOEt$$

catalyzed by (η^6-C$_6$H$_6$)Ru(CH$_2$=CHCO$_2$Et)$_2$/Na[C$_{10}$H$_8$] has been studied by workers at du Pont as a possible route to adipic acid, an important precursor for nylon. Suggest a mechanism. How might you use a slightly modified substrate to test your suggestion? (Na[C$_{10}$H$_8$] is simply a reducing agent.)

10. $(\eta^6\text{-}C_6H_6)Mo(CO)_3$ is a catalyst for the reduction of 1,3-dienes to cis monoenes with H_2; suggest how this might work, why the cis product is formed, and why the alkene is not subsequently reduced to alkane.

$$RCH{=}CH{-}CH{=}CHR + H_2 \longrightarrow cis\text{-}[RCH_2{-}CH{=}CH{-}CH_2R]$$

10

PHYSICAL METHODS IN ORGANOMETALLIC CHEMISTRY*

We now look at some of the main methods of identifying a new complex, assigning its stereochemistry, and learning something about its properties. We will look at some applications of the most commonly used spectroscopic and crystallographic methods to organometallic chemistry. Citations to both introductory and more advanced treatments of the methods themselves are also included.

10.1 ISOLATION

Before we can study the complexes, we have to isolate them in a pure form. The methods used resemble those of organic chemistry. Most organometallic complexes are involatile crystalline materials, although some are liquids [e.g., $CH_3C_5H_4Mn(CO)_3$], or even vapors [e.g., $Ni(CO)_4$] at room temperature and pressure. They normally have solubilities similar to those of organic compounds. The main difference from organic chemistry is that many organometallic compounds are "air sensitive," which usually means that they react with O_2 and sometimes with water. The electropositive f-block, and early d-block metals are the most reactive. Crystalline material is usually more stable than are solutions, but in many cases both must be kept under dry N_2 or Ar, and air and water must

*Undergraduates taking this course may not have had a physical chemistry course. The material on spectroscopy has therefore been gathered together here, so that instructors have the option of omitting all or part of it without losing the narrative flow of the rest of the book.

The Organometallic Chemistry of the Transition Metals, Fourth Edition, by Robert H. Crabtree
Copyright © 2005 John Wiley & Sons, Inc.

be completely removed from all the solvents used. One general method involves using flasks and filter devices fitted with ground joints for making connections and vacuum taps for removing air or admitting nitrogen. This so-called Schlenk glassware allows all operations to be carried out under an inert atmosphere on an ordinary benchtop. In an alternative setup, operations are carried out in a N_2-filled inert atmosphere box. This sounds more formidable than it is, and the details of the techniques used are available in an excellent monograph.[1a]

10.2 ^{1}H NMR SPECTROSCOPY

A variety of spectroscopic techniques are also available for structure determination.[1b] Organometallic chemists tend to rely heavily, perhaps too heavily, on NMR. The ^{1}H NMR technique[2] is perhaps most useful for metal hydrides, which resonate in an otherwise empty region of the spectrum (0 to -40δ) to high field of TMS.[3] This unusual chemical shift is ascribed to shielding by the metal d electrons, and the observed shifts generally become more negative on going to higher d^n configurations. The number of hydrides present may be determined by integration or, if phosphines are also present, from $^2J(P,H)$ coupling seen in the ^{31}P NMR (Section 10.4).[4] When we refer to $^nJ(X,Y)$ coupling, we mean the coupling of nucleus X and Y; n indicates the number of bonds that connect X and Y by the shortest route. $^2J(P,H)$ coupling to the phosphorus nuclei of cis or trans phosphines to the hydride proton in phosphine hydride complexes can also be seen in the ^{1}H spectrum. Trans couplings (90–160 Hz) are larger than cis ones (10–30 Hz), and this can be very useful in determining the stereochemistry of the molecule. Figure 10.1 shows the spectra of some octahedral iridium hydrides that illustrate how the stereochemistries can be deduced. The 5-, 7-, 8-, and 9-coordinate hydrides are often *fluxional*. That is to say, the molecules are nonrigid, so that the ligands exchange positions within the complex fast enough to become equivalent on the NMR timescale ($\sim 10^{-2}$ s). We will look at the consequences of fluxionality in more detail later (Section 10.5).

Virtual Coupling

Alkyl phosphines, such as PMe_3 or PMe_2Ph, also give important stereochemical information in the ^{1}H NMR. If two such phosphines are cis, they behave independently, and we usually see a doublet for the methyl groups, due to coupling to the $I = \frac{1}{2}$ ^{31}P nucleus. If the two phosphines are trans, the phosphorus–phosphorus coupling becomes so large that the ^{1}H NMR of the methyl substituents is affected. Instead of a simple doublet, we see a distorted triplet with a broad central peak. This behavior is called *virtual coupling*,[3] and means that the methyl group appears to be coupled to both its own and the trans phosphorus nucleus about equally, giving rise to the virtual triplet (Fig. 10.2a). This happens when $^2J(P,P)$ between equivalent P nuclei becomes large, as it is when the phosphines in question are trans. Intermediate values of the phosphorus–phosphorus coupling constant,

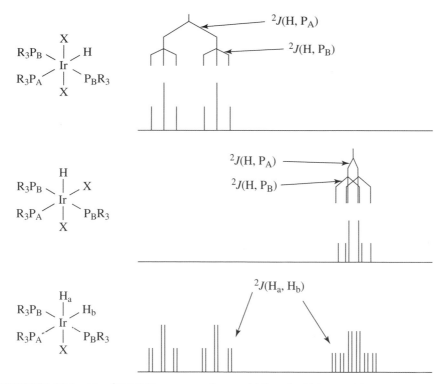

FIGURE 10.1 The ¹H NMR spectra of some iridium hydrides (hydride region). Each stereochemistry gives a characteristic coupling pattern.

which may occur for P–M–P angles between 90° and 180°, give patterns intermediate between a doublet and a virtually coupled triplet (Figs. 10.2*b*, and 10.2*c*).

Diastereotopy

The ¹H NMR spectrum of PMe₂Ph ligand in a metal complex gives useful information about the symmetry of the complex. Suppose we want to distinguish between **10.1** and **10.2** (Fig. 10.3) from the NMR alone. As shown in Fig. 10.3, **10.1** has a plane of symmetry containing X, Y, the PMe₂Ph phosphorus atom, and the metal. Note that in the rotamer of PMe₂Ph in which the Ph group also lies in the plane of symmetry, the mirror plane reflects one P–Me group into the other and makes them equivalent. In **10.2**, on the other hand, there is no such plane of symmetry and Me′ and Me″ are inequivalent. When this happens the two methyls are called *diastereotopic* groups.[2a] In general, two groups will be inequivalent if no symmetry element of the molecule relates the two groups. By far the most common situation is the presence of a plane of symmetry that contains the M–P bond; the presence of such a plane makes the two methyls equivalent. Diastereotopic groups are inequivalent and will generally resonate at different

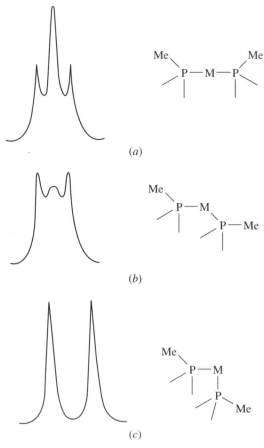

FIGURE 10.2 Virtual coupling in the PMe proton resonance of methylphosphine complexes. Each methyl group shows coupling both to its own phosphorus nucleus and to that of the second phosphine as long as $^2J(P, P')$ is large enough. As the P−M−P angle decreases from 180°, the virtual coupling decreases, until at an angle of 90°, the appearance of the spectrum is a simple doublet, owing to coupling of the phosphorus methyl protons only to their own phosphorus nucleus, not that of the second phosphine.

chemical shifts. We will therefore see a simple doublet (due to coupling to phosphorus) for **10.1** and a pair of doublets for **10.2**. Because each doublet comes at a different chemical shift, the appearance of the spectrum will be different at a different magnetic field, as shown in Fig. 10.3. The resonances for the diastereotopic groups differ by a certain chemical shift; the pattern therefore changes at higher field (also shown in Fig. 10.3), where there are more hertz per ppm (parts per million).[2a] The same inequivalence is found for any compound (e.g., **10.3**) in which no element of symmetry exists that will transform one of the two otherwise identical groups into the other. Structures **10.2** and **10.3** show inequivalent Me groups, regardless of whether the M−P or C−C bonds are freely rotating.

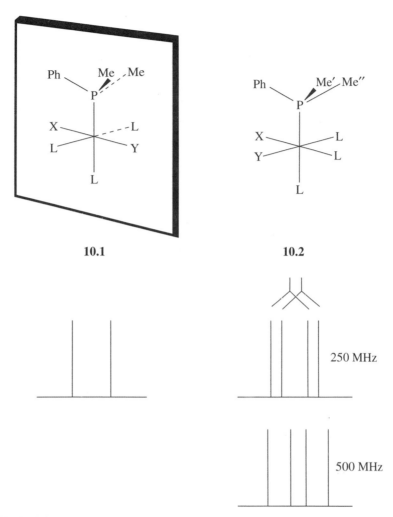

FIGURE 10.3 The methyl groups in **10.1** are equivalent in the proton NMR because of the presence of a mirror plane that contains the M–P bond; they appear as a single doublet as a result of ^2J(P, H) coupling. The methyl groups in **10.2** are inequivalent (diastereotopic) and so resonate at different frequencies. The two distinct doublets that result do not appear the same at a higher field and so are distinguishable from a doublet of doublets due to coupling, the appearance of which would be essentially invariant with field.

Chemical Shifts

In organic compounds, we are used to thinking of certain ranges of chemical shift values as being diagnostic for certain groups. We have to be more cautious in organometallic chemistry because the shifts are much more variable. For example, the vinyl protons of a coordinated alkene can resonate anywhere from 2 to 5δ (free alkene: $5-7\delta$). In the metalacyclopropane (X_2) extreme, the shifts are at the high-field end of the range, closer to those in cyclopropane itself; and in the opposite L extreme, they are closer to those in the free alkene, near 5δ. Hydride resonances are even more variable. In Ir(III) complexes, they tend to depend on the nature of the trans ligand and can range from -10δ, for high-trans-effect ligands, (e.g., H) to -40δ, for low-trans-effect ligands (e.g., H_2O).[4a] Structural assignments based on the value of a coupling constant tend to be more secure than ones based on the value of a chemical shift, although the shifts can be valuable in cases where their reliability has been well established, such as in the Ir(III) hydrides mentioned above. In general, protons attached to carbons bound to a metal show a coordination shift of 1–4 ppm to low field on binding; more remote protons usually show small coordination shifts (<0.5 ppm).

There are also special circumstances in which shifts can be affected by neighboring groups in predictable ways. In indenyl complexes, for example, the aromatic ring current of the benzo group induces high-field shifts in the protons of other ligands on the metal that spend a substantial amount of their time directly above the benzo ring. The ortho protons of the PPh_3 groups of **10.4** experience a

10.4

shift of -0.27 ppm relative to those of the analogous complex $CpIrHL_2^+$, which lacks the benzo ring. The preferred conformation of **10.4** in solution, shown below, was deduced from this evidence.[5]

Paramagnetic Complexes

Metal complexes can be paramagnetic, and this can lead to large shifts in the NMR resonances;[2d] for instance, $(\eta^6\text{-}C_6H_6)_2V$ has a ^1H NMR resonance at 290δ. More commonly, these resonances are broadened to such an extent that they become effectively unobservable. As we shall see in Section 10.5, other processes can also broaden resonances in diamagnetic molecules. A featureless NMR spectrum does not necessarily mean that no organometallic complexes are present.

10.3 ^{13}C NMR SPECTROSCOPY[2c]

Although ^{13}C $\left(I = \frac{1}{2}\right)$ constitutes only 1% of natural carbon, it is usually possible to get good proton decoupled ^{13}C NMR spectra from most organometallic complexes in a reasonable time. It is helpful to run an off-resonance decoupled spectrum as well; this introduces only 1-bond C,H couplings and reveals the number of protons to which each carbon is bound (CH$_3$ gives a quartet, CH$_2$ a triplet, etc.). The resulting spectra often allow the structure of a complex to be deduced, even in cases in which the proton NMR spectrum is too complex to decipher. The structures of **10.5** and **10.6**, which could be obtained only in an inseparable mixture, were deduced in this way.[6] After accounting for the PPh$_3$ resonances, each complex showed two quartets, two triplets, two doublets, and a singlet in the off-resonance decoupled spectrum. These were assigned as shown, ruling out any of the possible alternative structures that had been envisaged for the complexes.

| IrHL$_2^+$ | IrHL$_2^+$ |
| **10.5** | **10.6** |

Certain groups are found in characteristic resonance positions, for example, alkyls from -40 to $+20\delta$, π-bonded carbon ligands such as alkenes, Cp, and arenes from $+40$ to $+120\delta$, carbonyls around $150-220\delta$ (terminal) and $230-290\delta$ (bridging) and carbenes in the range $200-400\delta$. Relaxation (Section 10.7) of the ^{13}C nuclei, especially of carbonyls, may be slow, which makes them difficult to observe unless a relaxation reagent such as Cr(acac)$_3$ is added to the sample. Since the dynamic range of the method greatly exceeds that of ^{1}H NMR, the ^{13}C peaks for different carbons in a complex will normally be farther apart in frequency (hertz) than the corresponding ^{1}H peaks. This means that the spectra of complicated molecules are much easier to assign because overlapping of peaks is less likely and also that slower fluxional processes (Section 10.5) can be studied. Coupling is transmitted by the σ bonds of a molecule—the higher the s character of a bond, the higher is the coupling. This is the reason that ^1J(C,H) values depend on the hybridization of the C$-$H bond: sp^3, $\sim$125 Hz, sp^2, $\sim$160 Hz, and sp, $\sim$250 Hz. As in ^{1}H NMR of hydrides, trans couplings, for example, of methyl carbons to phosphorus are larger ($\sim$100 Hz) than cis couplings ($\sim$10 Hz).

Unfortunately, integration of carbon spectra is rarely reliable, in part because of the wide range of relaxation times encountered. Relaxation times of carbenes and other carbons lacking proton substituents are especially long. This means that the nuclei are easily saturated and intensities are low; again, a paramagnetic complex, such as Cr(acetylacetonate)$_3$, is added to help relax these carbons.

In polyene and polyenyl complexes, those carbons directly attached to the metal tend to be more shielded on binding, and a coordination shift (i.e., shift relative to the free ligand) of ~25 ppm to high field is observed. If the metal has a $\frac{1}{2}$ spin, coupling to the metal is also seen. This is very useful for determining the hapticity of the ligand. Coupling to other ligands is sometimes seen, but this is not reliable. The phenomenon of diastereotopy discussed in the last section also applies to carbon NMR and is shown by the diastereotopic P—Me carbons in complexes **10.1** and **10.2**.

10.4 ^{31}P NMR SPECTROSCOPY

Phosphorus-31 NMR spectroscopy^{2c,7} is very useful in studying phosphine complexes. Normally all the ligand protons are decoupled so as to simplify the spectra. The only common exception is the determination of n in $H_nM(PR_3)_m$. This can be done by decoupling only the PR$_3$ protons, while leaving the hydride protons undecoupled.[4] The phosphorus resonance will then appear as an $n + 1$ multiplet. MoH$_6$(PR$_3$)$_3$ could be identified only in this way because it could only be obtained in an impure form.[4b]

Different types of phosphorus ligand will normally resonate within different chemical shift ranges, so that phosphines and phosphites can be reliably distinguished, for example. Free and bound phosphorus ligands also show large coordination shifts. Of even more use is the chelation shift that is observed in chelating phosphines. If the phosphorus is part of a four-, five-, or six-membered ring, then it will resonate at a position shifted by -50, $+35$, and -15 ppm relative to a coordinated but nonchelating ligand having chemically similar R groups around phosphorus.[8] The origin of this shift is not yet understood, but it probably results from changing the hybridization at phosphorus as a consequence of changing the bond angles at phosphorus in different ways in rings of different sizes. This is useful for the detection of such species as cyclometallated phosphines and monodentate diphosphines, both of which are very difficult to characterize in any other way, except by crystallography.

Mechanistic Study of Wilkinson Hydrogenation

Tolman et al.[9] were able to deduce the initial events in the mechanism of Wilkinson hydrogenation (Eq. 10.1 and Section 9.2) from the ^{31}P NMR data shown in Fig. 10.4. Spectrum A shows the proton-decoupled ^{31}P NMR of RhCl(PPh$_3$)$_3$ itself. Two types of phosphorus are seen in a 2 : 1 ratio, P$_a$ and P$_b$ in **10.7**, each showing coupling to Rh ($I = \frac{1}{2}$, 100% abundance). P$_a$ also shows a cis coupling to P$_b$ and P$_b$ shows two cis couplings to the two P$_a$'s. On adding H$_2$ (spectrum B), the starting material almost disappears and is replaced by a new species, **10.8**, in which only P$_a$ now shows coupling to Rh and P$_b$ is a broad hump. Cooling to $-25°$ (spectrum B') restores the coupling pattern expected for the static molecule

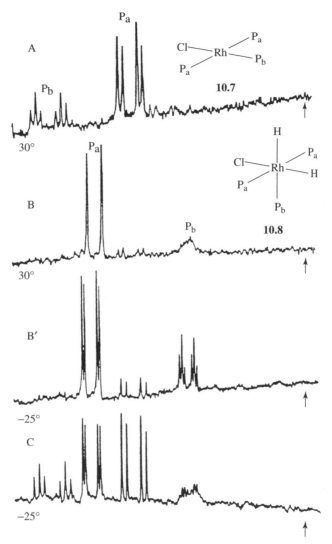

FIGURE 10.4 Proton-decoupled ³¹P NMR data for RhCl(PPh₃)₃: (A) dissolved in CH₂Cl₂; (B) after addition of H₂ at 30°; (B′) after addition of H₂ and cooling to −25°; (C) after sweeping solution B with nitrogen. The different P nuclei in the complex are seen, together with coupling to Rh (large) and couplings to other phosphines (small). In spectrum B, the loss of coupling to Rh and P for one of the two P resonances indicates that this ligand is reversibly dissociating. The most intense peaks are assigned to P_a. Free PPh₃ (arrow) is absent. (Reproduced from Ref. 7b with permission.)

10.8. The change from B to B′ is the result of P_b dissociating at a rate that is slow at −30° but comparable with the NMR timescale at 30° (Section 10.5). In a fluxional process in which two coupled nuclei always stay in the same molecule, couplings are retained in the NMR, but when dissociation of a ligand takes place we have crossover between molecules and couplings to that ligand are lost. In spectrum B, P_a retains full coupling to Rh, while P_b does not, so it is P_b which is dissociating. (The reason is that each of the two peaks of P_a doublet in spectrum B comes from a different population of molecules, one with the Rh spin α and the other with β spin. When P_b moves from molecule to molecule, it samples α and β Rh spins equally and so ends up resonating at an averaged chemical shift.) The amount of *free* PPh$_3$ always remains very small—the arrows show where free PPh$_3$ would appear. Passing nitrogen partially reverses the reaction and a mixture of **10.7** and **10.8** results (spectrum C).

10.7	**10.8**	minor species

$$(10.1)$$

NMR spectra can even be obtained from a number of the common transition metal nuclei,[2e] but this is not yet a routine procedure.

10.5   DYNAMIC NMR

Many organometallic species give fewer NMR resonances than would be predicted from their static structures. This is usually because the molecules are nonrigid,[10] and the nuclei concerned are exchanging places at a rate faster than the NMR timescale ($\sim 10^{-1}$–10^{-6} s).[2a] For example, Fe(CO)$_5$ gives only one carbon resonance at 25°, and yet its IR spectrum (a technique with the much faster timescale of $\sim 10^{-12}$ s) indicates a TBP structure with two types of carbonyl. The reason, proposed by Berry, is that the axial and equatorial carbonyls are exchanging by the *Berry pseudorotation* mechanism shown in Eq. 10.2. Ligands 1–4 become equivalent in the square pyramidal intermediate, and 1 and 4, which were axial in TBP, become equatorial in TBP′.

TBP	Sq. Pyramid	TBP′

$$(10.2)$$

Rate of Fluxionality

Sometimes the exchange takes place at a rate that is comparable with the NMR timescale. When this happens, we can usually slow the exchange down by cooling the sample until we see the static spectrum; this is called the *low-temperature limit*. On the other hand, if we warm the sample, the rate of exchange may rise to the extent that the fully averaged spectrum is observed (the *high-temperature limit*). In between these two extremes, broadened resonances are usually seen. For example, if we take a molecule with two sites A and B that are equally populated, on warming we will see the sequence of spectra illustrated in Fig. 10.5. The two sharp peaks broaden as the temperature rises. If we measure this initial broading at half peak height in units of hertz, and subtract out the natural linewidth that was present before broadening set in, then we have $W_{1/2}$, a measure by Eq. 10.3 of the rate at which the nuclei leave the site during the exchange process.

$$\text{Rate} = \pi(W_{1/2}) \tag{10.3}$$

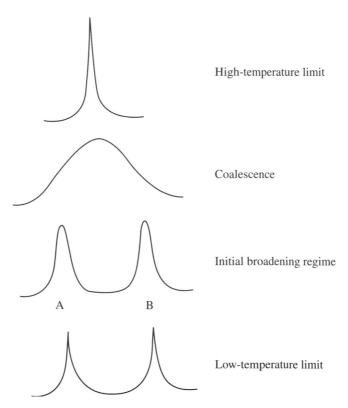

FIGURE 10.5 Changes in the ^1H NMR spectrum of a two-site system on warming of the H_A and H_B protons begin to exchange at rates comparable with the NMR timescale.

As we continue to warm the sample, the broadening increases until the two peaks *coalesce*. The exchange rate required to do this depends on how far apart the two peaks were initially; the appropriate equation is shown as Eq. 10.4, where Δv is the separation of the two resonances of the static structure.

$$\text{Rate} = \frac{\pi \Delta v}{\sqrt{2}} \qquad (10.4)$$

On further warming, the now single peak gets narrower according to Eq. 10.5, and we finally reach a point at which the signal is sharp once more.

$$\text{Rate} \approx \frac{\pi (\Delta v)^2}{2(W_{1/2})} \qquad (10.5)$$

This happens because the exchange is now much faster than the NMR timescale and only an averaged resonance is seen. Note that Eqs. 10.4 and 10.5 contain Δv, the separation of the two resonances measured in hertz. Since this will be different at different magnetic fields (two resonances 1 ppm apart will be 60 Hz apart as observed on a 60-MHz spectrometer, but 100 Hz apart as observed at 100 MHz), the coalescence temperature and the high-temperature limit are field dependent. On cooling the sample, the same changes occur in reverse, a process known as *decoalescence*. The position of the averaged resonance at the high-temperature limit is simply the weighted average of the resonance positions at the low-temperature limit. For example, if we have n_1 nuclei resonating at δ_1 and n_2 at δ_2, then at the high-temperature limit, the resonance position will be the weighted average δ_{av}, given by

$$\delta_{av} \approx \frac{n_1 \delta_1 + n_2 \delta_2}{n_1 + n_2} \qquad (10.6)$$

Dynamic NMR is a very powerful method for obtaining kinetic information about processes that occur at a suitable rate.[11]

Mechanism of Fluxionality

Fluxionality is very common for 5-coordinate TBP complexes, as it is for 7-, 8-, and 9-coordination; 4- and 6-coordinate complexes, on the other hand, tend to be rigid. There is also a second type of fluxionality that takes place irrespective of coordination number.[12] An example is $CpFe(CO)_2(\eta^1\text{-}C_5H_5)$ (Fig. 10.6), which shows only two proton resonances at room temperature, one for the $\eta^5\text{-}C_5H_5$, and one for the $\eta^1\text{-}C_5H_5$. The explanation is that the iron atom is migrating around the $\eta^1\text{-}C_5H_5$ ring at a sufficient rate to average all the proton resonances from the $\eta^1\text{-}C_5H_5$ ring. On going to lower temperature, separate resonances can be distinguished for the three different types of proton in the static $\eta^1\text{-}C_5H_5$ group. If we warm the sample from the low-temperature limit, there will be a different degree of initial broadening of the different proton resonances of the $\eta^1\text{-}C_5H_5$

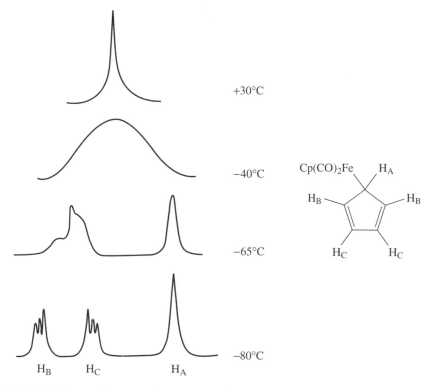

FIGURE 10.6 Fluxionality of $CpFe(CO)_2(\eta^1$-Cp), showing the faster collapse of the H_B resonance, indicating the operation of a 1,2, rather than a 1,3 shift. Only the resonances for the η^1 Cp group is shown, for greater simplicity.

group if the fluxionality involves 1,2 shifts rather than 1,3 shifts. This is because the H_C's are next to one another and so a 1,2 shift (which is indistinguishable from a 1,5 shift) will mean that one of the H_C's will still end up in an H_C site after a 1,2 shift. On the other hand, all the H_B's will end up in non-H_B sites. The exchange rate for H_C's will therefore appear to be one-half of the exchange rate for H_B's, and the resonance for H_C will show less initial broadening. Conversely, H_B's are three carbons apart and so 1,3 shifts will lead to the H_B's showing less initial broadening. Experimentally, it is found that a 1,2 shift is taking place.[13a] Note that we need to assign the spectrum correctly to obtain the correct mechanism and this is often the most difficult step.

In the case of the Cp ligand it is impossible to distinguish between a Woodward–Hoffmann allowed 1,5 shift and a least-motion 1,2 shift because they both give the same final observable result. In an η^1-C_7H_7 system, the two cases are distinguishable, Woodward–Hoffman giving a 1,4 and least motion a 1,2 shift. The appropriate compounds are difficult to make, but Heinekey and Graham were able to show that $(\eta^1$-$C_7H_7)Re(CO)_5$ follows a least motion and η^1-$C_7H_7SnMe_3$ a Woodward–Hoffmann path.[13b]

Another important case of fluxionality is bridge-terminal exchange in carbonyl complexes. The classic example is $[CpFe(CO)_2]_2$, which shows separate Cp resonances for cis and trans CO-bridged isomers in the proton NMR below $-50°C$, but one resonance at room temperature showing that fast exchange takes place. The Adams–Cotton mechanism of exchange (Eq. 10.7) invokes concerted opening of both CO bridges at once; indeed, 1% of the resulting nonbridged isomer can be detected in solution by NMR. The trans compound gives much faster exchange between bridging and terminal COs as shown by the greater initial broadening seen for the trans isomer by ^{13}C NMR. This is because only the nonbridged form of the trans compound, shown on the left in Eq. 10.7, has a choice of COs for re-forming the bridge. For example, if the starred COs were originally bridging, the compound can choose the unstarred pair to re-form the new bridge. This means that a terminal CO in the trans compound has a 50% chance of changing into a bridging carbonyl after one exchange event (going to the open form and back again to the trans form). In the open form of the cis compound, also shown, there is only one pair of trans COs, and the same ones that opened up also have to re-form the bridge unless a rotation takes place, which is thought to be slow.[14] This means that a terminal CO in the cis compound has a low chance of changing into a bridging carbonyl in any one exchange event.

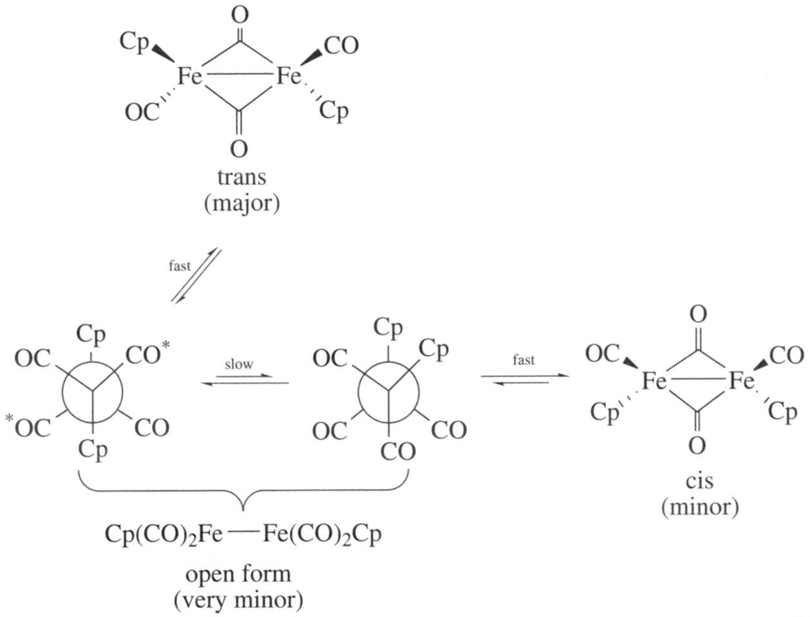

$$(10.7)$$

10.6   SPIN SATURATION TRANSFER

It sometimes happens that a fluxional exchange process is suspected on chemical grounds, but a low-temperature limiting (static) spectrum is seen at all accessible temperatures, and so the exchange is slow on the NMR timescale. An example is shown in Eq. 10.8, where we have to postulate exchange to account for the chemistry of the system, but it is too slow to affect the NMR lineshapes. In such circumstances, we can sometimes use *spin saturation transfer*.[15] The principle of the method is to irradiate one of the resonances in the spectrum of one of the two species and watch for the effects on the spectrum of the other species. If we irradiate the Me_A protons in **10.9a**, we see a diminution in the intensity of the resonance for Me_B in **10.9b**. This shows that Me_A in **10.9a** becomes Me_B in **10.9b** in the course of the exchange; likewise, irradiation at the frequency of H_C affects the intensity of the H_D. In this way we can obtain mechanistic information about the fluxional process.

$$(10.8)$$

10.9a   **10.9b**

The method works because by irradiating the Me_A protons we equalize the spin population in the α (lower-energy) and β (higher-energy) states. If the Me_A protons now become Me_B protons as a result of the exchange, then they carry with them the memory of the equalized populations. Since we need unequal α and β populations in order to observe a spectrum, the newly arrived Me_B protons do not contribute their normal amount to the intensity of the resonance. Now, a very important point is that the Me_A protons begin to lose their memory of the original, artificially equalized α- and β-spin populations by a process known as *relaxation*. There are several mechanisms for relaxation, one of which we will go into in detail in a moment. We need only recognize for now that the initially equal populations in the newly arriving protons relax back to the equilibrium population ratio with a rate $1/T_1(B)$, where $T_1(B)$ is the so-called spin lattice relaxation time, or T_1, of the Me_B site. This is commonly of the order of a few seconds and must be measured independently. The exchange rate has to be faster than about 10 times the T_1, or >0.1 s^{-1}, in order to give a measurable spin saturation transfer effect. This means that the exchange process must be taking place at a rate in the range of $\sim 0.1-1$ s^{-1} for useful information to be obtained; if the exchange is faster than this, line-broadening measurements usually give better rate data. If the initial intensity of the B resonance is I_0, the relaxation time of the B protons is $T_1(B)$, and the final intensity of the B resonance on irradiating the A resonance is I_f, then the exchange rate k is as given by Eq. 10.9.

$$\frac{I_0}{I_f} = \frac{\{T_1(B)\}^{-1}}{k + \{T_1(B)\}^{-1}} \qquad (10.9)$$

The most useful feature of the method is not so much the rate data but that it tells us which protons are exchanging with which, and so allows us to solve some

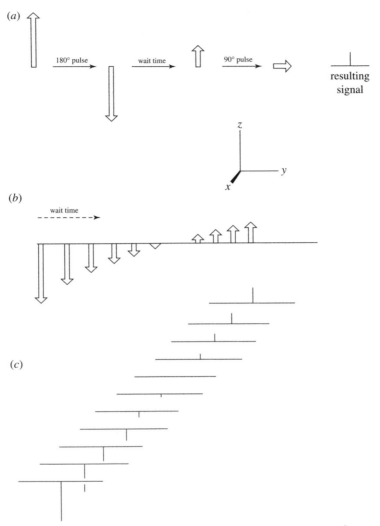

FIGURE 10.7   Inversion recovery method for determining T_1. (*a*) A 180° pulse inverts the spins. They partially recover during the wait time and are sampled by a 90° pulse. (*b*) Varying the wait time allows us to follow the time course of the recovery process, as seen in a stacked plot of the resulting spectra (*c*).

difficult mechanistic problems. In certain circumstances the nuclear Overhauser effect (NOE) (Section 10.7) can affect the experiment and must also be taken into account.[15]

10.7 T_1 AND THE NUCLEAR OVERHAUSER EFFECT

We now need to look at how we can determine the T_1 for any signal, something that we need to do in the spin saturation transfer experiment. If we imagine the sample in the magnetic field, where the z direction is the direction of the applied magnetic field, then the nuclei will line up with and against the field. The difference in energy between these two states is small, and so the excess of the more stable α spins is very slight. This excess we can consider as constituting a net magnetization of the sample pointing in the $+z$ direction (Fig. 10.7). The application of an RF (radio-frequency) pulse to the sample has the effect of rotating this vector out of the z direction into the xy plane, where it can be measured by sensitive detectors. A pulse that is of just the right strength to rotate the vector precisely into the xy plane is called a "90° pulse" because it has caused the vector to move through 90°. The reason we can measure it only in the xy plane is that the vector will now be rotating around the z axis at the Larmor frequency; this moving magnetic field generates a signal in the receiver coil of the instrument. This is the conventional Fourier transform (FT) NMR experiment.

One way to measure T_1 is to apply to the sample a pulse that precisely inverts the spins. This requires a so-called 180° pulse, which is twice the strength of the 90° pulse and rotates the vector from the $+z$ to the $-z$ direction. Originally, there was a slight excess of α spins because these are in a slightly more stable energy level in the magnetic field. A 180° pulse will now give us a slight excess of β spins. We now wait for relaxation to convert the new nonequilibrium distribution favoring β spins back to the old one favoring α. In separate experiments, we can sample the spins after, say, 0.1 s, then after 0.2 s and so on, to see how far they are along the path to recovery. Sampling simply requires a further 90° pulse to bring the spins back into the xy plane to be measured. This gives us the sort of result shown in Fig. 10.7. The negative peaks at short times are due to the inverted spin population at that time; after a sufficiently long time the resonances are all positive and the populations have therefore recovered. Relaxation is normally a first-order process with rate constant $1/T_1$.

T_1 and H_2 Complexes

A useful application of T_1 measurements is the distinction between molecular hydrogen complexes, **10.10**, and classical dihydrides, **10.11**. The reason is that two protons that are very close together can relax one another very efficiently by the so-called dipole–dipole mechanism. Dipole–dipole couplings are several orders of magnitude larger than the usual J couplings we see as splitting in the normal NMR spectrum. The reason we do not see the dipole–dipole splittings in the normal spectrum is that they average exactly to zero with the tumbling of the molecule in solution. Although we cannot see the effects of dipole–dipole coupling directly, the random tumbling of the molecule in solution causes one nucleus, say, H_A, to experience a randomly fluctuating magnetic field due to the magnetic field of a nearby nucleus, H_B. If the fluctuations happen to occur at the Larmor frequency, then H_A can undergo a spin flip by this means, and the α and β spins are eventually brought to thermal equilibrium, or relaxed, in this way. Relaxation is important because to see an NMR signal we need a difference in the populations of α and β spins—when the populations are equal in Fig. 10.7,

10.10   **10.11**

there is no signal. Observing the signal pumps energy into the spins and tends to equalize their populations—relaxation drains energy from the spins and tends to reestablish the population difference.

The rate of relaxation is given by Eq. 10.10, in which h is Planck's constant, γ is the gyromagnetic ratio of the nuclei involved, τ_c is the rotational correlation time (a measure of the rate of molecular tumbling in solution), ω is the Larmor frequency, I is the nuclear spin, and r is the distance between the two nuclei:

$$\text{Rate} = \frac{1}{T_1} = 0.4 \left(\frac{h^2}{2\pi}\right) \gamma^2 \{I(I+1)\} r^{-6} \left[\frac{\tau_c}{1 + \omega^2 \tau_c^2} + \frac{4\tau_c}{1 + 4\omega^2 \tau_c^2}\right] \quad (10.10)$$

The r^{-6} term makes the relaxation rate very sensitive to the distance r. In classical dihydrides, this distance would never be shorter than ~1.6 Å, leading to a relaxation time on the order of half a second. On the other hand, in unstretched molecular hydrogen complexes, this distance is ~0.85 Å and the relaxation time is tens of milliseconds at −80°C. Figure 10.8 shows how the method distinguishes between the classical and nonclassical hydride resonances in **10.12**.[16a]

10.12

Unfortunately, we do not know τ_c in Eq. 10.10. If we did, we could calculate the H—H distance. It turns out that on cooling the sample, T_1 passes through a minimum value. Equation 10.10 predicts that this should happen when $\tau_c = 0.62/\omega$. Since we know ω, we can calculate τ_c at the minimum and so estimate the H—H distance directly. A number of precautions need to be taken because rotation of the H$_2$ about the M—H$_2$ bond reduces the relaxation rate,[16c] and certain metals, notably Re, Nb, V, Mn, Co, and Ta, cause a substantial, but easily calculable, dipole–dipole relaxation of attached protons because both γ and I are high.[16d] We also assume isotropic (random) motion of the molecule,

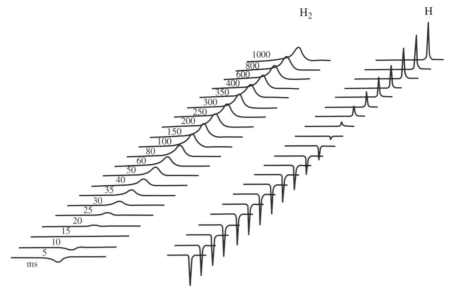

FIGURE 10.8 Differential relaxation of $M(H_2)$ and M–H resonances in **10.12**. The wait times in milliseconds are shown to the left. (We thank the American Chemical Society for permission to reproduce this figure from Ref. 16a.)

which is not the case for such systems as IrH_5L_2 and Cp^*ReH_6, where the MH_x unit has a low moment of inertia and so spins rapidly.

PHIP

A related phenomenon is PHIP,[17a] or para-hydrogen-induced polarization. On cooling a sample of H_2 in the presence of a suitable catalyst, the H_2 becomes enriched in p-H_2 in which the two nuclear spins are aligned. If a hydrogenation reaction is now carried out in an NMR tube under p-H_2, the two hydrogens may be transferred together to a substrate. Their spins are initially aligned in the product but the alignment decays with a rate of $1/T_1$. If decay is not too fast, this results in an extremely nonthermal distribution of spins in the product, and this in turn leads to very large enhancements of the resonances. Traces of a metal dihydride in equilibrium with H_2 are normally undetectable by NMR but can be seen using PHIP.[17b]

Nuclear Overhauser Effect

A valuable technique for determining the conformation of a molecule in solution is NOE. This is observed for any two nuclei in a molecule, say, H_A and H_B, that relax each other by the dipole–dipole mechanism. For this to be effective, the two nuclei need to be <3 Å apart, again as a result of the r^{-6} dependence shown in Eq. 10.10. Distance is the only criterion; the two nuclei do not have to have a bond between them.

The experiment consists of irradiating H_A, while observing H_B. NOE can lead to an increase in the intensity of the H_B resonance by as much as 50%, but usually the increase is 5–10%. In a typical application, NOE is observed in one isomer but not in the other. For example, H_A and H_B in **10.13**, but not **10.14**, show the NOE effect, leading to the assignments shown, which were later confirmed crystallographically ($R = C_2H_5$).[6]

10.13   **10.14**

The origin of the effect is described in Ref. 17c, but in essence by irradiating H_A, we equalize the α- and β-spin populations for this nucleus. Dipole–dipole relaxation then transfers some of the increased spin population in the upper β state of H_A to the lower α state of H_B and consequently increases the intensity of the H_B resonance. The enhancement is measured by the NOE factor, η, given by Eq. 10.11, where I_0 and I_f are the initial and NOE-enhanced intensities, respectively.

$$\eta = \frac{I_f - I_0}{I_0} \tag{10.11}$$

10.8   ISOTOPIC PERTURBATION OF RESONANCE

The isotopic perturbation of resonance (IPR) technique,[18] originally developed by Saunders, was first applied to organometallic chemistry by Shapley. IPR is useful where we are in the fast exchange limit of a fluxional system at all accessible temperatures. We might think that in such a case, we could never obtain information about what the spectrum would be at the low-temperature limit. For example, suppose we want to know whether the methyl group in a complex of 8-methylquinoline is agostic (**10.15**) or not (**10.16**). The usual ^{1}H NMR experiment does not help us because a singlet is expected for both structures. This is so because agostic methyl groups are fluxional, so that the terminal and bridging hydrogens are exchanging rapidly even at $-100°$C.

The IPR experiment consists of taking the proton spectrum (Fig. 10.9) of a mixture of isotopomers of the complex in which the methyl group has been partially substituted with deuterium. Instead of a single resonance, we see separate

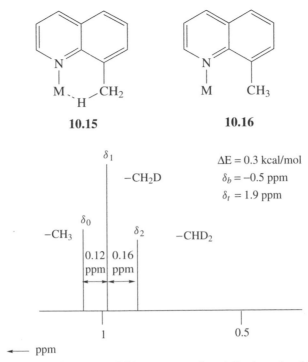

FIGURE 10.9 Schematic proton NMR spectrum of partially deuteriated **10.15** showing the different chemical shifts for each isotopomer.

resonances for each isotopomer. In the d_0 isotopomer (i.e., the isotopomer containing zero deuterium atoms), the observed chemical shift, δ_0, is the average of the shifts for the bridging and terminal positions, weighted by the fact that any given proton will spend twice as much time in terminal sites, as there are two of them, than in the bridging one.

In the d_1 isotopomer, there will be an isotopic preference for H to occupy the bridging sites. The reason is that the zero point energy of H is greater than that of D, and the stability difference depends on the strength of the C−(H,D) bond. The H/D zero-point energy difference is greater for the terminal C−(H,D)$_t$ than for the weaker bridging C−(H,D)$_b$ bond, and so there is an energy advantage for a hydrogen atom to be in a C−H$_b$ site. This population shift translates into a chemical shift in the ^{1}H NMR resonance of the methyl group; δ_1, the shift for the d_1 complex, will be an average that we can calculate by looking at the equilibrium shown in Fig. 10.10b. First we calculate the average shift that would be observed for each form in the absence of IPR. For example, **10.17** has one terminal and one bridging H and so the required average is $(\delta_t + \delta_b)/2$. We next apply a Boltzmann weighting, A, to the more stable form, **10.18**, with D in the bridge. Here, A is $\exp(-\Delta E/RT)$, and therefore always less than one, where ΔE is the energetic preference for D being in the bridge (this is usually about 150 cal/mol, but the exact value is extracted from the data), and T is the absolute temperature.

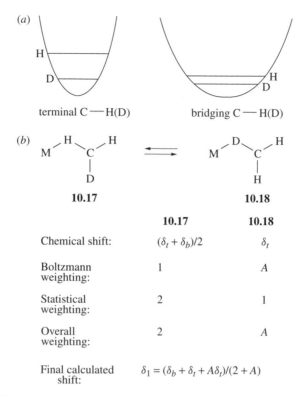

FIGURE 10.10   Origin of "isotopic perturbation of resonance." Zero-point energy differences between C–H and C–D bonds make H prefer the bridging position. A is the Boltzmann factor $(\exp -\Delta E/RT)$. (a) Zero-point energies are larger in the steeper well corresponding to the stronger terminal C–H(D) bond (left) as compared to the weaker bridging C–H(D) (right). (b) Calculation of the shifts and relative weightings for **10.17** and **10.18**.

Finally, we need a statistical weighting for **10.17** because there are two ways of having D terminal, since there are two terminal positions. Equation 10.13 gives the appropriate average. We can test that we have not made a mistake by putting $A = 1$, which should make the IPR go to zero and $\delta_0 = \delta_1 = \delta_2$:

$$\delta_0 = \frac{2\delta_t + \delta_b}{3} \tag{10.12}$$

$$\delta_1 = \frac{\delta_b + \delta_t + A\delta_t}{2 + A} \tag{10.13}$$

$$\delta_2 = \frac{\delta_b + 2A\delta_t}{2A + 1} \tag{10.14}$$

The best way to measure the shifts involved is to have all the isotopomers present in the same NMR tube. The shifts should be measured at different temperatures to confirm that they change in accordance with Eqs. 10.12–10.14. The mere fact of observing IPR only tells us the static structure is unsymmetric, but the results allow us to calculate δ_b, δ_t, and ΔE, and these values may help us find out what the static structure is. IPR has been applied[19] to a number of other hard structural problems involving hydrogen positions.

10.9   IR SPECTROSCOPY[20]

Bands in the IR spectrum[1b,20a] correspond to vibrational modes of a molecule. The position of the band, v, depends (Eq. 10.15) on the strength of the bond(s) involved as measured by a force constant k, and on the reduced mass of the system, m_r. Equation 10.16 shows the reduced mass calculated for a simple diatomic molecule, where m_1 and m_2 are the atomic weights of the two atoms:

$$v = \frac{1}{2\pi c}\left\{\sqrt{\frac{k}{m_r}}\right\} \tag{10.15}$$

(where c = the velocity of light)

$$m_r = \frac{m_1 m_2}{m_1 + m_2} \tag{10.16}$$

Carbonyl Complexes

Infrared spectroscopy is especially useful for carbonyl complexes because the C=O stretching vibration appears at 1700–2100 cm^{-1}, a region that is usually free of other ligand vibrations. The intensity is large because $d\mu/dr$, the dipole moment change during the vibration, is large, thanks to the polarization of the CO on binding to the metal. In complexes with more than one CO, the carbonyls do not usually vibrate independently but instead vibrate in concert, and are therefore said to be coupled together in a way that depends on the symmetry of the $M(CO)_n$ fragment.[20b]

The simplest case is an octahedral dicarbonyl, which may have the carbonyls cis or trans. If the carbonyls are trans, then coupling leads to the situation shown in Fig. 10.11. The COs may vibrate in phase, in which case both the carbonyls reach their maximum C–O extension at the same time (Fig. 10.11a), or they may vibrate out of phase (10.11b), in which case one carbonyl is at the maximum when the other is at the minimum C–O extension.

The in-phase, or symmetric vibration, v_s, appears at higher frequency because it is harder to stretch both COs at once. The reason is that on stretching, each CO becomes a better π acceptor; this is easier for the metal to satisfy if each CO stretches alternately, rather than at once. The intensity of the in-phase vibration is low because the dipoles of the two COs are opposed to each other. One might think that the absorbtion should have zero intensity, but there is usually enough

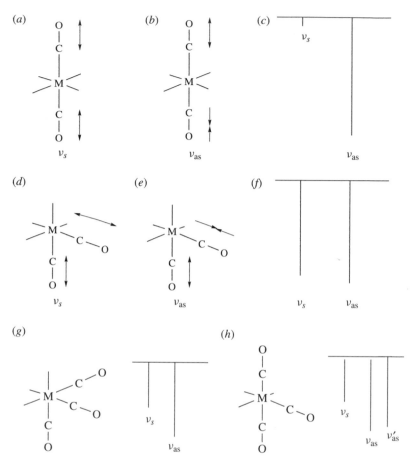

FIGURE 10.11   Effect of the structure of a metal carbonyl on the IR absorption pattern observed.

mixing with other, allowed vibrations in the molecule to lend it enough intensity to make it observable. The out-of-phase, or asymmetric, vibration, ν_{as}, has a very high intensity because the two opposed dipoles alternate in their stretching. The final spectrum, Fig. 10.11c, with an intense band at lower energy and a weak band at higher energy, is characteristic for a trans dicarbonyl. A cis dicarbonyl shows the same two bands, but now of approximately equal intensity, because ν_s now has a large $d\mu/dr$. The relationship between the ratio of the intensities and θ, the angle between the two COs, is shown in Eq. 10.17:

$$\frac{I(\text{sym})}{I(\text{asym})} = \cot^2 \theta \qquad (10.17)$$

Octahedral tricarbonyls can be facial (*fac*) or meridional (*mer*); tetracarbonyls can be cis or trans (the labels refer to the orientation of the noncarbonyl ligands);

but there are only single isomers of penta- and hexacarbonyls. In each case there is a characteristic pattern of IR bands that allow us to identify each type; Figs. 10.11g and h show the spectra expected for the two tricarbonyl isomers.

The pattern will be displaced to higher or lower frequency as the net ionic charge, or the noncarbonyl ligands, or the metal is changed. For example, a net negative charge, or more strongly donor ligands, or a more π-basic metal will give rise to more back bonding and so to a weaker C=O bond. This will shift the IR frequencies to lower energy, which means to lower wavenumber (Table 2.9).

Shifts in $v(CO)$ can also be used as a probe for changes at remote sites, as in Peris and co-workers' detection of hydrogen bonding to the uncoordinated pyridine nitrogen in **10.19**.[21a] Jaouen and co-workers' use of metal carbonyls for labeling proteins for FTIR detection has led to a new immunoassay method.[21b]

$(CO)_5W$—N⟨pyridyl⟩—CH=CH—⟨pyridyl⟩—N----HA   (HA = 2,4,6-trimethyl phenol; Ph$_2$NH; indole)

10.19

Other Ligands

The IR spectrum is also helpful in the identification of other ligands. Hydrides often show $v(M-H)$ bands, but the intensities can be very low as the polarity of the bond is usually small. Carboxylates can be chelating or nonchelating, and the IR data usually serves to distinguish the two cases. Complexes of CO_2, SO_2, NO, and other oxygen-containing ligands give intense bands that are often useful in their identification. Oxo ligands give very intense bands around $500-1000$ cm^{-1}, but the usual polyenes and polyenyls do not give very characteristic absorbtions. In an agostic C—H system, the bond is sometimes sufficiently weakened to give a band at ~ 2800 cm^{-1}. Dihydrogen complexes sometimes give a similar band at $2300-2700$ cm^{-1}, but in this case we again rely on mixing to obtain any intensity at all and indeed the band is completely absent in some cases. Metal–halogen stretching vibrations can be seen in the far IR at $200-450$ cm^{-1}, but since few spectrometers cover this range, they are rarely observed.

Identification of Bands

A common problem in IR work is the identification of a given absorption band as arising from a given ligand, rather than from some other vibration in the molecule. For example, a weak band at 2000 cm^{-1} might be a metal hydride, or there might be a small amount of a CO complex present. This kind of problem is solved by isotopic substitution. If we repeat the preparation with deuteriated materials, then we will either see a shift of the band to lower frequency, in which case it can be identified as the M—(H,D) stretch, or we will not, in which case the CO complex becomes a more likely alternative. If we can obtain the ^{13}CO analog, then the band should shift appropriately if it is due to CO stretching. The shift can be estimated by calculating the reduced masses of the normal and isotopically substituted systems from Eq. 10.16 (it is usual to assume that L_nM can be assigned infinite mass), and deducing the shift from Eq. 10.15. In the case cited above of an $M-H$ band at 2000 cm^{-1}, the $M-D$ band will come at about $2000/\sqrt{2} = 1414$ cm^{-1}.

Raman Spectroscopy

This is rarely applied to organometallic species, but the method is in principle useful for detecting nonpolar bonds, which do not absorb, or absorb only weakly in the IR. The intensity of the Raman spectrum depends on the change of polarizability of the bond during the vibration. It was used very early in its history to detect the Hg$-$Hg bond in the mercurous ion [ν(Hg$-$Hg) = 570 cm^{-1}], for example.

10.10 CRYSTALLOGRAPHY

Structure determination[1b,22] in the solid state is an extremely important part of organometallic chemistry. Two methods are generally used: X-ray and neutron diffraction. The whole three-dimensional structure of the crystal can be described in terms of a repetitive arrangement of the simplest unit of the structure called the *unit cell*, just as a single tile is often a unit cell for a two-dimensional repetitive pattern such as one might find in an Arabian courtyard. According to the space group of the three-dimensional arrangement of the unit cell of the structure, Bragg's law will be satisfied at certain orientations of the crystal, and a beam of X rays will flash out from the crystal at a certain angle to the incident beam. Bragg's law (Eq. 10.18, where λ is the wavelength of the radiation, 2θ is the angle between the incident and diffracted ray, n is an integer, and d is the spacing of the cells) requires that the diffracted radiation from different layers of unit cells be in phase.

$$2d \sin \theta = n\lambda \qquad (10.18)$$

In the X-ray method, a beam of monochromatic X rays is passed through a single crystal of the sample. The incident beam is diffracted at various angles; a photograph, for example, will show a pattern of spots. The intensity of this set of diffracted beams will depend on the nature and arrangement of the atoms in the unit cell. In short, the intensities carry the information about the locations of the atoms in the unit cell, while the relative positions of the spots on the film carry the information about the arrangement of the unit cells in space. The positions and intensities are seldom measured by film methods today, but by a computer-controlled device known as a *diffractometer*.

Limitations of the Method

The X rays are diffracted by the electron clouds around each atom. This means that the diffraction pattern is often dominated by the metal in a complex because it usually has a greater number of electrons than the other atoms present. Conversely, hydrogen atoms are very hard to find because they have few electrons. Where it is important to know the hydrogen positions (e.g., metal hydrides, dihydrogen complexes, or in determining the bond angles at carbon in ethylene complexes), neutron diffraction is used. Neutrons are diffracted from the nuclei of the atoms. All elements have broadly similar ability to diffract neutrons, so that the resulting intensities are not dominated by any one atom, and the positions of all the atoms can therefore be obtained. There are only a few laboratories around the world that are equipped to carry out neutron work; an added inconvenience is the much larger crystal size that has been required to obtain sufficient intensities of diffraction. In contrast, many large chemistry departments have an X-ray facility, and a substantial fraction of papers in organometallic chemistry include one or more X-ray structures.

The results of an X-ray structural determination are often represented as a diagram showing the positions of all the atoms in the molecule (e.g., Fig. 5.8). These have a deceptively persuasive appearance. As in all experimental methods, we have to be aware of the pitfalls. The first question is whether the crystal is representative of the bulk. It is not unusual to have several other compounds as minor impurities in a crystallizing sample, if only because the sample may be slowly decomposing. X-ray diffraction results are often based on work on one crystal. How do we know the rest of the material was the same? Usually it is possible to obtain an IR spectrum of the crystal on which the structural data were collected to check that it is the same material as the bulk of the sample. The more difficult question is whether the structure in the solid state is really the same as the structure of the same material in solution, to which the solution NMR data will correspond. Several organometallic complexes exist as one isomer in solution but as another in the solid state.[23a] If several isomers are interconverting in solution, then any crystals that form will generally consist of the *least soluble* (not the most stable) tautomer. A different tautomer may crystallize from a different solvent. Surprisingly large forces are present within the lattices, especially of ionic crystals; these may change the details of the structure compared with the solution state, in which most reactions take place. This is why it is so useful to have the NMR methods of structure determination in solution to compare with the X-ray results. IR spectroscopy is also very useful because we can obtain a spectrum both in solution and in the solid state. Sharp-line NMR spectra on solid-state samples by "magic angle" spinning can show how the NMR spectrum of the molecule under study changes on going from the solution to the solid state and, therefore, is a further check on the interpretation of any X-ray results. Co-crystallization with impurities can lead to deceptive artefacts such as erroneous bondlengths.[23b]

Interpreting the Results

In organic structures, it is generally always possible to describe the final structure obtained from X-ray work in simple valence bond terms. We know whether atom A is bonded to atom B, and we can make a very good estimate of the bond order, given the observed A−B distance. In organometallic structures, a similar interpretation of the results is not always easy. There are many examples of metal−ligand interactions that do not amount to a full bond and that are longer than the normal M⋯L covalent distance. We have seen agostic C−H bonds in Chapter 3; the M⋯H distance can be up to 1 Å longer than the sum of the covalent radii. Semibridging carbonyls can have the M⋯C distance 0.7 Å longer than the sum of the covalent radii. Binuclear bridged complexes are known that have almost all the possible M⋯M distances between the shorter ones appropriate for M−M bonding and the very long ones that unambiguously imply no bonding; in the midrange, of course, no clear-cut distinction is possible.

10.11   OTHER METHODS

Many other methods can be useful for the characterization of metal complexes, and we briefly discuss some of them here. Microanalysis of the products is standard practice, and the values found for C and H are normally acceptable if they fall within ±0.03% of the calculated figure. Solvent of crystallization can be present in the lattice and can alter the percentages observed; the presence of this solvent should be confirmed by another method such as NMR or IR. The molecular weight of a complex can be obtained by a method such as osmometry.

Conductivity measurements[24] in solution are useful for telling whether a given complex is ionic, and the measurements can also give the electrolyte type (A^+B^-, $A_2^+B^{2-}$, etc.).

The UV−visible spectrum of an organometallic complex is most commonly obtained when photochemical experiments are carried out, to help decide at which wavelength to irradiate (see Section 4.6). A detailed interpretation of the spectrum has been carried out for few organometallic complexes, a situation that contrasts with that in coordination chemistry, where UV−visible spectroscopy and the ligand field interpretation of the results has always been a major focus of attention.[25]

One other diffraction method that has proved useful for sufficiently volatile organometallic compounds is electron diffraction.[26] In this technique the organometallic compound is introduced into a vacuum chamber through a nozzle, and an electron beam is passed through the stream of molecules. The resulting diffraction pattern contains much less information than does an X-ray diffraction pattern, but by making simple assumptions about the structure of the molecule, valuable data can be obtained. A useful feature of the results is that they refer to the molecule in an isolated state in a vacuum, so solvation or crystal packing effects are absent.

Paramagnetic Organometallic Complexes

Once rare, these are much more commonly studied today.[27a] The magnetic moment is most conveniently determined by Evans's[27b] method. This involves measuring the chemical shift of a solvent resonance on going from the pure solvent (often present in the form of a sealed capillary tube placed in the sample) to a solution of the paramagnetic complex. A paramagnetic complex may give an EPR[1b,28] spectrum, which may be useful in characterizing the complex, particularly its symmetry, and in determining how the unpaired electron is delocalized. Paramagnetic complexes may give usable NMR spectra,[27c] but the resonance positions may be strongly shifted and broadened compared to a diamagnetic complex. If we oxidize a Ni(II) complex, LNi, we may make a paramagnetic species LNi$^+$. Sometimes the EPR of the product gives a resonance near $g = 2$ (the g scale is the equivalent of chemical shift in NMR) appropriate for an organic radical, in which case we assign the complex as Ni(II)(L•$^+$) with the oxidation having taken place at the ligand. In other cases the EPR shows $g \neq 2$ in which case a Ni(III)L formulation may be considered more appropriate. Assignment of the oxidation or reduction to M or L can be a contentious issue, however, because the real structure may not be purely Ni(II)(L•$^+$) or Ni(III)L. Electrochemical methods, especially cyclic voltammetry, are invaluable for studies on redox-active complexes. With this method the redox potentials and lifetimes of the oxidized or reduced species can be determined.[27a]

Volatile Species

Sufficiently volatile organometallic compounds can also be studied by mass spectrometry,[1b,29] or electrospray mass spectrometry for involatiles,[30] and photoelectron spectroscopy.[1b,31] Mass spectrometry often allows the molecular weight of a complex to be measured directly, if the molecular ion can be seen. Some ligands such as CO easily dissociate in the spectrometer and give false molecular ions.[29b] In addition, the isotopic pattern for many of the heavier elements (e.g., Mo, Cl, Br, Pd, Ru) is distinctive, and so the nature and number of these elements can usually be unambiguously identified in the molecular ion and in other fragments. Finally, thermodynamic data about the strength of bonds within the complex can sometimes be approximately estimated from the appearance potentials of certain fragments in the spectrum.[32] In another variant of the method, called *ion cyclotron resonance spectroscopy*,[33] the vapor-phase reactions of metal ions or of metal fragment ions with organic molecules can be studied. For example, it has been found that bare Fe$^+$ ions react readily with alkanes to break both C−H and C−C bonds.[34]

Photoelectron spectroscopy (PES)[35] is important because it gives us experimental data about the molecular energy levels within the complex. A solid sample is irradiated with X rays of a given frequency. If the X rays are of an appropriate energy, they can ionize even the core levels of the atoms; this is the electron spectroscopy for chemical analysis (ESCA) experiment. The photoelectrons emitted from the sample are detected and their energies analyzed. Each element in the sample emits at a characteristic energy, and so we have an elemental analysis. In addition, the energy observed shifts very slightly according to the charge on the particular element in the molecule; if the element is more positively charged in complex A than in complex B, the energy levels will be slightly stabilized and the photoelectron will be emitted with a slightly lower kinetic energy in complex A. Unfortunately, the data are not always sufficiently good to distinguish the small, chemically interesting differences between the charges on a metal in different environments. If the exciting radiation is less energetic [e.g., the He(I) lines at 21.22 or 40.8 eV], photoelectrons only from the valence orbitals of the molecule are observed. In this PES experiment, chemically interesting differences are found between different complexes. Each band can often be associated with a particular MO, and the effect of different substitution patterns on the MO energies can be studied. Vibrational fine structure can be seen in certain cases and this helps in assigning the bands.

Computational Methods

Molecular orbital (MO) theory[36] includes a series of quantum mechanical methods for describing the behavior of electrons in molecules by combining the familiar s, p, d, and f atomic orbitals (AOs) of the individual atoms to form MOs that extend over the molecule as a whole. The accuracy of the calculations critically depends on the way the interactions between the electrons (electron correlation) are handled. More exact treatments generally require more computer time, so the problem is to find methods that give acceptable accuracy for systems of chemical interest without excessive use of computer time. For many years, the extended Hückel (EH) method was widely used in organometallic chemistry, largely thanks to the exceptionally insightful contributions of Roald Hoffmann. The EH method allowed structural and reactivity trends to be discussed in terms of the interactions of specific molecular orbitals. Fenske–Hall methods also proved very useful in this period.[37]

Advances in computing power since the late 1990s have allowed more sophisticated methods to be implemented for organometallic molecules. These ab initio methods make fewer assumptions and are based more directly on the physics of the system. Once again the critical issue is handling electron correlation effects—very important in transition metal systems. A major step forward has been the widespread adoption of the present standard method, density functional theory (DFT),[38] in which the energy of a molecule is calculated from an expression involving the electron density distribution, the potential of the atomic nuclei, and a mathematical device called a *functional*. The inner electrons in each atom are not treated separately as was usual in prior methods but are replaced by a potential. The nuclear positions are varied until the energy of the molecule is minimized. The result is a structure and the corresponding energy. Typical errors are ±0.2 Å for bond lengths and ±5 kcal/mol for energies. One big advantage over EH is that even paramagnetic (open-shell) species seem to be successfully treated, but a disadvantage is that interpretations of trends in terms of specific orbitals and their interactions is no longer possible. The results of some EH and DFT studies are incorporated into the discussion in the other chapters of this book.

Molecular mechanics (MM)[39] is a method that has been very useful in organic chemistry by which the strain energy of a given structure can be evaluated by summing all the relevant interactions such as steric repulsions—this is much less expensive in computer time than an ab initio MO study. By minimizing the strain energy, favored conformations can be located. Attempts to use the method in transition metal chemistry have not always been completely successful because ligand field effects are not always considered. Four-coordinate carbon is always tetrahedral, for example, but 4-coordinate nickel can be square planar or tetrahedral, and one usually needs MO methods to determine which is more likely.

Combining MO and MM methods so that the metal and immediate ligand sphere is described by MO methods and the outer, purely organic part of the ligand by the much less expensive MM technique is also possible. One way to integrate molecular orbital and molecular mechanics is IMOMM, which allows steric effects of bulky ligands to be successfully modeled, for example.[40]

A very great advantage of computational methods is that results can be obtained for postulated transient intermediates and even for transition states. The accuracy of the computational results is often sufficient to rule out a postulated intermediate or decide between two competing mechanisms or structures even where there is no convincing experimental method for making the distinction.[41]

Interpretation of Results

Care always needs to be taken with interpreting physical data because Nature has a thousand ways to mislead. An approach to test your conclusion is to ask if there is any combination of events that could falsify it. Devising good control experiments is critical for testing alternate explanations of the data.

> • Crystallography and, for diamagnetic complexes, NMR are among the most useful physical methods.

REFERENCES

1. (a) G. S. Girolami, T. B. Rauchfuss, and R. J. Angelici, *Synthesis and Technique in Inorganic Chemistry*, University Science Books, Mill Valley, CA, 1999; D. F. Shriver, *The Handling of Air-Sensitive Compounds*, McGraw-Hill, New York, 1969; (b) E. A. V. Ebsworth, D. W. H. Rankin, and S. Cradock, *Structural Methods in Inorganic Chemistry*, Blackwell, Oxford, 1987.

2. (a) P. J. Hore, J. A. Jones, and S. Wimperis, *NMR, the toolkit*, OUP, New York, 2000, is a good nonmathematical introduction; a more rigorous treatment is given in (b) J. W. Emsley, J. Feeney, and L. H. Sutcliffe, *High Resolution Nuclear Magnetic Resonance*, Pergamon, Oxford, 1966; K. A. McLaughlan, *Magnetic Resonance*, Clarendon, Oxford, 1982; or J. A. Pople, W. G. Schneider, and H. J. Bernstein, *High Resolution Nuclear Magnetic Resonance Spectroscopy*, McGraw-Hill, New York, 1959; (c) P. S. Pregosin and R. W. Kunz, ^{31}P and ^{13}C *NMR Spectroscopy of Transition Metal Complexes*, Springer, Heidelberg, 1979; (d) G. N. Lamar, W. D. Horrocks, and R. H. Holm, *NMR of Paramagnetic Molecules*, Academic, New York, 1973; (e) W. v. Philipsborn, *Pure Appl. Chem.* **58**, 513, 1986; D. Rehder, *Chimia* **40**, 186,

1986; (f) M. Gielen, R. Willem, and B. Wrackmeyer, *Advanced Applications of NMR to Organometallic Chemistry*, Wiley, New York, 1996.

3. J. C. and M. L. H. Green, in *Comprehensive Organometallic Chemistry*, J. Bailar et al., eds., Pergamon, Oxford, 1973, Chapter 48, p. 355.

4. (a) J. Chatt, R. S. Coffey, and B. L. Shaw, *J. Chem. Soc.* 7391, 1965; (b) R. H. Crabtree and G. G. Hlatky, *Inorg. Chem.* **23**, 2388, 1984.

5. J. W. Faller, R. H. Crabtree, and A. Habib, *Organometallics* **4**, 929, 1985.

6. R. H. Crabtree, D. Gibboni, and E. M. Holt, *J. Am. Chem. Soc.* **108**, 7222, 1986.

7. D. G. Gorenstein, *Prog. NMR Spectrosc.* **16**, 1, 1983.

8. P. Garrou, *Chem. Rev.* **81**, 229, 1981.

9. P. Meakin, J. P. Jesson, and C. A. Tolman, *J. Am. Chem. Soc.* **94**, 3240, 1972.

10. J. W. Faller, *Adv. Organomet. Chem.* **16**, 211, 1977.

11. R. S. Drago, *Physical Methods in Chemistry*, Saunders, Philadelphia, 1977.

12. M. A. McKinney and M. A. Howarth, *J. Chem. Educ.* **57**, 110, 1980.

13. (a) F. A. Cotton, *J. Organomet. Chem.* **100**, 29, 1975; (b) D. Heinekey and W. A. G. Graham, *J. Am. Chem. Soc.* **101**, 6115, 1979.

14. L. J. Farrugia, *Organometallics* **11**, 2941, 1992, and references cited therein.

15. J. W. Faller, in *The Determination of Organic Structures by Physical Methods*, F. C. Nachod and J. J. Zuckerman, eds., Academic, New York, 1973, Vol. 5, Chapter 2.

16. (a) R. H. Crabtree, M. Lavin et al., *J. Am. Chem. Soc.* **108**, 4032, 1986; (b) R. H. Crabtree and D. G. Hamilton, *J. Am. Chem. Soc.* **108**, 3124, 1986; (c) R. H. Morris et al., *J. Am. Chem. Soc.* **113**, 3027, 1991; (d) J. Halpern et al., *J. Am. Chem. Soc.* **13**, 4173, 1991; X. L. Luo, H. Liu, and R. H. Crabtree, *Inorg. Chem.* **30**, 4740, 1991.

17. (a) R. Eisenberg et al., *J. Am. Chem. Soc.* **109**, 8089, 1987; *Acc. Chem. Res.* **24**, 110, 1991; (b) R. Eisenberg et al., *J. Am. Chem. Soc.* **115**, 5292, 1993; J. Bargon et al., *Mag. Res. Chem.* **40**, 157, 2001; (c) J. H. Noggle and R. E. Shirmer, *The Nuclear Overhauser Effect*, Academic, New York, 1971.

18. M. Saunders et al., *J. Am. Chem. Soc.* **99**, 8070, 1977. J. Shapley et al., *J. Am. Chem. Soc.* **100**, 7726, 1978.

19. M. L. Poveda, E. Carmona et al., *J. Am. Chem. Soc.* **121**, 346, 1999.

20. H. Günzler, *IR Spectroscopy*, Wiley-VCH, Weinheim, 2002; (a) Chapter 6 in Ref. 11, and Chapter 5 in Ref. 1b; (b) F. A. Cotton, *Chemical Applications of Group Theory*, Wiley-Interscience, New York, 1967.

21. (a) E. Peris et al., *J. Chem. Soc., Dalton* 3893, 1999; (b) G. Jaouen et al., *Chembiochem.* **5**, 519, 2001.

22. M. F. C. Ladd, *Structure Determination by X-ray Crystallography*, Kluwer, New York, 2003.

23. (a) K. G. Caulton, D. Eisenstein et al., *J. Am. Chem. Soc.* **121**, 3242, 1999; (b) G. Parkin, *Accts. Chem. Res.*, **25**, 455, 1992.

24. W. J. Geary, *Coord. Chem. Rev.* **7**, 81, 1971.

25. Chapter 6 in Ref. 1b, and B. N. Figgis, *An Introduction to Ligand Fields*, Wiley-Interscience, Chichester (UK), 1966.

26. Chapter 8 in Ref. 1b, and J. C. Brand and J. C. Speakman, *Molecular Structure*, Arnold, London, 1960, Chapter 9.

27. (a) M. Fourmigué, *Accts. Chem. Res.* **37**, 179, 2004; (b) J. Evans, *J. Chem. Soc.* 2003, 1960 (c) F. A. Walker, *ACS Symp. Ser.* **692**, 30, 1998.

28. Chapter 13 in Ref. 11, and Chapter 3 in Ref. 1b.

29. (a) Chapter 9 in Ref. 1b, J. R. Chapman, *Practical Organic Mass Spectrometry*, Wiley, New York, 1985; (b) W. A. Graham et al., *J. Am. Chem. Soc.* **95**, 1684, 1973.

30. D. P. Arnold et al., *Organometallics* **23**, 391, 2004; P. Chen, *Angew Chem. Int. Ed.* **42**, 2832, 2003.

31. Chapter 6 in Ref. 1b, and C. R. Brundle and A. D. Baker, eds., *Electron Spectroscopy, Theory, Techniques and Applications*, Academic, New York, 1977; J. L. Hubbard and D. L. Lichtenberger, *J. Am. Chem. Soc.* **104**, 2132, 1982; J. C. Green, D. M. P. Mingos, and E. A. Seddon, *Inorg. Chem.* **20**, 2595, 1981.

32. (a) R. L. Asher, E. H. Appelman, B. Rusic, *J. Chem. Phys. B*, **105**, 9781, 1996; (b) T. R. Spalding, *J. Organomet. Chem.* **149**, 371, 1978.

33. H. Hartmann and K. P. Wanczek, *Ion Cyclotron Resonance Spectroscopy*, Springer, Berlin, 1978, 1982.

34. D. P. Ridge et al., *J. Am. Chem. Soc.* **101**, 1332, 1979; J. L. Beauchamp et al., *Organometallics* **1**, 963, 1982.

35. D. L. Lichtenberger et al., *J. Am. Chem. Soc.* **108**, 2560, 1986; S. Huber, *Photoelectron Spectroscopy*, Springer, Berlin, 1995; J. C. Green, *Accts. Chem. Res.*, **27**, 131, 1994.

36. T. A. Albright, J. K. Burdett, and M.-H. Whangbo, *Orbital Interactions in Chemistry*, Wiley, New York, 1985.

37. M. B. Hall and R. F. Fenske, *Inorg. Chem.* **11**, 768, 1972.

38. W. Koch and R. H. Hertwig, *Encyclopedia of Computational Chemistry*, Wiley, New York, 1998, p. 689.

39. M. C. Baird, *Organometallics* **11**, 3712, 1992.

40. F. Maseras, N. Koga, and K. Morokuma, *Organometallics* **13**, 4008, 1994; G. Ujaque, F. Maseras, O. Eisenstein, L. Liable-Sands, A. L. Rheingold, and R. H. Crabtree, *New J. Chem.* **22**, 1493, 1998.

41. K. Morokuma, M. D. Fryzuk et al., *J. Am. Chem. Soc.* **121**, 523, 1999; M. Pavlov, P. E. M. Siegbahn, M. R. A. Blomberg, and R. H. Crabtree, *J. Am. Chem. Soc.* **120**, 548, 1998; S. Q. Niu and M. B. Hall, *J. Am. Chem. Soc.* **121**, 3992, 1999; O. Eisenstein et al., *Chem. Rev.* **100**, 601, 2000.

PROBLEMS

1. Sketch the ^1H NMR spectrum of (i) *cis-* and (ii) *trans-*$OsH_2(PMe_3)_4$. How could we go about finding the value of a trans $^2J(H,H)$ coupling by looking at the spectra of an isotopic modification of one of these complexes?

2. *trans-*$OsH_2(PMe_3)_4$ reacts with HBF_4 to give $[OsH_3(PMe_3)_4]^+$. What structures should we consider for this species, and how might ^1H NMR spectroscopy help you decide which structure is in fact adopted?

3. $(Indenyl)_2W(CO)_2$ is formally a 20e species. How might it achieve a more reasonable 18e configuration, and how could you use ^{13}C NMR spectroscopy to test your suggestion?

4. How could we distinguish between an $[(\eta^6\text{-benzene})ML_n]$ and an $[(\eta^4\text{-benzene})ML_n]$ structure for a given diamagnetic complex by ^{1}H NMR? Note that the observation of a single-benzene resonance does not prove the η^6-benzene structure because the η^4-benzene form might be fluxional at all accessible temperatures.

5. Two chemically inequivalent hydrides, H_a and H_b in a metal dihydride complex at 50°C, resonate at -5δ and -10δ, respectively, and are exchanging so that each resonance shows an initial broadening of 10 Hz at a field corresponding to 500 MHz. What is the rate of exchange? At 80°C we observe coalescence; what is the new rate of exchange?

6. Which of the methods (a) to (e) would be suitable for solving parts 1–6 mentioned below? (a) X-ray crystallography, (b) ^{1}H NMR spectroscopy, (c) ^{31}P NMR spectroscopy, (d) IR spectroscopy, (e) magnetic moment determination: (1) Characterizing a cyclometallated $Ph_2PC_6H_4$ complex, (2) characterizing a dihydrogen complex, (3) characterizing a CO_2 complex, (4) determining the stereochemistry of $M(CO)_2(dppe)_2$, (5) comparing the relative donor properties of a series of ligands L in $LNi(CO)_3$, (6) finding out whether a given complex $NiCl_2L_2$ were square planar or tetrahedral in solution and how would you interpret the data. If you cite more than one method, be sure to state which method you would use first.

7. $IrCl(CO)_2(PMe_3)_2$ has two solution IR bands in the CO region, for which I_{sym}/I_{asym} is 0.33. What is the preferred geometry of this complex in solution?

8. Why are the CO stretching bands of a bridging carbonyl at lower frequency in the IR spectrum than those of a terminal CO? What would you expect for a μ_3-CO?

9. How can a complex having an apparent formulation $[IrHCl(CO)(acetate)(PR_3)_2]$, as judged from analytical and NMR measurements, be formulated with (a) an η^1-acetate, (b) an η^2-acetate in solution? For each of your suggested formulations, state what methods of characterization would be useful to test your suggestions.

10. $[Ir(cod)(PMe_2Ph)(2\text{-methylpyridine})]^+$ shows a *pair* of doublets for the PMePh protons in the ^{1}H NMR; explain. (Coupling to the metal is not responsible; Ir does not have an $I = \frac{1}{2}$ nucleus.)

11

METAL–LIGAND MULTIPLE BONDS

We now look in detail at compounds with multiple bonds between metal and ligand. We are chiefly concerned with multiple bonds to carbon, as in metal carbene complexes $L_nM=CR_2$, which have a trigonal planar carbon and at least formally contain an M=C double bond, and metal carbyne complexes, $L_nM\equiv CR$, which are linear and contain an $M\equiv C$ triple bond, but we also look at complexes with multiple bonds to O and N. These are more often actor rather than spectator ligands.

11.1 CARBENES

A free carbene such as CH_2 has two spin states, singlet ($\uparrow\downarrow$) and triplet ($\uparrow\uparrow$). These are distinct spin isomers with different H−C−H angles and not just resonance forms, which always have the same spin. In the singlet, the electrons are paired up in the sp^2 lone pair, but in the triplet there is one electron in each of the sp^2 and p orbitals. (Fig. 11.1a). Unlike many of the ligands discussed in earlier chapters, carbenes are rarely stable in the free state. Methylene, :CH_2, for example, is a transient intermediate that reacts rapidly with a wide variety of species, even alkanes. This instability (both thermodynamic and kinetic) contributes to the very strong binding of carbenes to metal atoms by disfavoring dissociation.

The Organometallic Chemistry of the Transition Metals, Fourth Edition, by Robert H. Crabtree
Copyright © 2005 John Wiley & Sons, Inc.

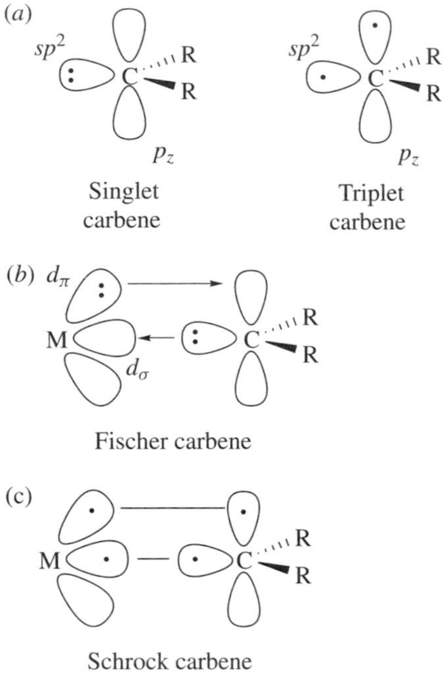

FIGURE 11.1 Singlet and triplet forms of a carbene (a) can be considered as the parents of the Fischer (b) and Schrock (c) carbene complexes. In the Fischer case, direct C → M donation predominates and the carbon tends to be positively charged. In the Schrock case, two covalent bonds are formed, each polarized toward the carbon, giving it a negative charge.

Fischer Versus Schrock Carbenes

Two alternative types of coordinated carbene can be distinguished: the Fischer[1] and the Schrock[2] type. Each represents a different formulation of the bonding of the CR_2 group to the metal. Carbenes, $L_nM=CR_2$, have Fischer character for low oxidation state, late transition metals, having π-acceptor ligands L, and π-donor substituents, R, such as $-OMe$ or $-NMe_2$, on the carbene carbon. Such a carbene behaves as if it carries a ∂^+ charge and is electrophilic. In the most common model, the CR_2 ligand in a Fischer carbene is considered to act as a 2e lone-pair donor (L type) derived from the singlet carbene. Schrock carbenes are bound to higher oxidation state, early transition metals, having non-π-acceptor ligands and non-π-donor R groups. In this case, the carbene behaves as a nucleophile, having a ∂^- carbon. The CR_2 ligand in a Schrock carbene is often considered to act like an X_2-type bis-alkyl, formally derived from the triplet carbene. Cases intermediate between the two extremes are especially common for $M=C(Hal)_2$ because the halide has intermediate π-donor strength between H and $-OMe$.[3a]

The reactivity of the carbene carbon is controlled by the bonding. A Fischer (singlet-derived) carbene is predominantly an L-type σ donor via the lone pair, but the empty p orbital on carbon is also a weak acceptor for π back donation from the $M(d_\pi)$ orbitals (Fig. 11.1b). This leads to an electrophilic carbene carbon because direct $C \rightarrow M$ donation is only partly compensated by $M \rightarrow C$ back donation. A Schrock carbene forms two covalent bonds via interaction of the triplet CR_2 fragment with a metal fragment having two unpaired electrons (Fig. 11.1c); it acts as an X_2 ligand. Each $M-C$ bond is polarized toward carbon because C is more electronegative than M, leading to a nucleophilic carbene carbon. Alternatively, the Schrock carbene can be considered as a Fischer carbene with very strong back donation. The later metals, being more electronegative, have stabler $M(d_\pi)$ orbitals. The presence of π-acceptor ligands L stabilizes the $M(d_\pi)$ levels even more, by the mechanism shown in Fig. 1.7. Therefore, π back donation is weak. The more electropositive early metals have less stable $M(d_\pi)$ levels (greater electropositive character implies that electron loss is easier, which in turn implies that the corresponding orbitals are less stable); d^2 metals are especially strong π donors. A change in oxidation state can alter the situation; for example, $RuCl_2COL_2(=CF_2)$ is predominantly Fischer type and $Ru(CO)_2L_2(=CF_2)$, with its higher-energy $M(d_\pi)$ orbitals, is borderline Schrock-type.[3a]

The electron-deficient Fischer carbene carbon is affected by the presence of the lone pair(s) of its π-donor substituents, denoted OR(lp). Structure **11.1** shows how the $M(d_\pi)$ and OR(lp) orbitals compete for π bonding to the carbene carbon. This can be described in VB (valence bond) language by resonance between **11.2** and **11.3**. The real structure often resembles **11.3** rather than **11.2** as shown by the long $M-C$ and short $C-O$ bonds found by X-ray studies. For electron counting purposes we regard the Fischer carbene as an L-type ligand such as CO. Note that the true $M=C$ bond order is less than 2, thanks to the contribution of **11.3**.

11.1	**11.2**	**11.3**

If we consider the Schrock carbene as a Fischer carbene with strongly enhanced back bonding, we can say that in the extreme case, the two electrons originally in $M(d_\pi)$ transfer to the $C(p_z)$ orbital, oxidizing the metal by two units and giving a CR_2^{2-} ligand. The system can therefore be seen as a metal-stabilized carbanion acting as both a σ donor and a π donor to the metal, not unlike phosphorus ylids such as $Ph_3P^+-CH_2^-$. This oxidation of the metal translates into the Schrock carbene acting as an X_2 ligand, just as the oxo group acts as O^{2-} in a complex such as $Re(=O)Cl_3(PPh_3)_2$ or $Re(=O)Me_4$.

In summary, we can think of the Fischer and Schrock extreme formulations as being L and X_2 models, respectively. This is not unlike the situation we saw in alkene complexes that are also 2e donors but can adopt either the L (alkene complex) or the X_2 (metalacyclopropane) extreme. In the latter case we also oxidize the metal by two units. In both cases, we expect all possible intermediate structures to exist. Table 11.1 summarizes the differences. Structures **11.4** and **11.5** show typical Fischer (**11.4**) and Schrock (**11.5**) carbenes.

$$(CO)_5W = \begin{array}{c} OR \\ \backslash \\ R \end{array} \qquad Cp_2(Me)Ta = \begin{array}{c} H \\ \backslash \\ H \end{array}$$

W(0), 18e   Ta(V), 18e

11.4   **11.5**

Structures **11.4** and **11.5** also show how oxidation states are assigned differently for the two types. Binding of a Fischer (singlet) carbene does not alter the oxidation state of the metal, but as an X_2 diyl ligand, a Schrock carbene is considered to increase the oxidation state of the metal by two units. This creates another oxidation state ambiguity in intermediate cases where the bonding is not clearly classifiable as Fischer or Schrock.

The term *alkylidene* refers to carbenes, CR_2 with alkyl substituents; for example, $MeCH=ML_n$ is an ethylidene complex but "alkylidene" is sometimes used as a synonym for "Schrock carbene" in the older literature because the first alkylidenes were of the Schrock type. There are electrophilic Fischer alkylidenes as well as nucleophilic Schrock ones, however, so the terms should be kept separate. For example, $[Cp_2W(=CH_2)Me]^+$ and $Cp_2Ta(=CH_2)Me$ are isoelectronic, but the former is electrophilic (Fischer) and the latter nucleophilic (Schrock) at the carbene carbon;[3b] the net positive charge on the tungsten complex must stabilize the $M(d_\pi)$ levels leading to much weaker back donation. Schrock

TABLE 11.1 Fischer and Schrock Carbenes $L_nM=CR_2$

Property	Fischer	Schrock
Nature of carbene carbon	Electrophilic	Nucleophilic
Typical R groups	π Donor (e.g., $-OR$)	Alkyl, H
Typical metal	Mo(0), Fe(0)	Ta(V), W(VI)
Typical ligands	Good π acceptor (e.g., CO)	Cl, Cp, alkyl
Electron count (covalent model)	2e	2e
Electron count (ionic model)	2e	4e
Oxidation state change on addition of CR_2 to L_nM	0	+2

carbenes with aryl substituents, such as $[Cp^*(Me_3P)(ArN)Nb=CHPh]$,[2] cannot be called alkylidenes.

N-Heterocyclic Carbenes

An important type of Fischer carbene is the *N*-heterocyclic carbene or NHC. The classic example, **11.6**, is derived from an *N,N'*-disubstituted imidazolium salt by deprotonating the acidic hydrogen at C-2.[4a] The resulting free carbene is sufficiently stable to be isolated if the R groups are bulky (e.g., adamantyl).

    Carbenes such as **11.6** are gaining increasing attention as spectator ligands in catalysis (Section 11.4)[4b] because they are strongly bound to the metal and successfully promote a variety of catalytic cycles much as do phosphines (Chapters 4 and 9). Like phosphines they can be turned both sterically and electronically.

$$RN \underset{\underset{\cdot\cdot}{C}}{\overset{\bigcirc}{\diagup\diagdown}} NR$$

11.6

Fischer Carbenes

Fischer made the first carbene complexes in 1964[1] by the attack of an alkyllithium on a metal carbonyl followed by methylation (Eq. 11.1). On our bonding picture, the methoxy substituent will also help stabilize the empty *p* orbital on the carbene carbon by π donation from one of the lone pairs on oxygen. Resonance form **11.3**, probably the dominant one in the heteroatom stabilized Fischer carbenes, shows the multiple character of this bond. This is responsible for the restricted rotation often observed for the heteroatom–carbene carbon bond in NMR studies. For example cis and trans isomers **11.8** and **11.9** of methoxymethyl carbenes are rapidly interconverting at room temperature (Eq. 11.2) but can be frozen out in the proton NMR at $-40°C$.[5]

$$(CO)_5M-CO \xrightarrow{\text{LiMe}} (CO)_5M^- \overset{Me}{\underset{O}{\diagdown\diagup}} \xrightarrow{\text{MeI}} (CO)_5M = \overset{Me}{\underset{OMe}{\diagdown\diagup}} \qquad (11.1)$$

11.7

$$(CO)_5M^- \underset{\underset{Me}{\diagup}}{\overset{Me}{\diagdown\diagup}}{O^+} \rightleftharpoons (CO)_5M^- \overset{Me}{\underset{O^+-Me}{\diagdown\diagup}} \qquad (11.2)$$

11.8   **11.9**

Preparation The key synthetic routes tend to fall into one of three general categories, illustrated by Eqs. 11.3–11.5. In Eq. 11.3, an acyl or similar species (often but not always formed by a nucleophilic attack on a CO or a similar ligand) is treated with an electrophile to give a Fischer carbene. In Eq. 11.4 an H^- (Fischer case) or an H^+ (Schrock case) is abstracted from the α position of an alkyl, and in Eq. 11.5 a carbene source is used. In Eqs. 11.4b and 11.5 the $L_n M$ fragment must be able to accept an extra pair of electrons during the reaction, and so the starting material must have <18e or be able to lose a ligand. Examples of Eqs. 11.3 and 11.4a are shown below, and examples of 11.4b and 11.5 are shown in the section on Schrock carbenes.

$$L_n M \!-\! C \!\equiv\! Y \xrightarrow{\;Nu^-\;} \left[L_n M \!-\! C \overset{\displaystyle \nearrow Y}{\underset{\displaystyle \searrow Nu}{}} \right]^{-} \xrightarrow{\;E^+\;} L_n M \!=\! C \overset{\displaystyle \nearrow Y\text{-}E}{\underset{\displaystyle \searrow Nu}{}} \tag{11.3}$$

$$(Y = O, \, NR;$$
$$Nu = OMe^-, \, NR_2^{\,-}, \, LiMe)$$

$$L_n M \!-\! CHR_2 \begin{cases} \xrightarrow{\;E^+\;} L_n M^+ \!=\! CR_2 \;+\; EH & (11.4a) \\[2em] \xrightarrow{\;Nu^-\;} L_n M^- \!=\! CR_2 \;+\; NuH & (11.4b) \end{cases}$$

$$L_n M + CH_2 N_2 \longrightarrow L_n MCH_2 + N_2 \tag{11.5}$$

Isonitriles are very sensitive to nucleophilic attack, and a wide range of bis-heteroatom-stabilized carbenes can be obtained:[6]

$$[(MeNC)_4 Pt]^{2+} + 4MeNH_2 \longrightarrow [Pt\{=C(NHMe)_2\}_4]^{2+} \tag{11.6}$$

Chugaev[7] obtained carbene complexes very similar to these as early as 1915 but was not able to assign the right structure, given the methods available at that time. Acetylides $L_n M \!-\! C \!\equiv\! CR$ are unexpectedly good bases[8a] via their canonical form $L_n M^+ \!=\! C \!=\! C^- R$. They can react with acid in alcohol solution to give the carbenes shown in Eq. 11.7. An intermediate vinylidene cation probably undergoes nucleophilic attack by the alcohol.[8b] In this case the usual order of attack shown in Eq. 11.3 (Nu^-, then E^+) is inverted.

$$ClL_2 Pt \!-\! C \!\equiv\! CR \xrightarrow{\;H^+\;} ClL_2 Pt^+ \!=\! C \!=\! CHR \xrightarrow{\;EtOH\;} ClL_2 Pt^+ \!=\! C \overset{\displaystyle \nearrow OEt}{\underset{\displaystyle \searrow CH_2 R}{}} \tag{11.7}$$

Electrophilic abstraction from an alkyl complex (Eq. 11.4a) is illustrated in the reactions of Eqs. 11.8 and 11.9; Eq. 11.9 is driven by the high Si−F bond strength.

$$Cp(CO)_2FeCH_2OMe + H^+ \longrightarrow \text{"}Cp(CO)_2Fe=CH_2^{+}\text{"} \longrightarrow \text{other products}$$

(11.8)

$$Cl(CO)(MeCN)L_2RuCF_3 + Me_3SiCl \longrightarrow Cl_2(CO)L_2Ru=CF_2 + Me_3SiF$$

(11.9)[9]

Alkylidenes can sometimes be made from organic carbene precursors such as diazo compounds or 1,1-diphenylcyclopropene.[10]

Spectroscopy ^{13}C NMR is very valuable for detecting carbene complexes because the very deshielded carbene carbon resonates at ~200–400 ppm to low field of TMS. It is tempting to ascribe this deshielding to the ∂^+ character of the carbene carbon, but as we shall see, Schrock carbenes, which are ∂^- in character at carbon, show similar shifts. In fact, the shift is probably a result of the existence of low-energy electronic excited states for the complex, which leads to a large "paramagnetic" contribution to the shift. A proton substituent at the carbene carbon resonates from +10 to +20δ.

Reactions Thermal decomposition of carbene complexes usually leads to one or both of two types of alkenes:[11a] one type is formed by rearrangement, and the other by dimerization of the carbene. Equation 11.10[11b] shows both types of product. The reaction does not go via the free organic carbene because cyclobutanone, which is known to be formed in the rearrangement of the free carbene, was not detected in the products.

(11.10)

Fischer carbenes without a heteroatom substituent are very reactive.[12a] The protonation of vinyl complexes is one route to these species (e.g., Eq. 11.11):[12b]

(11.11)

The addition of base reverses the first step by a nucleophilic abstraction. The ethylidene complex readily gives a 1,2 shift of the β proton to give the thermodynamically more stable alkene complex. Even carbenes that lack β hydrogens can be unstable: $[Cp(CO)_2Fe=CH-CMe_3]^+$ and $[Cp(CO)_2Fe=CH-CMe_2Ph]^+$

both rearrange by a 1,2 shift of a methyl or a phenyl group, respectively, to the electron-deficient carbene carbon (Eq. 11.12).[13] This reaction, analogous to the Wagner–Meerwein rearrangement in carbonium ions, is fast because of the electron-deficient character of the carbene carbon.

$$\text{Cp(CO)}_2\text{Fe} \Longrightarrow \overset{\text{Me}}{\underset{\text{Me}}{\big|}} \text{—R} \longrightarrow \text{Cp(CO)}_2\overset{+}{\text{Fe}}\text{—} \begin{array}{c} \text{Me} \quad \text{Me} \\ \big| \\ \big| \big| \\ \text{R} \end{array} \tag{11.12}$$

$$(R = \text{Me or Ph})$$

$[\text{Cp(Ph}_2\text{PCH}_2\text{CH}_2\text{PPh}_2)\text{Fe=CH–CMe}_3]^+$ does not rearrange, however, probably because the increased back donation by the more electron-rich phosphine-substituted iron decreases the electron deficiency at the carbene carbon.[14]

Where the carbene is sufficiently stabilized, an alkene can even rearrange to the corresponding carbene, the reverse of Eq. 11.12, as in the Ru example shown.[14b] In the Ir example, an equilibrium is seen between alkene and carbene forms.[14c]

$$\overset{\text{OEt}}{=\!\!\!=\!\!\!/} \xrightarrow{\text{RuHCl(PiPr}_3)_2} \quad (\text{PiPr}_3)_2\text{ClHRu}\!\!=\!\!\!\!\big\langle^{\text{OEt}}$$

$$\underset{\text{O}}{\overset{\text{H}}{\text{TpIr}}}\!=\!\text{C}\!\!\begin{array}{c}\text{CH}_3\end{array} \rightleftharpoons \underset{\text{O}}{\overset{\text{H}}{\text{TpIr}}}\!\!-\!\!\begin{array}{c}\text{CH}_2 \\ \big|\big| \\ \text{CH}\end{array}$$

Oxidative cleavage of a carbene ligand can be achieved with reagents such as Ce(IV) compounds, pyridine N-oxide, or DMSO, or even with air. The product is normally the ketone corresponding to the starting carbene. This reaction is useful not only for synthetic purposes but also for characterizing the original carbene (e.g., Eq. 11.13):[15]

$$(\text{CO})_5\text{Cr}\!=\!\text{C}\!\!\begin{array}{c}\text{O} \\ \\ \\ \text{CH}_2\end{array} \xrightarrow{\text{Ce(IV)}} \text{O}\!=\!\text{C}\!\!\begin{array}{c}\text{O} \\ \\ \\ \text{CH}_2\end{array} \tag{11.13}$$

11.10

The synthesis of **11.10** illustrates another useful reaction of Fischer carbenes, the abstraction of a proton β to the metal by a base such as an organolithium reagent. The resulting negative charge can be delocalized onto the metal as shown in Eq. 11.14[15] and is therefore stabilized. The anion can be alkylated by carbon electrophiles as shown.

11.10

$$(11.14)$$

Fischer carbenes readily undergo nucleophilic attack at the carbene carbon, as shown in Eq. 11.15.[16] The attack of amines can give the zwitterionic intermediate shown, or by loss of methanol, the aminocarbene. If we mentally replace the $(CO)_5Cr$ group with an oxygen atom, we can see the relation of this reaction to the aminolysis of esters to give amides (Eq. 11.16).

$$(11.15)$$

$$(11.16)$$

The addition of carbon nucleophiles or of alkenes can lead to the formation of metalacycles. These can break down to a carbene and an alkene (Eq. 11.17),[17] or reductive elimination may take place to give a cyclopropane (Eq. 11.18).[18] The formation of metalacycles from alkenes and carbenes is the key reaction in alkene metathesis (Section 11.4).

$$(11.17)$$

$$(11.18)$$

The reaction of carbenes with alkynes gives metalacyclobutenes, but these often rearrange. Equation 11.19 shows the Dötz reaction for the synthesis of naphthols.[19] Note that two naphthol haptomers are found.

Schrock Carbenes

High-valent metal alkyls, especially of the early metals, can undergo proton abstraction at the α carbon to give nucleophilic Schrock[2,20] carbenes. The first high-oxidation-state carbene was formed in an attempt to make $TaNp_5$ ($Np = CH_2CMe_3$, or neopentyl), by the reaction of $TaNp_3Cl_2$ with LiNp.* In fact, the product is $Np_3Ta=CH(t\text{-}Bu)$ (Eq. 11.20). The reaction probably goes via $TaNp_5$, which then loses neopentane by an α-proton abstraction from one (possibly agostic) Np ligand by another.[2,21] With Me_3SiCH_2, the intermediate TaR_5 species could be isolated at $-80°C$.[21]

One requirement for this α elimination is that the molecule be crowded. Substitution of a halide in Np_2TaCl_3 with a Cp group (Eq. 11.21)[22] is enough to do this, for example, as is addition of a PMe_3 (Eq. 11.22).[2] The corresponding

---

*Interestingly, Wittig was trying to make Ph_3PMe_2 when he discovered $Ph_3P=CH_2$.

benzyl complexes require one of the more bulky pentamethylcyclo-pentadienyls, Cp* (Eq. 11.23),[23] or two plain Cp groups (Eq. 11.24).[22]

$$(11.19)$$

$$(11.20)$$

$$(11.21)$$

$$(11.22)$$

$$\text{Cl}_2\text{Ta}\begin{array}{c}\diagup\text{CH}_2\text{Ph}\\-\text{CH}_2\text{Ph}\\\diagdown\text{CH}_2\text{Ph}\end{array}\xrightarrow[-\text{PhCH}_3]{\text{LiCp}^*}\text{Cp}^*\text{ClTa}\begin{array}{c}\diagup\text{CHPh}\\\diagdown\text{CH}_2\text{Ph}\end{array}\qquad(11.23)$$

$$\text{Cl}_2\text{Ta}\begin{array}{c}\diagup\text{CH}_2\text{Ph}\\-\text{CH}_2\text{Ph}\\\diagdown\text{CH}_2\text{Ph}\end{array}\xrightarrow[-\text{PhCH}_3]{\text{LiCp}}\text{Cp}_2\text{Ta}\begin{array}{c}\diagup\text{CHPh}\\\diagdown\text{CH}_2\text{Ph}\end{array}\qquad(11.24)$$

By adding two PMe_3 ligands, we see that the α proton of a benzylidene can undergo abstraction to give a benzylidyne (Eq. 11.25).

$$\text{Cp}^*\text{Ta}(=\text{CHPh})(\text{CH}_2\text{Ph})\text{Cl}\xrightarrow{2\text{PMe}_3}\text{Cp}^*\text{Ta}(\equiv\text{CPh})(\text{PMe}_3)_2\text{Cl}+\text{PhCH}_3\quad(11.25)$$

The methyl group is so sterically undemanding that it does not α-eliminate under the same conditions (Eq. 11.26). The synthesis of a methylene complex requires a deprotonation of a methyl complex by a strong base. By putting a net positive charge on the complex, we can activate the methyl for this reaction. Equation 11.27 shows how this can be done by an electrophilic abstraction of Me^-. Note that if this had been a low-valent metal, electrophilic abstraction of H^- by Ph_3C^+ to give an electrophilic (Fischer) methylene complex might have taken place.

$$\text{TaMe}_3\text{Cl}_2\xrightarrow{\text{LiCp}}\text{CpTaMe}_3\text{Cl}\xrightarrow{\text{LiCp}}\text{Cp}_2\text{TaMe}_3\qquad(11.26)^{[20]}$$

$$\text{Cp}_2\text{TaMe}_3\xrightarrow{\text{Ph}_3\text{C}^+}\text{Cp}_2\text{TaMe}_2{}^+\xrightarrow{\text{base}}\text{Cp}_2\text{Ta}(=\text{CH}_2)\text{Me}\qquad(11.27)^{[24]}$$

Structure and Spectra Few of the early metal complexes we have been looking at seem to be 18e. $TaMe_3Cl_2$ is ostensibly 10e, for example. This is not unusual for high-oxidation-state complexes, especially in the early metals, where the d orbitals are not so stabilized as in lower oxidation states or in later metals (Chapter 15). The halide has lone pairs that might be partially donated to the empty d_π orbitals, and the alkyls have C−H bonds that might become agostic, so the metal may be able to increase the electron count from these. Schrock carbene complexes that have <18e commonly have agostic C−H bonds. When this happens, the proton on the carbene carbon bends back toward the metal, the M=C bond becomes shorter, and the C−H bond becomes longer (**11.11**). In contrast, in late metals these d_π orbitals are usually full and the complex is often 18e, so we do not see agostic C−H bonds.

$$\text{M}=\text{C}\begin{array}{c}\diagup\text{H}\\\diagdown\text{R}\end{array}$$

11.11

Agostic binding leads to a high-field shift for this proton and a lowering of the C,H coupling constant in the ^{1}H NMR, together with a lowering of $v(C-H)$ in the IR. In 18e carbene complexes, such protons are not agostic and usually appear at 12δ with a $J(C,H)$ of 105–130 Hz; in the complexes with <18e, they can come as high as -2δ with a $J(C,H)$ of 75–100 Hz. At the same time a $v(C-H)$ band appears in the IR at a position indicating a weakened CH bond, for example, at 2510 cm^{-1} in CpTa{CH(t-Bu)}Cl$_2$. Crystal structures[25] show that the M=C−R angle can open up to as much as 175°, while the M=C−H angles fall to as little as 78°. The M=C bond length is always short (at least 0.2 Å shorter than an M−C single bond) in all cases but is even shorter in the complexes with <18e. The oxo alkylidene Cl$_2$(PEt$_3$)$_2$W(=O)(=CHCMe$_3$) has a much less distorted alkylidene group probably because the oxo lone pairs are more basic and so more available for the metal than the C−H bonding pair.[26]

The structure of Cp$_2$Ta(CH$_2$)Me (by neutron diffraction) shows the orientation of the methylene group is not the one predicted on steric grounds, with the CH$_2$ lying in the mirror plane of the molecule, but nearly at right angles (88°) to this plane, with the proton substituents pointing in the direction of the Cp groups. Whenever we see a countersteric conformation like this, an electronic factor is usually at work. Here, the filled p_z orbital of the CH$_2$ group is interacting with one of the empty orbitals on the metal. Since these orbitals are in the mirror plane of the molecule (see Section 5.4), this fixes the orientation of the CH$_2$ (Fig. 11.2). The larger CHR alkylidenes deviate only slightly from the orientation shown by CH$_2$, and so the two Cp groups become inequivalent. The ^{1}H NMR spectrum of the complexes shows this inequivalence but the two Cp groups become equivalent on warming. If we assume that the fluxional process is rotation about the M=CHR bond, then in the transition state, the alkylidene probably lies in the mirror plane and has no π interaction with the metal. The $\Delta G^\ddagger$ deduced from the data, 25 kcal/mol, therefore gives an estimate of the strength of the Ta=C π bond.

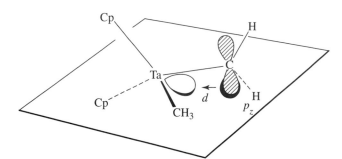

FIGURE 11.2 Orientation of the methylene group in Cp$_2$Ta(CH$_2$)Me is contrary to what would be expected on steric grounds and is controlled by the overlap of the C(p_z) with a metal d orbital that lies in the plane shown. Filled orbital hatched using CH$_2^-$ model.

Reactions   The reactions of Schrock carbenes illustrate their nucleophilic character. For example, they form adducts with the Lewis acid $AlMe_3$:

$$Cp_2Ta\overset{CH_2}{\underset{Me}{\diagup}} \xrightarrow{AlMe_3} Cp_2Ta^+\overset{CH_2-\overset{-}{A}lMe_3}{\underset{Me}{\diagup}} \qquad (11.28)$$

They also react with ketones like a Wittig ($Ph_3P=CH_2$) reagent (Eq. 11.29).[27]

$$Np_3Ta=CH(t\text{-}Bu) \xrightarrow{Me_2CO} \quad =CH(t\text{-}Bu) + [Np_3TaO]_x \qquad (11.29)$$

In their most important reaction, alkenes react with carbenes to give metalacycles. The alkene may coordinate to the metal first, if only transiently. The carbene carbon then attacks the coordinated alkene to give the product. The metalacycle can decompose in several ways (Eq. 11.30), either by reversal of the formation reaction to give alkene and a carbene, by reductive elimination to give a cyclopropane, or by β elimination to give an allyl hydride. The first route is the most important. Each time the $RCH=ML_n$ complex encounters an external alkene, it can exchange alkylidene ($RCH=$) groups between itself and the alkene. The final result is that alkylidene groups are catalytically exchanged between all the alkenes present. This alkene metathesis reaction[10] (Eq. 11.31) has proved to be of remarkably wide applicability in both organic and polymer chemistry and is discussed in detail in Section 12.1. Alkynes give a metalacyclobutene, which can rearrange as shown in Eq. 11.32.[20]

$$(11.30)$$

$$(11.31)$$

$$Cp_2Cl_2Ta = \underset{t\text{-}Bu}{\diagup} \quad \xrightarrow{RC \equiv CR} \quad Cp_2Cl_2Ta \underset{RC}{\overset{t\text{-}Bu}{\diagdown}} CR$$

(11.32)

$$Cp_2Cl_2Ta = \overset{R}{\underset{R}{\diagdown}} \diagup^{t\text{-}Bu}$$

As might be expected, the more electropositive Ti forms even more nucleophilic carbene complexes. One of the most interesting species of this class is Tebbe's reagent, formed from Cp_2TiCl_2 and $AlMe_3$ (Eq. 11.33):[28]

$$Cp_2Ti \overset{Cl}{\underset{Cl}{\diagdown}} \quad \xrightarrow{AlMe_3} \quad Cp_2Ti \overset{Cl}{\underset{CH_2-H}{\diagdown}} \quad Me-AlMe_2 \quad \xrightarrow{-CH_4} \quad Cp_2Ti \overset{Cl}{\underset{CH_2}{\diagdown}} AlMe_2$$

Tebbe's reagent

(11.33)

This is an example of a bridging carbene, but in its reactions it almost always loses Me_2AlCl first to give the mononuclear 16e $Cp_2Ti=CH_2$; a base is sometimes added to help remove the aluminum fragment by complexation. This reagent even gives a Wittig-type product with esters, substrates that are not methylenated with $Ph_3P=CH_2$. In addition, Tebbe's reagent does not racemize enolizable ketones as do the phosphorus ylids.[29a,b]

$$\text{"}Cp_2Ti=CH_2\text{"} + RCOOR' \longrightarrow R(OR')C=CH_2 \qquad (11.34)$$

Cases Intermediate Between Fischer and Schrock

In the Os complex **11.12**, Roper and co-workers[29c] have a carbene with character intermediate between the Fischer and Schrock extremes because it reacts both with electrophiles [e.g., SO_2 (Eq. 11.35)[29c] or H^+] and with nucleophiles [e.g., CO (Eq. 11.36)[29c] or CNR]. This is reasonable on the basis of our bonding picture. The osmium has π-donor (Cl) as well as π-acceptor (NO) ligands, the metal is in an intermediate oxidation state [Os(II) if we count the carbene as L, Os(IV) if X_2], and the carbene carbon has non-π-donor substituents (H).

$$Cl(NO)PPh_3Os=CH_2 \quad \xrightarrow{SO_2} \quad Cl(NO)PPh_3Os \overset{O}{\underset{CH_2}{\diagdown}} S=O \qquad (11.35)$$

11.12

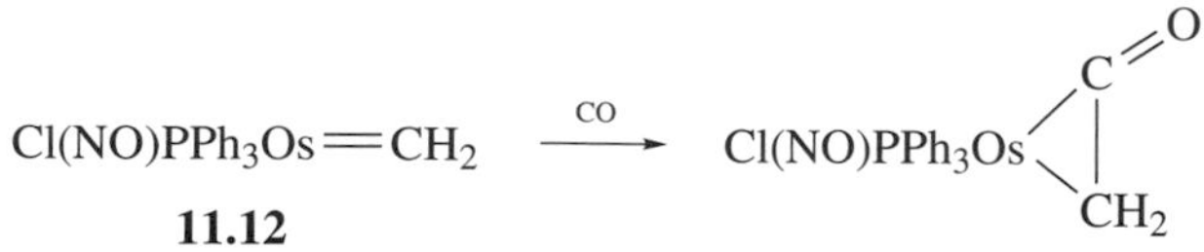

$$Cl(NO)PPh_3Os{=}CH_2 \quad \xrightarrow{CO} \quad Cl(NO)PPh_3Os\diagdown \qquad\qquad (11.36)$$

11.12

Boryl Complexes

The BR_2^- group is isoelectronic with CR_2 and a few boryl complexes have now been isolated, including $Cp_2WH(B\{cat\})$, $CpFe(CO)_2(B\{cat\})$[29d] (cat = catecholate {**9.30**}), and $RhHCl(B\{cat\})(PPh_3)_2$, which is one of the products formed from the oxidative addition of H–B(cat) with Wilkinson's catalyst.[29e] As in a carbene, an M=B multiple bond seems to be present; for example, in $Cp_2WH(B\{cat\})$, the B(cat) group is aligned in the least sterically favorable conformation, shown below, so the empty p orbital on boron can π bond with the filled metal d orbital shown. The π bond is not as strong as in a carbene, however, because the NMR spectrum shows that the B(cat) group is rapidly rotating.[29d]

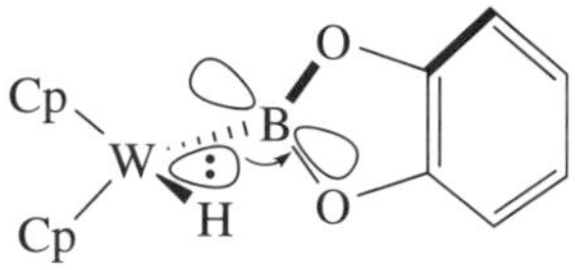

Vinylidene[29f]

The vinylidene ligand, M=C=CHR, readily formed from terminal acetylenes by a 1,2-migration of the H atom, is another intermediate character carbene. Although no heteroatom is present, the double bond adjacent to the carbene carbon clearly stabilizes the vinylidene relative to a Schrock alkylidene. A vinylidene is very prone to insertion: Eq. 11.37 shows a case where a double acetylene-to-vinylidene rearrangement is accompanied by an insertion into M–H. The final product, most likely formed by a further vinylidene insertion, is 5-coordinate, stabilized by a weak bond to a vinyl carbon (dotted line).

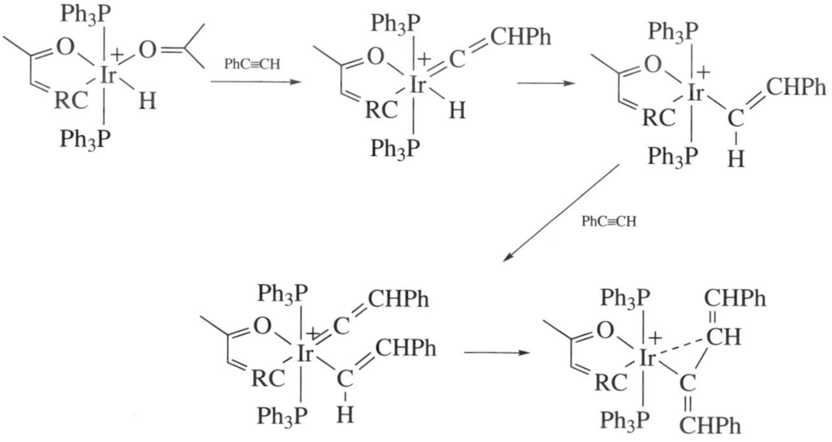

> • Carbenes can be heteroatom stabilized (Fischer) or not (Schrock).
> • Fischer carbenes have an electrophilic carbene carbon and Schrock carbenes have a nucleophilic carbene carbon.

11.2 CARBYNES[30]

Metal carbyne complexes M≡CR also have Fischer and Schrock extreme bonding formulations, although the distinction is less marked than for carbenes. In one bonding model, the free carbyne can be considered as doublet for Fischer and quartet for Schrock forms (Fig. 11.3a). A doublet carbene is a 2e donor via its *sp* lone pair and forms an additional covalent π bond (Fig. 11.3b). One *p* orbital on carbon remains empty and is able to receive back donation from the filled $M(d_\pi)$ orbital. We therefore have an LX ligand, 3e on the covalent model (ionic model: 4e). A quartet carbene can form three covalent bonds to a metal having three unpaired electrons, giving an X_3 ligand (Fig. 11.3c); this is also a 3e ligand on the covalent model (ionic model: 6e).

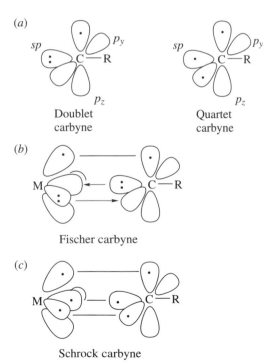

FIGURE 11.3 Doublet and quartet forms of (a) a carbyne can be considered as the parents of the (b) Fischer and (c) Schrock carbyne complexes.

Another common model considers the Fischer type as CR^+ bound to L_nM^- with weak back donation from two $M(d_\pi)$ orbitals. This leaves a net partial positive charge on carbon as in the Fischer carbene case. On this model, moving to very strong back donation converts the Fischer to the Schrock type with a net partial negative charge on carbon.

Oxidation state assignments again depend on the carbyne type. For example, the Fischer carbyne, $Br(Co)_4W{\equiv}CR$, is considered as W(II) and the Schrock carbyne, $Br_3L_2W{\equiv}CRe$, as W(IV). Once again, we have ambiguity in intermediate cases.

Synthesis

Fischer first prepared carbyne complexes (1973) by electrophilic abstraction of methoxide ion from a methoxy methyl carbene.

$$L(CO)_4M{=}C(OMe)Me + 2BX_3 \longrightarrow [L(CO)_4M{\equiv}CMe]^+BX_4{}^-$$

$$+ BX_2(OMe) \longrightarrow X(CO)_4M{\equiv}CMe \qquad (11.37)^{13}$$

If L is CO, then the halide ion (Cl, Br, or I) displaces the CO trans to the carbyne in the intermediate cationic complex; this shows the high trans effect of the carbyne. On the other hand, if L is PMe_3, then the cationic species is the final product.

By carefully controlled oxidation, Mayr and McDermott[31a] have been able to remove the carbonyl ligands in a Fischer carbyne to give a Schrock carbyne, thus making a direct link between the two types. This also allows synthesis of Schrock carbenes and carbynes with substituents other than the ones that can be obtained by the standard methods. In Eq. 11.39, we can think of the Br_2 oxidizing the metal by two units. This destabilizes the metal d_π orbitals relative to the carbon p orbitals, and so switches the polarity of the metal–carbon multiple bond. Note how the coligands change on going from the soft carbonyls in the W(II) starting material to the hard dme ligand in the W(VI) product. Schrock carbynes are nearly always d^0 (counting the carbyne as an X_3 ligand) as here.

$$Br(CO)_4W{\equiv}CMe \xrightarrow{Br_2, \text{ dme}} Br_3(dme)W{\equiv}CMe \qquad (11.38)^{31a}$$

$$(dme = MeOCH_2CH_2OMe).$$

Otherwise, Schrock carbynes can be made by deprotonation of an α-CH (Eq. 11.39); by an α elimination, in which this CH bond in effect oxidatively adds to the metal (Eq. 11.40); or in rare cases by a remarkable metathesis reaction[32] (Eq. 11.41). This reaction fails for coligands other than t-butoxide, showing the sensitivity of the different reaction pathways to the electronic and steric environment of the metal. MeCN is cleaved in the same way to a carbyne and a nitride

$(t\text{BuO})_3W\equiv N$. These reactions can be considered as triple binuclear oxidative additions with the O S rising by three units.

$$CpCl_2Ta=CHR \xrightarrow{\text{(i) PMe}_3,\text{ (ii) Ph}_3P=CH_2} CpCl(PMe_3)Ta\equiv CR \qquad (11.39)$$

$$Cp^*Br_2Ta=CHt\text{-Bu} \xrightarrow{\text{(i) dmpe, (ii) Na/Hg}} Cp^*(dmpe)HTa\equiv Ct\text{-Bu} \qquad (11.40)$$

$$(t\text{-BuO})_3W\equiv W(Ot\text{-Bu})_3 + t\text{-BuC}\equiv CtBu \longrightarrow 2(t\text{-BuO})_3W\equiv Ct\text{-Bu} \quad (11.41)$$

Structure and Spectra

The carbyne ligand is linear, having *sp* hybridization, and the $M\equiv C$ bond is very short (first row, 1.65–1.75 Å; second and third rows, 1.75–1.90 Å). The ^{13}C NMR shows a characteristic low-field resonance for the carbyne carbon at +250 to +400 ppm.

Reactions

A carbyne can couple[31b] with another carbyne to give an alkyne or alkyne complex.[33] For instance, $Br(CO)_4Cr\equiv CPh$ reacts with Ce(IV) to give free $PhC\equiv CPh$. In the Fischer series, the carbyne carbon is electrophilic and subject to nucleophilic attack, for example, by PMe_3, pyridine, RLi, or isonitrile (= Nu) to give a carbene of the type $L_nM=CR(Nu)$.[30] Alternatively, the nucleophile may attack the metal in $L_n(CO)M\equiv CR$ and produce a ketenyl complex $L_n(Nu)M(\eta^2\text{-OC}=CR)$ or $L_n(Nu)_2M(\eta^1\text{-OC}=CR)$. On the other hand, Schrock carbynes are nucleophilic and subject to attack by electrophiles, for instance, $(t\text{-BuO})_3W\equiv C(t\text{-Bu})$ reacts with HCl to give $(t\text{-BuO})_2Cl_2W=CH(t\text{-Bu})$.

Carbides

Acetylenes $RC\equiv CH$ are readily deprotonated. Some carbynes, $L_nM\equiv CH$, also having an acidic CH proton, can be deprotonated to give terminal carbide complexes, $[L_nM\equiv C]^-$. For example, Peters and co-workers[34a] have deprotonated $[(ArN\{t\text{-Bu}\})_3Mo\equiv CH]$ with KCH_2Ph. After the counterion was sequestered by complexation to a crown ether, $[(ArN\{t\text{-Bu}\})_3Mo\equiv C][K(crown)]$ was isolated and structurally characterized. Bridging acetylides, $L_nM-C\equiv C-ML_n$, and poly-acetylides, $L_nM-\{C\equiv C\}_x-ML_n$, are also well known.[34b]

11.3 BRIDGING CARBENES AND CARBYNES

Like CO, carbenes can act not only as terminal ($M=CH_2$) but also as bridging ligands.[35] When they bridge, a metal–metal bond is usually present as well (**11.13** and **11.14**). In bridging, carbenes lose some of their unsaturation, and therefore the very high reactivity of their mononuclear analogs. Fischer methylenes are

$$
\begin{array}{cc}
\underset{\textbf{11.13}}{\overset{\displaystyle \mathrm{CH_2}}{M\!\!-\!\!M}} &
\underset{\textbf{11.14}}{\overset{\displaystyle \mathrm{CH_2}}{M\qquad M}}
\end{array}
$$

very reactive and barely isolable, while bridging methylenes are well known and relatively stable.

One of the most valuable synthetic routes to bridging carbenes involves the use of diazomethane (Eq. 11.42) and related compounds (Eq. 11.43)[36], which are precursors for free carbenes in organic chemistry.

$$
\mathrm{CpMn(CO)_2(thf)} \xrightarrow{\mathrm{CH_2N_2}} \underset{\mathrm{Cp(CO)_2Mn\!\!-\!\!Mn(CO)_2Cp}}{\overset{\mathrm{CH_2}}{\diagup\!\diagdown}} \qquad (11.42)
$$

$$
\mathrm{CpRh(CO)_2} \xrightarrow[h\nu]{\mathrm{CE_2N_2}} \underset{\mathrm{Cp(CO)Rh\!\!-\!\!Rh(CO)Cp}}{\overset{\mathrm{CE_2}}{\diagup\!\diagdown}} \qquad (11.43)
$$

$$(E = COOMe)$$

Diazomethane adds not only to monomeric metal complexes but also to compounds containing metal–metal double bonds, a reaction somewhat analogous to the addition of a free carbene to a C=C double bond to give a cyclopropene. This analogy suggested itself to three groups at the same time, and, remarkably, they tried exactly the same reaction, Eq. 11.44:[37]

$$(11.44)$$

Note how loss of CO regenerates the Rh=Rh double bond in what is really a substitution of CO by CH$_2$. Insertion of CH$_2$ into a metal–metal single bond is seen in the synthesis of the platinum "A-frame" (so called because the structure resembles the letter A) complex **11.15** in Eq. 11.45,[38] a rare example of a bridging methylene complex without an M–M bond.

$$(11.45)$$

The second general method of bridging carbene complexes involves the analogy between C=C and M=C double bonds. Since many metal complexes react with C=C double bonds to give alkene complexes, Stone investigated the reactions of the same metal complexes with compounds containing an M=C bond (Eq. 11.46). This is a very powerful method of making a variety of homo- and heterometallic complexes and can be extended to the M≡C triple bond as well.

$$L_nM{=}CR_2 \ + \ \Big\|{\longrightarrow}Pt(PPh_3)_2 \ \xrightarrow{-C_2H_4} \ \begin{array}{c} R_2C \\ | \\ L_nM \end{array}\!\!{>}Pt(PPh_3)_2 \qquad (11.46)$$

Structure and Spectra

The ^{13}C NMR resonance positions of the carbene carbon for terminal and bridging carbenes reflects the greater unsaturation of the terminal type. Terminal groups resonate at a range from 250 to 500δ, while bridging groups appear from 100 to 210δ if an M–M bond is present, and between 0 and 10δ if not; for comparison, simple metal alkyls resonate at −40 to 0δ. These values probably reflect the change in hybridization required for the carbon atom to form bonds at the angles required by the geometry of the complex. If no metal–metal bond is present (**11.14**), then these angles will be close to 109° apart and no special rehybridization will be required. If an M–M bond is present, the two M–C bonds are usually 75°–85° apart. In a terminal carbene, the two bonds are, of course, formed with the same metal atom.

Reactions

Bridging carbenes are highly reactive toward alkynes, which give insertion as shown in Eq. 11.47:[39]

$$\text{(11.47)}$$

Hydride abstraction from a bridging carbene can give a μ_2-bridging carbyne, which is unsaturated, is very reactive, and shows pronounced carbonium ion character. The bonding scheme resembles the one we saw for Fischer carbenes, except that this is a bis-metal-stabilized carbonium ion, **11.16**.

Carbynes can also bridge three metals, as in the long-known and very stable tricobalt complex **11.17**; these are much less reactive than the unsaturated μ_2-carbynes discussed above.

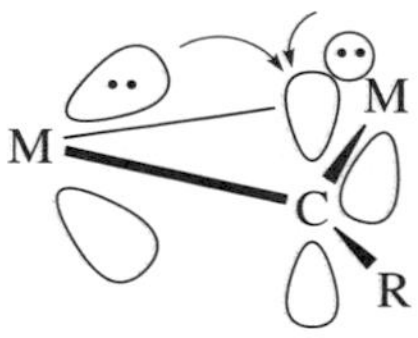

11.16

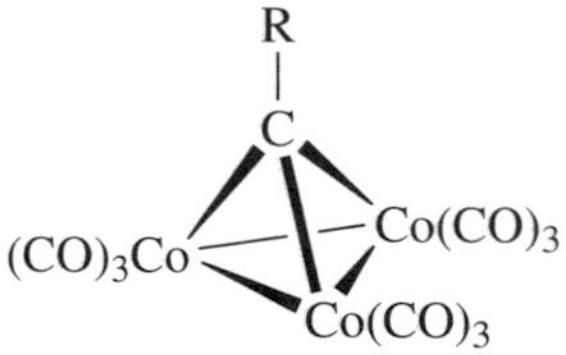

11.17

11.4   *N*-HETEROCYCLIC CARBENES

The M=C bonds in the majority of the terminal carbenes described up to now are reactive. Such carbenes are actor ligands because the M=C bonds are broken in typical reactions. *N*-heterocyclic carbenes,[40–42] or NHCs, are an exception—their M=C bonds are so stable that they often act as spectator ligands. They are gaining increasing attention in that role because, like phosphines, they are electronically and sterically tunable. Also like phosphines, they promote a wide series of catalytic reactions. For the moment we lack the same level of detailed understanding of their steric and electronic properties that is provided for phosphines by the Tolman map (Fig. 4.4).

Although they have sometimes been regarded as phosphine analogs, NHCs differ from PR_3 in important ways. The thermodynamic instability of free NHCs strongly disfavors simple dissociation, but reductive elimination can occur with loss of the imidazolium salt (Eq. 11.48).[43] Many catalysts containing NHCs are nevertheless stable for thousands of turnovers[42] or more,[43] so productive chemistry can be much faster than decomposition via Eq. 11.48.

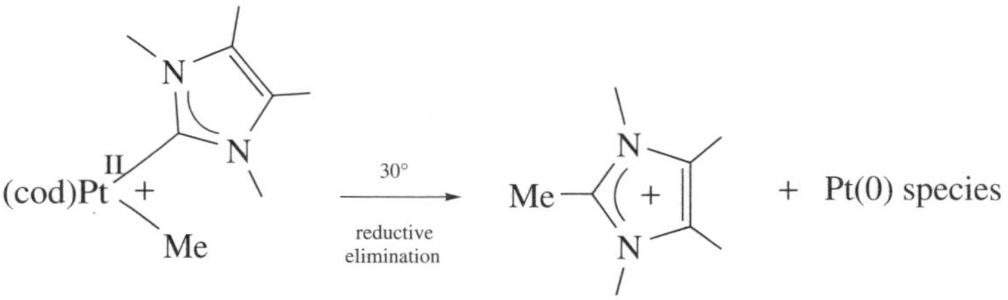

$$(11.48)$$

N-heterocyclic carbenes tend to be more donor than PR_3—the Tolman electronic parameters (TEP) for typical NHCs show they are more donor than even the most donor phosphines: PMe_3, 2064 cm^{-1}; **11.18**, 2054 cm^{-1}; **11.19**,

2050 cm^{-1}. Unlike PR_3, the nature of the R groups at N1 and N3 do not have a very large effect on the TEP; instead, a change in the nature of the ring is the factor that has the most effect on TEP. The R groups do influence the steric effect of the NHC, but the ligand is fan shaped, not cone shaped, like PR_3. Since rotation about the formal M=C bond is usually easy, the NHC normally orients to avoid steric clashes with other ligands making the NHC less bulky in reality than might at first appear.

Bonding

The free NHC (**11.18**) has a lone pair at C2 that can act as the donor to the metal. An empty *p*-orbital at C2 acts as acceptor for the π lone pairs at N1 and N3. This π-bonding strongly disfavors the triplet carbene because occupation of the C2 *p* orbital would interfere with the π bonding, so the singlet-triplet gap for the NHC is huge (ca. 82 kcal/mol by calculation[40]). Like an alkyl, the NHC is a very high trans effect ligand.

On binding to a d^2 or higher d^n-configuration metal, the metal at d_π electron density can now in principle engage in back bonding to the NHC. As this is a Fisher carbene like **11.1**, the importance of back bonding is probably small,[46] so the structure may be represented as a metal-substituted imidazolium **11.20**. In practice structure **11.21** is often seen, although, taken literally, this implies that C2 bears a hydrogen (because one assumes tetravalency for carbon).

Synthesis

The commonest route goes via the free carbene, formed via deprotonation of the imidazolium salt with strong base such as BuLi (Eq. 11.49). This normally requires bulky R groups like mesityl so the free carbene has at least transient stability and forbids the presence of other labile protons in the structure that would also react with BuLi. These limitations have led to the development of milder routes that avoid the free carbene.

Simplest among these is direct oxidative addition (Eq. 11.50), where the outcome can be complicated by subsequent reactions of the hydride formed in the oxidative addition step.

Direct metallation can be assisted by weak bases such as acetate because it is no longer necessary to deprotonate the free imidazolium ion.[48] A very useful method[49] is initial formation of a silver carbene using Ag_2O, followed by transmetallation to give the final product (Eq. 11.51).

$$\text{(11.49)}$$

$$\text{(11.50)}$$

isolable intermediate

$$\text{(11.51)}$$

N-heterocyclic carbenes were known for many years.[40–42] Öfele[50] and Wanzlick and Schonherr[51] made the first ones in 1968, and Lappert[52] made a whole series in the 1970s by cleavage of the C=C bond in the electron-rich olefin **11.22**.

Applications

After the initial activity in the 1960s and 1970s, Arduengo[42] drew attention back to the area in 1991 with the isolation of the first NHC in the free state, where bulky R groups stabilize the carbene center. From 1994, Herrmann[41] developed the use of NHCs as spectator ligands in homogeneous catalysis. Perhaps the most dramatic success in the area came from the modification of the Grubbs[10] catalyst with an NHC to give a much improved version, **11.23**. The very high trans effect NHC labilizes the PCy$_3$ ligand, loss of which is necessary for activity, and the rates go up by a factor of 10^2 to 10^3 relative to the prior bis-PCy$_3$ complex, **11.24**. Air stability is also improved on NHC substitution in this and other[44] catalysts.

Other NHCs, such as **11.25–28**, are readily accessible by similar routes, starting from the corresponding azoles: **11.25**, deriving from 1,2,4-triazole, and **11.26**, from thiazole. The abnormal C5 binding mode[53] **11.27**, seen in an increasing number of examples, is among the most strongly electron-donating NHCs. Chelating and pincer carbenes are also common, as illustrated by **11.28**.[44]

11.23
(R = mesityl)

11.24

11.25

11.26

11.27

11.28

There are numerous catalytic applications[40,41] of NHCs (hydrogenation, hydrosilation, metathesis, coupling chemistry, etc.) in which NHCs can have advantages over phosphines. Rates can be faster, and the catalysts usually do not need protection from air during catalysis. Imidazoles are also more readily synthesized in a variety of structural modifications, although subsequent formation of the M−C bond can be somewhat more difficult than in the case of PR$_3$.

> • *N*-heterocyclic carbenes (**11.18**) can be useful spectator ligands, tunable both electronically and sterically.

11.5   MULTIPLE BONDS TO HETEROATOMS

Related to carbenes and carbynes are species with multiple bonds to heteroatoms, of which the most important are terminal oxo M=O, nitrido M≡N, and imido M=NR. The high electronegativity of O and N give these ligands "Schrock" character; that is, they can be regarded as O^{2-}, NR^{2-}, and N^{3-}, respectively.[54] Stable compounds of these types tend to be found along a diagonal of the periodic table that runs from V to Os, where Mo is the element with the most examples; the great majority of examples have electron configurations from d^0 to d^2. Oxo groups have a high tendency to form M−O−M bridges; for some metals, such as Zr, terminal oxo complexes are rare.

For M=O in an octahedral complex, there are strong interactions between two of the M d_π orbitals and the O lone pairs (Fig. 11.4). When the two d orbitals are empty (d^0, d^1, or d^2), the interaction is bonding, and the M=O group has triple-bond character **11.29** with the LX_2 O atom as a 6e donor. This can be represented as **11.29a** or **11.29b**.

With more electrons on the metal, the bond order drops and electron–electron repulsions between $M(d_\pi)$ electrons and heteroatom lone pairs destabilize the

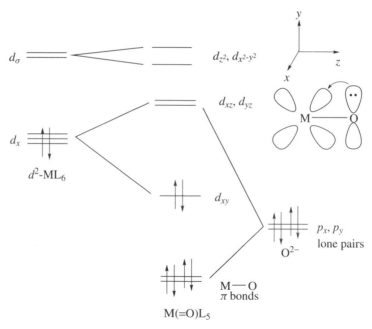

FIGURE 11.4 π Bonding in metal oxo complexes. After the σ bonds have been considered, a $d^2ML_6{}^{2+}$ species has a two-above-three orbital pattern characteristic of an octahedron. As long as they remain empty, two of the three d_π orbitals (xz and yz) can accept electrons from the O^{2-} lone pairs; one of these interactions is shown at the top right. This is a special case of the situation shown in Fig. 1.8. With one σ bond and two π bonds, the net $M^-\equiv O^+$ bond order is three.

$$M^- \equiv O^+ \quad \text{or} \quad M \overset{\leftarrow}{=} \ddot{O}$$

11.29a   **11.29b**

system and stable octahedral oxo complexes with d^4 or higher configurations are unknown. The Mayer and co-workers[55] d^4 oxo species, $Re(=O)X(RC\equiv CR)_2$, adopts a tetrahedral structure and the d^6 (η^6-$C_6H_4(i$-$Pr)Me)Os^- \equiv NAr^+$ and (η^5-$C_5Me_5)Ir^- \equiv NAr^+$ of Bergman and co-workers[56] are linear, thus avoiding the destabilization that would arise in an octahedral ligand field. Otherwise, octahedral late metal species normally have bridging oxo structures. A rare terminal oxo in $[py(porph.^+)Fe^{IV}=O]$ (porph = bulky porphyrin ligand) makes this species extremely reactive, even with alkane C—H bonds, and it is only observable at low temperatures.[57] This means that species such as $d^8(Me_3P)_3Pt=O$ are not plausible in a mechanistic scheme; $L_3Pt^+-O^-$ is not forbidden and a rare d^6 Pt(IV) oxo of this type has recently been observed.[58] Similar ideas hold for $M^- \equiv NR^+$ and $M \equiv N$. $M^- \equiv NR^+$ is linear at nitrogen, as expected for a $M \equiv N$ triple bond. A rare bent M—NR double-bonded structure is found in **11.30**, where the M=NR bond length of 1.789 Å can be compared with the adjacent $M^- \equiv NR^+$ at 1.754 Å. The reason for the unusual structure is that since =NR is an X_2 and $\equiv NR$ is an LX_2 ligand, if both imides were linear the Mo would have 20 electrons.

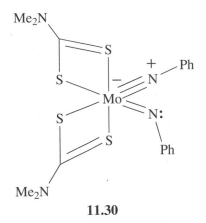

11.30

Synthesis

The complexes are often formed by oxidation, hydrolysis, or aminolysis (Eqs. 11.52–11.57). Equation 11.51 shows an unusual and very interesting route that forms multiple bonds to O and to C at the same time.[59]

$$[Os^{III}(NH_3)_6]^{3+} \xrightarrow{Ce^{4+}} [N\equiv Os^{VI}(NH_3)_5]^{3+} \qquad (11.52)$$

$$WCl_6 + t\text{-BuNH}_2 \longrightarrow [(t\text{-BuNH})_4W^-(\equiv N^+t\text{-Bu})_2] \qquad (11.53)$$

$$Np_3W\equiv Ct\text{-Bu} + H_2O \longrightarrow Np_2W^-(\equiv O^+)(=CHtBu) + NpH \quad (11.54)$$

$$(Np = Me_3CCH_2)$$

$$(11.55)$$

$$(L = PMe_2Ph)$$

$$WF_6 + 2MeOMe \longrightarrow WOF_4(OMe_2) + 2MeF \qquad (11.56)$$

The most oxophilic elements are even able to extract O from organic compounds, which prevents use of oxygenated solvents in many of these systems (Eq. 11.56). The nitride ligand has a lone pair that can sometimes be alkylated in a synthesis of an imido complex (Eq. 11.57):[60]

$$[R_4Os\equiv N]^- + Me_3O^+ \longrightarrow R_4Os^-\equiv N^+Me \qquad (11.57)$$

$$(R = Me_3SiCH_2)$$

Spectra and Structure

The $M^-\equiv O^+$ band at 900–1100 cm^{-1} in the IR spectrum is characteristic of the terminal oxo group; $M^-\equiv NR^+$ appears at 1000–1200 cm^{-1} and $M\equiv N$ at 1020–1100 cm^{-1}. The assignment can be confirmed by ^{18}O or ^{15}N substitution. An exception is $Cp_2M=O$ (M = Mo,W) with $\nu(M-O)$ frequencies below 880 cm^{-1}; electron counting shows that these must be $M=O$, not $M^-\equiv O^+$ species, however. The long $M=O$ bond length of 1.721 Å in $(MeC_5H_4)_2Mo=O$ is consistent with this idea. Low frequencies are also seen in bis-oxo species where the two oxo groups probably compete for electron donation into the empty $M(d_\pi)$ orbital(s). Useful NMR spectra can be obtained with ^{17}O- and ^{15}N-substituted species (both $I = \frac{1}{2}$), and these can be used to assign a bridging or terminal mode for the ligands present.

The presence of two *distortional isomers* was suggested for a number of metal oxo species, such as $MoOCl_2(PR_3)_3$ (**11.31**). The blue and green "isomers" of this series were found to have different $M=O$ bond lengths. Parkin and co-workers[61] have found that $MoCl_3(PR_3)_3$ (**11.32**) can co-crystallize with **11.31** in such a way as to cause both the color change and an apparent lengthening of the crystallographically determined $M-O$ distance, and so this distortional isomerism is not real. This is an illustration of how easy it is to miss alternative interpretations of the data.

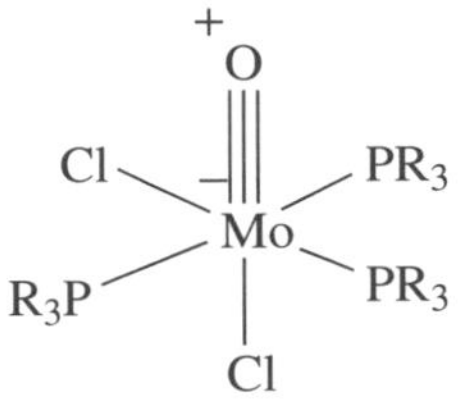

11.31

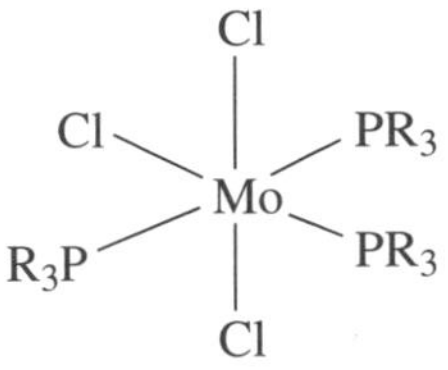

11.32

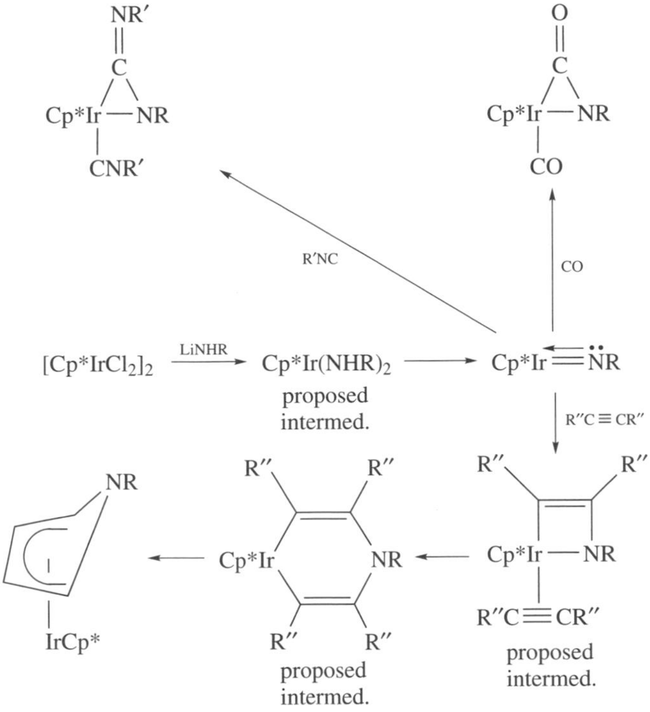

FIGURE 11.5 Some reactions of one of the Bergman and co-workers[56] late metal imido complexes.

Reactions

Two general reactivity principles apply. As the electronegativity of M increases on moving to the right in the periodic table, the orbital energies move from situation (*c*) in Fig. 11.1 to one where the $M(d_\pi)$ and O or $N(p)$ orbitals have comparable energy. The basic character of the O or N therefore falls. High-valent oxo, imido, or nitrido species are often stable enough to be isolated, but low-valent ones tend to be much more reactive. For example, $(CO)_5Mo=NPh$ has been implicated by McElwee-White and co-workers as a transient intermediate in a variety of reactions.[62] Bergman and co-workers[56] $(\eta^6\text{-}C_5Me_5)Ir{\equiv}NAr$ is isolable but very reactive (Fig. 11.5). $Cp_2^*Zr{=}O$, made by deprotonating $Cp_2^*Zr(OH)(O_3SCF_3)$ with the strong base $K[N(TMS)_2]$, reacts with acetylenes and nitriles to give metalla-cycles (Eq. 11.58).[63]

(11.58)

REFERENCES

1. J. W. Herndon, *Coord. Chem. Rev.* **248**, 3, 2004; **243**, 3, 2003.
2. R. R. Schrock, *JCS Dalton* 2541, 2001; *Chem. Rev.* **102**, 145, 2002.
3. (a) P. J. Brothers and W. R. Roper, *Chem. Rev.* **88**, 1293, 1988; (b) N. J. Cooper, *Pure Appl. Chem.* **56**, 25, 1984.
4. (a) A. J. Arduengo et al., *J. Am. Chem. Soc.* **113**, 361, 1991; (b) W. A. Herrmann et al., *J. Organomet. Chem.* **557**, 93, 1998; (c) G. Bertrand et al., *Chem. Rev.* **100**, 39, 2000.
5. E. O. Fischer and A. Maasbol, *Chem. Ber.* **100**, 2445, 1967.
6. J. S. Miller and A. L. Balch, *Inorg. Chem.* **11**, 2069, 1972.
7. L. Chugaev, *J. Russ. Chem. Soc.* **47**, 776, 1915; J. H. Enemark, A. L. Balch et al., *Inorg. Chem.* **12**, 451, 1973.
8. (a) M. Bullock et al., *J. Am. Chem. Soc.* **109**, 8087, 1987; (b) M. H. Chisholm and H. C. Clark, *Inorg. Chem.* **10**, 1711, 1971; M. H. Chisholm et al., *Inorg. Chem.* **16**, 677, 1977.
9. M. A. Gallop and W. R. Roper, *Adv. Organomet. Chem.* **25**, 121, 1986.
10. T. M. Trnka and R. H. Grubbs, *Accts. Chem. Res.* **34**, 18, 2000.
11. (a) E. O. Fischer, U. Schubert, and H. Fischer, *Pure Appl. Chem.* **50**, 857, 1978; (b) C. P. Casey and R. L. Anderson, *Chem. Commun.* 895, 1975.
12. T. Bodner and A. R. Cutler, *J. Organomet. Chem.* **213**, C31, 1981.
13. R. S. Bly and R. K. Bly, *Chem. Commun.* 1046, 1986.
14. (a) A. Davidson and J. P. Selegue, *J. Am. Chem. Soc.* **102**, 2455, 1980; (b) E. R. Davidson, O. Eisenstein, and K. G. Caulton, *J. Am. Chem. Soc.* **120**, 9388, 1998; (c) E. Carmona, A. Lledos et al., *Angew. Chem. Int. Ed.*, **43**, 3708, 2004.
15. C. P. Casey and R. L. Anderson, *J. Am. Chem. Soc.* **96**, 1230, 1974; K. H. Doetz et al., *J. Organomet. Chem.* **182**, 489, 1979.

16. H. Werner, E. O. Fischer et al., *J. Organomet. Chem.* **28**, 367, 1971.

17. E. O. Fischer et al., *Chem. Ber.* **105**, 3966, 1972.

18. C. P. Casey et al., *J. Am. Chem. Soc.* **101**, 7282, 1979.

19. K. H. Dötz et al., *Angew. Chem., Int. Ed.* **23**, 97, 1984.

20. R. R. Schrock, *Acc. Chem. Res.* **12**, 98, 1979; *Science* **219**, 13, 1983.

21. (a) R. R. Schrock et al., *J. Am. Chem. Soc.* **100**, 359, 1978; (b) Z. Xue et al., *J. Am. Chem. Soc.* **117**, 12746, 1995.

22. R. R. Schrock, L. W. Messerle, C. D. Wood, and L. J. Guggenberger, *J. Am. Chem. Soc.* **100**, 3793, 1978.

23. R. R. Schrock, W. J. Youngs, M. R. Churchill et al., *J. Am. Chem. Soc.* **100**, 5962, 1978.

24. R. R. Schrock, *J. Am. Chem. Soc.* **97**, 6577, 1975; R. R. Schrock and P. R. Sharp., *J. Am. Chem. Soc.* **100**, 2389, 1978.

25. J. M. Williams, R. R. Schrock, et al., *J. Am. Chem. Soc.* **103**, 169, 1981.

26. J. H. Wengrovius and R. R. Schrock, *Organometallics* **1**, 148, 1982.

27. R. R. Schrock, *J. Am. Chem. Soc.* **98**, 5399, 1976.

28. F. N. Tebbe, G. W. Parshall, and G. S. Reddy, *J. Am. Chem. Soc.* **100**, 3611, 1978.

29. (a) R. H. Grubbs et al., *Pure Appl. Chem.* **55**, 1733, 1983; (b) A. J. Fairbanks et al., *Tet. Lett.* **44**, 3631, 2003; (c) W. R. Roper et al., *Adv. Organomet. Chem.* **25**, 121, 1986; (d) J. Hartwig, *J. Am. Chem. Soc.* **115**, 4908, 1993; (e) K. Burgess, R. T. Baker et al., *J. Am. Chem. Soc.* **114**, 9350, 1992; (f) P. H. Dixneuf, *Acc. Chem. Res.* **32**, 311, 1999.

30. H. Fischer et al., *Carbyne Complexes*, VCH, Weinheim, 1988.

31. (a) A. Mayr and G. A. McDermott, *J. Am. Chem. Soc.* **108**, 548, 1986; (b) A. Mayr et al., *J. Am. Chem. Soc.* **109**, 580, 1987.

32. (a) J. S. Murdzek and R. R. Schrock, Chapter 5 in Ref. 30; R. R. Schrock et al., *J. Am. Chem. Soc.* **104**, 4291, 1982.

33. G. Huttner et al., *Angew. Chem. Int. Ed.* **88**, 649, 1976.

34. (a) J. C. Peters, A. L. Odom, and C. C. Cummins, *Chem. Commun.* 1997, 1995; (b) J. A. Gladysz et al., *Chem. Eur. J.* **4**, 1033, 1998; *New J. Chem.* **21**, 739, 1997.

35. J. W. Herndon, *Coord. Chem. Rev.* **181**, 177, 1999.

36. K. K. Mayer and W. A. Herrman, *J. Organomet. Chem.* **182**, 361, 1979.

37. (a) J. R. Shapley et al., *J. Organomet. Chem.* **201**, C31, 1980; (b) N. M. Boag, M. Green, F. G. A. Stone et al., *Chem. Commun.* 1171, 1980; (c) W. A. Herrmann et al., *Angew. Chem., Int. Ed.* **20**, 183, 1980.

38. R. J. Puddephatt, K. R. Seddon et al., *Inorg. Chem.* **18**, 2808, 1979.

39. S. A. R. Knox, *Chem. Commun.*, 2803, 1980.

40. D. Bourissou, O. Guerret, F. P. Gabbai, and G. Bertrand, *Chem. Rev.* **100**, 39, 2000.

41. W. A. Herrmann, *Angew. Chem. Int. Ed.* **41**, 1291, 2002.

42. A. J. Arduengo, *Acc. Chem. Res.* **32**, 913, 1999.

43. (a) D. S. McGuinness, N. Saendig, B. F. Yates, and K. J. Cavell, *J. Am. Chem. Soc.* **123**, 4029, 2001; (b) C. M. Crudden and D. P. Allen, *Coord. Chem. Rev.* **248**, 2247, 2004.

44. E. Peris, J. A. Loch, J. Mata, and R. H. Crabtree, *Chem Comm.* 201, 2001; J. A. Loch, M. Albrecht, E. Peris, J. Mata, J. W. Faller, and R. H. Crabtree, *Organometallics* **21**, 700, 2002.

45. L. Perrin, E. Clot, O. Eisenstein, J. Loch, and R. H. Crabtree, *Inorg. Chem.* **40**, 5806, 2001.

46. C. Boehme and G. Frenking, *Organometallics* **17**, 5801, 1998.

47. S. Grundemann, M. Albrecht, A. Kovacevic, J. W. Faller, and R. H. Crabtree, *JCS Dalton,* 2163, 2002.

48. M. Albrecht, J. R. Miecznikowski, A. Samuel, J. W. Faller, and R. H. Crabtree, *Organometallics* **21**, 3596, 2002.

49. H. M. J. Wang, and I. J. B. Lin, *Organometallics* **17**, 972, 1998; A. R. Chianese, X. W. Li, M. C. Janzen, J. W. Faller, R. H. Crabtree, *ibid*, **22**, 1663, 2003.

50. K. Öfele, *J. Organometal. Chem.* **12**, P42, 1968.

51. H. W. Wanzlick and H. J. Schonherr, *Angew. Chem. Int. Ed.* **7**, 141, 1968.

52. M. F. Lappert, *J. Organometal. Chem.* **358**, 185, 1988.

53. A. R. Chianese, A. Kovacevic, B. M. Zeglis, J. W. Faller, and R. H. Crabtree, *Organometallics* **23**, 2461, 2004.

54. J. M. Mayer, *Comments Inorg. Chem.* **8**, 125, 1988; W. A. Nugent and J. M. Mayer, *Metal Ligand Multiple Bonds*, Wiley, New York, 1988.

55. J. M. Mayer, D. L. Thorn, and T. H. Tulip, *J. Am. Chem. Soc.* **107**, 7454, 1985.

56. R. G. Bergman et al., *J. Am. Chem. Soc.* **113**, 2041, 5100, 1991; (b) *Organometallics* **11**, 761, 1992.

57. R. Ortiz de Montellano, *Cytochrome P-450: Structure, Mechanism and Biochemistry*, Plenum, New York, 1986. S. J. Lippard and J. M. Berg, *Principles of Bioinorganic Chemistry*, University Science Books, Mill Valley, CA, 1994.

58. T. M. Anderson, W. A. Neiwert, M. L. Kirk, P. M. B. Piccoli, A. J. Schultz, T. F. Koctzle, K. Morokuma, R. Cao, and C. L. Hill, *Science*, **306**, 2074, 2004.

59. J. M. Mayer, *J. Am. Chem. Soc.* **109**, 2826, 1987.

60. P. A. B. Shapley et al., *Organometallics* **5**, 1269, 1986; *J. Organometal. Chem.* **335**, 269, 1987.

61. G. Parkin et al., *J. Am. Chem. Soc.* **113**, 8414, 1991.

62. L. McElwee-White et al., *J. Am. Chem. Soc.* **113**, 4871, 1991.

63. R. G. Bergman et al., *Organometallics* **11**, 761, 1992.

PROBLEMS

1. How could you use Tebbe's reagent to convert cyclohexanone to 1,1-dimethylcyclohexane?

2. Provide a plausible mechanism and experimental mechanistic tests for

3. Can you suggest a mechanism for the reactions of Eq. 4.31?

4. (a) We can view $Ph_3P=CH_2$ as a carbene complex of a main-group element. Does it show Fischer- or Schrock-like behavior? Using arguments of the type shown in Fig. 11.1, explain why it behaves as it does. (b) Metal oxo complexes, such as $Re(=O)Cl_3(PPh_3)_2$, might also be regarded as carbenelike if we make the isoelectronic substitution of O for CH_2. Do the same arguments of Fig. 11.1 give any insight into whether an M=O group will have greater or lesser nucleophilic character than the corresponding $M=CH_2$ species?

5. Propose a mechanism for

$$(CO)_5M = CR(OR)$$

6. Would you expect changes in the formal orbital occupation to effect the orientation of a CH_2 group? Given the orientation shown in Fig. 11.2, draw the appropriate diagram for the isoelectronic $[Cp_2W(=CH_2)Me]^+$, which has an electrophilic methylene. What about the hypothetical $[Cp_2W(=CH_2)Me]^-$? What would be the CH_2 orientation, and would you expect the complex to be stable?

7. Why is the NHC ligand regarded as a 2e neutral donor (L ligand) when its M−C bond resembles that in M−Ph, an undoubted 1e ligand on the covalent model? **11.33** is also an L ligand but if it is deprotonated (**11.34**), is it still an L ligand?

11.33   **11.34**

12

APPLICATIONS OF ORGANOMETALLIC CHEMISTRY

An important series of catalytic reactions involve some of the intermediates and pathways discussed in previous chapters. Alkene metathesis (Eq. 12.1), now gaining wide acceptance in organic and polymer synthesis, goes via metal carbene intermediates. Alkene polymerization, a key modern development in polymer synthesis, uses unsaturated alkyl complexes. This catalytic reaction allows an exceptional level of control over the molecular structure and therefore over the polymer properties. The water–gas shift and related reactions are of commercial importance in providing a simple route to H_2 and to acetic acid. C−H activation refers to a class of catalytic reactions in which unactivated C−H bonds are broken. Finally, we look at some organometallic materials.

12.1 ALKENE METATHESIS

The alkene metathesis reaction[1] is one of the most original and unusual transformations in chemistry (Eq. 12.1). Remarkably, the strongest bond in the alkene, the C=C double bond, is broken during the reaction. The resulting RHC= fragments are exchanged between the alkenes. Metathesis was originally developed in industry[2] and only applied to simple alkenes because the catalysts then used were intolerant of functionality, and the reaction itself was not widely known to organic chemists. With the development of much more tolerant and versatile catalysts and the wider diffusion of information on the reaction, the number of

The Organometallic Chemistry of the Transition Metals, Fourth Edition, by Robert H. Crabtree
Copyright © 2005 John Wiley & Sons, Inc.

applications to functionalized alkenes in industrial chemistry and organic and polymer synthesis continues to increase.[1,3]

$$RCH{=}CHR + R'CH{=}CHR' \rightleftharpoons 2RCH{=}CHR' \qquad (12.1)$$

The final products are statistical unless the reaction can be driven in some way such as by continuous removal of a volatile product like C_2H_4 (Eq. 12.2).

$$2RCH{=}CH_2 \Rightarrow 2RCH{=}CHR + C_2H_4{\uparrow} \qquad (12.2)$$

A critical step in making metathesis broadly useful was finding catalysts more tolerant of organic functionality. This required moving to the right in the periodic table. The early titanium catalysts are the least useful because they react preferentially with heteroatom functionalities ($RCOOH > ROH > R_2CO > RCO_2R > C{=}C$), in line with the high oxophilic character of early metals. Molybdenum and tungsten catalysts are intermediate in character, while ruthenium catalysts prefer $C{=}C$ bonds over heteroatoms ($C{=}C > RCOOH > ROH > R_2CO > RCO_2R'$). Rhodium is apparently too far to the right, because it fails to give metathesis cleanly—the key carbene intermediate tends to cyclopropanate the alkene. Grubbs' Ru catalysts[1] (**12.1**) have proved to be the easiest to handle, but for some applications Schrock's Mo catalysts[3] are needed. Chiral metathesis catalysts[3] are discussed in Chapter 14.

Grubbs catalyst   Schrock catalyst

Metathesis can usefully be divided into a number of types, depending on the nature of the substrates and products in the catalytic reaction. A reaction such as Eq. 12.1 is a *simple metathesis*. With two substrates we have the reverse version, a *cross metathesis* (Eq. 12.3). With some choices of R and R', the cross product can be favored kinetically. This happens in Eq. 12.4 where one alkene, $t\text{-BuCH}{=}CH_2$, present in excess, is too bulky to metathesize with itself, and the cross product is formed in 93% yield.[1]

$$RCH{=}CHR + R'CH{=}CHR' \rightarrow 2RCH{=}CHR' \qquad (12.3)$$

$$AcOCH_2CH{=}CHCH_2OAc + t\text{-BuCH}{=}CH_2 \rightarrow AcOCH_2CH{=}CH_2 \qquad (12.4)$$

With a conjugated diene, ring-closing metathesis (RCM) is often possible, particularly where the product ring is unstrained (Eq. 12.5).[4] The reverse of

Eq. 12.4 is a ring-opening metathesis (ROM), favored by the presence of a large excess of C_2H_4.

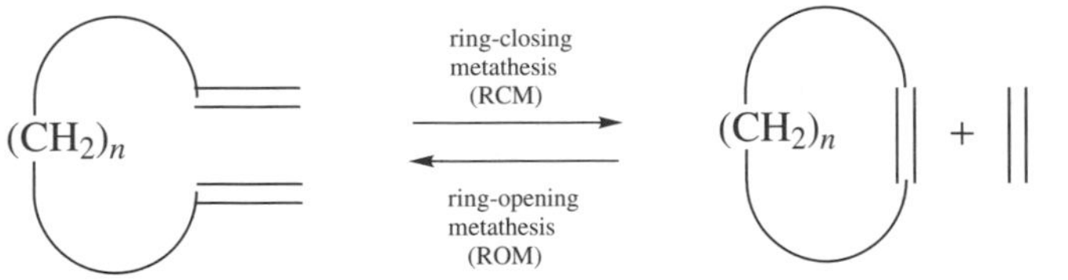

$$(12.5)$$

The efficiency of the catalysts is high enough for application to polymerization. The two best known cases[1] are acyclic diene metathesis (ADMET, Eq. 12.6) and ring-opening metathesis polymerization, or ROMP, where ring strain (ca. 15 kcal/mol in Eq. 12.7) drives the ring opening. These reactions are considered to be living polymerizations because the resting state of the catalyst is the fully active $[Cl_2L_2Ru=CH-\{P\}]$, where $\{P\}$ is the polymer chain. This means that when one monomer, A, is used up, a second monomer, B, can be added. The reaction then continues with the result that a block copolymer (...AAAABBBB...) is obtained. Such a material has very different properties from a simple mixture of homopolymers A_n and B_n or of a random copolymer (...AABABB...). Once again, the reaction is very tolerant of functional groups in the monomer.

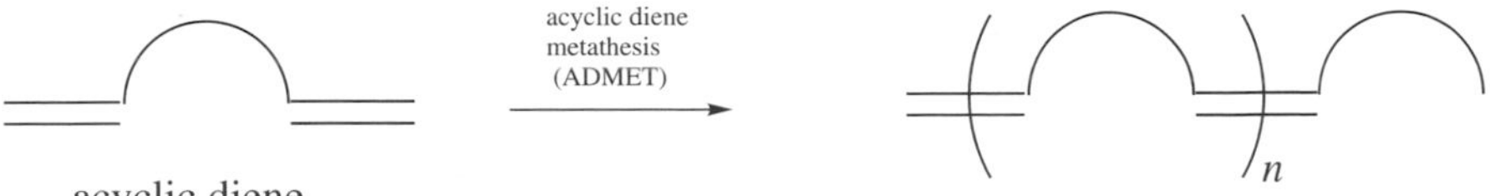

acyclic diene

$$(12.6)$$

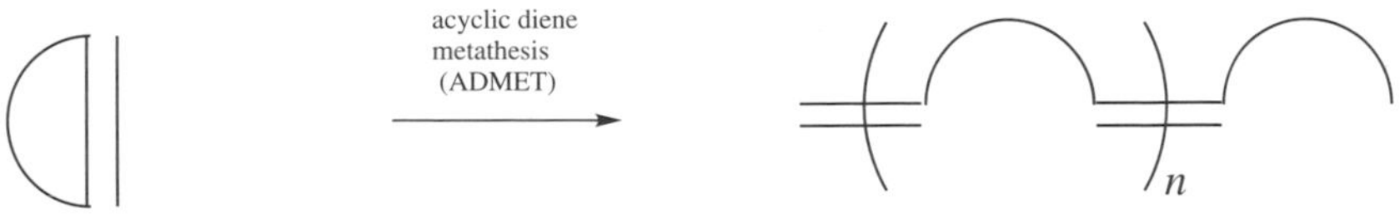

strained
cycloalkene

$$(12.7)$$

Mechanism

After the initial discovery, several suggestions were made for the mechanism during the 1970s (Eq. 12.8). A cyclobutane metal complex was considered, but cyclobutanes were not formed in the reaction and added cyclobutane did not participate in the reaction. Pettit proposed a tetracarbene complex, for example $M(=CH_2)_4$ from ethylene, but that seemed to require an unreasonably large number of available sites on the metal. Grubbs proposed a metallocyclopentane intermediate, formed from oxidative coupling of the two alkenes, but it was not clear how this species could rearrange appropriately. All these mechanisms proved misconceived. In an earlier (1971) article that had completely escaped the attention of the organometallic community—no doubt because it was published in a polymer journal—Hérisson and Chauvin[5] suggested the correct solution. This came out of a series of well-chosen "double-cross" experiments designed to test whether the two alkenes simultaneously bound to the metal (*pairwise mechanisms*), as in all the mechanisms of Eq. 12.8, or whether the alkenes were converted one at a time (*nonpairwise mechanisms*). The nonpairwise *Chauvin mechanism* of Eq. 12.9 shows how a metalacyclobutane can be formed from an initial carbene reacting with an incoming alkene and then cleaving in a different way (along the dotted line in the equation) to give the new alkene and a different carbene.

$$(12.8)$$

$$(12.9)$$

The critical experiment to decide between these two routes, the double cross shown in Eq. 12.10, is a more elaborate form of the crossover experiment. If the reaction is pairwise, then at the beginning of the reaction we will see products from only two of the alkenes (e.g., the C_{12} and C_{16} products in Eq. 12.10), not the double-cross product containing fragments of all three alkenes (C_{14} in Eq. 12.10), which would be expected on the nonpairwise mechanism. The pairwise mechanism requires that no C_{14} form initially; later on in the reaction, double-cross products are bound to form, whatever the mechanism, by subsequent metathesis of C_{12} with C_{16}.

$$(12.10)$$

The amounts of C_{12}, C_{14}, and C_{16} were measured as a function of time and the $[C_{14}]/[C_{12}]$ and $[C_{14}]/[C_{16}]$ ratios extrapolated back to time zero. These ratios should be zero for the pairwise and nonzero for the nonpairwise routes. The results showed that a nonpairwise mechanism operates: $[C_{14}]/[C_{12}]$ extrapolated to 0.7 and $[C_{14}]/[C_{16}]$ to 8.35 for one of the best-known metathesis catalysts, $MoCl_2(NO)_2(PPh_3)_2$ and $Me_3Al_2Cl_3$.[6a] Staunch adherents of the pairwise mechanism suggested the "sticky olefin" hypothesis. This held that the alkene, once metathesized by a pairwise mechanism, was retained by the metal at the active site, rather than being immediately released into solution. While it remains at the site, the single-cross product might metathesize several times, and so only the double-cross product would be released into solution and detected. This salvages the pairwise mechanism and requires a more sophisticated experiment to test the new hypothesis.

To test this, we need a system in which the metathesis products do not themselves metathesize, so that we can be sure that we are seeing the *initial* products. Perhaps the best example is shown in Eq. 12.11,[6b] in which **12.2** is converted into ethylene and phenanthrene, neither of which metathesize further with the particular Mo catalyst chosen. The initial isotope distributions in the products will then truly reflect a single catalytic cycle. The results of this reverse double cross showed a $1:2:1$ mixture of the d^0, d^2, and d^4 isotopomers of the resulting ethylene, which successfully defends the nonpairwise mechanism against the sticky olefin idea. It was only at the end of the 1970s that consensus was established, however.

12.2   **12.2-d$_4$**   phenanthrene

\+

$$H_2C=CH_2 \quad + \quad D_2C=CH_2 \quad + \quad D_2C=CD_2$$

$$1 \quad : \quad 2 \quad : \quad 1$$

$$(12.11)$$

Types of Catalyst

The early catalysts, such as $MoCl_2(NO)_2(PPh_3)_2$, needed an $Al_2Me_3Cl_3$ co-catalyst to generate an alkylidene, probably via α elimination of L_nMoMe_2 to give $L_nMo=CH_2$ as active catalyst, but the active species were never directly observed. The first isolable carbene shown to be a metathesis catalyst was $(CO)_5W(=CPh_2)$,[7] but activity was not very high. Two types of catalyst are in common use, the Schrock and Grubbs catalysts; neither requires a co-catalyst because the carbene is already present.

Commercial Applications

Metathesis plays a key role in the SHOP process discussed in the next section and in ROMP polymerization (Eq. 12.7).

The neohexene process starts with the acid-catalyzed dimerization of isobutene, followed by metathesis with ethylene, to give neohexene, an intermediate in the manufacture of synthetic musk, and regenerate isobutene.

The commercial synthesis of the housefly pheromone **12.3** illustrates the technique of driving the metathesis reaction by removing the more volatile alkene product, in this case, ethylene; undesired noncross products can easily be separated by distillation. Unfortunately, the presence of the alkylaluminum co-catalyst severely limits the range of functional groups tolerated by this system.

$$Me(CH_2)_7CH=CH_2 + Me(CH_2)_{12}CH=CH_2 \longrightarrow$$

$$Me(CH_2)_7CH=CH(CH_2)_{12}Me + C_2H_4 + \text{other products}$$
$$\textbf{12.3} \qquad\qquad\qquad (12.13)$$

Commercial synthesis of unusual polymers has also been possible with Grubbs metathesis catalyst. Polydicyclopentadiene can be formed from dicyclopentadiene by ROMP; the presence of the second C=C bond in the monomer allows some cross-linking to occur, giving exceptional strength to the material. It can even stop bullets within a few centimeters of penetration! Direct reaction injection molding has proved possible, in which the monomer and catalyst are injected into the heated mold and the item formed in place. The polymer is being used to fabricate sports equipment and several other applications are being considered.

dicyclopentadiene

Alkynes can be metathesized by the complex[8] $(t\text{-BuO})_3W\equiv C(t\text{-Bu})$ (**12.4**), apparently via the tungstenacyclobutadiene species **12.5** in Eq. 12.14.

$$(t\text{-BuO})_3W\equiv C(t\text{-Bu})$$
$$\textbf{12.4}$$

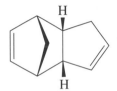

$$R'C\equiv CR + M\equiv CR'$$

12.2 DIMERIZATION, OLIGOMERIZATION, AND POLYMERIZATION OF ALKENES

All three reactions rely on repeated alkene insertion into an M—C bond to form new C—C bonds via the Cossee mechanism.[9,10] The three types differ in the relative rates of chain growth (k_g) by insertion to termination (k_t), normally by β elimination. If chain termination is very efficient, alkene dimerization may be seen. If it is very inefficient, a polymer will result, as in Ziegler–Natta and metallocene catalysis. In the intermediate case, oligomeric α olefins can be formed (Fig. 12.1), as in the SHOP process. Even though we discuss these reactions separately, they are nevertheless closely related.

Alkene polymerization[9] is one of the most important catalytic reactions in commercial use and an important contribution to polymer and materials science. The Ziegler–Natta catalysts, for which Ziegler and Natta won the Nobel Prize in 1963, account for more than 15 million tons of polyethylene and polypropylene annually. These catalysts are rather similar to the early metathesis catalysts in that mixtures of alkylaluminum reagents and high-valent early metal complexes are used. The best known is $TiCl_3/Et_2AlCl$, which is active at 25°C and 1 atm; this contrasts with the severe conditions required for thermal polymerization (200°C, 1000 atm). Not only are the conditions milder, but the product shows much less branching than in the thermal method. Propylene also gives highly crystalline stereoregular material, in which long sequences have the same stereochemistry at adjacent carbons in a head-to-tail polymer; this is called an isotactic polymer

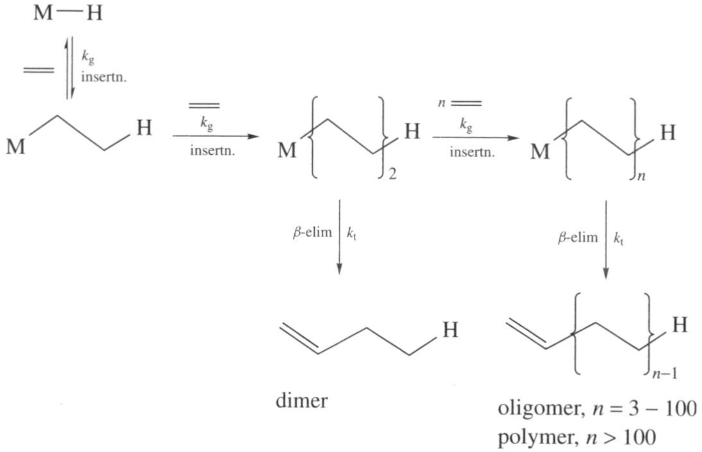

FIGURE 12.1   Relative rates of insertion and β elimination determine the value of n in the products of di-, oligo-, and polymerization reactions in the Cossee mechanism. Slower β elimination implies higher n.

(**12.6**). The commercial catalysts are heterogeneous in the sense that the active centers are on crystallites of $TiCl_3$ supported on $MgCl_2$.

Homogeneous versions of the Ziegler–Natta catalysts were soon developed. The most common are of general form $[LL'MCl_2]$ (M = Ti, Zr, or Hf), where L and L' are a series of C- or N-donor ligands. Initially, L and L' were cyclopentadienyl groups, so the catalysts were at first termed metallocene catalysts. Later improvements incorporated a much wider range of ligands, and so the term *single-site catalysts* is now also used.

These catalysts have had a revolutionary impact on the polymer industry because variation of L and L' allows delicate control over the microstructure of the polymer—how the atoms are connected in the chains—and over the polydispersity—the distribution of chain lengths. Control of the microstructure and the polydispersity in turn gives control of the physical properties of the final polymer material. Metallocene polymers can be designed to be very tough, or act as elastomers, or be easily heat-sealed, or have excellent optical properties, or have excellent processability and they have therefore displaced higher-cost polymers, such as polyurethanes, in certain applications. Syndiotactic polypropylene (**12.7**), unobtainable pure before metallocene catalysis, is softer but tougher and clearer than other forms. It is in films for food storage and in medical applications. The global production of metallocene polymers exceeds 2×10^9 lb.

Cp_2ZrCl_2 must first be activated with methylalumoxane (MAO, $[MeAlO]_n$), a species with an ill-defined polymeric structure formed from the partial hydrolysis of $AlMe_3$. Initial methylation by MAO gives Cp_2ZrMe_2 followed by Me^- abstraction from Zr–Me by MAO to form the active 14e species, $[Cp_2ZrMe]^+$, stabilized by the noncoordinating $[Me\{MeAlO\}_n]^-$ counterion.

| **12.6** isotactic | **12.7** syndiotactic |

Microstructure

Metallocenes produce polyethylene that is strictly linear, without side branches, termed LLDPE (linear low-density polyethylene). Other processes tend to produce branches and give a lower quality product.

Polypropylene has an almost perfectly regular head-to-tail structure when produced with metallocenes. The arrangement of the methyl groups in isotactic polypropylene (**12.6**) gives the polymer chain a helical rod structure. This is reminiscent of the α helix of proteins. The rods are chiral and catalysts that form isotactic polypropylene are also chiral. Since both hands of the catalyst are normally present, rods of both left- and right-handed forms are present in equal amounts.

Syndiotactic polypropylene has no chirality and is formed by catalysts lacking chirality. It tends to adopt a planar zig-zag conformation (12.7) of the main chain.

Types of Catalyst

Metallocene catalysts of type 12.8 have been found to be highly selective for the formation of isotactic polypropylene. Structure 12.8 is chiral because it lacks a plane of symmetry. Like the asymmetric hydrogenation catalysts discussed in Chapter 9, 12.8 also has a C_2 symmetry axis, so both binding sites, both occupied by Cl ligands in 12.8 and 12.9 have the same chirality. Each new propylene monomer that is incorporated is therefore expected to enter into the polymer chain with the same chirality, giving isotactic polymer, whichever binding site is operative for any given step.

12.8   12.9

To explain the selective formation of syndiotactic propylene by metallocene polymerization catalysts, it has proved necessary to assume that polymerization proceeds stepwise, with the polymer chain moving from one metallocene binding site to the other at each step much like a windshield wiper. The nth alkene to insert therefore occupies the opposite binding site from the $(n-1)$th and $(n+1)$th alkene. This is reasonable because once the insertion takes place the newly formed M—C bond automatically finds itself in the other binding site (see Fig. 12.2). The model may be oversimplified, however, because one would expect $[Cp_2ZrR]^+$ to be trigonal with the R group in the symmetry plane of the molecule, not pyramidal as in Fig. 12.2. Ion pairing with the counterion may pyramidalize the metal. In a low dielectric solvent such as is used for polymerization, ion pairing energies can be surprisingly large (>10 kcal/mol) and so may have a more important role in organometallic chemistry than currently appreciated.[9c]

Catalyst 12.9 is highly selective for forming syndiotactic polypropylene, (12.7). Each binding site is locally chiral, but, because the whole molecule has a plane of symmetry, the two binding sites have opposite local chirality. Because each successive propylene occupies opposite sites, they can therefore be incorporated into the growing chain with opposite chirality, leading to syndiotactic polymer.

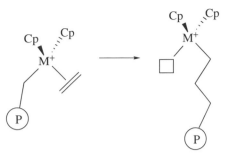

FIGURE 12.2 Windshield wiper model for alkene polymerization by metallocene catalysts. The insertion causes the M—C bond to the polymer chain to move from one side to the other. The open square represents an open site.

Molecular mechanics indicate probable structures for the key intermediate propylene complexes in the two classes of catalyst. In the chiral isotactic catalyst, **12.8**, the methyl group tends to be located as shown in Fig. 12.3 (upper), so that successive propylenes enter with the same chiralities and bind via the same face (*re* in the figure). In the achiral syndiotactic catalyst, **12.9**, in contrast, successive propylenes enter with opposite chiralities and bind via alternating faces (*re* then *si*).

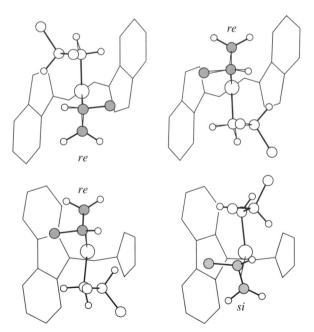

FIGURE 12.3 Chiral metallocene catalyst **12.8** (upper) leads to alternate propylenes (shaded) binding via the same *re*-face to give isotactic polymer. The achiral catalyst **12.9** (lower) leads to alternate propylenes binding via the opposite faces, *re* then *si*, to give syndiotactic polymer. (Adapted from Ref. 9b with permission.)

Green–Rooney Proposal

Why is the C=C insertion step into the M–C bond so quick in these catalysts, while in isolable 18e alkyl olefin complexes it tends to be very slow? The reason seems to be that the reaction can be strongly accelerated by coordinative unsaturation. This can allow the alkyl to become agostic and so turn toward the alkene, facilitating insertion, as proposed by Green and Rooney.[11] Theoretical work by Ziegler et al.[12] has indicated that in the model intermediate $[Cp_2ZrMe(C_2H_4)]^+$ the CH_3 group is agostic (Fig. 12.4, left), as allowed by the formally 16e count for this species. The principal (C_3) axis of the methyl group is rotated by 40%, turning the CH_3 sp^3 hybrid orbital toward the alkene. At the transition state for insertion (Fig. 12.4, right) this value has increased to 46%.

It is difficult to test the Green–Rooney proposal, but the challenge was taken up by Grubbs et al.,[13a] Kraulendat and Britzinger,[13b] and Piers and Bercaw,[13c] who developed an elegant mechanistic experiment to determine whether agostic species were involved. Figure 12.5 shows one version[13b] of the experiment in which the polymerization of *trans-n*BuCH=CHD is halted after one insertion by hydrogenolysis with H_2. Cp_2Zr–H first inserts to give **12.10**. This can then insert in one of two ways to give **12.11**. The approach shown, with the alkyl RCH_2 and alkene R groups pointing away from one another, is not only reasonable but known to be favored from other work. Either **12.11a** or **12.11b** can be formed. On the pure *Cossee* mechanism a 50:50 ratio is expected, but on the *Green–Rooney* mechanism, the ratio will depend on whether C–H or C–D prefers to be agostic. As we saw in Fig. 10.10, C–H prefers to be agostic, so we expect the erythro product to predominate (Fig. 12.5). Experimentally, erythro is favored by 1.3:1 (by ^2H NMR), and so the *Green–Rooney* mechanism is followed in this case.

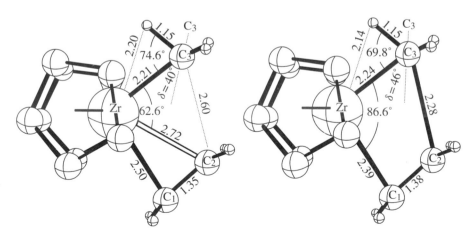

FIGURE 12.4 Structures of a model intermediate $[Cp_2ZrMe(C_2H_4)]^+$ (left), showing the agostic methyl. The methyl leans over even more at the transition state (right). The results were obtained by Ziegler and co-workers by density functional theoretical calculations. (Reproduced from Ref. 9b with permission.)

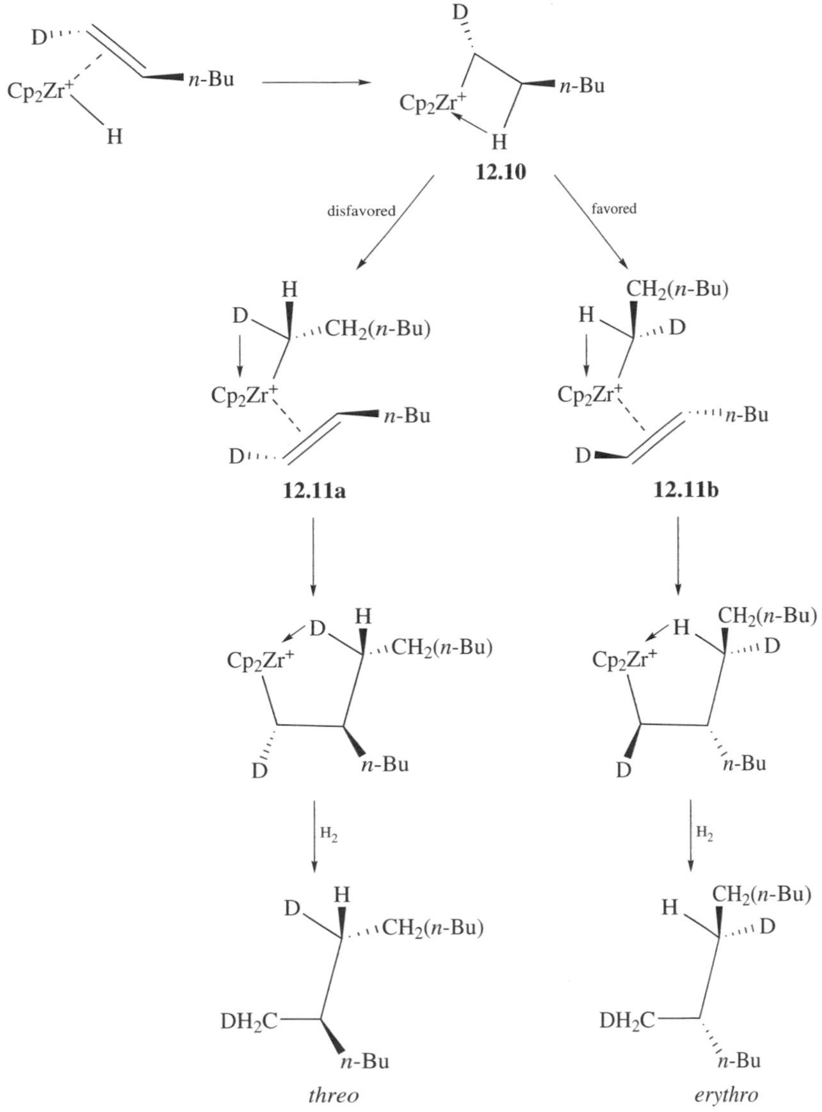

FIGURE 12.5 Grubbs experiment. Since the α-CH bonds of the metal alkyl are not involved in the Cossee mechanism (Fig. 12.1), we expect a 50:50 mixture of isotopomers, as observed in some situations. On the modified Green–Rooney mechanism shown here, we would expect a preferential binding of C–H over C–D in the agostic intermediate, which leads to a non-50:50 ratio as observed for certain systems.

In her studies on the f-block metals, Watson and Roe[14] found a remarkable system in which successive alkene insertions into a Lu–R bond can be observed step by step (Eq. 12.15). Not only do the alkenes insert but the reverse reaction, β elimination of an alkyl group, as well as the usual β elimination of a hydrogen, are also observed. For the *d* block this β elimination of an alkyl group would normally not be possible; it is probably the larger M$-$R bond dissociation energies in the *f* block that make the thermodynamics of the overall process favorable.

$$\text{Cp}_2\text{LuH} \underset{\substack{\beta\text{-H} \\ \text{elim.}}}{\rightleftharpoons} \text{Cp}_2\text{Lu} \underset{\substack{\beta\text{-alkyl} \\ \text{elim.}}}{\rightleftharpoons} \text{Cp}_2\text{Lu} \qquad (12.15)$$

Most late metals only dimerize or oligomerize alkenes (oligomers are trimers, tetramers, and other short-chain molecules), rather than polymerize them. This is because β elimination very soon stops the chain from growing. One of the best known systems is $NiCl_2/EtAlCl_2$, in which a nickel hydride is believed to be the active catalyst. If we consider ethylene, the first insertion gives an ethyl complex, this can either β-eliminate or insert another ethylene, the same is true for the *n*-butyl product of the second insertion. The product distribution therefore depends on the ratio of the rates of insertion and β elimination.[15]

Brookhart Catalysts

Once it was recognized that coordinative unsaturation could facilitate insertion, the widely accepted rule of thumb that only early metals could give efficient polymerization catalysts seemed shortsighted. Brookhart et al.[16a,b] and Gibson et al.[16c] have exploited highly unsaturated, electrophilic late metal systems. Electrophilicity is also important because it slows down β elimination by limiting the extent of back donation into the β-C$-$H σ^* orbital, required for the reaction to proceed. In early work, $[Cp^*Co\{P(OMe)_3\}Et]^+$, having an agostic ethyl group was shown to slowly polymerize ethylene. In later work, more unsaturated systems were developed that allow the alkyl alkene intermediate to be agostic; this cannot be the case for 18e $[Cp^*LCo(Et)(C_2H_4)]^+$.

Catalysts of the general type shown, once activated by MAO were found to be extremely effective, with the exact properties depending on the steric environment of the site, governed by the nature of the Ar group.[16b,c]

(M = Fe, Co; Ar = 2,6-iPr$_2$-C$_6$H$_4$)

These catalysts can produce polymers of quite different microstructures than previously seen and could become commercially important in future.

SHOP Oligomerization

The Shell higher-olefins process (SHOP) is an industrial process based on homogeneous nickel catalysts of the type shown in Fig. 12.6 and discovered by Keim et al.[17a] These oligomerize ethylene to give 1-alkenes of various chain lengths (e.g., C_6-C_{20}). Insertion is therefore considerably faster than β elimination. The $C_{10}-C_{14}$ fraction is a desirable feedstock; for example, hydroformylation gives $C_{11}-C_{15}$ alcohols that are useful in detergent manufacture. The non-$C_{10}-C_{14}$ fraction consists of 1-alkenes with longer (e.g., C_{16}) and shorter (e.g., C_8) chain lengths. Figure 12.6 shows how isomerization and metathesis can be combined to manipulate the chain lengths so as to produce more $C_{10}-C_{14}$ material from the longer and shorter chains. The fact that internal $C_{10}-C_{14}$ alkenes are formed does not matter because hydroformylation gives linear alcohols even from internal alkenes, as discussed in Section 9.3. Homogeneous catalysts were strong contenders for the isomerization and metathesis steps of SHOP, but in practice heterogenized catalysts were adopted. There are now several plants operating.

Replacing the PR_3 ligand of the SHOP Ni catalyst with an ylid ligand, $^-CH_2-P^+R_3$, gives highly active alkene polymerization catalysts with tunable properties.[17b]

Nickel complexes are also used for the oligomerization of butadiene where Ni(0) mediates the oxidative coupling of two butadienes to give the bis-π-allyl complex **12.12** (Fig. 12.7). According to the exact conditions, the dimers, cyclooctadiene, vinylcyclohexane, and even divinylcyclobutane, can be formed by reductive elimination from **12.12**.[16] Alternatively, a third molecule of butadiene can add to give 1,5,9-cyclododecatetraene. Only naked Ni(0) can give the trimerization, addition of PR_3 diverts the reaction to give dimers by occupying the site to which the third butadiene would otherwise bind.

Another commercially important reaction is du Pont's synthesis of 1,4-hexadiene. This is converted to synthetic rubber by copolymerization with ethylene and propylene, which leaves the polymer with unsaturation. Unsaturation is also present in natural rubber, a 2-methylbutadiene polymer **12.13**, and is necessary for vulcanization.

12.13

The 1,4-hexadiene is made by codimerization of ethylene and butadiene, with a $RhCl_3$/EtOH catalyst (Eq. 12.16).[18] The catalyst is about 80% selective for the *trans*-1,4-hexadiene, a remarkable figure considering all the different dimers that could have been formed. The catalyst is believed to be a rhodium hydride formed by reduction of the $RhCl_3$ with the ethanol solvent (Section 3.2). This must react with the butadiene to give mostly the *anti*-methylallyl (crotyl) intermediate, which selectively inserts an ethylene at the unsubstituted end. The cis/trans ratio of the product probably depends on the ratio of the two isomers of the crotyl

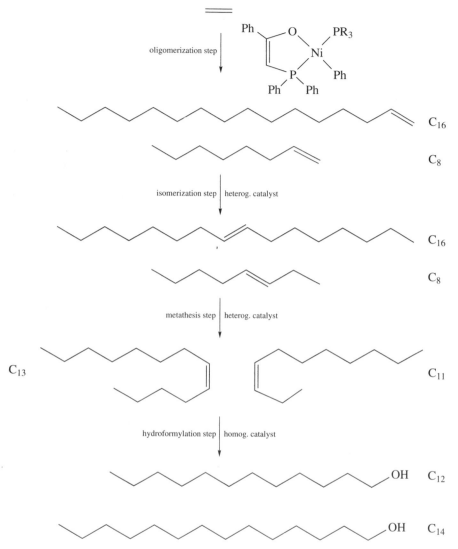

FIGURE 12.6 In the Shell higher olefins process (SHOP), Keim's nickel catalyst gives 1-alkenes of various chain lengths. The subsequent steps allow the chain lengths to be manipulated to maximize the yield of $C_{10}-C_{14}$ products. Finally, SHOP alkenes are often hydroformylated, in which case the internal alkenes largely give the linear product, as discussed in Chapter 9.

intermediate. Adding ligands such as HMPA to the system greatly increases the selectivity for the trans diene. By increasing the steric hindrance on the metal, the ligand probably favors the anti isomer of the crotyl ligand over the more hindered syn isomer. The rhodium hydride is also an isomerization catalyst, and so the 1,4-hexadiene is also converted to the undesired conjugated 1,3 isomers.

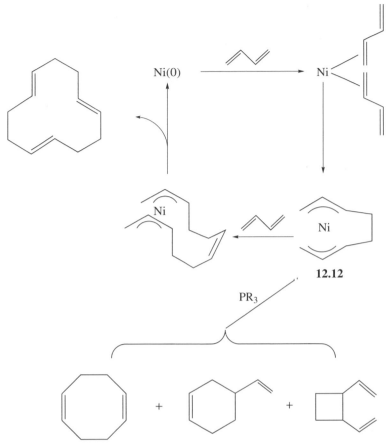

FIGURE 12.7   Wilke oligomerization of butadiene. "Naked" nickel catalysts give cyclo-dodecatriene, while the presence of ligands such as PPh$_3$ causes the reaction to produce the dimers shown.

The usual way around a problem like this is to run the reaction only to low conversion, so that the side product is kept to a minimum. The substrates, which are more volatile than the products, are easily recycled.

$$\text{(12.16)}$$

$$\text{(12.17)}$$

12.3   ACTIVATION OF CO AND CO_2

Most organic chemicals are currently made commercially from ethylene, a product of oil refining. It is possible that in the next several decades we may have to shift toward other carbon sources for these chemicals as depletion of our oil reserves continues. Either coal or natural gas (methane) can be converted into CO/H_2 mixtures with air and steam (Eq. 12.18), and it is possible to convert such mixtures, variously called "water–gas" or "synthesis gas" to methanol (Eq. 12.18) and to alkane fuels with various heterogeneous catalysts. In particular, the Fischer–Tropsch reaction (Eq. 12.19) converts synthesis gas to a mixture of long-chain alkanes and alcohols using heterogeneous catalysts.[19]

$$C + H_2O \xrightarrow{\text{heat}} \underset{\text{synthesis gas}}{H_2 + CO} \xrightarrow{\text{het. catal.}} CH_3OH \tag{12.18}$$

$$H_2 + CO \xrightarrow{\text{het. catal.}} CH_3(CH_2)_nCH_3 + CH_3(CH_2)_nOH + H_2O \tag{12.19}$$

Water–Gas Shift

It is often useful to change the $CO:H_2$ ratio in synthesis gas, and this can be accomplished by the water–gas shift reaction (Eq. 12.20), which can be catalyzed heterogeneously (Fe_3O_4 or Cu/ZnO) or by a variety of homogeneous catalysts, such as $Fe(CO)_5$[20] or $Pt(i\text{-}Pr_3P)_3$.[21] The reagents and products in Eq. 12.20 have comparable free energies; the reaction can therefore be run in either direction and this can be regarded as both CO and CO_2 activation.

$$H_2O + CO \rightleftharpoons H_2 + CO_2 \tag{12.20}$$

In the mechanism proposed for the homogeneous iron catalyst (Fig. 12.8), CO binds to the metal and so becomes activated for nucleophilic attack by OH^- ion at the CO carbon. Decarboxylation of the resulting metalacarboxylic acid probably does not take place by β elimination because this would require prior

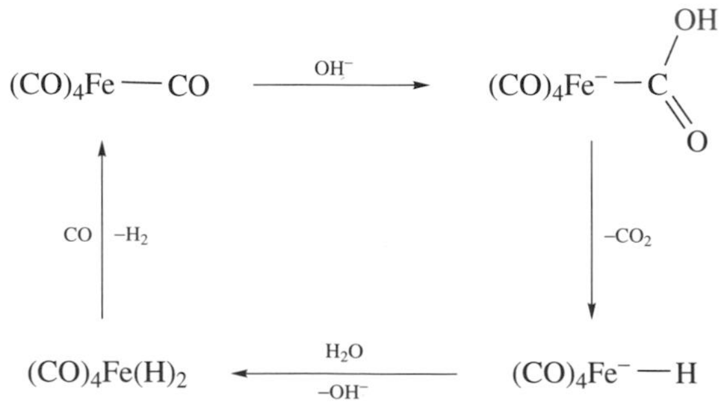

FIGURE 12.8   Cycle proposed for the $Fe(CO)_5$-catalyzed water–gas shift reaction.

loss of CO to generate a vacant site; instead, deprotonation may precede loss of CO$_2$, followed by reprotonation at the metal to give HFe(CO)$_4^-$. Protonation of this anionic hydride liberates H$_2$ and regenerates the catalyst. The platinum catalyst (Fig. 12.9) is perhaps more interesting in that it activates both the water and the CO, so no added base is needed. This happens because the platinum complex is sufficiently basic to deprotonate the water, leading to a cationic hydride complex. The cationic charge activates the CO for nucleophilic attack by hydroxide ion to give the metalacarboxylic acid (M$-$COOH). Such a compound is seen as a stable intermediate when water reacts with TpIr(CO)$_2$ (Eq. 12.21, Tp = tris(pyrazolyl)borate). The final product, TpIr(H)$_2$(CO), does not lose H$_2$, so this system is not a catalyst.[22] We look at a biological analog of the water$-$gas shift in Section 16.4.

$$\text{TpIr(CO)}_2 \xrightarrow{\text{H}_2\text{O}} \text{TpIr(H)(COOH)(CO)} \xrightarrow{\text{heat, } -\text{CO}_2} \text{TpIr(H)}_2\text{(CO)} \qquad (12.21)$$

Reppe Reaction[20a]

This uses the water$-$gas shift to generate H$_2$/CO for subsequent hydroformylation of the substrate alkene to give an aldehyde, followed by hydrogenation to give an alcohol, as shown in Eq. 12.22. With the Fe(CO)$_5$/base catalyst mentioned above, the product is the linear alcohol.

$$\text{RCH=CH}_2 + 3\text{CO} + 2\text{H}_2\text{O} \xrightarrow{\text{KOH, Fe(CO)}_5\text{, 15 atm, 100}^\circ\text{C}}$$

$$\text{RCH}_2\text{CH}_2\text{CH}_2\text{OH} + 2\text{CO}_2 \quad (12.22)$$

The alkene is believed[23] to insert into an Fe$-$H bond of the active catalyst, H$_2$Fe(CO)$_4$, formed as in Fig. 12.8, followed by migratory insertion to give (RCH$_2$CH$_2$CO)FeH(CO)$_3$, which in turn reductively eliminates the aldehyde

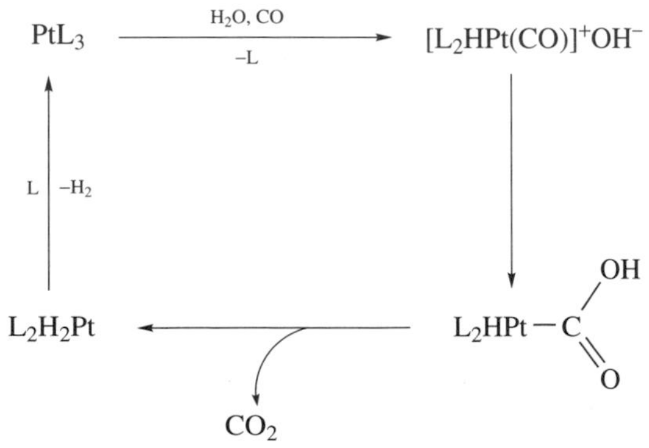

FIGURE 12.9 Cycle proposed for the PtL$_3$-catalyzed water$-$gas shift reaction {L = P(i-Pr)$_3$}.

RCH_2CH_2CHO. This aldehyde is then hydrogenated to the alcohol with $HFe(CO)_4^-$ as catalyst. By itself, $Fe(CO)_5$ is not a hydroformylation catalyst because H_2 cannot displace CO to form $H_2Fe(CO)_4$, hence the need for the base to remove the CO.

Monsanto Acetic Acid Process[24a]

Over two million tons of acetic acid a year are produced by carbonylation of methanol, which happens in >99% selectivity with a rhodium catalyst. The active catalyst is

$$[Rh^II_2(CO)_2]^-$$

12.14

The net effect is the cleavage of the methanol C$-$O bond and insertion of a CO. To be carbonylated, the methanol has to bind to the catalyst and this requires adding a certain amount of HI to the system,

$$MeOH + CO \xrightarrow{\text{180°C, 30 atm, MeI, } 10^{-3} M \text{catal.}} MeCOOH \qquad (12.23)$$

$$MeOH + HI \rightleftharpoons MeI + H_2O \qquad (12.24)$$

which produces an equilibrium concentration of MeI, which can in turn oxidatively add to the metal in the turnover limiting step. Once we have the rhodium methyl, migratory insertion can take place with CO to give an acetylrhodium iodide. Reductive elimination of the acyl iodide completes the cycle (Fig. 12.10). The free acyl iodide is hydrolyzed by the methanol to give methyl acetate and can be ultimately converted to acetic acid with water. The resulting acetic acid can be entirely derived from synthesis gas if the methanol comes from the reaction shown in Eq. 12.18. In a very closely related reaction, CH_3COOMe can be carbonylated to acetic anhydride $(CH_3CO)_2O$.[24b] The Monsanto process for making acetic acid is replacing the older route that goes from ethylene by the Wacker process to acetaldehyde, which is then oxidized to acetic acid in a second step. This example shows how important it is that chemical companies carry out research into possible alternative ways to make a compound, even though the current route is working well; otherwise their competitors may discover a better one. A biological analog of this reaction is discussed in Section 16.4, and an application in organic synthesis, the Heck reaction, is discussed in Section 14.4. An improved process based on iridium has been developed by BP-Amoco.[24c]

A related process, CO_2 activation,[25] may be of considerable future importance. Carbon dioxide, as a growing constituent of the atmosphere, has been implicated as a factor in global warming by the greenhouse effect. CO_2 is transparent to the incoming solar radiation but not to the infrared frequencies at which the Earth reradiates heat into space during the night. CO_2 is so thermodynamically stable that only a very few potential products can be made from CO_2 by exothermic processes. One could reduce it to CO with hydrogen by the water$-$gas shift, and

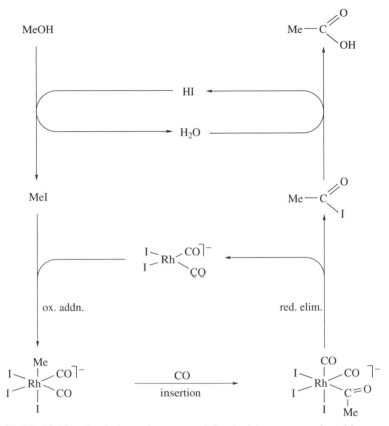

FIGURE 12.10 Catalytic cycle proposed for the Monsanto acetic acid process.

then use CO chemistry to make various carbon compounds, except that H$_2$ is very expensive. Indeed, the current methods of making H$_2$ involve the consumption of either coal or natural gas, which are valuable carbon sources:

$$CH_4 + H_2O \longrightarrow H_2 + CO \tag{12.25}$$

$$C + H_2O \longrightarrow H_2 + CO \tag{12.26}$$

$$H_2 + CO_2 \xrightarrow{\text{water-gas shift}} CO + H_2O \tag{12.27}$$

The most important CO$_2$ activation process is photosynthesis, in which solar photons drive a reaction that would otherwise be uphill thermodynamically: the reduction of CO$_2$ to carbohydrates coupled to water oxidation to O$_2$. Many metalloenzymes are involved in these processes; the one that "fixes" CO$_2$ is ribulose diphosphate carboxylase, in which an enolate anion of the sugar nucleophilically attacks the CO$_2$ carbon. Cu(II), Mn(II), and Mg^{2+} are all present in the active enzyme, and one of these probably plays a role in polarizing the CO$_2$, perhaps via an η^1-OCO complex.

CO_2 insertion into M—H bonds is probably involved in the catalytic reduction of CO_2 with H_2 to give HCOOH. Although this is "uphill" thermodynamically ($\Delta G° = +8$ kcal/mol) the reaction becomes favorable under gas pressure and in the presence of base to deprotonate the formic acid formed. The best catalyst to date is [Rh(cod)Cl]$_2$/Ph$_2$P(CH$_2$)$_4$PPh$_2$, which gives 45 turnovers per hour at room temperature at 40 atm pressure.[26]

12.4 CH ACTIVATION[27]

The goal is the conversion of an alkane RH into the more valuable species, RX, where X is any of a variety of useful groups. Classical organic procedures tend to give branched products, such as i-PrX from propane, but linear products are very desirable, hence the potential of transition metal complexes, where formation of n-PrX is favored. For the moment no system has proved economically viable, but steady progress is being made.

The term *CH activation*[27] emphasizes the selectivity difference between low-valent metal complexes and classical organic reagents. In a classical electrophilic or radical route, radicals such as •OH abstract an H atom from alkanes, PrH, but always to give the branched radical i-Pr•. Superacids, abstract H⁻ ion from PrH, but always to give the branched ion i-Pr⁺. By such classical routes, the ultimate functionalization product, i-PrX, is branched.

An alkane CH bond can oxidatively add to a variety of low-valent transition metals preferentially to give the linear product, n-Pr-M-H, however, and in any subsequent functionalization, the linear product, n-PrX, is often obtained. In addition, methane activation holds promise as methane seems likely to become a more important feedstock for the chemical industry. Methane conversion to methanol or a derivative (e.g., MeOCH$_2$OMe) would make a conveniently transportable fuel. Partial oxidation such as this is particularly hard. Methanol is much more easily oxidized than methane, so classical oxidation procedures give CH$_2$O, CO and CO$_2$. In the CH activation route, the CH bond of methanol is not much more reactive than in methane, so the overoxidation problem is less severe.

Already in the 1960s, Chatt and Davidson[28] showed that [Ru(dmpe)$_2$] cyclometalates and also reacts with naphthalene (ArH) to give [Ru(H)(Ar)(dmpe)$_2$] (Eq. 12.28). The dmpe has limited back-bonding capability, so low-valent [Ru(dmpe)$_2$] has a very high tendency to give oxidative addition.

$$(12.28)$$

During the 1970s, Shilov[27] saw preferential activation of primary CH bonds in H/D exchange in alkanes catalyzed by Pt(II) in $D_2O/DOAc$. This was the first indication of the special reactivity pattern associated with oxidative addition. Moving to $[Pt(IV)Cl_6]^{2-}$ as oxidant, alkanes were oxidized to ROH and RCl with the same Pt(II) catalyst with linear product still preferred, so the Pt(IV) clearly intercepts the same intermediate alkyl that led to RD in the deuteriation experiments. With methane as substrate, a methylplatinum intermediate was seen.[29] Labinger and Bercaw[30] revisited the system in the 1990s using a series of mechanistic probes that confirmed Shilov's main points as well as extending the picture. Figure 12.11a shows the current mechanistic view. An alkane complex either leads to oxidative addition of the alkane and loss of a proton or the alkane σ complex loses a proton directly (Eq. 12.29). In isotope exchange, the resulting alkyl is cleaved by D^+ to give RD. In the alkane functionalization, oxidation of the Pt(II) alkyl by Pt(IV) gives a Pt(IV) alkyl by electron transfer. The Pt(IV) now becomes a good leaving group, and Cl^- or OH^- can nucleophilically attack the R-Pt(IV) species with departure of Pt(II) to regenerate the catalyst.

Periana et al.[31a] made Shilov-like chemistry much more efficient in a series of methane conversion catalysts. With Hg(II) salts in H_2SO_4 at 180°, the acid is both solvent and mild reoxidant (Eq. 12.30). Methane was converted to the methanol ester, methyl bisulfate, $MeOSO_3H$, in which the $-OSO_3H$ provides a powerful deactivating group to prevent overoxidation. At a methane conversion of 50%, 85% selectivity to methyl bisulfate (ca. 43% yield) was achieved with the major side product being CO_2. The expected intermediate $MeHg^+$ cation was seen by NMR spectroscopy, and a Shilov-like mechanism (Fig. 12.11b) proposed. Since Hg(II) is not expected to give oxidative addition, Hg(IV) being unknown, the initial activation step must occur via deprotonation of a σ complex. Similar selectivity is seen as for Pt(II) and indeed deprotonation of a Pt(II) σ complex still cannot be excluded for this case.

$$\text{(12.29)}$$

$$CH_4 \xrightarrow[180°]{\text{Hg(II), } H_2SO_4} CH_3OSO_3H \qquad \text{(12.30)}$$

$$\text{(12.31)}$$

$$CH_4 \xrightarrow[180°]{} CH_3OSO_3H$$

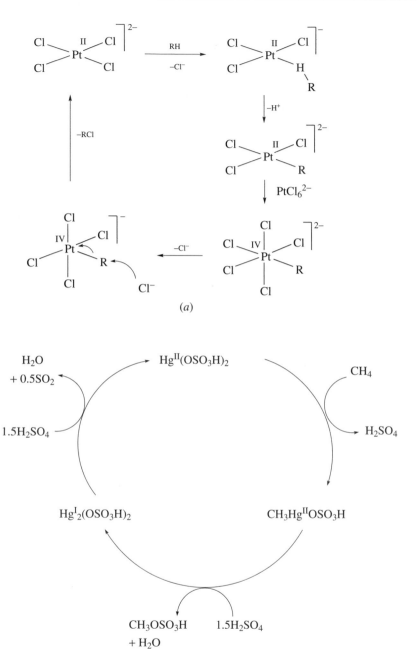

FIGURE 12.11 (*a*) The proposed mechanism of the Shilov reaction; (*b*) Periana's related methane oxidation.

Periana et al.[31b] also developed a similar Pt(II) catalyzed process in H_2SO_4 at 180° (Eq. 12.31), as well as a direct, selective, catalytic oxidative condensation of two methane molecules to acetic acid at 180° in liquid sulfuric acid with Pd(II) salts as catalyst.[31c] Both carbons of acetic acid originate from the methane as shown by isotope labeling. The results are consistent with methane C—H activation to generate Pd—CH_3, followed by oxidative carbonylation with methanol, generated in situ from methane, to produce acetic acid. Sen[32] has reported an intriguing series of catalytic systems for conversion of methane and other alkanes that are probably similar in character.

CH activation pathways may go via an intermediate σ complex with the alkane bound to the metal. In $M(CO)_6$ photolysis in an alkane matrix, formation of $[M(CO)_5(alkane)]$ was identified by Perutz and Turner[33] in 1975. More recently, $CpRe(CO)_2(n\text{-heptane})$ was detected by FTIR at room temperature in heptane,[34] and $CpRe(CO)_2(cyclopentane)$ was detected by NMR spectroscopy.[35]

By reversing transition metal catalyzed hydrogenation of alkenes to alkanes, stoichiometric dehydrogenation of alkanes was reported in 1979.[36] For example, cyclopentane reacts with $[IrH_2(Me_2CO)(PPh_3)_2]^+$ to give $[CpIrH(PPh_3)_2]^+$ with t-BuCH=CH_2 as "hydrogen acceptor" (Eq. 12.32). Oxidative addition of an alkane CH bond was proposed as the initial step of the pathway.

$$[IrH_2(Me_2CO)_2(PPh_3)_2]^+ + 3t\text{-BuCH=}CH_2 \longrightarrow \underset{\underset{Ph_3P}{Ph_3P}}{\overset{\oplus}{Ir}}\diagdown H \qquad (12.32)$$

$$+ \; 3t\text{-BuCH}_2CH_3$$

The key oxidative addition was directly observed by Janowicz and Bergman[37a] in 1982, via photogeneration of $Cp^*Ir(PMe_3)$ from the dihydride and reaction with the alkane solvent, RH, to give a variety of $Cp^*Ir(R)(H)(PMe_3)$ species (e.g., Eq. 12.33).

$$Cp^*Ir(PMe_3)H_2 + n\text{-}C_5H_{12} \rightarrow Cp^*Ir(PMe_3)H(n\text{-}C_5H_{11}) + H_2 \qquad (12.33)$$

High selectivity was seen for attack at a terminal CH bond in linear alkanes, analogous to Shilov chemistry, providing a mechanistic link with the prior Pt work. Strong CH bonds, present in linear alkanes or cyclopropane or benzene, were preferentially activated, suggesting that the Ir—R bond strength tracks the H—C bond strength. $Cp^*Ir(R)(H)(PMe_3)$ thermally eliminates RH under methane to give the very stable $Cp^*Ir(Me)(H)(PMe_3)$,[38] so photolysis is not essential to success. Analogous iridium carbonyls and rhodium phosphine complexes gave similar chemistry.[37b]

Flash infrared kinetics of the photochemistry of $Tp^*Rh(CO)_2$ and $Bp^*Rh(CO)_2$ in liquid xenon solution [Tp^* = hydridotris(3,5-dimethylpyrazolyl)borate; Bp^* = dihydridobis(3,5-dimethylpyrazolyl)borate] give evidence for the formation of xenon complexes $(\eta^3\text{-Tp*})Rh(CO)\cdot Xe$ and $(\eta^2\text{-Tp*})Rh(CO)\cdot Xe$ before the formation of the cyclohexyl hydride from the cyclohexane also present. $C-H$ oxidative addition can even be preferred over $C-Cl$ addition, as is the case for alkyl chlorides with the Tp*Rh(NCR) fragment.[39]

Felkin et al.[40a] saw selective catalytic conversion of a variety of cycloalkanes into cycloalkenes with using $ReH_7(PR_3)_2$ with $t\text{-BuCH}=CH_2$ as hydrogen acceptor (Eq. 12.34). Several alkane dehydrogenation catalysts were soon found,[40b-c] although some lack robustness and deactivate rather readily. The buildup of alkene product naturally causes problems since this is an alternative substrate for the catalyst. To avoid using $t\text{-BuCH}=CH_2$, the reaction can either be run photochemically[41a] or under reflux.[41b] In each case unfavorable thermodynamics of alkane dehydrogenation is overcome, in the first by input of light energy and in the second by continuous removal of the product H_2.

Under irradiation, $RhCl(CO)(PMe_3)_2$ gives catalytic alkane carbonylation,[42] in which the usual alkyl hydride intermediate undergoes CO insertion, followed by reductive elimination of RCHO. Once again, terminal selectivity is seen in linear alkanes.

Goldman and co-workers[43,44] have reported an acceptorless PCP Ir(III) pincer complex that is among the most efficient to date for alkane dehydrogenation (Eq. 12.34). It has been applied to the introduction of C=C double bonds into aliphatic polymers and to the formation of enamines by dehydrogenation of tertiary amines.

Brookhart's[45] bis(phosphinite) [(PCP)IrHCl] pincer complexes. NaO-t-Bu in cyclooctane with $t\text{-BuCH}=CH_2$ as acceptor generates species with exceptional catalytic activity for Eq. 12.34. Turnover numbers up to 2200 and initial turnover frequencies between 1.6 and 2.4 s^{-1} were seen at 200°C.

Hartwig and co-workers'[46] transition-metal-catalyzed terminal borylation of linear alkanes with $Cp*Rh(\eta^4\text{-}C_6Me_6)$ gives linear alkylboranes from commercially available borane reagents under thermal conditions in high yield (Eq. 12.35). The hydrogen transfer to boron occurs by a boron-assisted, metal-mediated σ-bond metathesis. The "unoccupied" p orbital of boron lowers the energy of the transition state and the intermediates by accepting electron density from the metal. Smith and co-workers'[47] iridium catalysts are also highly effective for metal-catalyzed terminal borylation.

$$\text{(12.34)}$$

catalysts $= ReH_7(PPh_3)_2,\ R_2P-\underset{\underset{H\ \ H}{/\ \ \backslash}}{Ir}-PR_2$

$$\text{(alkene drawing)} \xrightarrow[\text{HBcat}]{\text{Cp*Rh}(\eta^4\text{-C}_6\text{Me}_6)} \text{(alkene-Bcat drawing)} \quad (12.35)$$

When alkanes bind to metals as σ complexes, the net donation of charge from the CH to the metal in forming the M$-$alkane bond is not compensated by back donation because C$-$H σ^* is high in energy and only one lobe of the C$-$H σ^* orbital is available for back donation—the other is remote from the metal. This induces a depletion of electron density on the CH bond that acidifies the CH proton. We have already seen how deprotonation of an alkane complex has been implicated in Periana's Hg(II) catalyst and cannot be excluded for Shilov's Pt(II) system. Alkane acidification on binding is relevant to σ-bond metathesis pathways of alkane activation where the alkane σ complex transfers a proton to a basic group such as an alkyl M$-$R$'$ already present on the metal. This can lead to exchange of the R group from the RH substrate with M$-$R$'$. Watson's[48] methane exchange between Cp$_2^*$LuMe and CH$_4$ was verified with C-13 methane (C*H$_4$). The oxidation state of the metal does not change, so this route is available to redox-inactive metals like d^0 ions and the f block. Indeed, Fendrick and Marks[49] found similar reactions with actinides.

In the late metals it is hard to completely eliminate the possibility of oxidative addition/reductive elimination as an alternative redox pathway to the same final products, but Bergman and Arndtsen[50] proposed a σ-bond metathesis pathway for the reaction of alkanes including methane with the dichloromethane complex [Cp*IrR(PMe$_3$)(ClCH$_2$Cl)]$^+$ where a variety of alkanes R$'$H gave RH and [Cp*IrR$'$(PMe$_3$)(ClCH$_2$Cl)]$^+$.

An alkane RH can also be activated by addition across a metal$-$heteroatom bond, as shown by Wolczanski[51] Sadow and Tilley[52] have seen a number of σ-bond metathesis reactions of methane with Cp$_2$*ScR (R = alkyl).

C$-$C Oxidative Addition

Breaking the C$-$C bonds of alkanes is worse both thermodynamically and kinetically than breaking the C$-$H bond because we make two relatively weak M$-$C bonds (together worth $\sim$70 kcal/mol), for the loss of a C$-$C bond ($\sim$85 kcal/mol) and a C$-$C bond is less sterically accessible than a C$-$H bond. Direct alkane C$-$C bond breaking has been observed only in very strained alkanes in which the relief of strain provides a substantial extra driving force. The first example dates from 1955, when Tipper[53a] observed the reaction between cyclopropane and PtCl$_2$ (Eq. 12.36); the correct metalacyclobutane structure of the product was suggested by Chatt et al.[53b] in 1961.

$$\triangle \xrightarrow{\text{PtCl}_2} [\text{(metalacyclobutane drawing)} \text{PtCl}_2]_n \quad (12.36)$$

The product in Eq. 12.37 seems to be a C$-$C bond-breaking product of 1,1-dimethylcyclopentane, but isolation of the intermediate shows that the reaction goes via prior C—H bond breaking.[54] The system is set up so that the unfavorable C—C cleavage is accompanied by the formation of a thermodynamically very stable Cp—M bond.

$$(12.37)$$

Ligand precursors of type **12.15** (X = H) readily cyclometallate to give species of type **12.16** (X = H), commonly called "pincer" complexes. Milstein[55] has shown how ligand precursors of this type can undergo C—C bond cleavage if the precursor contains a suitably placed alkyl group, X. Where X = CH$_3$, a concerted oxidative addition pathway was proposed. Where X = CF$_3$, the CC bond cleaved (BDE Ph—CF$_3$ = 109 kcal/mol) is among the strongest known but M—Ph and M—CF$_3$ bonds are also known to be very strong (Section 3.6). In none of these cases were the alternative CH or CF oxidative addition pathways seen. In a striking extension of this work, it even proved possible to catalytically dealkylate ligand **12.15** (X = Et) using H$_2$ (Eq. 12.39) although only four turnovers were possible.

$$(12.38)$$

12.15
R = t-Bu;
X = H, CH$_3$, CF$_3$.

12.16

$$(12.39)$$

In spite of these advances in alkane chemistry, the development of a series of robust and selective catalysts for different alkane conversion reactions remains a continuing challenge in organometallic chemistry today.

12.5 ORGANOMETALLIC MATERIALS AND POLYMERS

Inorganic chemistry provides many of the materials—from concrete to silicon chips—that are indispensable to modern life. The need for designed materials with special properties will continue to grow in the new century. Organometallic chemistry is beginning to contribute in several ways. For example, in metal organic chemical vapor deposition (MOCVD),[56] a volatile metal compound is decomposed on a hot surface to deposit a film of metal. A typical example is the use of $Cr(C_6H_6)_2$ to deposit Cr.[57]

Porous materials like zeolites,[58] with well-defined structures having voids in the interior, have proved exceedingly valuable as catalysts. These have aluminosilicate lattices with acidic protons in the pores as the catalytically active group. While these are not organometallic, organosilicon compounds can be precursors in their synthesis. Reaction only happens in the interior of the structure within a defined cavity, so only compounds having certain sizes can enter or leave, depending on the exact zeolite structure. Exxon-Mobil's acidic ZSM-5 zeolite catalyst, for example, converts MeOH to gasoline range hydrocarbons and water.[59a] Once again, main-group chemistry is predominant in this area, but hybrid materials with transition metal catalytic sites are also known. In recent examples, a Pd(0)/zeolite hybrid material acts as a very efficient catalyst for the Heck reaction,[59b] and a zirconium metallocene catalyst/zeolite hybrid acts as a propene polymerization catalyst.[59c]

Another type of organometallic material is formed by crystallizing organometallic precursors that have hydrogen bonding groups capable of establishing a network of hydrogen bonds throughout the lattice.[60a]. This is sometimes called crystal engineering and serves to orient the molecules in the lattice. A related development is Yaghi et al.'s[60b] use of linearly bridging ligands such as 4,4′-dipyridine as rigid rods to connect metal ions into open lattices that possess large cavities; these *mesoporous* materials can be crystallized and their structures determined. So far, none is organometallic, however, although such lattices might be more stable and useful.

Organometallic chemistry has contributed strongly to the polymer industry by providing polymerization catalysts (Section 12.2), but polymers derived from organometallic monomers are also attracting attention,[61] although for the moment they remain laboratory materials. Among the best studied polymers of this type are the poly(ferrocenylsilanes),[62] formed by ring-opening polymerization (ROP) of ferrocenophanes (ferrocenes with bridges between the rings, e.g., **12.17**). The strain of ca. 16–20 kcal/mol present in the bridge, evident from the ring tilt angle of $16°-21°$, serves to drive the polymerization. The thermal route of Eq. 12.40 gives very high-molecular-weight material (polymer from **12.17** M_w ca. 10^5-10^6). The polymer is processable and films can be formed by evaporating a solution on a flat surface. The nature of the R groups can also be readily changed (e.g., OR, NR_2, alkyl), allowing a range of materials to be accessed.

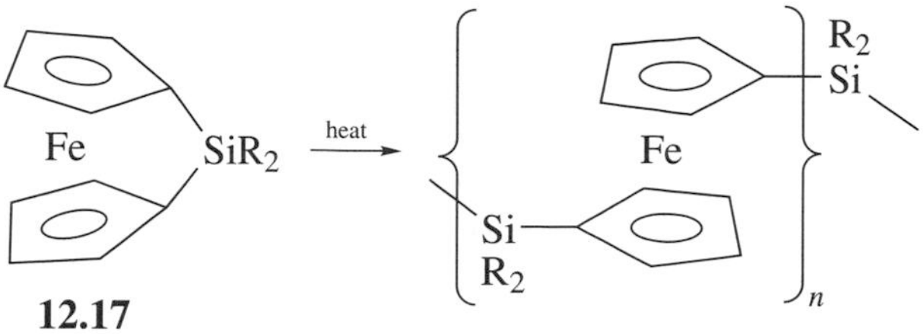

(12.40)

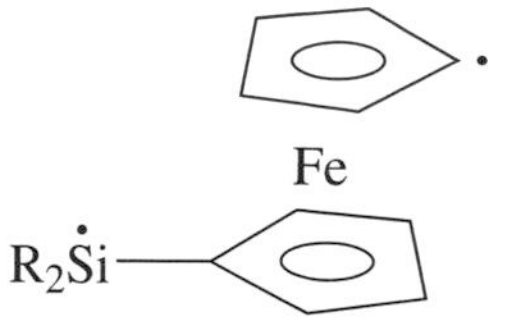

12.18

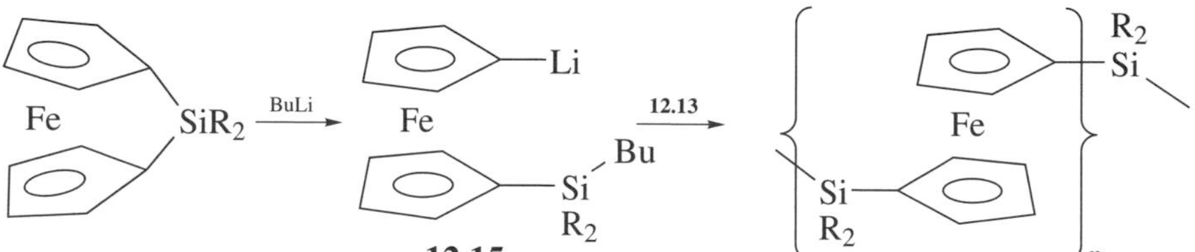

(12.41)

The thermal ROP presumably goes via the diradical **12.18**, but the polymerization can also be initiated by BuLi in THF, when the intermediate is likely to be **12.19** This is more easily controllable and gives better polydispersity (less deviation of molecular weights of individual chains from the average molecular weight). The polymer is also *living*, meaning that the Li remains at the end of the chain, allowing chain-end functionalization or the introduction of a second monomer to form a new block of a second polymer. Transition-metal-catalyzed ROP of **12.17** is also possible and this has the advantage that the process is less affected by impurities than the BuLi-initiated version.

Organic light-emitting diodes (OLEDs) constitute another materials application of organometallic compounds—the term "organic" is used because some OLEDs have no metals—emit light in response to a voltage. They may replace the familiar liquid crystal displays (LCDs) of computers or of DVD players in the next generation of these devices because OLEDs are much more resistant than traditional LCDs to bending and shock.

An applied voltage injects electrons and holes into the material, and when these find each other, energy is emitted that corresponds to the HOMO–LUMO gap of the molecule.[63] Cyclometalated iridium(III) complexes,[64] such as **12.19**–**12.21**, have proved very effective in allowing the emission wavelength to be tuned by variation of the structure.

12.19   **12.20**   **12.21**

REFERENCES

1. R. H. Grubbs, *Handbook of Metathesis*, Wiley-VCH, New York, 2003.

2. R. L. Banks, *CHEMTECH*, **16**, 112, 1986.

3. A. H. Hoveyda and R. R. Schrock, *Chem. Eur. J.* **7**, 945, 2001 and references cited.

4. R. H. Grubbs et al., *Pure Appl. Chem.* **75**, 421, 2003.

5. J. L. Hérisson and Y. Chauvin, *Makromol. Chem.* **141**, 161, 1970.

6. (a) T. J. Katz and J. McGinnis, *J. Am. Chem. Soc.* **99**, 1903, 1977; (b) T. J. Katz and R. Rothchild, *J. Am. Chem. Soc.* **98**, 2519, 1976.

7. T. J. Katz et al., *Tetrahedron Lett.* 4247, 1976.

8. R. R. Schrock et al., *J. Am. Chem. Soc.* **103**, 3932, 1981.

9. (a) W. Kaminsky, *Pure Appl. Chem.* **70**, 1229, 1998; (b) H. H. Brintzinger, D. Fischer, R. Mülhay, B. Rieger, and R. M. Waymouth, *Angew. Chem., Int. Ed.* **34**, 1143, 1995; (c) E. Y. X. Chen and T. J. Marks, *Chem. Rev.* **100**, 1391, 2000.

10. E. J. Arlman and P. Cossee, *J. Catal.* **3**, 99, 1964.

11. M. L. H. Green, J. J. Rooney et al., *Chem. Commun.* 604, 1978.

12. T. Ziegler et al., *Organometallics* **14**, 2018, 1995.

13. (a) R. H. Grubbs et al., *J. Am. Chem. Soc.* **107**, 3377, 1985; (b) H. Kraulendat and H. H. Brintzinger, *Ang. Chem., Int. Ed.* **29**, 1412, 1990; (c) W. E. Piers and J. E. Bercaw, *J. Am. Chem. Soc.* **112**, 9406, 1990.

14. P. Watson and D. C. Roe, *J. Am. Chem. Soc.* **104**, 6471, 1982.

15. B. Bogdanovic, *Adv. Organomet. Chem.* **17**, 105, 1979; P. W. Jolly and G. Wilke, *The Organic Chemistry of Nickel*, Academic, New York, 1975.

16. (a) G. F. Schmidt and M. Brookhart, *J. Am. Chem. Soc.* **107**, 1443, 1985; M. Brookhart and E. Hauptman, *J. Am. Chem. Soc.* **114**, 4437, 1992; (b) M. Brookhart et al., *J. Am. Chem. Soc.* **120**, 4050, 1998; (c) V. C. Gibson et al., *J. Am. Chem. Soc.* **121**, 8728, 1999.

17. (a) W. Keim et al., *Organometallics* **2**, 594, 1983; K. Hirose and W. Keim, *J. Mol. Catal.* **73**, 271, 1992; (b) K. A. Ostoja-Starzewski and J. Witte, *Angew. Chem. Int. Ed.*, **27**, 839, 1988.

18. A. C. L. Su, *Adv. Organometal. Chem.* **17**, 269, 1979.

19. P. M. Maitlis, *J. Organomet. Chem.*, **689**, 4366, 2004.

20. J. W. Reppe, *Annalen* **582**, 121, 1953.

21. T. Yoshida, Y. Ueda, and S. Otsuka, *J. Am. Chem. Soc.* **100**, 3941, 1978.

22. L. A. Oro et al., *J. Organomet. Chem.* **438**, 337, 1992.

23. R. Pettit et al., *J. Am. Chem. Soc.* **99**, 8323, 1977.

24. (a) D. Forster, *Adv. Organomet. Chem.* **17**, 255, 1979; (b) J. R. Zoeller et al., *Adv. Chem. Ser.* **230**, 377, 1992; (c) M. L. Fernandez et al., *J. Organomet. Chem.* **438**, 337, 1992; (d) G. J. Sunley et al., *Sci. Technol. Catal.* **121**, 61, 1999.

25. X. L. Yin and J. R. Moss, *Coord, Chem. Rev.* **181**, 27 1999.

26. E. Graf and W. Leitner, *Chem. Commun.* 623, 1992.

27. A. E. Shilov and G. B. Shul'pin, *Activation and Catalytic Reactions of Saturated Hydrocarbons,* Kluwer, Dordrecht, 2000; R. H. Crabtree, *JCS Dalton,* 2437, 2001.

28. J. Chatt and J. M. Davidson, *J. Chem. Soc.* 843 1965.

29. L. A. Kushch, V. V. Lavrushko, Y. S. Misharin, A. P. Moravsky, and A. E. Shilov, *New J. Chem.* **7**, 729, 1983.

30. J. A. Labinger and J. E. Bercaw, *Nature* **417**, 507 2002.

31. (a) R. A. Periana, D. J. Taube, E. R. Evitt, D. G. Loffler, P. R. Wentrcek, G. Voss, and T. Masuda, *Science* **259**, 340, 1993; (b) R. A. Periana, D. J. Taube, S. Gamble, H. Taube, T. Satoh, and H. Fujii, *Science* **280**, 560, 1998; (c) R. A. Periana, O. Mironov, D. Taube, G. Bhalla, and C. J. Jones, *Science* **301**, 814, 2003.

32. A. Sen, *Acc. Chem. Res.* **31**, 550 1998.

33. R. N. Perutz and J. J. Turner, *J. Am. Chem. Soc.* **97**, 4791 1975.

34. G. I. Childs, D. C. Grills, X. Z. Sun, and M. W. George, *Pure Appl. Chem.* **73**, 443, 2001.

35. S. Geftakis and G. E. Ball, *J. Am. Chem. Soc.* **120**, 9953 1998.

36. R. H. Crabtree, J. M. Mihelcic, and J. M. Quirk, *J. Am. Chem. Soc.* **101**, 7738 1979.

37. (a) A. H. Janowicz and R. G. Bergman, *J. Am. Chem. Soc.* **104**, 352 1982; (b) R. G. Bergman, *Science* **223**, 902 1984; (c) W. D. Jones and F. J. Feher, *J. Am. Chem. Soc.* **106**, 1650 1984; J. K. Hoyano and W. A. G. Graham, *J. Am. Chem. Soc.* **104**, 3723 1982.

38. M. J. Wax, J. M. Stryker, J. M. Buchanan, C. A. Kovac, and R. G. Bergman, *J. Am. Chem. Soc.* **106**, 1121 1984.

39. A. J. Vetter and W. D. Jones, *Polyhedron* **23**, 413, 2004.

40. (a) D. Baudry, M. Ephritikhine, H. Felkin, and R. Holmes-Smith, *Chem. Commun.* 788, 1983; (b) D. Baudry, M. Ephritikhine, H. Felkin, and J. Zakrzewski, *Chem. Commun.* 1235, 1982; (c) M. J. Burk and R. H. Crabtree, *J. Am. Chem. Soc.* **109**, 8025 1987; (d) K. C. Shih and A. S. Goldman, *Organometallics* **12**, 3390 1993; (e) C. M. Jensen, *Chem. Commun.* 2443 1999.

41. (a) M. J. Burk, R. H. Crabtree, and D. V. McGrath, *Chem. Commun.* 1985, 1829; (b) T. Aoki and R. H. Crabtree, *Organometallics* **12**, 294, 1993.

42. T. Sakakura, T. Sodeyama, K. Sasaki, K. Wada, and M. Tanaka, *J. Am. Chem. Soc.* **112**, 7221 1990.

43. K. B. Renkema, Y. V. Kissin, and A. S. Goldman, *J. Am. Chem. Soc.* **125**, 7770, 2003 and references cited.

44. (a) A. S. Goldman, personal communication; (b) X. Zhang, A. Fried, S. Knapp, and A. S. Goldman, *Chem. Commun.* **2003**, 2060; (c) K. Krogh-Jespersen, M. Czerw,

N. Summa, K. B. Renkema, P. D. Achord, and A. S. Goldman, *J. Am. Chem. Soc.* **126**, 7192, 2004.

45. I. Gottker-Schnetmann, P. White, and M. Brookhart, *J. Am. Chem. Soc.* **126**, 1804, 2004.

46. (a) K. M. Waltz and J. F. Hartwig, *Science* **277**, 211 1997; H. Y. Chen, S. Schlecht, T. C. Semple, and J. F. Hartwig, *Science*, **287**, 1995, 2000; (b) C. E. Webster, Y. B. Fan, M. B. Hall, D. Kunz, and J. F. Hartwig, *J. Am. Chem. Soc.* **125**, 858, 2003.

47. J. Y. Cho, M. K. Tse, D. Holmes, R. E. Maleczka, and M. R. Smith, *Science*, **295**, 305 2002; R. E. Maleczka, F. Shi, D. Holmes, and M. R. Smith, *J. Am. Chem. Soc.*, **125**, 7792, 2003.

48. P. L. Watson, *J. Am. Chem. Soc.* **105**, 6491 1983.

49. C. M. Fendrick and T. J. Marks, *J. Am. Chem. Soc.* **106**, 2214 1984.

50. B. A. Arndtsen and R. G. Bergman, *Science* **270**, 1970 1995.

51. C. P. Schaller, C. C. Cummins, and P. T. Wolczanski, *J. Am. Chem. Soc.* **118**, 591 1996.

52. A. D. Sadow and T. D. Tilley, *J. Am. Chem. Soc.* **125**, 1971, 2003.

53. (a) C. F. H. Tipper, *J. Chem. Soc.* 2043, 1955; (b) J. Chatt et al., *J. Chem. Soc.* 738, 1961.

54. R. H. Crabtree, R. P. Dion, D. V. McGrath, and E. M. Holt, *J. Am. Chem. Soc.* **108**, 7222, 1986.

55. M. E. van der Boom and D. Milstein, *Chem. Rev.* **103**, 1759, 2003.

56. J. C. Hierso, P. Feuerer, and P. Kalck, *Coord. Chem. Rev.* **180**, 1811, 1998; W. C. J. Wei and M. H. Lo, *Appl. Organomet. Chem.* **12**, 201, 1998; M. Bochmann, *Chem. Vapor Depos.* **2**, 85, 1996; G. B. Stringfellow, *Mater. Sci. Eng.* **87B**, 97, 2001.

57. F. Maury et al., *J. Electrochem. Soc.* **146**, 3716, 1999.

58. *Zeolites: A refined tool for designing catalytic sites*, L. Bonneviot and S. Kaliaguine, eds., Elsevier, New York, 1995.

59. (a) M. Jayamurthy, *Ber. Bunsen Phys. Chem.* **99**, 1521, 1995; (b) J. Y. Ying et al., *J. Am. Chem. Soc.* **120**, 12289, 1998; (c) D. O'Hare et al., *Chem. Commun.* 603, 1997.

60. (a) D. Braga, F. Grepioni, and G. R. Desiraju, *Chem. Rev.* **98**, 1375, 1998; (b) O. M. Yaghi et al., *Accs. Chem. Res.* **31**, 474, 1998.

61. R. J. Puddephatt, *Chem. Commun.* 1055, 1998.

62. I. Manners et al., *Chem. Rev.* **99**, 1515, 1999; *J. Polym. Sci. A* **40**, 179, 2002.

63. M. A. Baldo, S. Lamansky, P. E. Burrows, M. E. Thompson, and S. R. Forrier, *Appl. Phys. Lett.* **74**, 4, 1999; A. Tsuboyama et al., *J. Am. Chem. Soc.* **125**, 12971, 2003.

64. S. Sibley, M. E. Thompson, P. E. Burrows, and S. R. Forrest, *Optoelectronic Properties of Inorganic Complexes*, J. Fackler, ed., Plenum, New York, 2003.

65. J. Burdeniuc, B. Jedlicka, and R. H. Crabtree, *Chem. Ber.* **130**, 145, 1997.

PROBLEMS

1. Given the mechanisms of the water–gas shift reaction starting from CO and H_2 shown in Fig. 12.9, what can you deduce about the mechanism of the reaction in the reverse sense, starting from CO_2 and H_2O?

2. The attempted metathesis of ethyl vinyl ether, EtOCH=CH$_2$, with Grubbs's catalyst [RuCl$_2$(PCy$_3$)$_2$(=CHPh)], gives only a stable metal complex and one equivalent of a free alkene as product. Predict the structures of these products and explain why the reaction is only stoichiometric, not catalytic.

3. The reaction shown below appears to be a cyclometallation, but is there anything unusual about it that might excite suspicion that it does not go by a conventional oxidative addition mechanism? Suggest an alternative.

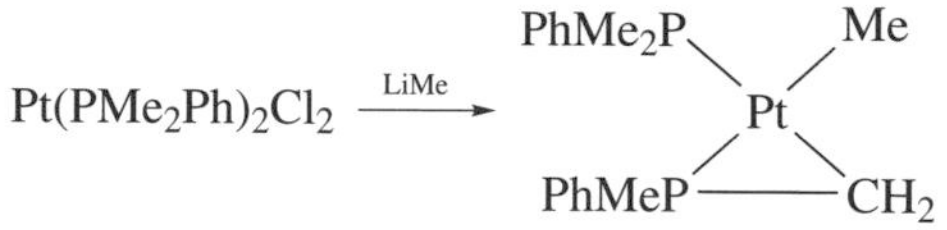

$$\text{Pt(PMe}_2\text{Ph)}_2\text{Cl}_2 \xrightarrow{\text{LiMe}} \begin{array}{c} \text{PhMe}_2\text{P} \quad\quad \text{Me} \\ \diagdown \diagup \\ \text{Pt} \\ \diagup \diagdown \\ \text{PhMeP} \text{———} \text{CH}_2 \end{array} \qquad (12.42)$$

4. Suppose that you were about to study the following complexes to see if any of them bind CO$_2$. Describe what type(s) of product you would anticipate in each case: Re(PR$_3$)$_5^-$, (η^5-Indenyl)Ir(PR$_3$)$_2$, and CpMo(CO)$_3$H. Given that you had samples of all three, which would you try first as the most likely to bind CO$_2$ (R = Me)?

5. Suggest a plausible mechanism for

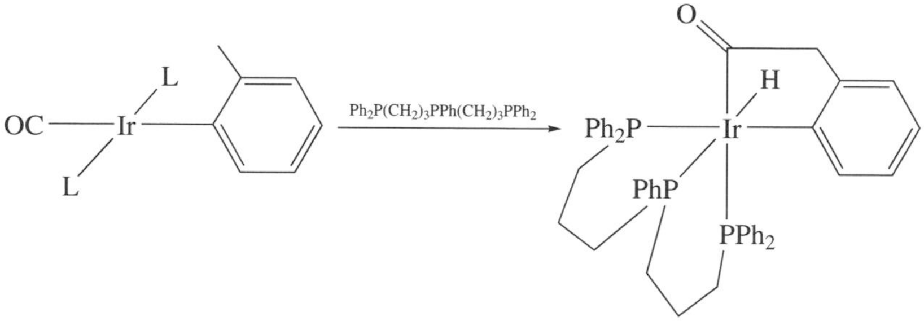

$$(12.43)$$

6. Suggest a plausible mechanism and mechanistic tests for

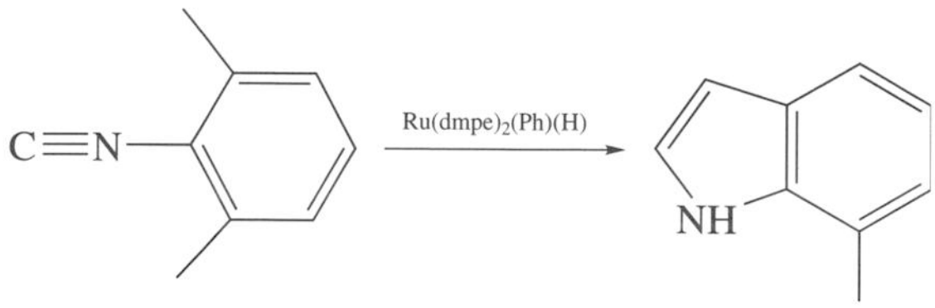

$$(12.44)$$

7. Suggest a plausible mechanism and mechanistic tests for

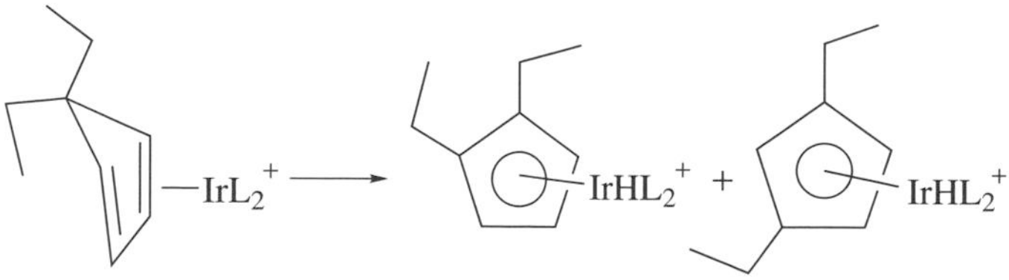

$$(12.45)$$

50 : 50 mixture

8. Suggest a plausible mechanism for Eq. 12.45 and some ways of testing your suggestion.

$$P_2Pt \diamond \rightleftharpoons P_2Pt \diamond -Ph$$

Ph

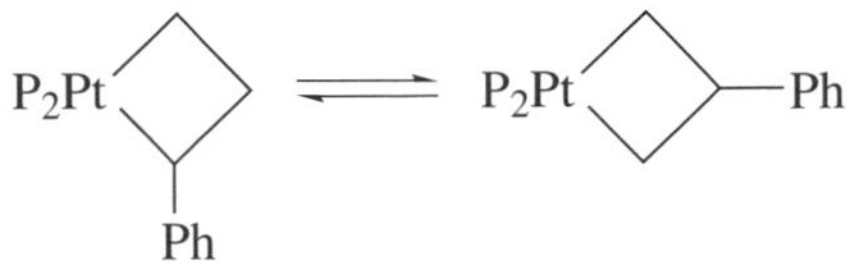

(12.46)

9. Suggest a plausible mechanism for Eq. 12.47 and some ways of testing your suggestion.

$$(cod)IrL(thf)^+ + HCOONa \longrightarrow (cod)IrLH \qquad (12.47)$$

10. Account for the product formed in Eq. 12.48.

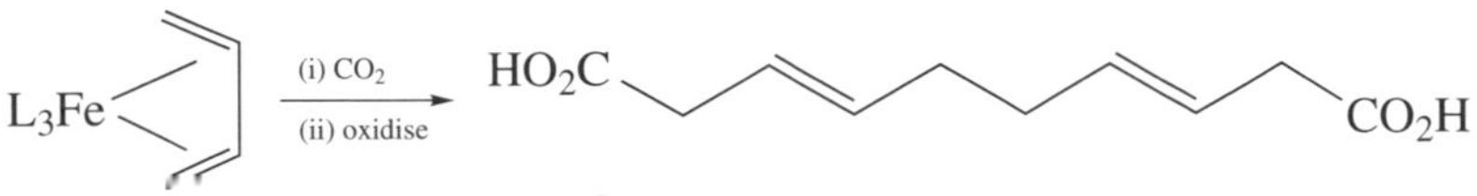

$$L_3Fe \diagup \xrightarrow[\text{(ii) oxidise}]{\text{(i) } CO_2} HO_2C \diagup\diagdown\diagup\diagdown\diagup CO_2H$$

(12.48)

11. Hydrosilation (shown below) is mediated by a variety of catalysts, both homogeneous and heterogeneous. Write a plausible mechanism for a generalized homogeneous catalyst L_nM.

$$RCH\equiv CH + R_3Si-H \longrightarrow RCH=CH-SiR_3 + RC(SiR_3)=CH_2. \quad (12.49)$$

12. If methanol/HI is carbonylated in a system resembling the Monsanto acetic acid process, but with [(dpe)RhI(CO)] as catalyst and H_2 present, ethanol is formed from methanol. Provide two reasonable mechanisms and suggest an experimental test to distinguish between them (see Ref. 65).

13

CLUSTERS AND THE METAL−METAL BOND

We now see what happens when several metal atoms are bound together in a *cluster*. Rather than form chains like carbon, they tend to agglomerate so as to form the maximum number of M−M bonds—the structures resemble the close-packed structures of the elemental metals themselves.[1] The reason is that clusters contain unsaturated L_nM fragments. The triangular cluster $Os_3(CO)_{12}$ can be regarded as the stable trimer of the unsaturated 16e fragment $Os(CO)_4$. The 15e $Rh(CO)_3$ fragment forms $Rh_4(CO)_{12}$, with a tetrahedron of metal atoms. The condensed structures of clusters allow the few available electrons to be maximally shared over the cluster as a whole. We also study the new bonding and reactivity patterns possible for organic fragments bound to a cluster.

Organometallic clusters are almost always rich in carbonyl ligands, probably because $M(CO)_n$ fragments are sufficiently unhindered to approach to within M−M bonding distance of each other. It is surprising that so few stable *homoleptic** clusters of other small high-field ligands, such as hydride, silyl, methyl, or methylene, are as yet known.

An early stimulus to cluster chemistry was the cluster−surface analogy,[2] which proposed that cluster chemistry would resemble the surface chemistry of metals because both surfaces and clusters consist of arrays of metal atoms. Supported metals such as Pd/C are very active catalysts. Carbonyl clusters have so far not shown the high catalytic activity of either metal surfaces or mononuclear homogeneous catalysts, probably because clusters are "poisoned" by the presence

*A homoleptic compound has only one type of ligand.

The Organometallic Chemistry of the Transition Metals, Fourth Edition, by Robert H. Crabtree
Copyright © 2005 John Wiley & Sons, Inc.

of a monolayer of CO. Bare metal nanoparticles (Section 13.5) are highly active, however. Organic compounds do bind to clusters differently than to single metals, and these new structures[3] provide important clues for surface chemistry, where direct structural data for surface-bound species are still very hard to obtain.

A second point of interest in cluster chemistry is the gradual evolution of cluster structure, magnetic behavior, and ionization potential with increasing cluster size. In principle, these properties should approach that of the bulk metal, but some may do so faster than others.

The term *cluster* has been applied to a vast range of chemical systems involving aggregation of simpler units. Main-group clusters such as the boron hydrides played an important role in clarifying bonding theory in the 1960s. Stable transition metal clusters are sometimes divided into organometallic and inorganic. Organometallic clusters, such as $Os_3(CO)_{12}$, tend to be low-oxidation-state (≤ 2) species, often with carbonyl ligands. Inorganic clusters,[4,5] such as $(RO)_3Mo\equiv Mo(OR)_3$, $Cl_4Re\equiv ReCl_4$ (**13.1**), or $Mo_6(\mu^3\text{-}Cl)_8{}^{4-}$ (**13.2**), are often higher valent and have ligands such as Cl^-. Hybrid organometallic main-group clusters such as $[\{Fe(CO)_3\}_2\{BH\}_2(\mu^2\text{-}H)_4]$ also exist.[4b] There are also a number of naked metal clusters[5] of the posttransition elements, such as $Sn_9{}^{2-}$.

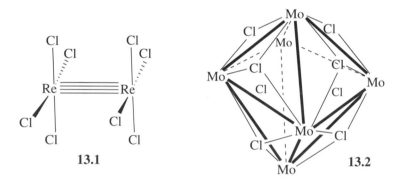

13.1   **13.2**

The term *cluster* was once reserved for complexes containing at least three metals, bound by metal–metal bonds, but is now normally used for any aggregate, including di- and polynuclear complexes bound together only by bridging ligands. In this chapter, we emphasize organometallic, M–M bonded species.

13.1 STRUCTURES

Cluster chemistry usually requires X-ray crystallography for characterization, and perhaps for this reason, structural questions have tended to be given most attention. Once a given structure has been determined, it is sometimes possible to use spectroscopic methods to deduce the structures of closely related species. In particular, IR studies are often useful in showing whether the CO ligands have been affected during a reaction, and ^{1}H NMR studies are often used to look at the organic ligands.

The M—M single-bond lengths are often comparable to those found in the elemental metal, but the attractive interaction between the metals is often increased by the presence of bridging ligands such as CO. Not all M—M bonds are bridged; $[(CO)_5Mn—Mn(CO)_5]$ is an example, but this bond[6] is weak (28 ± 4 kcal/mol) and unusually long, at 2.93 Å versus 2.46 Å in $[(CO)_3Fe(\mu\text{-}CO)_3Fe(CO)_3]$. With a bond strength of only 17 kcal/mol, the unsupported M—M bond of $[CpCr(CO)_3]_2$ undergoes spontaneous breaking and reforming even at room temperature.

Effective Atomic Number (EAN) Rule

Only the simpler clusters can be described in terms of the 18e rule. For example, each 16e $Os(CO)_4$ group in $Os_3(CO)_{12}$, **13.3**, can be considered as achieving 18e by forming two M—M bonds, one with each of the other metal atoms. Since each metal has the same electronegativity, the bond is considered as contributing nothing to the oxidation state. The complex contains 18e, Os(0). It is usually more convenient to count the electrons for the cluster as a whole, rather than attempt to assign electrons, especially electrons from bridging ligands, to one metal rather than another. On this counting convention, $Os_3(CO)_{12}$ is a $3 \times 8e$ (Os is in group 8) $+ 12 \times 2e = 48e$ cluster. This is the appropriate number of electrons for a triangular cluster. We have $3 \times 9 = 27$ orbitals, which you might think ought to require 54e, but this assumes that we count each metal individually, and then sum the totals from each metal. By doing this, we would count the M—M bonding electrons twice over because in counting Os^1, we count 1e "originating" (from a bookkeeping point of view) from Os^2. In counting Os^2, we would count these M—M bonding electrons again. Six M—M bonding electrons are involved so we expect $54 - 6 = 48e$ to account for the three M—M bonds.

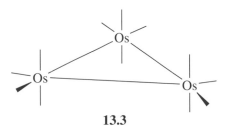

13.3

Since we always deal with electron counts that are > 18, it is more convenient in cluster chemistry to use the alternative name of the 18e rule, the effective atomic number, or EAN, rule. The closed-shell configuration resembles that of the noble gases [Rn (radon) in the case of Os], and so the Os in the complex is said to have the same effective atomic number as Rn and is *coordinatively saturated*.

The EAN electron count for a cluster of nuclearity x and having y metal–metal bonds is defined by

$$\text{EAN count} = 18x - 2y \tag{13.1}$$

For $Mo(CO)_6$, for example, y is 0 and we expect an 18e count from Eq. 13.1. For $(CO)_5Mn-Mn(CO)_5$, y is 1 and we expect a count of $(2 \times 18) - 2$, or 34e. This is indeed the case because 2 Mn contribute 14e and 10 COs contribute 20e, so this is an EAN cluster. The 48e $Os_3(CO)_{12}$ case ($y = 3$) was discussed above. For tetrahedral $Rh_4(CO)_{12}$ ($y = 6$), we expect 60e from Eq. 13.1 for an EAN cluster, as are indeed present. For TBP $Os_5(CO)_{16}$ ($y = 9$), we expect 72e for an EAN cluster, as found; note that the bridging CO counts as 2e just like a terminal CO.

$$Rh_4(CO)_{12} \qquad\qquad Os_5(CO)_{16}$$

These ideas can be extended to more complex clusters as shown in Fig. 13.1.

The CO bonding mode is unimportant: whether a CO is terminal or bridging, it still contributes 2e to the cluster as a whole, so we cannot predict by counting electrons whether a given molecule will have any bridging COs or not. Of the isoelectronic group 8 $M_3(CO)_{12}$ clusters, only the iron analog, **13.4**, has bridging COs; the others, like **13.3**, have only terminal carbonyls. Note that in the diagrams in this chapter a single unlabeled line drawn from the metal denotes a terminal carbonyl substituent and a bent line connecting two metals denotes a bridging CO; only non-CO ligands are shown explicitly.

$Re_4H_4(CO)_{12}$	$4 \times Re$	$= 28$		$Fe_6C(CO)_{16}^{2-}$	$6 \times Fe$	$= 48$
	$4 \times H$	$= 4$			$1 \times C$	$= 4$
	$12 \times CO$	$= \underline{24}$			$16 \times CO$	$= 32$
		56			$2 \times e^-$	$= \underline{2}$
						86
$Os_3H_2(CO)_{10}$	$3 \times Os$	$= 24$		$Fe_3(\mu\text{-}CO)_2(CO)_{10}$	$3 \times Fe$	$= 24$
	$2 \times H$	$= 2$			$2 \times \mu\text{-}CO$	$= 4$
	$10 \times CO$	$= \underline{20}$			$10 \times CO$	$= \underline{20}$
		46				48

FIGURE 13.1   Electron counting in clusters. For the structure of $Fe_6C(CO)_{16}^{2-}$, see Fig. 13.7.

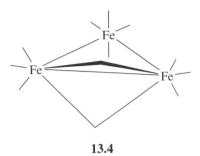

13.4

$Os_3H_2(CO)_{10}$ behaves as an unsaturated cluster in that it is much more reactive than $Os_3(CO)_{12}$. One way of looking at this is to say that, as a 46e cluster, it lacks 2e from the EAN count of 48e. It is often viewed as containing an Os=Os "double bond" because the EAN count for a system with four M—M bonds in a three-atom cluster is 46e. We would then regard an Os=Os double bond, like a C=C double bond, as being unsaturated. Structure **13.5** shows that there are two Os—H—Os bridges.

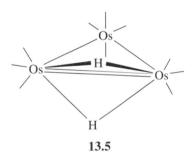

13.5

In our discussion of M—H—M bonding (Section 3.2), we saw that the presence of such a bridge implies M—M bonding. The representation shown in **13.5** is the conventional one, but it should not be taken to mean that there are separate M—M and M—H—M bonds. In fact, each M—H—M unit constitutes a 2e, three-center bond as shown in **13.6**. This means that the Os=Os "double bond" is really a reflection of the presence of the two hydride bridges. The bridge can open and generate a vacant site. This makes the dihydride far more reactive than $Os_3(CO)_{12}$ itself and therefore a very useful starting material in triosmium cluster chemistry.

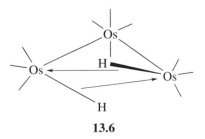

13.6

The tetranuclear group 9 clusters $M_4(CO)_{12}$ have 60e. Equation 13.1 shows that six M−M bonds must be present if the cluster is to conform with the EAN rule. As expected, a tetrahedral cluster framework with six M−M bonds is adopted. In summary, we can deduce whether the molecule has the EAN count if we know how many M−M bonds are present, or we assume that the molecule is an EAN one and deduce the number of M−M bonds we expect to find.

Face (μ_3) bridging is a bonding mode unique to polynuclear complexes. If we have a face bridging CO (**13.7**), we count only the 2e of the carbon lone pair as contributing to the cluster. On the other hand, some ligands have additional lone pairs they can bring into play. A Cl ligand is 1e when terminal, **13.8**, but 3e when edge (μ_2) bridging, **13.9**, and has 5e to donate to the cluster if it is face bridging (**13.10**), as two of its lone pairs come into play (the corresponding numbers for the ionic model are 2e, 4e, and 6e, respectively, but this model is not commonly used in cluster chemistry).

| **13.7** | **13.8** | **13.9** | **13.10** |

As shown in Fig. 13.1, $Re_4H_4(CO)_{12}$ (**13.11**) has 56e. This requires the presence of eight M−M bonds, rather than the six normally implied by a tetrahedral arrangement of four metals. The distortions that would be expected for a static structure of type **13.12** with two localized M=M double bonds are not found, and so the extra M−M bonds are conventionally considered to be delocalized over the metal framework, so as to make each M−M bond slightly shorter. An alternative picture comes from our discussion of the nature of the hydride bridge

| **13.11a** | **13.11b** |

in Section 3.2. Each H in **13.11** is found to be face bridging (μ_3-H). We can regard the 2e of the M_1−H bond to be donated to both M_2 and M_3 as shown in **13.13**. This gives an EAN cubanelike structure (**13.11a**) for $Re_4H_4(CO)_{12}$. In this model, the delocalized M−M bonds are included in the μ_3-H bridging.[7] In this way, each μ_2-H reduces the EAN by 2e, and each μ_3-H reduces it by 4e. Note that on the conventional model the position of the hydrogen (whether terminal, μ_2-H, or μ_3-H) is irrelevant to the EAN count. The alternative picture successfully predicts the position of the hydrogen in a large number of clusters.*

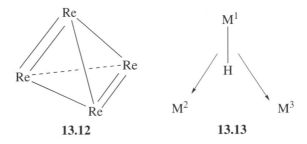

13.12   **13.13**

In M_5L_n clusters, we can have a trigonal bipyramid (TBP) of metals with nine M−M bonds or a square pyramid (SP) with eight. By Eq. 13.1, the TBP is adopted for a 72e system like $Os_5(CO)_{16}$ and the SP for a 74e cluster like $M_5(CO)_{15}C$ (**13.14**). Note how all four valence electrons of the C are counted as contributing to the cluster.

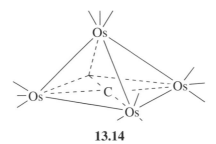

13.14

Wade's Rules

When we get to six-metal clusters and beyond, the EAN picture can start to fail. For example, the octahedral $Os_6(CO)_{18}^{2-}$, **13.15** is an 86e cluster. On the basis of Eq. 13.1, and assuming there are 12 M−M bonds, the EAN should be 84e. Yet the cluster shows no tendency to lose electrons or expel a ligand. $Os_6(CO)_{18}$, **13.16**, which is an authentic 84e cluster, does not adopt the octahedral framework at all but does have 12 M−M bonds.

---

*Students sometimes ask which model is "right"—models are only mental constructs that reflect some aspect of reality. One model may work for one compound, a second model for another. Model **13.12** is certainly more widely used.

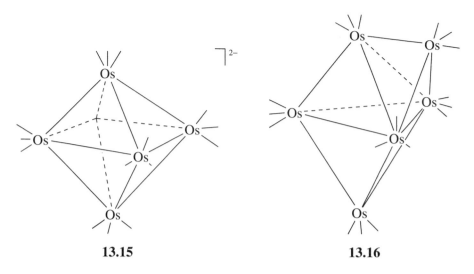

13.15   13.16

The cluster counting model that applies to these non-EAN clusters is the polyhedral skeletal electron pair theory, sometimes known as *Wade's rules*.[8] On this picture, an analogy is drawn between the metal cluster and the corresponding boron hydride cluster. Elements like C and H, which have the same number of electrons and orbitals, can form closed-shell molecules, such as CH_4. Elements to the right of carbon, such as N, have more electrons than orbitals and so give molecules with lone pairs, like NH_3. Like transition metals, boron has fewer electrons than orbitals, and so it forms compounds in which the BH_x units cluster together to try and share out the few electrons that are available by using 2e, three-center bonds, such as in B_2H_6. The higher borane hydride anions $B_nH_n^{2-}$ ($n = 6–12$) form polyhedral structures, some of which are shown in Fig. 13.2; these form the basis for the polyhedral structures adopted by all molecules covered by Wade's rules. The shape of the cluster is decided purely by the number of cluster electrons (called "skeletal" electrons), not by any other factor.

The number of skeletal electrons appropriate to the borane clusters, $B_nH_n^{2-}$, can be deduced as follows. First, we assume that each B–H bond is a normal 2e covalency, requiring 1e from H and 1e from B. As boron starts with 3e, it has 2e left to contribute to the cluster, and this means that $B_nH_n^{2-}$ has $2n + 2$ cluster electrons, $2n$ electrons of which come from the n BH groups, and the remaining two electrons come from the 2- net charge. In order to see where these electrons go, we consider that each BH unit has an sp orbital pointing directly toward the center of the cluster, and a p_x and a p_y orbital, pointing along the surface (Fig. 13.3). The MO analysis of this arrangement predicts that the sp orbitals contribute to one low-lying orbital, when they are all taken with the same sign (in phase). Other combinations are high lying and empty. The p orbitals, $2n$ in number, combine to give n filled bonding MOs and n empty antibonding MOs.

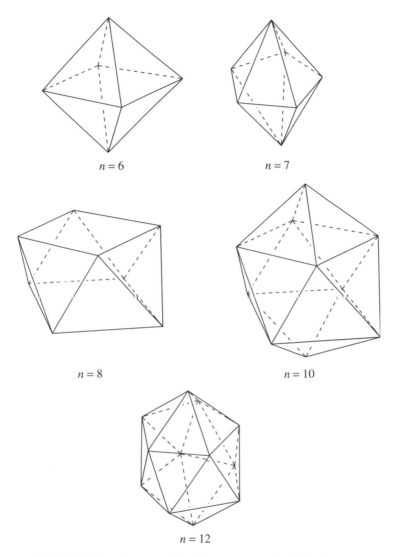

$n = 6$

$n = 7$

$n = 8$

$n = 10$

$n = 12$

FIGURE 13.2 Some polyhedral structures adopted by boranes.

This picture provides $n + 1$ orbitals, which offer an appropriate home for $2n + 2$ skeletal electrons.

Since the cluster shape depends only on the number of skeletal electrons, we should be able to remove a vertex group, say, BH, from the cluster without changing the cluster structure, as long as we leave behind the two skeletal electrons that the vertex BH group was contributing. This means we must remove a BH^{2+}, not a BH unit, in order to leave one vertex of the cluster empty. If we remove a

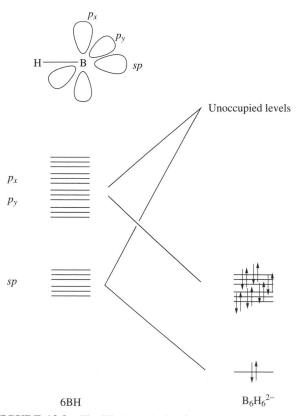

FIGURE 13.3 The Wade analysis of a close borane cluster.

BH^{2+} unit in the case of $B_6H_6^{2-}$, we get the hypothetical $B_5H_5^{4-}$ fragment.

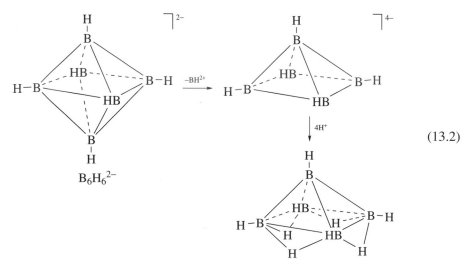

(13.2)

This will have exactly the same polyhedral structure because the electron count has not changed, but one vertex is now empty. To make the known neutral borane, B_5H_9, we add four protons, which, as zero-electron species do not alter the electron count. Note that the protons bridge the faces of the polyhedron, which include the missing vertex; they could be said to sense the electron density left behind in the cluster faces when we removed the BH^{2+} group. As a species with one empty vertex, B_5H_9 is given the descriptor *nido*. Molecules that have every vertex occupied are designated *closo*. In general, a species $B_xH_y^{z-}$ will have $\frac{1}{2}(x + y + z)$ skeletal electron pairs. The appropriate number of vertices, v, is

$$v = \tfrac{1}{2}(x + y + z) - 1 \tag{13.3}$$

The number of BH groups we have to find vertices for is x. If the number of vertices v called for by Wade's rules also happens to equal x, then each vertex can be occupied and we will have a closo structure. On the other hand, if x happens to be one less than v, one vertex will be empty and a nido structure will result. If x is two or three units less than v, then the structures are called *arachno* and *hypho* with two or three empty vertices, respectively. Normally adjacent (rather than nonadjacent) vertices are left empty.

Wade's rules can also apply to other main-group elements: the 14 skeletal electron octahedral Sn_6^{2-} has been isolated as $[SnCr(CO)_5]_6^{2-}$ in which all the exo-lone pairs on Sn are bound to the 16-valence-electron fragment, $\{Cr(CO)_5\}$.[9a]

Surprisingly, the same model also describes many transition metal clusters, including many of the non-EAN ones. In order to see how we can do this, we first have to find a way of replacing the BH groups by transition metal equivalents that donate the same number of skeletal electrons. Since transition metals have nine orbitals but only three are required for cluster bonding on the Wade picture, we first have to fill the six orbitals not required for cluster bonding and see how many electrons remain for the cluster bonding orbitals. If we take the $Os(CO)_3$ fragment, we have to assign the nine orbitals as follows: (1) three orbitals are filled with the three CO lone pairs; (2) three more orbitals are filled with six electrons out of the eight electrons appropriate for a group 8 element such as Os—these electrons back-bond to the COs; and (3) two metal electrons are now left for the remaining three orbitals, which are the ones that bond to the cluster (Fig. 13.4). This implies that $Os(CO)_3$ contributes the same number of skeletal electrons (two) as does a BH group. We can therefore replace all the BHs in $B_6H_6^{2-}$ with $Os(CO)_3$ groups without altering the structure. We end up with $Os_6(CO)_{18}^{2-}$, exactly the cluster we could not explain on the EAN model.

There also exist many clusters, called *metalaboranes*,[9b] in which some of the vertices of the polyhedron have a boron atom and others a transition metal [e.g., *closo*-$(CpCo)_2(BH)_4(\mu_3 - H)_2$, **13.17**].

For the fragment MX_aL_b, the Wade analysis leads us to predict that the cluster electron contribution, F, of that fragment will be

$$F = N + a + 2b - 12 \tag{13.4}$$

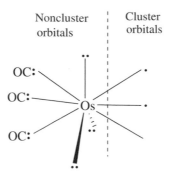

FIGURE 13.4   Applying Wade's rules to a transition metal fragment. The three CO groups of the $Os(CO)_3$ fragment supply 6e, and these electrons occupy three of the metal's nine orbitals. Six of the eight metal electrons occupy the d_π orbitals and back-bond to the CO groups. Two metal electrons are left to fill the three cluster bonding orbitals shown to the right of the dotted line.

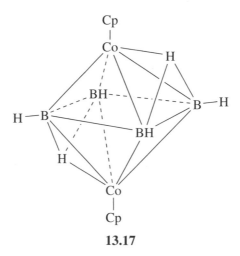

13.17

(where N = group number of metal). To find the total number, T, of cluster electrons, we then sum the contribution from all the fragments in the cluster, add the sum of the contributions from the bridging ligands (ΣB) to account for any electrons donated to the cluster by edge bridging, face bridging, or encapsulated atoms (see example below), and adjust for the total charge, z^-, on the cluster as a whole:

$$T = \Sigma F + \Sigma B + z \tag{13.5}$$

(where $B = 1$ for bridging H, 2 for bridging CO, 3 for η^2-Cl, etc.). The number of vertices, v, in the cluster will then be given by

$$v = \frac{T}{2} - 1 \tag{13.6}$$

We have seen what happens in a borane cluster if there are not enough BH fragments to fill the vertices: We get a nido structure with an empty vertex. The same is true for transition metal clusters, for example, in $Fe_5(CO)_{15}C$, the carbon atom, which is not considered as a vertex atom, is encapsulated within the cluster and gives all its four valence electrons to the cluster. The $Fe(CO)_3$ fragment contributes two cluster electrons as it is isoelectronic with $Os(CO)_3$. The total count is therefore $(5 \times 2) + 4 = 14$, and the number of vertices is $\frac{14}{2} - 1 = 6$. This requires the structure shown as **13.18**, as is observed for this and the analogous Ru and Os species.

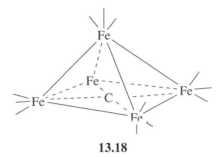

13.18

What happens when there are more atoms than vertices into which they can fit? For example, $Os_6(CO)_{18}$ is a $(6 \times 2) = 12$ cluster electron species. This means that the number of vertices required by Wade's rule is $\frac{12}{2} - 1 = 5$. The structure found for the molecule, **13.16**, shows that the extra metal atom bridges to a face of the five-vertex base polyhedron, and so is able to contribute its electrons to the cluster, even though it cannot occupy a vertex.

Only when we move up to clusters of nuclearity 6–12, do the EAN and Wade predictions become different. Often the Wade structure is the one observed, but sometimes we find that both a Wade's rule and an EAN cluster are stable. Adams and Yang[10] have shown how in such situations there can be facile interconversion between the two forms by gain or loss of a ligand:

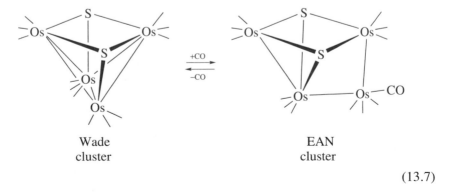

Wade
cluster

EAN
cluster

(13.7)

Linear Clusters If suitable ligands can be provided, it has proved possible to stabilize linear clusters made of chains of metal atoms.[11]

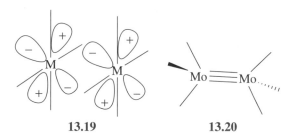

M–M Multiple Bonds Multiply bonded species, such as $Cl_4Re{\equiv}ReCl_4{}^{2-}$ (**13.1**), were first recognized by Cotton[4] and tend to be formed from the middle transition elements, the same elements that give strong $M{\equiv}O$ multiple bonds (Section 11.5).[3,4] For $\{L_nM\}_2$ to form a bond of order n, the L_nM fragment has to have a d^n or higher configuration because it needs a minimum of n electrons, just as the CH fragment needs three available electrons to form $HC{\equiv}CH$. In **13.1**, two square planar $d^4ReCl_4{}^-$ fragments face each other in the unusual eclipsed (Cl atoms face-to-face) geometry with a very short Re–Re distance. Taking the M–M direction as the z axis, the quadruple bond is formed from overlap of the d_{z^2} (the σ bond), the d_{xz} and d_{yz} (which form two π bonds), and of the d_{xy} on each Re, which forms the so-called δ bond. It is this last δ bond that causes the eclipsed geometry because only in this geometry is overlap possible, as illustrated in **13.19**. The electronic structure of **13.1** is often represented as $\sigma^2\pi^4\delta^2$, which indicates how many electrons are present in each type of bond. $(RO)_3Mo{\equiv}Mo(OR)_3$ has an M–M triple bond of the $\sigma^2\pi^4$ type, in which good overlap is still possible in the staggered geometry, **13.20**.

13.19   **13.20**

M–M multiple bonds are short; for example, typical values for Mo are 2.1 Å, $Mo{\equiv}Mo$; 2.2 Å, $Mo{\equiv}Mo$; 2.4 Å, $Mo{=}Mo$; 2.7 Å, Mo–Mo; and 2.78 Å, Mo metal. Bond strengths are known for few systems, but for $Re{\equiv}Re$ in **13.1** it is 85 ± 5 kcal/mol, of which only ca. 6 kcal/mol is assigned to the δ bond. (This δ bond strength is comparable to that of a hydrogen bond.)

- Cluster formation is a result of electron deficiency in $M(CO)_n$ fragments.
- The EAN and Wade's rules are alternate models that help predict cluster structures; where they give different predictions, both structures may exist.

13.2   THE ISOLOBAL ANALOGY

Hoffmann's[12] isolobal analogy is a general unifying principle that goes far beyond the confines of cluster chemistry. Nevertheless it has found most application in this area, and so we will look at it now. The idea is very simple; the backbone of most organic compounds is made up of the familiar groups CH_3, CH_2, and CH, which we can put together at will. What is the special property of a methyl radical that makes it univalent? Clearly, it is the singly occupied sp^3 orbital. We will consider this fragment as having one orbital and one "hole"; a hole for this purpose simply means that the fragment has one electron less than the 8e closed-shell configuration CH_3^-. As far as the rest of the molecule is concerned, a methyl radical can be considered as providing a hole and an orbital. Hoffmann points out that any fragment with a half-filled orbital of a σ type may be able to form structures similar to those found for the methyl group. $Mn(CO)_5\bullet$ is an example of such a radical. We can imagine that it is formed by removing a CO from the 18e species $Mn(CO)_6^+$ to give $Mn(CO)_5^+$, a 16e species with an empty orbital (two holes) pointing toward the missing ligand. To make the 17e radical, we merely have to add 1e to this orbital. The resulting $Mn(CO)_5\bullet$ can replace one methyl group in ethane to give $MeMn(CO)_5$, or both of them to give $(CO)_5Mn-Mn(CO)_5$, for example. The two fragments are not isoelectronic because $Mn(CO)_5\bullet$ has far more electrons than $CH_3\bullet$, but the significant orbital by which the two fragments form bonds to other groups, are the same both in symmetry and in occupancy. The isolobal analogy is expressed by a double-headed twirly arrow, as follows:

$$Me-Me \xleftrightarrow{\hspace{1cm}} Me-Mn(CO)_5 \xleftrightarrow{\hspace{1cm}} (CO)_5Mn-Mn(CO)_5 \qquad (13.8)$$

Suppose that we moved one element to the left. How could we treat $Cr(CO)_5$, a fragment that, like $Mn(CO)_5$ has one orbital, but that is empty (two holes)? Clearly, CH_3^+ is the appropriate organic fragment because it too has an unfilled σ-type orbital. As we know, $Cr(CO)_5$ reacts with CO to give $Cr(CO)_6$. The linear acetyl cation CH_3CO^+, an important intermediate in Friedel–Crafts reactions, can now be seen as a CO complex of CH_3^+. This is not a conventional way of looking at this species and illustrates how the isolobal principle can give new insights in organic as well as in inorganic chemistry.

The CH_2 fragment has two orbitals and two electrons with which to make bonds; in other words, CH_2 has two orbitals and two holes. If the CH_2 fragment is to bond to two H atoms to give methane, we will hybridize these two orbitals in such a way as to have two sp^3 lobes. If two CH_2 fragments are to dimerize to give ethylene, then we will rehybridize the system to give an sp^2 and a p orbital, so that we can form a σ and a π bond. The question is to discover what metal fragments are isolobal with CH_2. It turns out that $Mo(CO)_5$ is one such fragment. This is not so obvious until one recognizes that the key point in the isolobal analogy is that the number of holes has the fixed value of (18 minus the electron count of the ML_n fragment). The number of orbitals can vary according to the hybridization. For example, we can hybridize the single empty orbital of $Mo(CO)_5$ with one of the filled d_π orbitals to give a fragment that still has two holes but now has two orbitals. This picture in turn implies that CH_3^+ is isolobal with CH_2. Hoffmann has called this the *deprotonation analogy*. This extension of the analogy is more useful for organometallic rather than organic fragments because in the organic case we can only take a C–H bonding orbital for the rehybridization; this, which is more stable than the nonbonding d_π orbital of the organometallic fragment, is more reluctant to cooperate. We can see the $Mo(CO)_5$ fragment acting as isolobal with CR_2 in the Fischer carbenes $(CO)_5Mo=CR_2$. Just as $Mo(CO)_5$ forms a carbonyl complex, $Mo(CO)_6$, so does CH_2, in the form of $CH_2=C=O$, ketene.

Table 13.1 shows how the analogy works. We need to calculate n_H, the number of holes in our metal fragment (Eq. 13.9 shows this explicitly for the $MX_aL_b^{c+}$, where N is the group number of the metal).

$$n_H = 18 - N - a - 2b + c \qquad (13.9)$$

This shows us at once which organic fragments are isolobal with the organometallic fragment in question. The most direct analogy will be with the organic fragment that has the same number of orbitals. For the metal fragments, the number of orbitals, n_o, is calculated on the basis of an octahedral model. If there are three ligands in the fragment, three orbitals of the octahedron are available; Eq. 13.10 shows the general expression

$$n_o = 6 - a - b \qquad (13.10)$$

TABLE 13.1 Isolobal Relationships[a]

Inorganic Fragment	n_H	n_o	Organic Fragment	Complex	Isolobal with
$Mn(CO)_5$	1	1	CH_3	$Me-Mn(CO)_5$	$Me-Me$
$Mo(CO)_5$	2	1	CH_3^+	$Me_3P-Mo(CO)_3$	Me_3P-Me^+
	2	2^b	CH_2	$OC=Mo(CO)_5$	$OC=CH_2$
	2	3^b	CH^-	—	—
$Fe(CO)_4$	2	2	CH_2	$(C_2H_4)-Fe(CO)_4$	Cyclopropane
$Cp(CO)_2Mo$	3	2^b	CH_2^+	—	—
	3	3^b	CH	$Cp(CO)_2Mo\equiv CR$	Acetylene
$CpRh(CO)$	2	2	CH_2	$\{CpRh(CO)\}_2(\mu\text{-}CH_2)$	Cyclopropane
$PtCl_3^-$	2^c	1^d	CH_3^+	$Cl^--PtCl_3^-$	$Cl^--CH_3^+$
	2^c	$2^{b,d}$	CH_2	$(C_2H_4)-PtCl_3^-$	Cyclopropane

[a] n_H and n_o are the number of holes and of orbitals.
[b] After rehybridizing to include one or more d_π orbitals. Note that on the deprotonation analogy, CH_3, CH_2^-, and CH^{2-} are isolobal, as are CH_3^+, CH_2, and CH^- and CH_3^{2+}, CH_2^+, and CH.
[c] On the basis of a 16e closed shell.
[d] On a square planar basis.

By the deprotonation analogy, metal fragments can make up to three more orbitals available by using their d_π set; reference to Table 13.1 will show how we often have to resort to using the d_π set. For example, $Mo(CO)_5$ in Table 13.1 is isolobal with CH_3^+ by Eqs. 13.9 and 13.10 ($n_H = 2$, $n_o = 1$). If we bring in an extra filled d_π orbital, we move to ($n_H = 2$, $n_o = 2$), which makes the fragment isolobal with CH_2. This means that the $Me_3P-Mo(CO)_5$ or $Me-Mn(CO)_5$ bonds are formed without a significant contribution from a d_π orbital, while the $OC=Mo(CO)_5$ double bond with its strong Mo-to-CO π back-bonding component requires a strong contribution from a d_π orbital. The deprotonation analogy gets its name from the fact that CH_2 can be formed by deprotonation of CH_3^+.

Because CH has three orbitals and three holes, the most direct analogy is therefore with the group 9 $M(CO)_3$ fragments, such as $Co(CO)_3$. Figure 13.5 shows the conversion of the hydrocarbon tetrahedrane into a tetrahedral $M_4(CO)_{12}$ cluster by the isolobal replacement of $M(CO)_3$ groups by CH. $Co_4(CO)_{12}$ has a bridged structure, and only the Rh and Ir analogs are all-terminal; since the all-terminal structure can only be unstable with respect to the real structure by a few kilocalories per mole for Co, we must not hold it against the isolobal analogy, or any counting rule for not being able to predict the pattern of CO bridges. Structure **13.24**, best known for Co, is normally considered as μ_3-carbyne cluster.

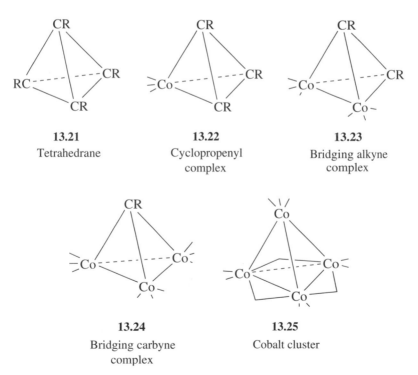

13.21	**13.22**	**13.23**
Tetrahedrane	Cyclopropenyl complex	Bridging alkyne complex

13.24	**13.25**
Bridging carbyne complex	Cobalt cluster

FIGURE 13.5 Stepwise isolobal replacement of CH by $Co(CO)_3$ in tetrahedrane. $Co_4(CO)_{12}$ has the CO bridged structure shown.

Structure **13.23** is usually considered as a bridging alkyne complex of $Co_2(CO)_8$, and **13.22** as a cyclopropenyl complex of $Co(CO)_3$. The all-carbon compound, **13.21**, is unstable and reverts to two molecules of acetylene, but stable tetrahedranes C_4R_4 have been made by using very bulky R groups.

Those metals that prefer to be 16e, such as Pt(II), can also be treated on isolobal ideas, but the number of holes is determined on the basis of a closed shell of 16e, not 18e. The argument is that the fifth d orbital, although empty, is too high in energy to be accessible, and so its two holes do not count. For example, the 14e $PtCl_3^-$ fragment is considered as having two holes, not four. The number of orbitals is also calculated on the basis of a square planar structure, so that $PtCl_3^-$ has one orbital, and is therefore isolobal with CH_3^+. Both species form a complex with NH_3, for example, $(NH_3)PtCl_3^-$ and $CH_3NH_3^+$. An extra nonbonding orbital on Pt can also be considered to contribute, giving two orbitals and two holes, which makes $PtCl_3^-$ isolobal with CH_2. Both fragments form complexes with ethylene—$(C_2H_4)PtCl_3^-$ and cyclopropane, respectively.

Any bridging hydrides can be removed as protons; for example, the dinuclear hydride in Eq. 13.11 is isolobal with acetylene because the 15e $IrHL_2^+$ fragment has three holes and three orbitals. CO ligands contribute in the same way whether they are bridging or terminal (e.g., Eq. 13.12), but the rhodium dimer (Eq. 13.13) has bridging CO groups.

$$HL_2Ir \overset{H}{\underset{H}{\diamond}} IrHL_2^+ \xrightarrow{-3H^+} HL_2Ir \equiv IrHL_2^{2-} \longleftarrow \bigcirc \longrightarrow HC \equiv CH \quad (13.11)$$

$$CpRh(CO) \longleftarrow \bigcirc \longrightarrow CH_2 \quad (13.12)$$

$$2CpRh(CO) \longrightarrow CpRh \overset{CO}{\underset{CO}{\diamond}} RhCp \quad (13.13)$$

As we see in the next section, we can even use the isolobal analogy to plan synthetic strategies, but we must guard against expecting too much from such a simple model. There are many cases in which molecules isolobal with stable organic compounds have not been made. This may be because the right route has not yet been found, or it may be that another structure is more favorable. C–C multiple bonds are stronger than M–M multiple bonds, and so a species like $(CO)_3Co \equiv Co(CO)_3$ is unlikely, although it is isolobal with acetylene. Similarly, we saw that acetylene is more stable than tetrahedrane. Finally, the isolobal analogy is a structural one; we cannot expect it to predict such things as reaction mechanisms, for example.

> • Isolobal ideas give us a way to draw on analogies between organic and inorganic structures.

13.3 SYNTHESIS

Many metal cluster complexes were originally synthesized by unplanned routes or as by-products in other reactions. Only recently have systematic procedures been developed for making metal–metal bonds and building up clusters.

Clusters are formed efficiently in a number of ways:

1. By pyrolysis of mononuclear carbonyl complexes[13] (it appears that CO is lost first, and the unsaturated fragment then attacks the original carbonyl):

$$Ru(CO)_5 \xrightarrow{50°C} Ru_3(CO)_{12} \qquad (13.14)$$

Photolysis can also be used to expel the CO.[14]

$$Fe(CO)_5 \xrightarrow{h\nu} Fe_2(CO)_9 \qquad (13.15)$$

2. By nucleophilic attack of a carbonyl anion:[15,16]

$$Mn(CO)_5^- + BrRe(CO)_5 \longrightarrow (CO)_5Mn–Re(CO)_5 \quad (13.16)[15]$$

$$Ru_3(CO)_{12} \xrightarrow{Fe(CO)_4^{2-}} FeRu_3(CO)_{12}^{2-} \xrightarrow{2H^+}$$

$$H_2FeRu_3(CO)_{12}$$

$$(13.17)$$

3. By binuclear reductive elimination:[17]

$$HMn(CO)_5 + MeAuL \longrightarrow (CO)_5Mn–AuL + MeH \qquad (13.18)$$

We saw some other examples of this reaction in Section 6.5.

4. By addition of a coordinatively saturated cluster to an unsaturated one via a bridging group (Eq. 13.19). In this method, we rely on a bridging ligand, such as hydride, to link the coordinatively saturated species to an

unsaturated cluster. In the example shown,[18] MeCN is introduced by the use of the Me_3NO reagent, which oxidizes a CO to CO_2 (Section 8.1). Ready dissociation of the MeCN provides the unsaturation, which allows an Os–H bond to bind to give a "spike" structure with one metal bound to the cluster by a single bond. The last thermal step shows the high tendency for cluster to agglomerate in such a way as to produce the maximum number of M–M bonds. In this and some later high nuclearity systems, the CO groups have been omitted for clarity.

$$Os_5(CO)_{16} \qquad Os_5(CO)_{15}(MeCN) \qquad Os_6H_2(CO)_{19}$$

$$Os_6H_2(CO)_{18}$$

$$(13.19)$$

5. By addition of an M–C multiple bond to a metal (Eq. 13.20). This method was developed by Stone[19] on the basis of the isolobal analogy. Because the M=C double bond is isolobal with the C=C double bond, those metals

$$(13.20)$$

that form alkene complexes might also be expected to form complexes with metal carbenes. This reaction is a very rich source of clusters.

6. By addition of an M—M multiple bond to a metal (Eq. 13.21). Green and co-workers[20] have taken the isolobal analogy one step further by invoking an analogy between the M=M multiple bond and an alkene. Both of these methods are likely to be very powerful.

$$(13.21)$$

7. By the use of bridging ligands. The common diphosphine $Ph_2PCH_2PPh_2$ has a high tendency to bring two metals close together, rather than chelate to a single metal (Eq. 13.22).[21a] This is presumably the result of geometric factors associated with the different ring sizes in the two cases. A large number of related ligands, such as $CN(CH_2)_3NC$ can behave similarly.[21b]

$$(13.22)$$

8. By using main-group elements to bring about cluster formation or expansion:

$$CpMn(CO)_2(thf) + PbCl_2 \longrightarrow Cp(CO)_2Mn=Pb=Mn(CO)_2Cp$$

$$(13.23)^{22}$$

13.4 REACTIONS

Clusters give a rich reactivity pattern with the usual organometallic ligands, often involving bridging of the ligand to several metals. Unfortunately, it is still a difficult area in which to try to rationalize or to predict.

With Electrophiles

Perhaps the simplest reaction of a cluster is the addition of a zero-electron electrophilic reagent such as H^+ because this should take place without any change in the cluster geometry. Anionic clusters are especially easy to protonate, and the resulting hydrides tend to be bridging (Eq. 13.24). Note that a μ without a subscript means that the ligand is bridging to two metals (i.e., $\mu = \mu_2$); bridging to three metals is shown as μ_3.

$$(13.24)$$

Electrophiles more bulky than the proton often add to the carbonyl oxygen, as we saw in Eq. 4.15. The same is true for clusters; for example, $Ru_3(CO)_{12}$ is converted from the normal CO-unbridged structure to a bridged $Ru_3(\mu\text{-}COAlR_3)_2(CO)_{10}$ structure with $AlBr_3$ (Eq. 13.25).[23] This structure resembles that of $Fe_3(CO)_{12}$, which is really $Fe_3(\mu\text{-}CO)_2(CO)_{10}$. On rare occasions, the proton may also add to a CO oxygen, as in the protonation product of $Fe_3(CO)_{11}{}^{2-}$, which is $(\mu\text{-}H)Fe_3(\mu\text{-}COH)(CO)_{10}$.[24] Carbon electrophiles may also add to a sufficiently nucleophilic vertex atom such as a sulfur, such as in $Os_3(CO)_9(\mu\text{-}H)_2(\mu_3\text{-}S)^-$ (Eq. 13.26),[25] which shows that the sulfur has a lone pair not involved in cluster bonding, and therefore this S should be considered as contributing only four of its six valence electrons to the cluster.

$$(13.25)$$

$$(13.26)$$

With Nucleophiles

The addition of nucleophiles adds 2e to the cluster, and so it must either rearrange or lose a 2e ligand. Equation 13.27 shows an interesting example of the reversible conversion of the trigonal bipyramidal $Os_5(CO)_{16}$ to the "bow tie" cluster $Os_5(CO)_{19}$ with CO,[26] and Eq. 13.28 shows rearrangement of the dicapped tetrahedral $Os_6(CO)_{18}$ to the raft cluster $Os_6(CO)_{17}L_4$ with $P(OMe)_3 (=L)$.[27] In each case the addition of CO or of L, which adds 2e to the cluster, causes breakage of an Os−Os bond, which "absorbs" the two electrons.

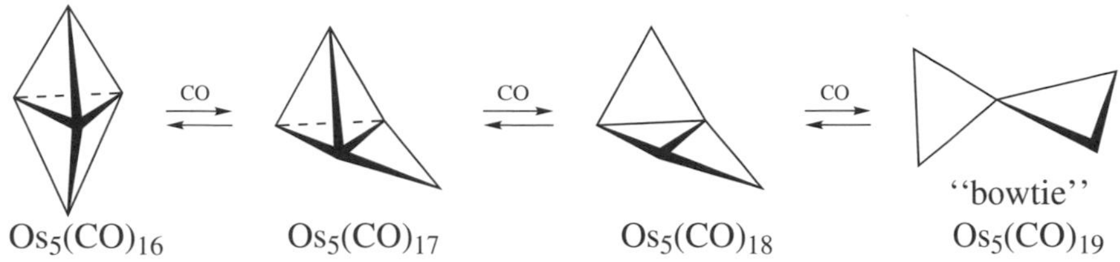

$$Os_5(CO)_{16} \quad\rightleftharpoons\quad Os_5(CO)_{17} \quad\rightleftharpoons\quad Os_5(CO)_{18} \quad\rightleftharpoons\quad \text{"bowtie"}\ Os_5(CO)_{19}$$

(13.27)

"raft" $Os_6(CO)_{18}L_3$ $Os_6(CO)_{18}L_2$

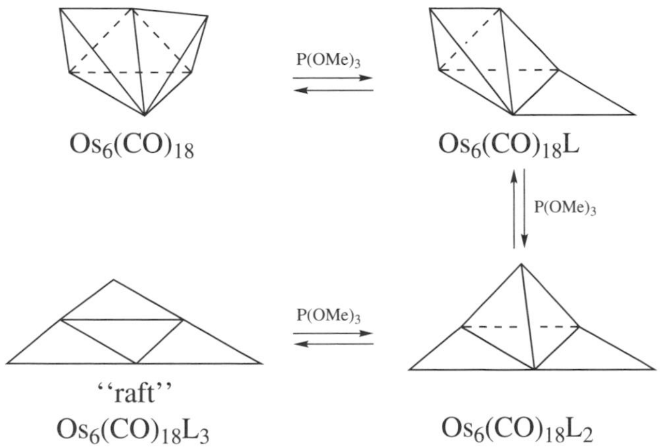

(13.28)

An "unsaturated" cluster, such as $(\mu_2\text{-}H)_2Os_3(CO)_{10}$ does not have to lose a ligand on addition of a nucleophile because one of the M−H−M bridges can open up and generate a vacant site. This is why the triosmium dihydride is such a popular starting material in cluster studies. For example, CO adds to give a product, $(\mu\text{-}H)HOs_3(CO)_{11}$, in which one of the two M−H−M bridges has opened and the hydride has become terminal. This turns the Os=Os "double bond" into an Os−Os single bond and means that the cluster is still an EAN one. This reaction can lead to substitution if a CO is expelled, as shown in Eq. 13.29.[28] Cluster breakdown into smaller fragments is also a possible outcome of substitution. The less stable cluster $Ru_3(CO)_{12}$ gives not only $Ru_3(CO)_9L_3$ but also $Ru(CO)_3L_2$ and $Ru(CO)_4L$ as substitution products with PPh_3. For osmium, mononuclear products are observed only under forcing conditions.

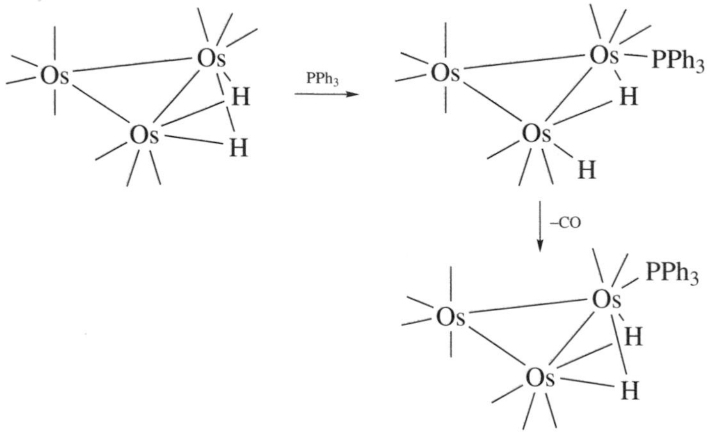

(13.29)

Nucleophiles may also attack the ligands. The use of Me_3NO to liberate CO from clusters has already been mentioned. Mayr and Kaesz[29] have shown that when amines attack a CO in $Os_3(CO)_{12}$, the metala-amide that is formed can labilize other COs in the molecule by bridging:

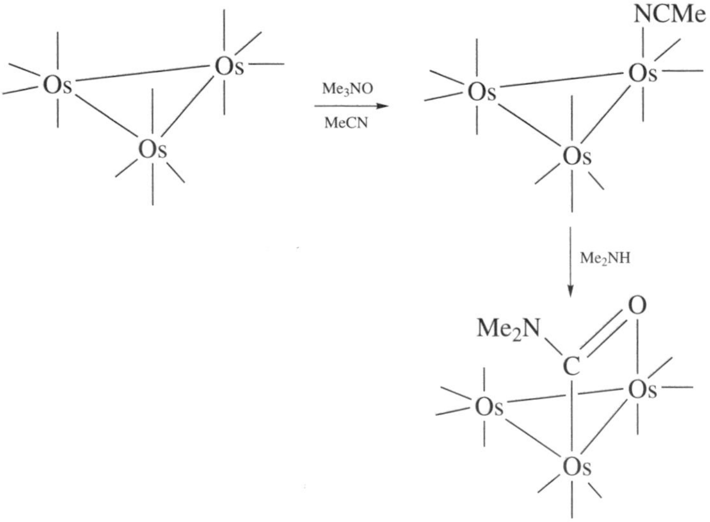

(13.30)

Oxidative Addition

As this reaction adds 2e to the cluster, subsequent loss of CO is required if the structure is not to change. The addition of H_2 to $Os_3(CO)_{12}$ probably takes place by loss of CO. The initial product is believed to be $(\mu\text{-H})HOs_3(CO)_{11}$, which then loses another CO to go to the final product, $(\mu\text{-H})_2Os_3(CO)_{10}$. As in the case of oxidative addition to mononuclear metal centers, there are many different mechanisms at work in oxidative addition. For example, Cl_2 addition does not require prior CO dissociation. The Cl_2 directly oxidizes the cluster by taking two electrons from a metal–metal bond (Eq. 13.31). This leads to a linear cluster in which only two M–M bonds are left. Pyrolysis of this complex leads to a chloro-bridged cluster $(\mu_2\text{-}Cl)_2Os_3(CO)_{10}$.[30] This is not unsaturated like $(\mu_2\text{-}H)_2Os_3(CO)_{10}$ because Cl is a 3e, not a 1e, donor, and so the cluster has 50e. By the EAN rule, we only require two M–M bonds; this means that the Os atoms bridged by the chlorides are not also metal–metal bonded.

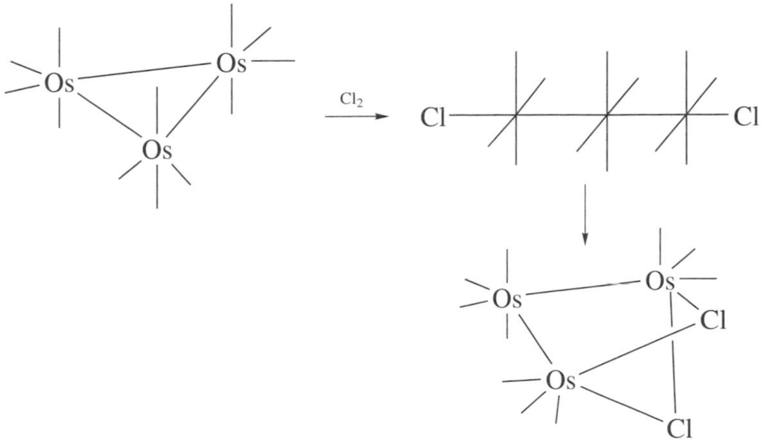

(13.31)

One striking difference between clusters and mononuclear systems is the difference in selectivity for C–H oxidative addition in ligands. For example, a mononuclear species will activate the allylic C–H bond of a coordinated alkene to give an allyl hydride; a cluster, in contrast, breaks the vinyl C–H bond. An alkyl in a mononuclear system gives β elimination of hydride, an alkyl in a cluster usually gives α elimination (Eq. 13.32).[31] In each case, the bond broken by the cluster is one atom closer to the point of attachment of the ligand to the metal than in the mononuclear case. This is probably because, in the cluster, the C–H bond is broken not by the metal to which the ligand is bound, but by the adjacent metal. This is shown in Fig. 13.6.

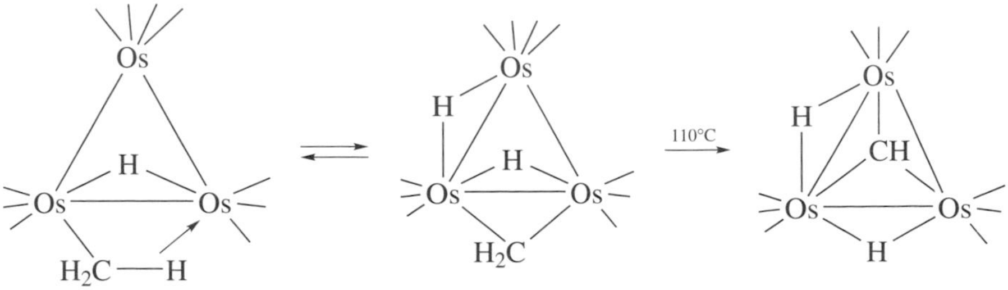

(13.32)

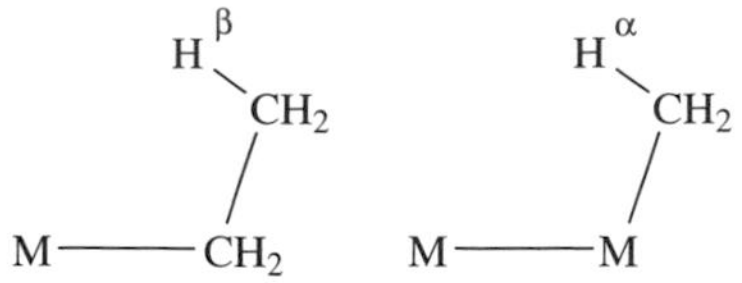

FIGURE 13.6 Geometric analogy between a β-CH in a mononuclear complex and an α-CH in a cluster.

Ethylene can undergo two successive C–H bond scissions at the same carbon (Eq. 13.33):[32]

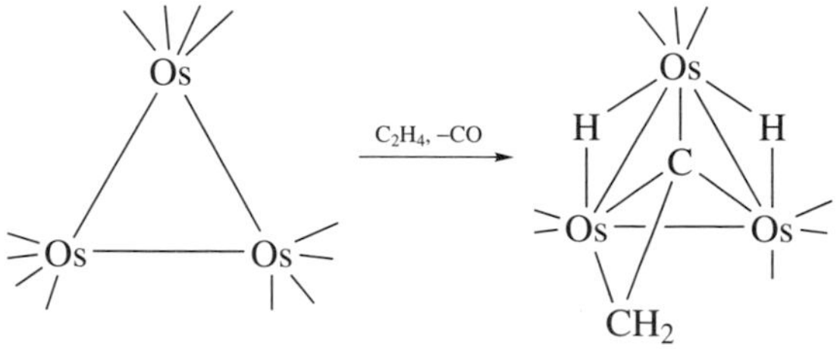

$$(13.33)$$

C–H bond breaking in γ and δ positions is also possible if dictated by the structure of the ligand (Eq. 13.34).[33] Further bond scissions can also occur (Eq. 13.35):[34]

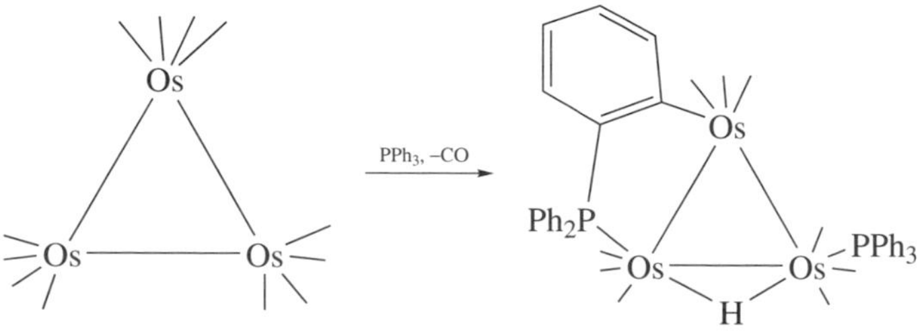

$$(13.34)$$

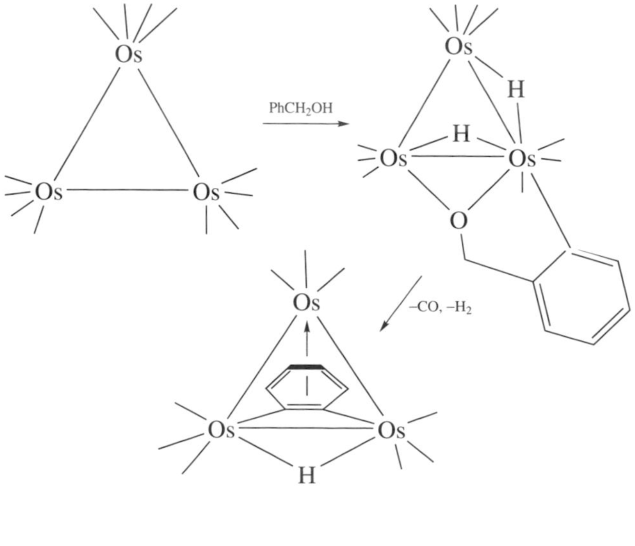

$$(13.35)$$

Another interesting bond cleavage reaction is the scission of the C$-$C triple bond in alkynes. We have already seen how metal$-$metal triple bonds can do this to give metal carbyne complexes (Section 11.2, Eq. 11.41). This reaction is not unusual in clusters and can be encouraged by using an alkyne that forms a specially stabilized carbyne. $Et_2NC{\equiv}CNEt_2$ has even been used as a source of the Et_2NC fragment in a reaction that generates a cluster from a mononuclear cobalt complex (Eq. 13.36):[35]

$$CpCo(CO)_2 \xrightarrow{\;Et_2NC{\equiv}CNEt_2\;} \text{[cluster structure]} \tag{13.36}$$

The cluster structure shows a bipyramidal framework with NEt$_2$ groups at top and bottom carbons, and CpCo, CpCo (CoCp), CoCp vertices.

Reactions Involving CO

One of the objects of cluster carbonyl chemistry has been to find ways of reducing CO and incorporating it into organic compounds. As we saw when we looked at CO activation in Section 12.3, the heterogeneously catalyzed Fischer$-$Tropsch reaction is an interesting route from CO to long-chain alkanes and alcohols. This is believed to go by scission of the CO on metallic iron to give a surface-bound oxo group and a surface-bound carbide. Hydrogenation of these surface species then leads to H_2O and CH_2, which is believed to polymerize to give the long chains observed. Interestingly, carbide clusters, like $Fe_6(\mu_6\text{-}C)(CO)_{16}{}^{2-}$, can be made by reduction of metal carbonyls.[36] These carbide clusters were known for many years, but the reactivity of the carbide could not be studied because it was buried in the cluster. Later work (Fig. 13.7) has shown how the cluster can be opened up to give an Fe$_4$ "butterfly" by controlled oxidation. In spite of its name, the "carbide" reacts more like a carbonium ion. This carbon binds a CO, polarizing it so that the solvent methanol can attack to give the ester derivative **13.26**, hydrogenation of which gives methyl acetate.[36a] Related work (Fig. 13.7) has shown how a μ_3-CO can be dissected to a carbide with loss of water. Note the interesting tetrahedral to butterfly rearrangement on protonation. Structure **13.27** is unusual in that it is a carbyne ligand with an agostic C$-$H bond, the longest such bond yet discovered; further protonation leads to CH_4. Similar reductive transformations of other unsaturated groups, such as isonitriles and NO, are also known.[36b] Heterobimetallic systems such as $Cp_2(Me)Zr-Ru(CO)_2Cp$

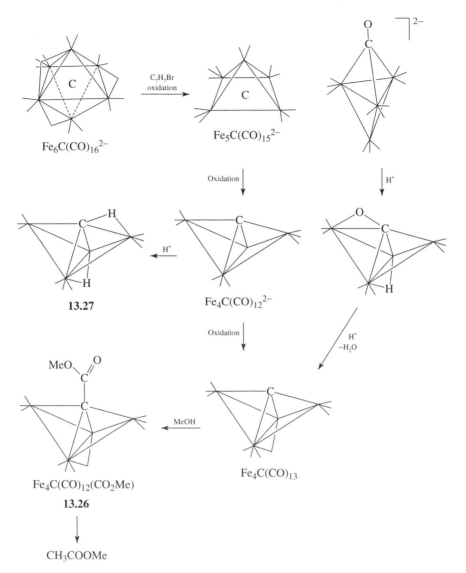

FIGURE 13.7 Some interesting chemistry of carbide clusters.

have been prepared by Casey in the course of attempts to design clusters to reduce CO.[36c]

Catalytic activity is sometimes seen in metal clusters, but it is sometimes difficult to tell if this arises from cluster breakdown to mononuclear fragments or whether the active catalyst is polynuclear. Some examples[37] are $[HRu_3(CO)_{11}]^-$ and $[Rh_2\{(Et_2PC_2H_4)_2PCH_2\}_2]$, which are both active for hydroformylation (Section 9.3) and in which the active catalysts are believed to be trinuclear and dinuclear, respectively.

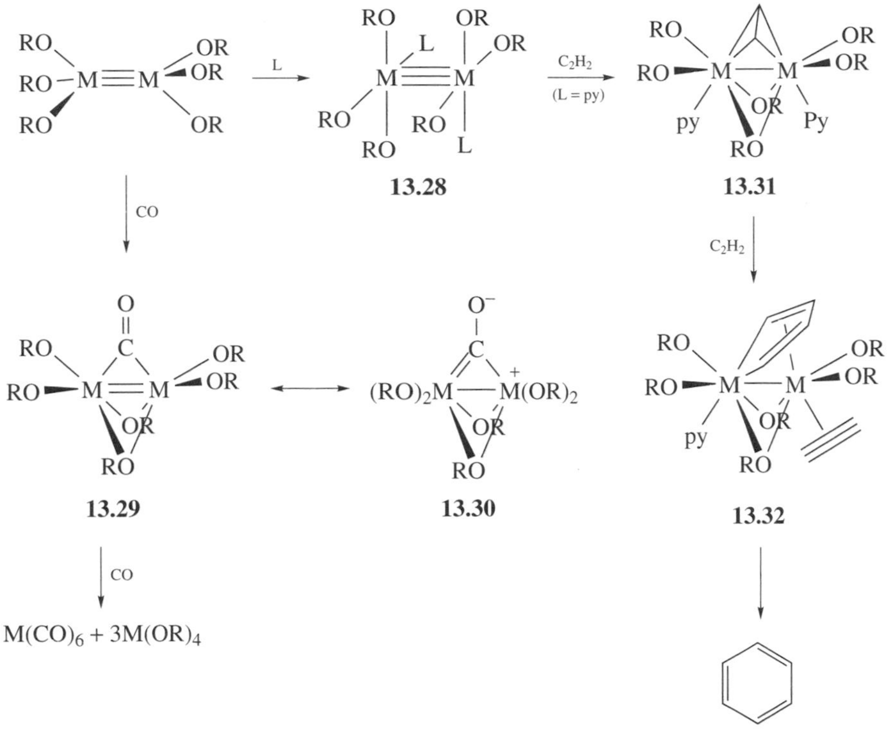

FIGURE 13.8 Some reactions of $M_2(OR)_6$. L = pyridine or PMe_3; M = Mo or W.

M–M Multiple Bonds

Chisholm[5] has studied the reaction of M–M multiple bonds as shown in Fig. 13.8. In forming **13.28**, it is not an M–M bond that is lost as would be the case for an Os carbonyl cluster. Instead two RO-to-metal π-bonding interactions are lost (the lone pairs on O are 2e donors), and this allows the 2e of the two incoming nucleophiles to be accommodated. The carbonyl complex **13.29** is interesting because the $\nu(CO)$ frequency is very low ($\sim$1600 cm^{-1}), and this is attributed to contributions from resonance forms of type **13.30**. The system is an alkyne cyclotrimerization catalyst, probably via the sequence **13.31** $\rightarrow$ **13.32**.

13.5 GIANT CLUSTERS AND NANOPARTICLES

Small metal particles, called colloids or nanoparticles, have found uses for many years—medieval red stained glass contains colloidal gold, for example. Very striking advances have been made in recent years in their controlled synthesis, better characterization, and in the identification of new commercial applications. This has contributed to nanotechnology—the applications of material objects in the 100 Å to 1000 Å range.

Traditional aqueous metal colloids were formed by reducing a metal salt in the presence of protective polymer such as polyvinyl alcohol (PVA), which absorbs on the surface. More recently large ligand-stabilized particles have been prepared that are intermediate between clusters and nanoparticles.

Chaudret has shown that the metal–organic precursor $Fe[N(SiMe_3)_2]_2$ can be reduced by H_2 in the presence of $n\text{-}C_{16}H_{33}\,NH_2$ to give iron nanoparticles of very similar shape and size—cubes of 7 Å edge length. They even "crystallize" into a cubic superlattice, as indicated by the electron micrographs in Fig. 13.9. For example, Moiseev and co-workers[40] have used dipyridyl to protect a Pd colloid formed from H_2 and $Pd(OAc)_2$ and have synthesized a *giant cluster* that is believed to have an isocosahedral close-packed structure of *approximate* formulation "$[Pd_{561}(phen)_{60}](OAc)_{180}$." Electron microscopy (Fig. 13.10) shows that the 25 Å particle size distribution is very narrow and X-ray absorption spectroscopy shows the Pd–Pd distances are very close to those in metallic Pd but that the packing is probably icosahedral. The crystallites are catalytically active for O_2 or peroxide oxidation of ethylene, propylene, and toluene to vinyl acetate, allyl acetate, or benzyl acetate.

None of the nanoclusters described in this section should be considered as having a precisely defined stoichiometry. The number of metals cited usually comes from the idealized formulations shown in Fig 13.11, which show hexagonal close-packed structures with one to five shells of atoms.[40c]

Gold colloids are stabilized with $P(m\text{-}C_6H_4SO_3Na)_3$ to the extent of making them isolable as red solids.[41] When two different metals are reduced, alloy or "onion" structures can be formed. In the latter case a colloid of one metal is used as the seed particles for growing a second metal: Au encapsulated by Pt is an example. Lewis co-workers[42] have evidence that the active catalyst in Speier hydrosilation (Section 9.5) of $RCH{=}CH_2/R_3SiH$ to $RCH_2CH_2SiR_3$ with $H_2PtCl_6/i\text{-}PrOH$ as catalyst is a Pt colloid. The surface may be capped with SiR_3 groups that act as protectant, the role taken by dipyridyl or PVA in the systems mentioned above. A 35 Å Pd colloid stabilized by a polymeric hydrosilane has

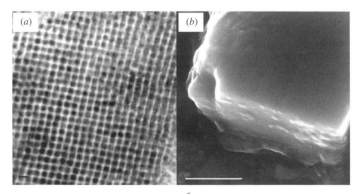

FIGURE 13.9 (*a*) The crystallization of 7 Å nanoparticulate iron cubes in a cubic lattice. (*b*) The morphology of the resulting crystallite. [Illustrations kindly provided by B. Chaudret.]

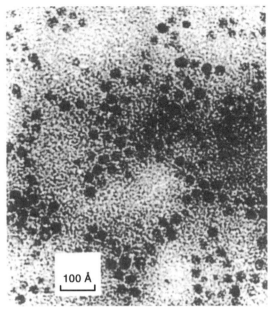

FIGURE 13.10   Electron micrograph of Moiseev's giant palladium clusters on a carbon support. (Reproduced from Ref. 40a with permission of the Royal Society of Chemistry © 1985.)

Full-Shell "Magic Number" Clusters				
Number of shells	1	2	3	4	5
Number of atoms in cluster	M_{13}	M_{55}	M_{147}	M_{309}	M_{561}
Percentage surface atoms	92%	76%	63%	52%	45%

FIGURE 13.11   Idealized nanoclusters of close-packed atoms with one to five shells of atoms, together with the numbers of atoms (magic numbers) in these clusters. (Reproduced with permission from Ref. 40c.)

substantially different selectivity than either Pd/C or homogeneous Pd catalysts in hydrogenation and hydrogenolysis reactions.[43]

Since metal complexes can sometimes decompose to give catalytically active colloidal preparations that maintain the appearance of a normal solution, there is always the possibility that the true catalyst in such a case is the metal surface and that the system is not a true homogeneous catalyst.

Giant clusters can be obtained as pure compounds. The largest clusters that can still be crystallized for X-ray studies and are found to be of a defined nuclearity

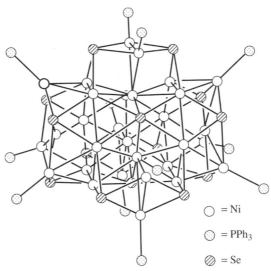

FIGURE 13.12 Molecular structure of $Ni_{34}Se_{22}(PPh_3)_{10}$. (Reproduced from Ref. 39 with permission.)

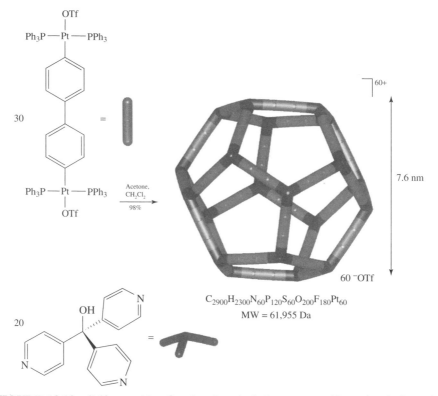

FIGURE 13.13 Self-assembly of a giant icosahedral organometallic molecule from the 30 rod and 20 vertex units.

are now in the M_{100} range. Examples are the close-packed $Pd_{69}(CO)_{36}(PEt_3)_{18}$ and $Pd_{145}(CO)_x(PEt_3)_{30}$. $Ni_{34}Se_{22}(PPh_3)_{10}$ is interesting in that the core is a particle of nickel selenium alloy, not of metallic nickel (Fig. 13.12).

Unusual physical properties are sometimes seen for these particles. For example, "$Pt_{309}(phen)_{36}O_{30}$" shows two ^{195}Pt NMR resonances that are assigned to surface and bulk Pt. The latter show the so-called Knight shift, which is a shift in the resonance position as a result of metallic character.[44] $[Au_{55}(PPh_3)_{12}Cl_6]$ has been used in microelectronic devices.[45]

> • Nanoparticles, often highly reactive, have interesting potential for the future.

13.6 GIANT MOLECULES

Self-assembly of simple units into complex structures is a key feature of biochemistry. Multiple copies of one or more proteins can assemble with a strand

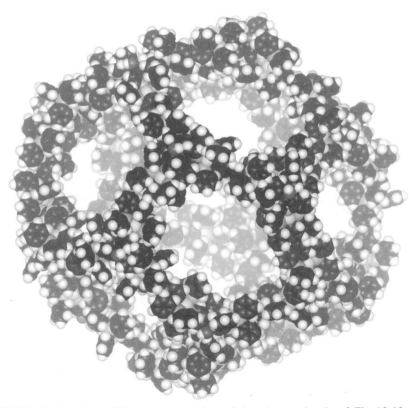

FIGURE 13.14 Space-filling representation of the giant molecule of Fig 13.13. (Art kindly provided by Prof. Stang.)

of RNA or DNA to form rod-shaped or globular viruses, for example. A similar principle has been used by Stang and co-workers[46a] to synthesize giant molecules of definite shape and composition. These are large enough to be easily visualized by transmission electron microscopy (TEM).

For example, Fig. 13.13 shows how 30 diplatinum biphenyl units (rods) can react with 20 molecules of a tripyridine (corners) to give an icosohedron of diameter 76 Å. A space-filling model of the resulting giant molecule is shown in Fig 13.14. A TEM picture of the molecule is shown in Fig. 13.15. A wide variety of other shapes are or should be accessible in this way—for example, molecular rods, squares, and cubes.[46b]

Several important questions remain unanswered in cluster chemistry. Can clusters be synthesized with other high-field ligands than CO, and will they have reactivity patterns different from those of the carbonyl clusters we have been looking at in this chapter? In particular, can a wider range of catalytically active clusters be prepared, by choosing more labile ligands than CO? Can

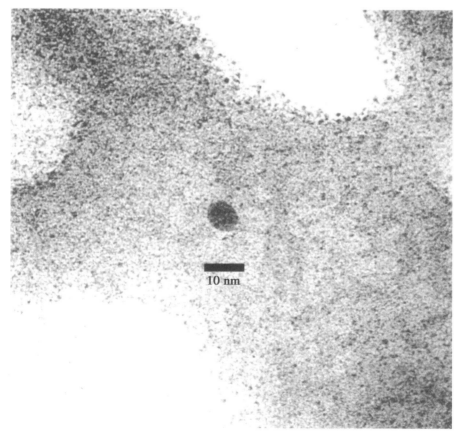

FIGURE 13.15 Electron micrograph of the giant molecule. (Art provided by Prof. Stang.)

cluster fragmentation be controlled, perhaps by using ligands that keep the cluster together in some way? A related question concerns mechanism: How do we know whether a given stoichiometric or catalytic reaction is a reaction of the intact multimetal cluster unit or of dissociated, even mononuclear intermediates that subsequently re-form a cluster once again?

REFERENCES

1. D. M. P. Mingos, *Pure Appl. Chem.* **63**, 807, 1991; J. B. Kiester, in *Encyclopedia of Inorganic Chemistry*, R. B. King, ed., Wiley, New York, 1994 p. 3348.

2. J. M. Basset, F. Lefebvre, and C. Santini, *Coord. Chem. Rev.* **180**, 1703, 1998.

3. M. Bowker, *Basis and Applications of Heterogeneous Catalysis*, OUP, New York, 1998.

4. (a) F. A. Cotton and R. A. Walton, *Multiple Bonds between Metal Atoms*, OUP, New York, 1993; (b) T. P. Fehlner ed., *Inorganometallic Chemistry*, Plenum, New York, 1992.

5. M. H. Chisholm, *Acc. Chem. Res.* **33**, 53, 2000.

6. J. R. Pugh and T. J. Meyer, *J. Am. Chem. Soc.* **114**, 3784, 1992.

7. M. L. H. Green, N. J. Cooper et al., *J. Chem. Soc., Dalton* 29, 1980.

8. K. Wade, *Adv. Inorg. Organometal. Chem.* **18**, 1, 1976.

9. (a) G. Huttner, *Angew. Chem., Int. Ed.* **32**, 297, 1993; (b) C. E. Housecroft and T. P. Fehlner, *Adv. Organometal. Chem.* **21**, 57, 1982.

10. R. D. Adams and L. W. Yang, *J. Am. Chem. Soc.* **105**, 235, 1983.

11. Y. Tatsumi, T. Naga, H. Nakashima, T. Murahashi, and H. Kurosawa, *Chem. Comm.*, 1430, 2004.

12. R. Hoffmann, *Angew. Chem., Int. Ed.* **21**, 711, 1982; L. C. Song, D. S. Guo, and Y. B. Dong, *Trans. Met. Chem.* **25**, 37, 2000.

13. F. Calderazzo, R. Ercoli, and G. Natta, in *Organic Syntheses via Metal Carbonyls*, I. Wender and P. Pino, eds., Wiley-Interscience, New York, 1968, Chapter 1.

14. B. F. G. Johnson and J. Lewis, *Adv. Inorg. Chem. Radiochem.* **24**, 225, 1981.

15. J. E. Ellis, *J. Organomet. Chem.* **86**, 1, 1975.

16. G. L. Geoffroy, *Acc. Chem. Res.* **13**, 469, 1980.

17. F. G. A. Stone and C. M. Mitchell, *J. Chem. Soc., Dalton* 102, 1972.

18. B. F. G. Johnson et al., *Chem. Commun.* 507, 1986.

19. F. G. A. Stone, *Pure Appl. Chem.* **58**, 529, 1986.

20. M. Green, F. G. A. Stone et al., *J. Chem. Soc., Dalton* 869, 1981.

21. (a) M. P. Brown, R. J. Puddephatt, M. Rashidi, and K. R. Seddon, *J. Chem. Soc., Dalton* 951, 1977; (b) D. M. Hoffman and R. Hoffmann, *Inorg. Chem.* **20**, 3543, 1983; K. R. Mann, N. S. Lewis, H. B. Gray et al., *Inorg. Chem.* **17**, 828, 1978; A. S. Balch, *Inorg. Chem.* **98**, 8049, 1978.

22. W. A. Herrmann, *Angew. Chem., Int. Ed.* **24**, 1062, 1985.

23. D. F. Shriver et al., *Inorg. Chem.* **13**, 499, 1974.

24. D. F. Shriver et al., *J. Am. Chem. Soc.* **100**, 5239, 1978.

25. P. R. Raithby et al., *Chem. Commun.* 551, 1978.

26. D. H. Farrar, B. G. F. Johnson, J. Lewis, P. R. Raithby, and M. J. Rosales, *J. Chem. Soc., Dalton* 2051, 1982.

27. R. J. Goudsmit, B. F. G. Johnson, J. Lewis, P. R. Raithby, and K. H. Whitmire, *Chem. Commun.* 640, 1982.

28. A. J. Deeming and S. Hasso, *J. Organometal. Chem.* **114**, 313, 1976.

29. A. Mayr and H. D. Kaesz, *J. Organomet. Chem.* **272**, 207, 1984.

30. A. J. Deeming, B. F. G. Johnson, and J. Lewis, *J. Chem. Soc. A* 897, 1970.

31. J. R. Shapley et al., *J. Am. Chem. Soc.* **99**, 5225, 1977.

32. A. J. Deeming and M. Underhill, *J. Chem. Soc., Dalton* 1415, 1974.

33. C. W. Bradford and R. S. Nyholm, *J. Chem. Soc., Dalton* 529, 1973.

34. A. J. Deeming et al., *J. Chem. Soc., Dalton* 1201, 1978.

35. R. B. King and C. A. Harmon, *Inorg. Chem.* **15**, 879, 1976.

36. (a) J. S. Bradley, *Adv. Organometal. Chem.* **22**, 1, 1983; (b) R. D. Adams and I. T. Horvath, *Prog. Inorg. Chem.* **33**, 127, 1985; (c) C. P. Casey, *J. Organometal. Chem.* **400**, 205, 1990.

37. G. Süss-Fink, *Adv. Chem. Ser.* **230**, 419, 1992; S. A. Laneman and G. G. Stanley, *Adv. Chem. Ser.* 349, 1992.

38. J. D. Aitkin and R. G. Finke, *J. Mol. Catal.* **145**, 1, 1999.

39. N. T. Tran and L. F. Dahl, *Angew. Chem. Int. Ed.* **42**, 3533, 2003.

40. (a) I. Moiseev et al., *Chem. Commun.* 937, 1985; (b) I. Moiseev, K. I. Zamaraev et al., *J. Mol. Catal.* **53**, 315, 1989.

41. C. Larpent and H. Patin, *J. Mol. Catal.* **44**, 191, 1988.

42. L. N. Lewis, N. Lewis, and R. J. Uriarte, *Adv. Chem. Ser.* **230**, 541, 1992.

43. L. Fowley, D. Michos, and R. H. Crabtree, *Tetrahedron Lett.* **34**, 3075, 1993.

44. M. A. Marcus et al., *Phys. Rev.* **B42**, 3312, 1990.

45. A. Schmid et al., *Angew. Chem. Int. Ed.* **32**, 250, 1993.

46. (a) P. J. Stang et al., *Nature* **398**, 796, 1999; *J. Am. Chem. Soc.* **121**, 10434, 1999; (b) P. J. Stang et al., *Acc. Chem. Res.* **35**, 972, 2002.

PROBLEMS

1. Given the existence of cyclopropenone, suggest two cluster complexes that are isolobal with this species, and how you might try to synthesize them.

2. Give the cluster electron counts (see Fig. 13.1) of the following: $Cp_3Co_3(\mu_3\text{-CS})(\mu_3\text{-S})$; $Fe_3(CO)_9(\mu_3\text{-S})_2$; $Fe_3(CO)_{10}(\mu_3\text{-S})_2$. In deciding how to count the S atoms, take account of the fact that these seem to have one lone pair not engaged in cluster bonding, as shown by their chemical reactivity in methylation with Me_3O^+, for example.

3. For the species listed in problem 2, how many M–M bonds would you expect for each? Draw the final structures you would predict for these species.

4. $Co_4(CO)_{10}(EtC≡CEt)$ has structure **A** shown below. What is the cluster electron count? Does it correctly predict the number of M—M bonds? How would you describe the structure on a Wade's rule approach?

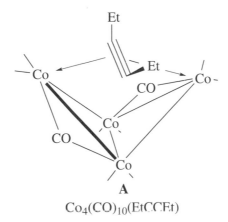

A

$Co_4(CO)_{10}(EtCCEt)$

5. What light do the isolobal ideas throw on structures **B** and **C** (below)?

$Cp_4Fe_4(\mu_3-CO)_4$

B

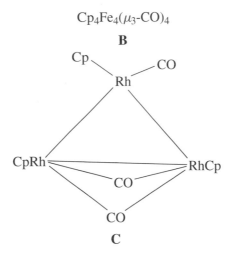

C

6. What structures would you predict for $Fe_4CO_{13}^{2-}$, $Ni_5CO_{12}^{2-}$, and $Cr_2(CO)_{10}(Ph_2PCH_2PPh_2)$?

7. Pt(0) forms an RC≡CR complex $Pt(C_2R_2)_n$. Predict the value of n based on an isolobal relationship with structure **D** (below). Why are the two W—C vectors orthogonal in **D**?

$\{Cp(CO)_2WCR\}_2Pt$

D

8. Predict the structure of **E** (below), making it as symmetric as possible. With what organoiron complex is **E** isolobal?

$$Fe(CO)_3\{B_4H_4\}$$
E

9. Why do boron and transition metal hydrides tend to form clusters, when carbon and sulfur hydrides tend to form open-chain hydrides $Me(CH_2)_nMe$, and $HS(S)_nSH$? Why is sulfur able to form clusters in the compounds mentioned in problem 2?

10. $Os_3(CO)_{10}(\mu_2\text{-}CH_2)(\mu_2CO)$ reacts with CO to give structure **F** (below), which reacts with H_2O to form acetic acid, Suggest a structure for **F**.

$$Os_3(CO)_{12}(CH_2CO)$$
F

14

APPLICATIONS TO ORGANIC SYNTHESIS

It is hard to find an organic synthesis today that does *not* use organometallic reagents, The Grignard reagent dates from the early twentieth century, and by the 1970s, main-group species such as MeMgI and BuLi became workhorse reagents in synthesis. Today, transition metal catalyzed reactions such as cross-coupling (Pd) and hydrogenation (Rh, Ir, Ru) are standard in both organic and medicinal chemistry. Alkene metathesis (Ru, Mo) is also being used more and more for ring closure and cross-coupling. Oxidation is also a key type of catalytic reaction (Ti, Os, Ru, Mn), but coordination compounds are involved, not organometallics. Many of these reactions can be made asymmetric with the use of suitable homochiral ligands. Both transition and main-group elements are involved, and so we take the opportunity to look at some main-group chemistry here.

We saw in Chapter 9 how organometallic chemistry has responded to the challenge of synthesizing organic compounds on an industrial scale. Such commodity chemicals as ethylene or acetic acid are not expensive and so practical syntheses must use catalytic, rather than stoichiometric, amounts of organometallic compounds. The organic compounds we look at now are synthesized on a smaller scale.[1a,b] These fine chemicals are usually additives, plasticizers, drugs, or other high-value items. Here, stoichiometric quantities of one of the cheaper metal reagents, and in some cases even of the precious-metal reagents, can be used. Nevertheless, with the continuing rise in environmental concerns and green chemistry, pressure has grown to maximize the ratio of product to waste (Q factor) in industrial processes.[2] This has, in turn, led to increasing interest in catalytic reactions, where the metal catalyst is present in minimal quantity, and the

The Organometallic Chemistry of the Transition Metals, Fourth Edition, by Robert H. Crabtree
Copyright © 2005 John Wiley & Sons, Inc.

selectivity of the reaction is normally much enhanced, so waste by-products are minimized. The pharmaceutical industry is also under pressure to produce chiral drugs in their biologically active enantiomeric forms rather than as racemic mixtures as in the past. This has enhanced the importance of asymmetric catalysis, a procedure that can produce essentially only the desired enantiomer in favorable cases.

- Organometallic methods, both main group and transition metal, have risen to dominance in organic synthesis.
- Catalytic chemistry is increasingly being adopted over stoichiometric chemistry, where possible.

14.1 METAL ALKYLS ARYLS, AND HYDRIDES

Metal alkyls tend to be polarized M^+-R^-, especially for electropositive metals, and so the R group often acts as a nucleophile. By changing metal, we alter the polarization of the bond as we alter the electronegativity of M. LiR is very reactive, as Li is electropositive, but the R group cannot contain halo, keto, or carboxymethyl functionality or RLi will decompose by reacting with itself. For the electronegative Hg, in contrast, R can vary widely and still form a stable species RHgX, but the reactivity of RHgX is much lower than that of RLi. A different reactivity–stability compromise and therefore a different metal may be needed for different applications.

Alkyls QR_n of elements (Q) to the left of carbon are electron poor in the sense that they have fewer electron pairs than orbitals, and the octet is not achieved (LiMe, 2e; RMgX, 4e; $AlMe_3$, 6e), and so they show a strong tendency to associate with themselves (e.g., Al_2Me_6 or Li_4Me_4) or with electron donors (e.g., $Me_2O: \rightarrow AlMe_3$). Self-association allows easy exchange of R groups between metals; for example, although $Me_2Al(\mu\text{-Me})_2AlMe_2$ has two types of methyl group, a single methyl resonance is seen in the proton NMR at room temperature. Elements to the right of carbon are electron rich, having more electron pairs than orbitals, so they form alkyls $:QR_n$ having one or more lone pairs (PMe_3, 1 lone pair, SMe_2, 2 lone pairs) and act as lone-pair donors (ligands). Elements of the carbon group form electron-precise alkyls QR_4 that lack both empty orbitals and lone pairs. This is the origin of the unreactivity of alkanes CR_4.

Lithium and Magnesium

The metal alkyls with the longest history of organic applications are the Grignard reagents, RMgX, and alkyllithiums, RLi.[3] These act as sources of R^- and are highly reactive carbon nucleophiles toward $R_2'CO$ and $RCOOR'$, for example. Alkyls of the more electropositive elements, such as Na [Pauling electronegativity (EN): 0.9], are less suitable because they are less stable. Li^+ and Mg^{2+}

(EN: Li, 1.0 and Mg, 1.3), as small and therefore highly polarizing ions, also tend to coordinate the substrate, such as a ketone, and polarize it so as to favor nucleophilic attack by the R group. RLi and RMgX are usually very air and moisture sensitive and are made and used under an inert gas.

Organolithium or organomagnesium[4] reagents are prepared from the metal and an alkyl halide or from an alkylmetal reagent and a compound with a labile X–H proton such as cyclopentadiene and RC≡CH (Eqs. 14.1 and 14.2). Specially activated "Rieke" magnesium is useful for less reactive halides such as vinyl halides and alkyl fluorides.[4b] In Grignard synthesis from Mg and RX, electron transfer to give RX·⁻ is thought to be followed by loss of X⁻ and recombination of R· with the surface, which then releases RMg⁺.[4c]

$$EtBr + Li \longrightarrow EtLi + LiBr \qquad (14.1)$$

$$EtMgBr + CpH \longrightarrow EtH + LiCp \qquad (14.2)$$

A very useful feature of the deprotonation route is that heteroatoms on the substrate can bind the organolithium reagent and direct the deprotonation to the ring C–H bond ortho to the heteroatom. For example, –OMe, –CONMe$_2$, –NMe$_2$, –SO$_2$Me, and even –F substituents on a benzene ring act in this way:

$$(14.3)$$

Organolithium reagents and aryl bromides and iodides readily undergo metal–halogen exchange by nucleophilic attack of R on the halide (Eq. 14.4). This very rapid reaction is often carried out at very low temperature ($-80°C$) where other processes do not compete.

$$(14.4)$$

Organolithium reagents are oligomers (i.e., dimers, trimers, and higher species) in nondonor solvents such as alkanes: LiMe is a tetramer with a cubane structure **14.1**, for example. {RLi}$_n$ forms solvates with THF. Addition of the chelating ligand Me$_2$NCH$_2$CH$_2$NMe$_2$ (TMEDA) leads to formation of a monomer, and this increases the reactivity. n-BuLi can deprotonate toluene to form PhCH$_2$Li only if TMEDA is present. The organomagnesium reagents are usually prepared from the alkyl or aryl halide and magnesium metal in ether or THF. The products are not usually isolated but used directly in the ethereal solvent. Their constitution has been a subject of debate for many years, but the Schlenk equilibrium

(Eq. 14.5) probably describes the situation well in most cases. The addition of dioxan complexes and precipitates the MgX_2 and leaves R_2Mg in solution.

$$2RMgX \rightleftharpoons R_2Mg + MgX_2 \qquad (14.5)$$

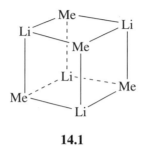

14.1

The following are some of the numerous classical reactions of Grignard reagents:[3,4]

$$RMgX + R'R''C=O \longrightarrow RR'R''C-OMgX \xrightarrow{\text{hydrolysis}} RR'R''C-OH \qquad (14.6)$$

$$(R' \text{ and } R'' = \text{aryl, alkyl, or H})$$

$$RMgX \xrightarrow[\text{(ii) hydrolysis}]{\text{(i) ethylene oxide,}} RCH_2CH_2OH \quad (14.7)$$

$$2RMgX + R'COOEt \longrightarrow R_2R'C-OMgX \xrightarrow{\text{hydrolysis}} R_2R'COH \qquad (14.8)$$

$$RMgX + CO_2 \longrightarrow RCO_2MgX \xrightarrow{\text{hydrolysis}} RCO_2H \qquad (14.9)$$

An alternative pathway via a single-electron transfer mechanism has also been invoked in some cases (Eq. 14.10).[4d] Chiral auxiliaries such as binaphthols can make Grignard and related reactions asymmetric:[4e]

$$RMgX + R_2CO \longrightarrow R\cdot + R_2CO\cdot^- + MgX^+ \longrightarrow R_3COMgX \qquad (14.10)$$

While organomagnesium reagents only very rarely add to C=C double bonds, the more reactive EtLi can add to dienes. The resulting allyllithium can continue adding to further diene molecules in the *anionic polymerization* reaction.

Boron and Aluminum

Organoboranes are of special importance because they are easily formed in borane addition to C=C bonds (*hydroboration*). The high electronegativity of B (2.0) means that the B−C bond is not very polar and BR_3 species are usually water, although not air stable but sufficiently reactive to be useful. In contrast to most other reagents, a B−H bond adds in an anti-Markownikov manner to an alkene to give the corresponding organoboron reagent (Eq. 14.11). This can be converted to a variety of useful organic compounds (e.g., alcohols, alkanes, and alkyl bromides; Eqs. 14.12–14.14) in a subsequent step. This hydroboration procedure has an important place in organic synthesis:[5a]

$$RCH{=}CH_2 + [BH_3]_2 \longrightarrow (RCH_2{-}CH_2)_3B \qquad (14.11)$$

$$(RCH_2{-}CH_2)_3B + H_2O_2 \longrightarrow RCH_2{-}CH_2OH \qquad (14.12)$$

$$(RCH_2{-}CH_2)_3B \xrightarrow{\text{acid}} RCH_2{-}CH_3 \qquad (14.13)$$

$$(RCH_2{-}CH_2)_3B \xrightarrow{\text{Br}_2,\ \text{base}} RCH_2{-}CH_2Br \qquad (14.14)$$

Organoaluminum reagents are important in Ziegler–Natta catalysts (Section 12.2) but are not widely used in organic synthesis. They can be violently pyrophoric and water sensitive and can add readily to alkenes. The *Aufbau* reaction (Eq. 14.15) is a commercial synthesis of C_{12}–C_{16} linear alcohols that are useful in detergents.

$$AlEt_3 + 3nC_2H_4 \longrightarrow Al\{(C_2H_4)_n Et\}_3 \xrightarrow{O_2} Al\{OC_2H_4(C_2H_4)_{n-1}Et\}_3$$

$$\longrightarrow HOC_2H_4(C_2H_4)_{n-1}Et \qquad (14.15)$$

Trimethylaluminum is a methyl-bridged dimer Al_2Me_6. In contrast to transition metals, the bridge contains Al$-$C$-$Al bonds only and is not agostic, presumably because the metal is incapable of back donation into the C$-$H σ^*. The small Al$-$C$-$Al angle at the bridging C suggests a direct Al$-$Al interaction similar to the M$-$M bonding present in M$-$H$-$M transition metal systems. NMR studies in solution show bridge–terminal alkyl exchange. In alkylaluminum hydrides, such as $[Me_2AlH]_3$, the hydrides prefer the bridge positions.

Up to now we have looked at metal alkyls from groups to the left of carbon (RLi, R_2Al, ...). These are electron deficient as monomers (RLi, 2e; R_3B, 6e) and are commonly found as dimers or polymers. As electron-precise (8e, all bonding) species, R_4Si are monomeric and do not coordinate extra ligands as avidly as the electron deficient alkyls.

Si, Ge, Sn, and Pb

Organosilicon reagents[5b] are of special importance in organic synthesis because they share some but not all of the properties of alkanes. The Si$-$C bond is strong and relatively nonpolar, and SiR_4, like CR_4, is electron precise so the reagents are stable and are not strong nucleophiles. Their usefulness is a result of a number of special properties: (1) the R_3Si-O (108 kcal/mol) and R_3Si-F (135 kcal/mol) bonds are unusually strong, (2) the "$SiMe_3{}^+$" group behaves like a proton that can be readily cleaved from carbon, and (3) Si stabilizes a carbonium ion in the β position. The first property is a result of the R_3Si group being an electron acceptor. This is clearly shown in the bond angles of silicon compounds such as $Me_3Si-O-SiMe_3$ (Si$-$O$-$Si = 148°). This is far larger than the sp^3 angle of Me$-$O$-$Me (109°) because there is partial O$-$Si double-bond character that in the extreme would lead to a linear molecule.

The acceptor orbital on the Me_3Si group is the Si$-$Me σ^*, just as we saw for PR_3 in Fig. 4.3. The third property is an interesting one and its origin is still debated, but one possibility is that the "carbonium ion" (**14.2**) has some of the character of an alkene complex of a Si cation (**14.3**). Equation 14.18 shows how electrophilic cleavage of a $SiMe_3$ (= TMS) group can occur with retention of configuration and so be used in the stereospecific cleavage of a vinylsilane by DCl.

The stabilization of a β-carbonium ion is also involved in the reaction of an allylsilane with an electrophile (Eq. 14.16). An advantage of silicon over other metals in this context is that it does not undergo 1,3 shifts, and so the point of attachment of the electrophile can be reliably predicted (Eq. 14.17).[5b] This β stabilization of the carbonium ion also has stereochemical implications; Eq. 14.18 shows how the stereochemistry of a vinylsilicon reagent can be retained on protonation. A TMS group on carbon has been described as a "superproton" in that it leaves easily, especially with fluoride ion as nucleophile (Eq. 14.19) consistent with the high Si$-$F bond strength.

$$(14.16)$$

$$(14.17)$$

$$(14.18)$$

$$PhCHCl-C(SiPh)=CH_2 \xrightarrow{KF} PhCH=C=CH_2 + Ph_3SiF + KCl \quad (14.19)$$

The fact that Si—O bonds are strong is used in stabilizing enol forms of various carbonyl compounds. Generally, a base such as i-Pr$_2$NLi (LDA) is used to deprotonate the carbonyl compound, and Me$_3$SiCl then gives the silyl enol ether, which can react with a wide variety of carbon electrophiles, such as aldehydes, ketones, 3° alkyl halides, and α,β-unsaturated ketones, for example:

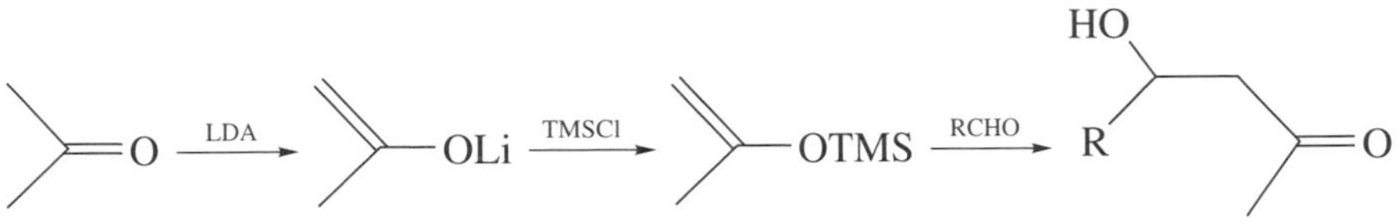

$$(14.20)$$

A synthesis of dihydrojasmone that uses some of these principles is shown in Eq. 14.21:[6]

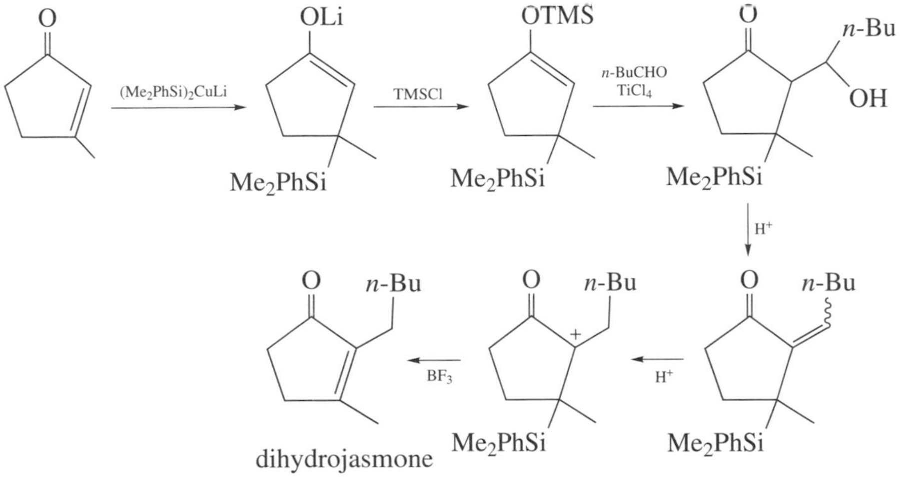

$$(14.21)$$

Cationic Si in R$_3$Si$^+$ is very unfavorable, so nucleophilic displacement of a group at Si never goes by an S$_N$1 route, but by attack at Si (S$_N$2). This can take place with or without inversion because the 5-coordinate intermediate is fluxional by the Berry process (Eq. 10.2).

The most important application[6b] of organostannanes is the initiation of radical reactions with n-Bu$_3$Sn—H, such as the replacement of halide (X) by H shown below:

$$R_3Sn-H \longrightarrow R_3Sn\cdot + H\cdot \quad \text{(initiation)} \quad (14.22)$$

$$R_3Sn\cdot + RX \longrightarrow R_3SnX + R\cdot \quad \text{(chain propagation)} \quad (14.23)$$

$$R\cdot + R_3Sn-H \longrightarrow R_3Sn\cdot + RH \quad \text{(chain propagation)} \quad (14.24)$$

Lead alkyls such as PbEt$_4$ with their weak Pb—C bonds (~36 kcal/mol) were used in gasoline to promote combustion by thermolytic release of Et• radicals, but environmental concerns have led to its abandonment in most places.

Zinc

Organozinc reagents can be prepared from RLi or RMgX and ZnCl$_2$ from RI and Zn. R$_2$Zn is monomeric, but bases readily associate, for example, R$_2$Zn(TMEDA). In the Reformatsky reaction (Eq. 14.25) the Zn—C bond is sufficiently unreactive to tolerate the ester group of the substrate but sufficiently reactive to nucleophilically attack the ketone.

$$\text{BrCH}_2\text{COOR} \xrightarrow{\text{(i) Zn (ii) Me}_2\text{CO, (iii) H}^+} \text{Me}_2\text{C(OH)CH}_2\text{COOR} \qquad (14.25)$$

The use of chiral ligands has allowed asymmetric carbonyl additions with zinc reagents.[7a]

In the Simmons–Smith reaction, the zinc forms a carbenoid reagent[7b] (Eq. 14.26), which acts as a carbene equivalent in the cyclopropanation of Eq. 14.27.

$$\text{CH}_2\text{I}_2 + \text{Zn(Cu)} \longrightarrow \text{IZnCH}_2\text{I} \qquad (14.26)$$

$$\qquad (14.27)$$

The advantage of zinc reagents, their greater tolerance of functional groups and greater configurational stability, is lost if they are prepared from RLi or RMgBr. Direct oxidative addition to metallic zinc in reactive Rieke form, aided in some cases by ultrasound or transmetallation from a hydroboration product can give a wide variety of organozinc reagents. The lower electronegativity of Zn versus Li or Mn means that reactivity of the resulting organozinc is also lower. Knochel[7c] has reviewed the use of a variety of transition metal catalysts to speed up organozinc reactions.

Mercury

Organomercury reagents[7d] are of interest because the Hg—C bond is relatively nonpolar, so that the compounds are much less reactive than the group 1 and 2 alkyls but more closely resemble organosilicon compounds. Most organomercurials are stable to water and even to acids. This means that a much wider variety of organic functionality can be incorporated into the R group than is the case for the organolithiums or magnesiums. Environmental concerns have limited their use, however.

$$\text{PhCHBrCO}_2\text{R} \xrightarrow{\text{Hg, } h\upsilon} \text{PhCH(HgBr)CO}_2\text{R} \qquad (14.28)$$

The direct mercuration of arenes by electrophilic attack with $Hg(OAc)_2$ or $HgCl_2$ is perhaps the most useful synthetic route.

$$C_6H_6 \xrightarrow{\text{(i) Hg(OAc)}_2, \text{ (ii) NaCl}} PhHgCl \tag{14.29}$$

Another useful preparative procedure is transalkylation from the corresponding organoborane, which can be prepared from the alkene (see also below).

$$RCH=CH_2 \xrightarrow{B(OR)_2H} RCH_2CH_2B(OR)_2 \xrightarrow{HgCl_2} RCH_2CH_2HgCl \tag{14.30}$$

Oxymercuration of alkenes probably involves formation of a transient cationic alkene complex, which undergoes nucleophilic attack by solvent (Eq. 14.31) and gives the Markownikov product and so complements Eq. 14.30.

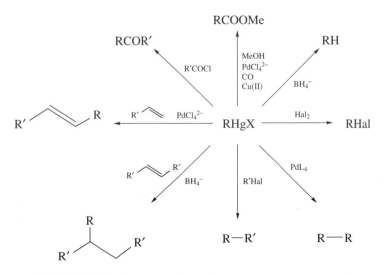

$$\tag{14.31}$$

Organomercurials give the reactions shown in Fig. 14.1. Halogenation is useful not only to prepare organic halides but also to determine the position of attachment of the mercury atom in the original compound. Mercury-bound R groups are easily transferred to Pd, and if they resist β elimination, can be used in a variety of transformations shown in Fig. 4.1.

FIGURE 14.1 Some reactions of organomercury compounds.

The relatively low reactivity of the RHgX reagent is shown by acylation with RCOCl; this gives a ketone that is stable to further attack by the organometallic species. The organomercury reagents with their very weak Hg–C bonds are a useful source of radicals. Once the first R• is released, the remaining R is very weakly bound (Eq. 14.31), and so both radicals are effectively released at the same time.

$$HgMe_2 \xrightarrow{\Delta H \ 51 \ kcal/mol} Me\bullet + \bullet HgMe \xrightarrow{\Delta H \ 7 \ kcal/mol} 2Me\bullet + Hg \qquad (14.32)$$

Mercury also gives the synthetically useful transformation of R–H to R_2 in the photochemical Mercat reaction[7e]; the weakest C–H bond in the molecule is selectively cleaved. The product shown in Eq. 14.33 is a useful ligand but very difficult to make by conventional routes.

$$Me_2CHNH_2 \xrightarrow{Hg, \ NH_3, \ h\nu} H_2N\text{-}CMe_2CMe_2\text{-}NH_2 \qquad (14.33)$$

Copper

Although copper is a transition metal, it is sufficiently far to the right in the periodic table so that it begins to show main-group characteristics, especially in the d^{10} Cu(I) state. Organocuprates $Li[CuR_2]$,[8] prepared by reaction of the organolithium compound with a Cu(I) salt such as CuI, do not β-eliminate and are sufficiently nucleophilic, thanks to the net anionic charge, to attack a usefully wide variety of organic electrophiles. The structures of these reagents is still a matter of discussion, but oligomeric forms are likely to be present. As shown in Eq. 14.34, the reagents suffer from the disadvantage that only one of the two R groups is transferred to the electrophile, E. The electrophile may be an alkyl iodide, or even a vinyl halide (Eq. 14.35), for which most nucleophiles are ineffective; perhaps the extra activating effect of the copper reagent comes from the coordination by the metal of the halide (page 119) or of the C=C group of the vinyl halide.

$$2LiR + CuI \xrightarrow{-LiI} Li[CuR_2] \xrightarrow{E} R\text{-}E + CuR + LiI \qquad (14.34)$$

$$trans\text{-}PhCH\text{=}CHBr \xrightarrow{LiCuMe_2} trans\text{-}PhCH\text{=}CHMe \qquad (14.35)$$

One of the most important applications of organocuprates is their addition to α, β-unsaturated carbonyl compounds, in which exclusive 1,4 addition is observed.

$$R'CH\text{=}CH\text{-}COOEt \xrightarrow{LiCuMe_2} R'CHMe\text{-}CH\text{=}C(OLi)OEt$$

$$\xrightarrow{hydrolysis} R'CHMe\text{-}CH_2COOEt \qquad (14.36)^9$$

With alkynes, insertion is observed to give a vinyl cuprate, which can then be quenched with an electrophile (Eq. 14.37).[10]

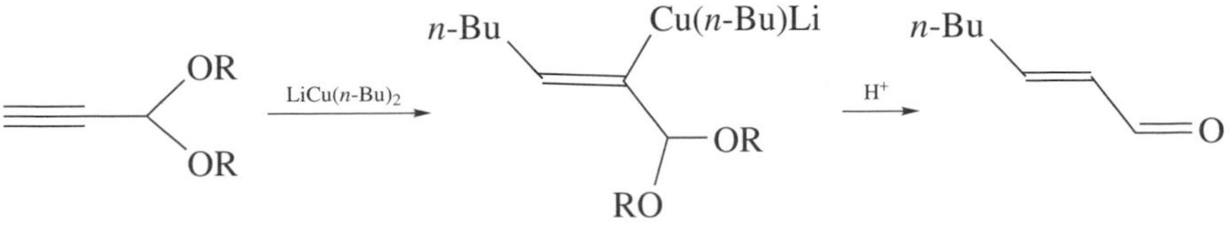

$$(14.37)$$

Ce

Cerium(III) reagents,[11] formed from RLi and $CeCl_3$ at $-80°$, have the advantage of reacting more cleanly with organic carbonyl groups than does RLi. Ce being less electropositive, the cerium reagents have less tendency to abstract protons α to the C=O or to add to adjacent C=C bonds in α, β-unsaturated ketones (Eq. 14.38). The reagents may be of the form $Li_n[CeR_{(3+n)}]$.

$$RCH{=}CH\ COR' \mid MeLi/CeCl_3 = RCH{=}CH\ CMe(OH)R' \qquad (14.38)$$

Other Metals

In hydrozirconation with Schwartz's[12a] reagent, Cp_2ZrHCl, addition to alkenes leads to the anti-Markovnikov alkyl (Eq. 14.39). Remarkably, 1-, 2-, and 3-hexene all give the same n-hexyl product. The reason must be that the initially formed alkyls rapidly β-eliminate. This moves the C=C bond along the chain in an alkene isomerization reaction (Section 9.1), until the n-hexyl complex is formed. This must therefore be the thermodynamically most stable alkyl. In general, primary alkyls tend to be most stable, probably because they are the least bulky isomer. In hydroboration, in contrast, no isomerization is observed.

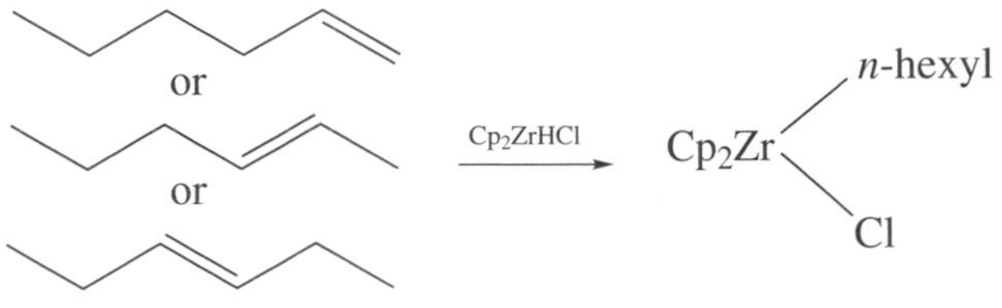

$$(14.39)$$

Addition to an alkyne takes place stereospecifically to the cis vinyl complex:

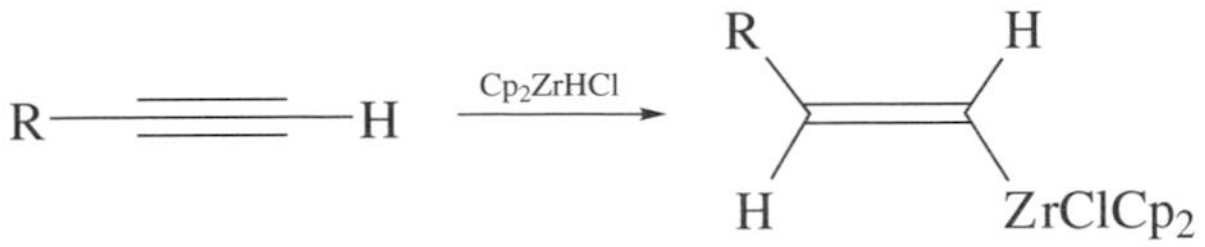

$$(14.40)$$

Transfer of the alkyl formed by hydrozirconation to another metal, such as Cu, can be useful; in the reaction shown below, the cyclohexylzirconium species, formed by hydrozirconation, is trapped with an acyl halide.

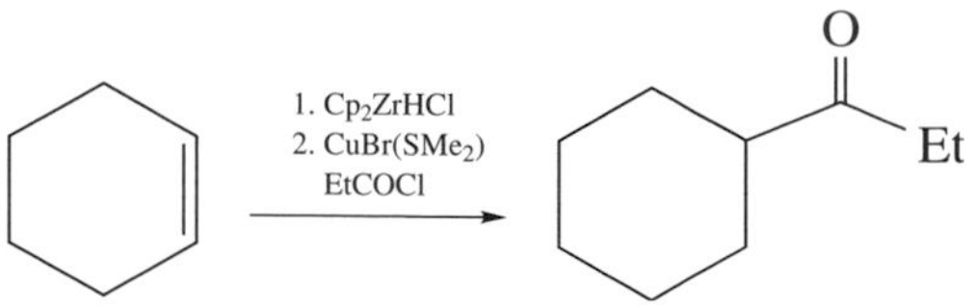

Oxidative coupling of dienes to "Cp$_2$Zr," formed from Cp$_2$ZrCl$_2$ and BuLi, gives a metallacycle that can be further functionalized, as by carbonylation in the tecomanine synthesis of Eq. 14.41.[12b]

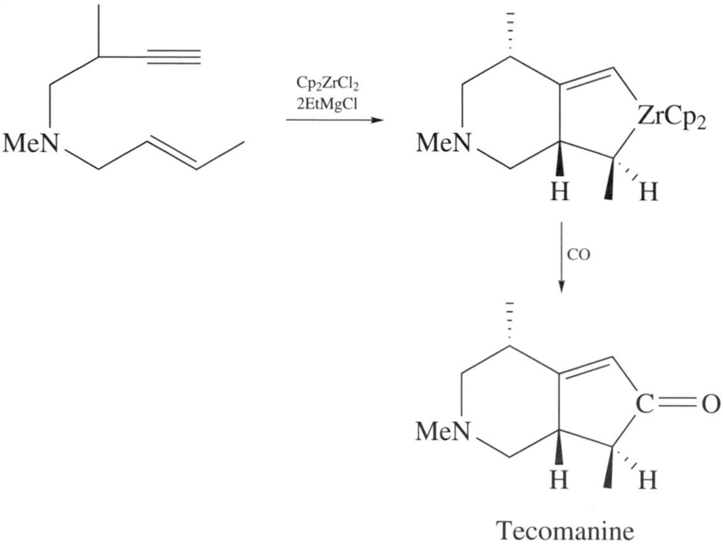

(14.41)

Tecomanine

Cyclometalation has been applied to synthesis in the *Murai reaction*. The substrate has to contain a group, a ketone in Eq. 14.42, that binds the metal complex. Cyclometalation gives an intermediate aryl that undergoes insertion with a vinyl silane followed by reductive elimination to give the product; the result is a C–H addition across a C=C bond.[12c]

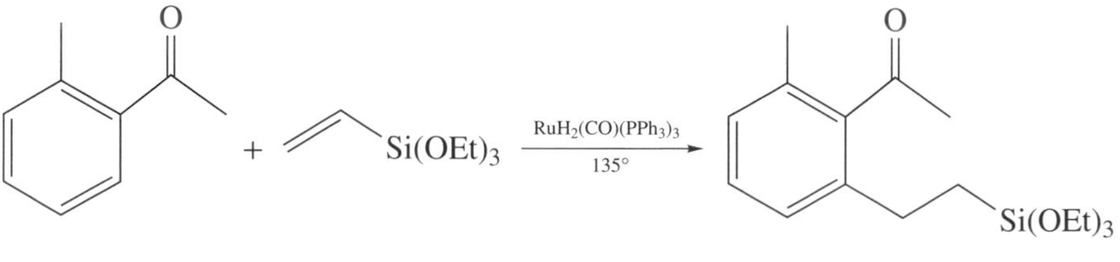

(14.42)

In the *Nozaki–Hiyama–Kishi reaction* (Fig. 14.2), an organic halide undergoes two-center oxidative addition to Cr(II), followed by addition of the resulting Cr(III) alkyl to an aldehyde. The advantages are a tolerance for a variety

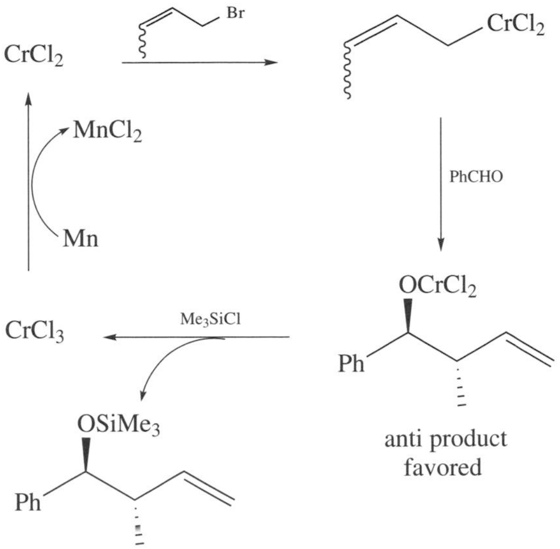

FIGURE 14.2   Nozaki–Hiyama–Kishi reaction in which a Cr(III) alkyl intermediate adds to a carbonyl compound.

of functional groups in the substrates and the high anti diastereoselectivity of the addition.[12d]

14.2   REDUCTION, OXIDATION, AND CONTROL OF STEREOCHEMISTRY

Organometallic compounds tend to be reducing in character and so tend to be applied in reduction. High-valent coordination compounds tend to be used in oxidation. Even in oxidation the intermediacy of species with M−C bonds has been proposed, which makes it difficult to maintain the somewhat artificial distinction between organometallic and coordination compounds in this area.

It is in the area of oxidation and reduction that directed and asymmetric reactions have been particularly successful. Organic synthesis is vitally concerned with the stereochemical outcome of a given reaction. A typical synthetic target (e.g., **14.4**; the cyclopentenone ring of prostaglandin A) will have more than one asymmetric (or stereogenic) center; these are starred in **14.4**. In a racemic synthesis, still common, the racemate of the target is formed, in this case **14.4** and **14.5** in a 50 : 50 mixture. The stereocenters have the right relative configuration but the compound is not a single enantiomer as in the natural product itself. In such a synthesis we will need reactions that selectively create new asymmetric centers with a defined stereochemistry with respect to preexisting centers. Increasing emphasis has been placed on asymmetric syntheses in which both the relative and absolute configurations of the target molecule are reproduced. If the target is a drug, then we prefer to synthesize the active enantiomer only. In this way, we avoid giving the patient the inactive enantiomer along with the active drug.

$$(14.43)$$

14.4 **14.5**

Directed and Asymmetric Oxidation

The traditional method of asymmetric synthesis involves modifying the substrate with a resolved chiral auxiliary and finding a reagent that introduces an asymmetric center in a defined way relative to the auxiliary. The auxiliary is then removed, ideally leaving a single enantiomer of the product. This method requires a mole of auxiliary per mole of product formed. A more sophisticated approach is to mimic Nature's own solution: the use of an enantiomerically pure catalyst. In this case the handedness of the product is decided by the handedness of the catalyst, and only a small amount of resolved catalyst produces a large amount of asymmetric product.

OsO_4 is the best reagent for the *cis*-dihydroxylation of alkenes (Eq. 14.44).[13] Of great practical importance, use of a chiral amine as L with an unsymmetric alkene (RCH=CHR') can lead to high asymmetric induction in the product diol. One enantiomer predominates as measured by the enantiomeric excess (e.e.) of the reaction. The percent e.e. is defined on page 246. It is most convenient to carry out the reaction with catalytic quantities of osmium and excess N-methylmorpholine-N-oxide to reoxidize the Os back to Os(VIII).[13a] Free OsO_4 reacts with 0% e.e., so we need a system in which reaction via $[L^*OsO_4]$ is preferred. This is the case here because the chiral amine strongly promotes the oxidation rate.

$$(14.44)$$

The Sharpless[13b] epoxidation provides good examples of both directed and asymmetric catalytic reactions. It has long been known that alkenes can be epoxidized with peracids, which deliver an electrophilic oxygen atom, as shown in Eq. 14.45. Sharpless showed that alkyl hydroperoxides in the presence of high-valent metal catalysts, such as VO(acac)$_2$, can also epoxidize alkenes. Equation 14.46 shows a suggested mechanism for the Sharpless reaction; comparison with Eq. 14.45 shows the mechanistic analogy between the two processes: just as RCOOH is a good leaving group in the first case, departure of ROH and an M=O group delivers the electrophilic oxygen in the second. The oxophilicity of the early metals used as catalysts clearly plays a role in stabilizing the M=O group.

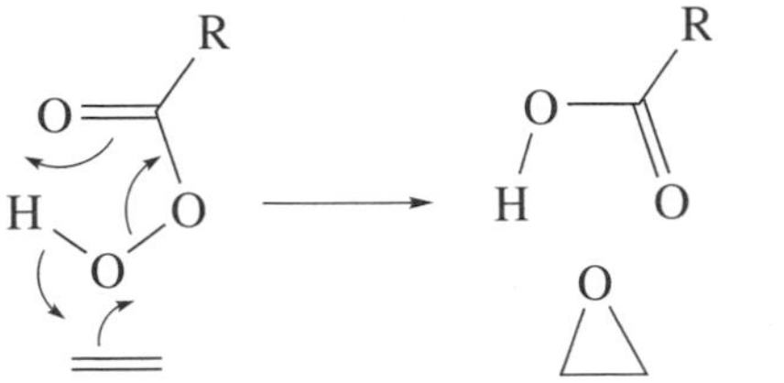

(14.45)

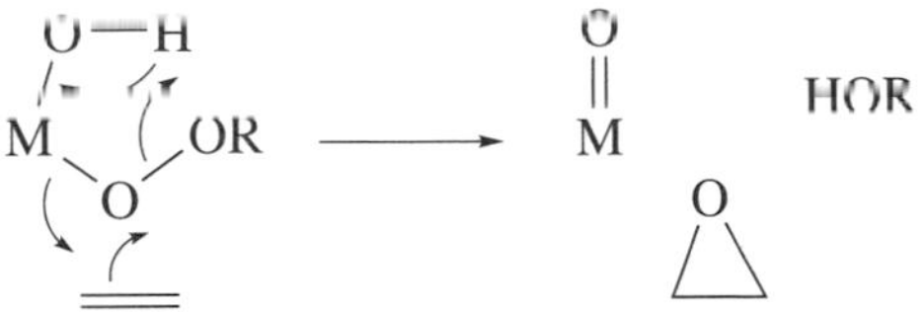

(14.46)

Normally, the most basic, and therefore the most highly alkyl-substituted alkene reacts first, but the vanadium catalyst shows strong directing effects that allow the catalyst to overcome the usual selectivity order if an allylic or homoallylic −OH group is present (e.g., Eq. 14.47).[13c] In cyclic compounds the stereochemistry of the final epoxide is determined by the directing effect of the −OH group to which the catalyst binds (Eq. 14.48). Peracids tend to give the other isomer of the product, by a simple steric effect.

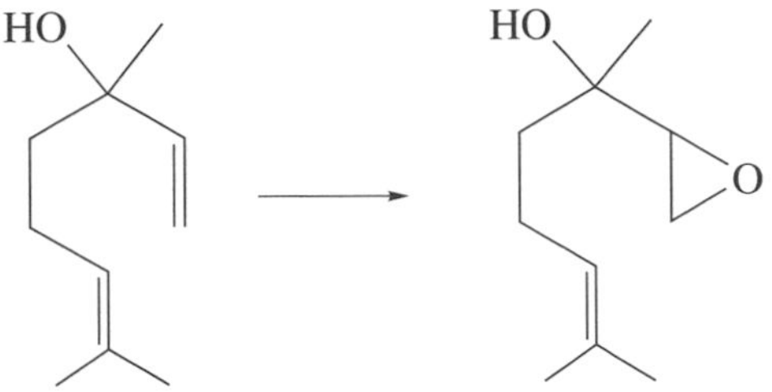

(14.47)

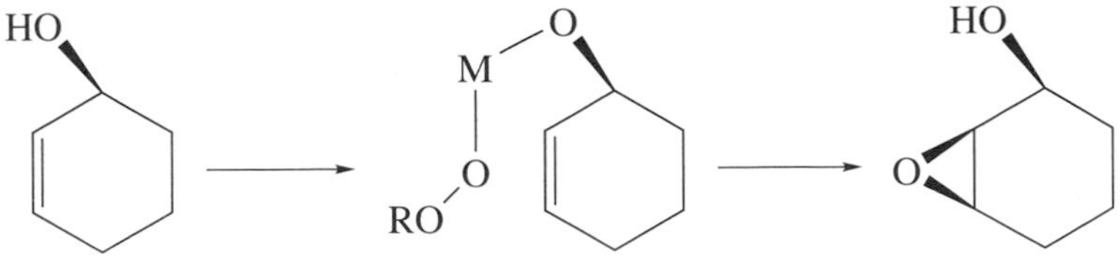

(14.48)

One of the most useful applications of the chemistry of transition metals in organic synthesis is the Sharpless asymmetric epoxidation.[14] By using one or other enantiomer of diethyl tartrate (DET) as a ligand, Ti(IV) as the catalyst, and *t*-BuOOH as the oxidant, allylic alcohols can be epoxidized to give chiral epoxy alcohols of *predictable* stereochemistry. The product stereochemistry observed for each enantiomer of DET used as ligand is shown in Eq. 14.49. This means that the stereochemistry of the reaction is imposed by the reagent ("reagent control"), rather than the much more common situation in which it is a result of the substrate structure and conformation ("substrate control"). The attractive features of the system are the simplicity of the reagents used and the synthetic versatility of the epoxy alcohols obtained.

(14.49)

Jacobsen and co-workers[15a] have found a system using **14.6** that catalyzes asymmetric epoxidation of alkenes with ArIO as oxidant and does not require that the substrate contain a hydroxy group. For example, Z−PhCH=CHMe is converted to the epoxide with an 84% e.e.

The ability of the catalyst to form asymmetric epoxides led Jacobsen to ask whether the same chiral salen ligands that discriminate between the enantiotopic faces of an approaching olefin also create an effective dissymmetric environment for nucleophilic attack at a bound epoxide. Indeed, Jacobsen et al.[15b] were also able to use very similar salen catalysts for the asymmetric ring opening of epoxides by nucleophilic attack on an epoxide activated by binding to a chiral, Lewis acidic Cr(III) metal salen complex.

Directed and Asymmetric Reduction

The principles of directed and asymmetric reactions were first developed for hydrogenation, as discussed in Section 9.2. Asymmetric hydrosilation of ketones can now be carried out catalytically with rhodium complexes of diop (**9.22**). The widely used chiral ligand Et-duPHOS, made by Burk[16] at du Pont, allows chiral amination of ketones via Eq. 14.50. Note how the use of the hydrazone generates an amide carbonyl to act as a ligand, as is known to favor high e.e. (see Section 9.2). Noyori's[17] powerful BINAP ligand has been applied to a very large number of asymmetric reactions.

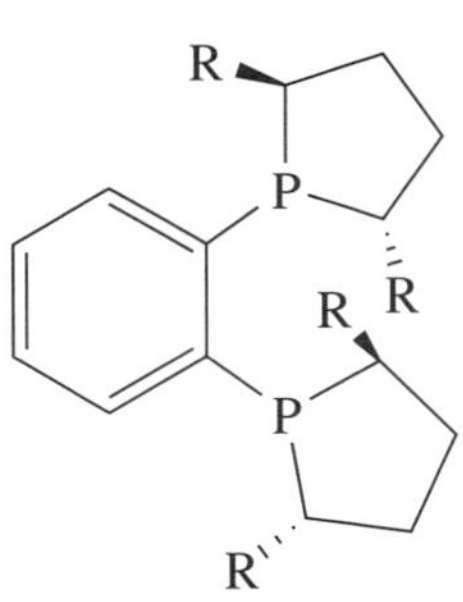

duPHOS

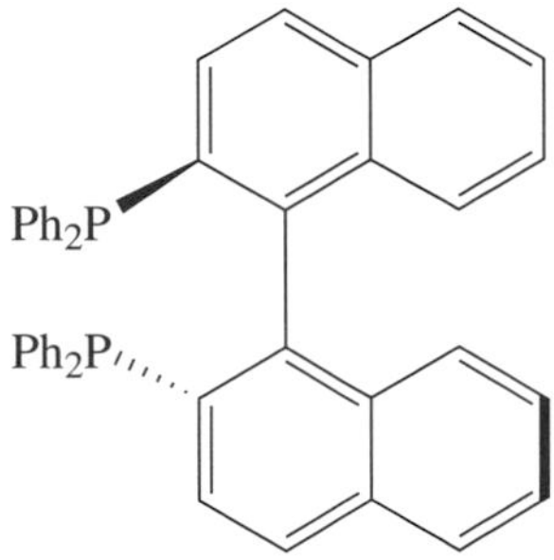

BINAP

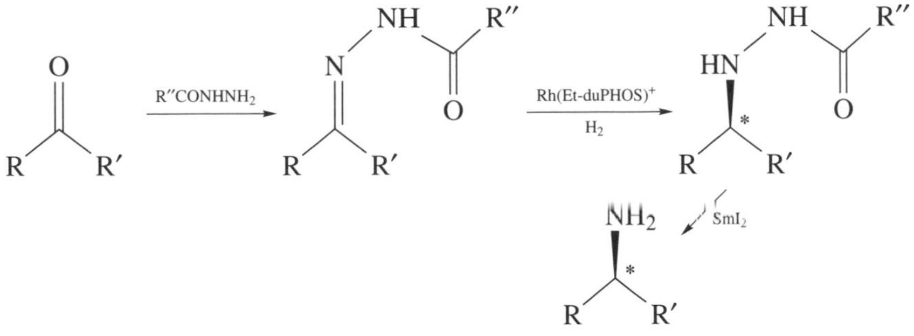

72–97% e.e.

$$(14.50)$$

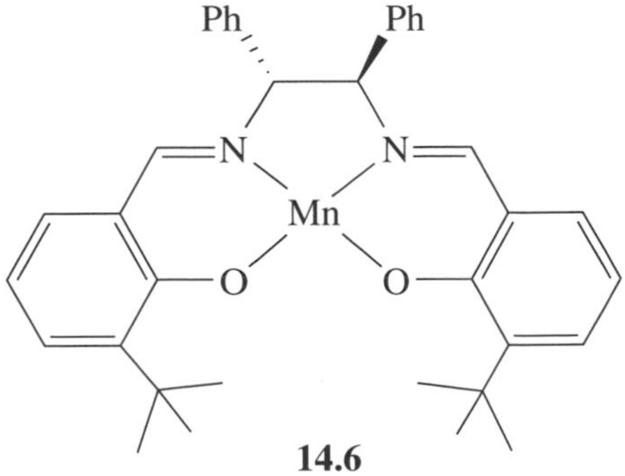

14.6

For example, in the reaction shown in Fig. 14.3, the catalyst reacts 100 times faster with one of the two equilibrating enantiomers of the starting material, leading to a kinetic resolution with a 99 : 1 ratio of threo : erythro product. In addition, the hydrogenation is asymmetric with an e.e. of 98%.[17]

Even though borane addition to alkenes happens without a catalyst, the catalytic version is important because it has usefully different chemo-, regio-, and

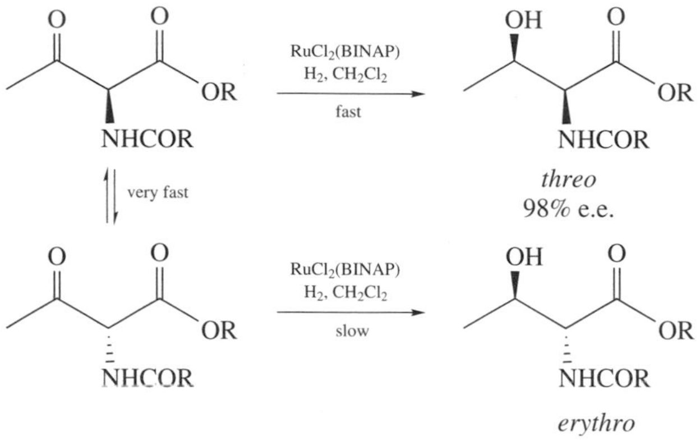

FIGURE 14.3 Asymmetric hydrogenation by Noyori et al.[17] that includes a kinetic resolution step. The threo : erythro ratio is 99 : 1.

stereoselectivities (Section 9.5).[18a] Enantiomeric excesses as high as 96% can be obtained with a Rh({R}-binap)$^+$ catalyst in the conversion of norbornene to *exo*-norborneol,[18b] and additions to allylic alcohols, which give a 10 : 90 ratio of syn : anti product in the absence of a catalyst switch to a 96 : 4 ratio with Wilkinson's catalyst.[18c]

The samarium(II) reagent, SmI_2, formed from Sm and ICH_2CH_2I, is a powerful 1e reductant that allows a variety of reductions (Eq. 14.51).[18d]

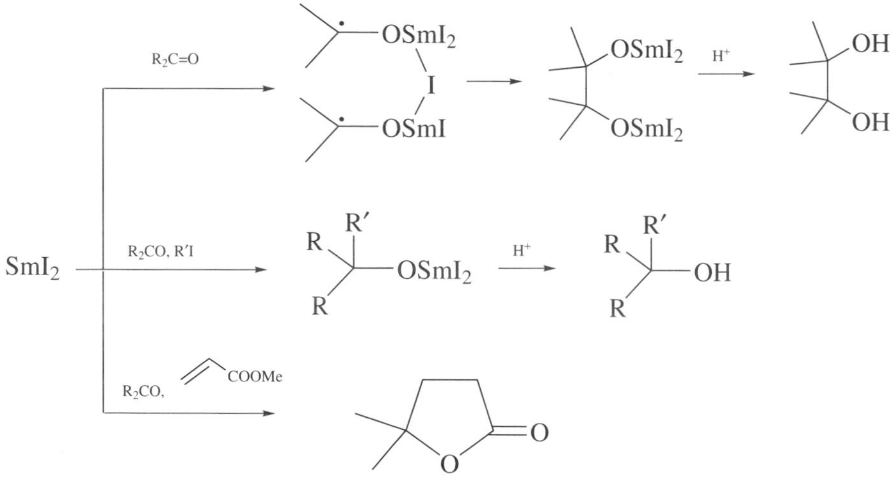

$$(14.51)$$

- Decrease of the electronegativity of the metal, M, usefully raises the selectivity and lowers the reactivity of alkyl M—R.

14.3 PROTECTION AND DEPROTECTION

One role that a metal reagent plays is simply to act as a protecting group. Conventional protection works best for heteroatom functionalities, but alkene alkyne, diene, and arene groups are perhaps best protected by organometallic reagents.

Cyclopentadienyliron Alkene Reagents

The best-known reagent for alkenes is the $Cp(CO)_2Fe$ fragment, which is often designated simply as Fp (pronounced "fip"). Rosenblum[19] has shown how the isobutylene group in $Fp(CH_2=CMe_2)^+$ can be displaced by less bulky alkenes to give the Fp complex of the new alkene, which protects it from hydrogenation and from electrophilic attack. Protection of norbornadiene in Eq. 14.52 allows clean bromination without the usual carbonium ion rearrangements taking place. If there are several C=C double bonds in a molecule, the Fp group selectively complexes the least hindered or the most strained. Such C=C groups are usually the most reactive, and so it is particularly useful to be able to protect them selectively. Deprotection takes place readily with iodide ion in acetone:[19]

(14.52)

Alkyne Cobalt Carbonyl

Alkynes are protected as the tetrahedrane-like clusters **14.7**.[20] In this case, deprotection is carried out oxidatively with a reagent such as $FeCl_3$ or Et_3NO; as we saw in Section 4.3, oxidation often increases substitution rates at metal complexes and also reduces back donation to an unsaturated ligand, like an alkyne, which now dissociates more easily. The protecting group binds a C≡C selectively over a C=C group, and the complex is stable to the conditions required for the conversion of any free C=C group in the molecule to an alcohol by acid-catalyzed hydration or by hydroboration–oxidation, and to an alkyl group by diimine reduction (Eq. 14.53):

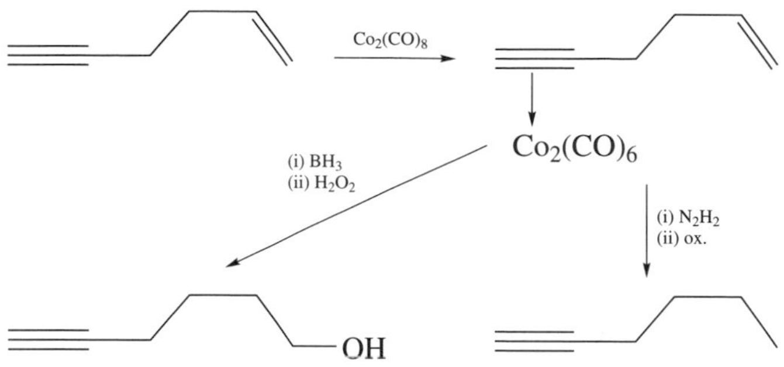

$$
\text{(14.53)}
$$

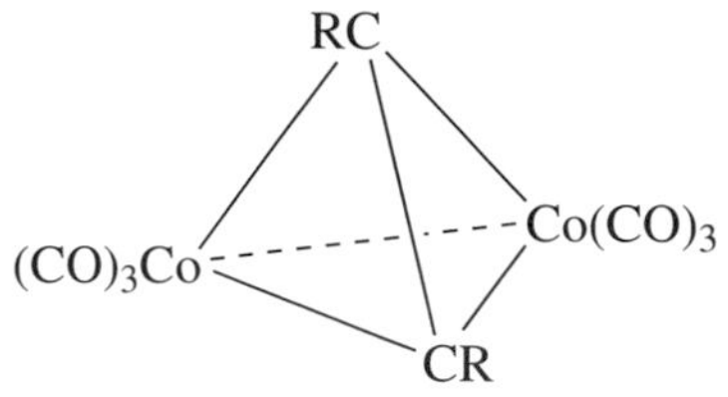

14.7

In the Nicholas[20b] reaction, carbonium ions α to the alkyne carbon are stabilized in the Co complex and can react with a variety of nucleophiles, such as the allylsilane in Eq. 14.54. The positive charge is probably stabilized by delocalization into the cluster by some such resonance form as **14.8**.

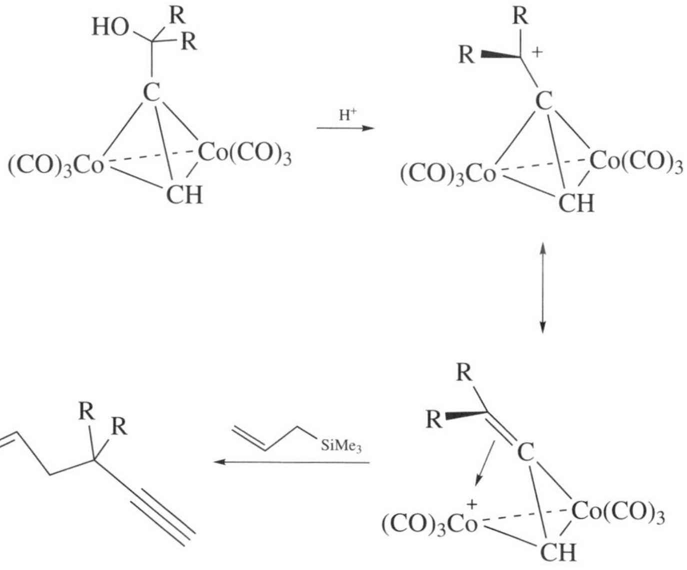

$$
\text{(14.54)}
$$

14.8

Diene Iron Carbonyl[21]

Dienes are most commonly protected with the $Fe(CO)_3$ group. Once again, an oxidative deprotection step with $FeCl_3$ is often used. One important application is the protection of a diene in the B ring of certain steroids (e.g., **14.9**). Under these circumstances, the side chain C=C groups can be successfully converted into a number of useful derivatives by osmylation, hydroboration, or hydrogenation, without affecting the diene.

Fe(CO)₃

14.9

Arene Chromium Carbonyl

Arenes are generally protected with the $Cr(CO)_3$ group, but as this complexation leads to a number of other important changes in the chemical properties of the arene, in particular making it much more susceptible to nucleophilic attack.[22]

Stabilizing Highly Reactive Species

Complexation has also been used to trap highly reactive species that might otherwise decompose. An early example was cyclobutadiene, not isolable except in the complexed form, such as the $Fe(CO)_3$ complex. Ce(IV) oxidation releases the free diene.[23a] In the case of **14.10**, trapping as the $Pt(PPh_3)_2$ complex allowed this unusually strained and reactive alkene to be purified and stored. The alkene itself, which is stable for short periods under ambient conditions, is released by treatment of the complex with CS_2.[23b]

14.10

- Oxidation mainly involves metal–organic catalysts.

14.4   REDUCTIVE ELIMINATION AND COUPLING REACTIONS

Early coupling processes, such as the Ullmann reaction (1901) for the conversion of ArHal to Ar−Ar with Cu powder, were very inefficient. In the 1990s, the discovery of a broad series of coupling reactions, often catalyzed by Pd complexes (Section 9.6), has transformed the situation, and these reactions now find very widespread use in organic synthesis. C−C, C−N, and C−O bonds can all be formed. Bromo- and iodoarenes are the most reactive, but the use of bulky, basic phosphines[24a] such as P(t-Bu)$_3$ can even allow use of the much more available, but less reactive, chloroarenes. Examples[24b] are shown in Eq. 14.55, where Pd$_2$(dba)$_3$ is a common precursor of Pd(0), lightly stabilized by the labile dba ligand.

dba = {(PhCH=CH)$_2$C=O})

$$(14.55)$$

The same strategy makes the Heck reaction efficient for chloroarenes (Eq. 14.56).[25] A feature of all these reactions is that the R groups to be coupled must in general resist β elimination so that they survive unchanged during the catalytic reaction while attached to Pd; this means that aryls, vinyls, and benzyls are useful, and this type of reaction has become a standard method in the synthesis of complex organic molecules:

$$(14.56)$$

Many variants of these palladium coupling reactions have been developed. For example, in a synthesis of Pumilitoxin, Kibayashi et al.,[26a] have used the sequence shown in Eq. 14.57 involving an organozinc reagent and a vinyl iodide.[26b] This example is unusual in that the organozinc-derived R group has a β hydrogen, yet reductive elimination to give coupling is preferred over β elimination.

$$(14.57)$$

Nicolaou et al.[27] developed the vinylphosphate route shown in Eq. 14.58 for the synthesis of Brevetoxin. Equation 14.59 shows how the I > Br > Cl reactivity order can be used to differentiate positions on an arene ring; this was a key step of the synthesis of a napyradiomycin antibiotic.[28]

$$(14.58)$$

$$(14.59)$$

Cyclotrimerization of Alkynes

We saw in Section 6.7 that the oxidative coupling of two acetylenes is a common process for a variety of low-valent metal complexes. The metalacyclic product can go on to an arene with excess alkyne, leading to a catalytic cyclotrimerization of the alkyne (Eq. 14.60).[29]

$$(14.60)$$

$$(14.61)$$

Vollhardt[29a] has adapted this reaction for the organic synthesis, by using the strategy shown in Eq. 14.61. The bis alkyne component is thought to form a metalacycle, which then reacts with the free mono alkyne. This alkyne is chosen so as to be too bulky to cyclotrimerize but reactive enough to convert the metalacycle to the arene: $Me_2SiC{\equiv}CSiMe_3$ and related alkynes fulfill these conditions and have the added advantage that the TMS groups can be easily removed or used to introduce further functionality. Palladium catalysts are also useful.[29b] Equation 14.62 shows the system applied to the synthesis of the protoberberine alkaloids.[30]

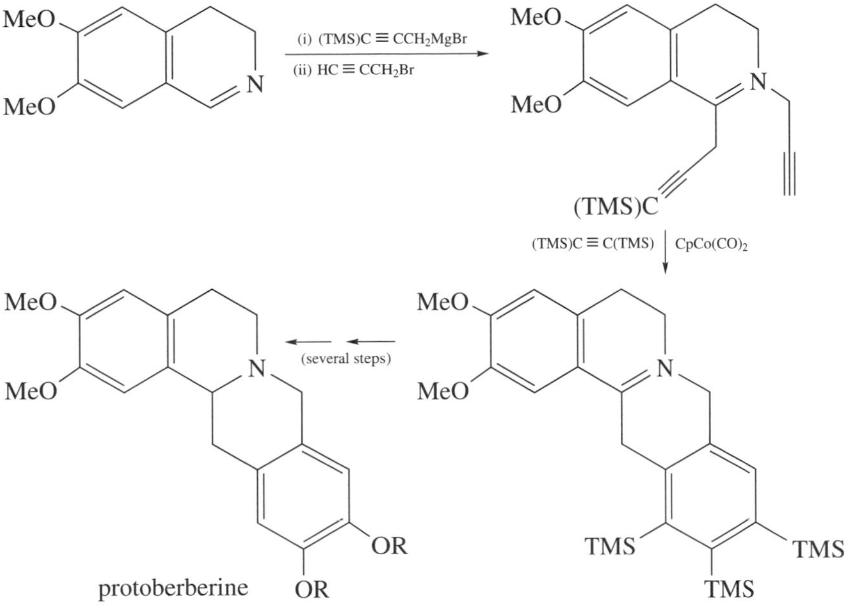

(14.62)

The skeletons of the steroids and the anthracyclines can also be constructed in a similar way. The strategy used for the steroids is exemplified in Eq. 14.63:

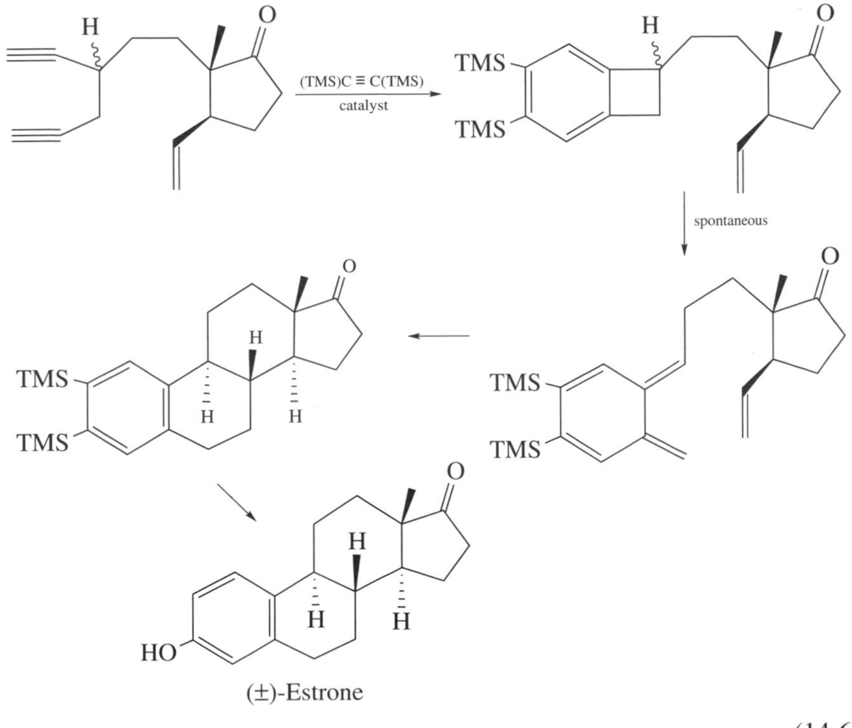

(±)-Estrone

(14.63)

which shows the key step. The usual cobalt-catalyzed [2 + 2 + 2] reaction gives a reactive benzocyclobutane; this spontaneously opens to the o-quinodimethane, which undergoes an internal Diels–Alder reaction to give the steroid skeleton. The formation of the arene has enough thermodynamic driving force to make the very strained benzocyclobutane. Some of the exothermicity of this first step, stored in the strained C_4 ring, then drives the subsequent ring opening leading to the final product. The desired trans–anti-trans product of the Diels–Alder step is thought to result from a "chair" transition state. Two further steps lead to estrone.[31a]

The reaction can be extended to the case in which two alkynes and a nitrile are trimerized to give a pyridine, or two alkynes and an isocyanate are trimerized to give an α-pyridone, also shown in Eq. 14.60, as exemplified in syntheses of vitamin B_6, a pyridine derivative,[31b] and the antitumor agent, camptothecin,[31] an α-pyridone.

Pauson–Khand Reaction

As shown in Eq. 14.64, this reaction[20] leads to substituted cyclopentanones in which the bulkiest substituent of the alkyne usually ends up α to the carbonyl.[32a] In the following application by Schreiber et al.[32b] (Eq. 14.65), a complex tricyclic natural product is constructed. [W(CO)$_5$(thf)] is also a useful catalyst for the Pauson–Khand reaction.[20]

$$(14.64)$$

$$(14.65)$$

McMurry Reaction

A coupling reaction of great interest is McMurry's[33] titanium-mediated synthesis of alkenes from two ketones (Eq. 14.66). This involves a reduced form of titanium, perhaps Ti(0), which may give the sequence of reactions shown in Eq. 14.67. These ideas are supported by the fact that 1,2-diols are also reduced to the alkene. Whatever the mechanism, the reaction shows the strongly oxophilic character of this early metal. A large number of reactions of this type are known with SmI_2.[18d]

$$R_2C{=}O \xrightarrow{\text{TiCl}_3,\ \text{LiAlH}_4} R_2C{=}CR_2 + TiO_2 \qquad (14.66)$$

$$(14.67)$$

Diels–Alder reactions between α, β-unsaturated carbonyl compounds and dienes can be catalyzed by Lewis acids, such as $Cp_2M(OSO_2CF_3)_2$[34a] or $[\{HC(2\text{-pyridyl})_3Mo(NO)_2\}]^{2+}$.[34b]

14.5   INSERTION REACTIONS

Mizoroki–Heck Reaction

From the point of view of the alkene or alkyne, an alkene insertion into an M—R bond is a *carbometallation* of the alkene or alkyne by the M—R group. The most important insertion reactions involve alkenes, alkynes, and CO. The first is exemplified in the Mizoroki–Heck coupling reaction,[35a] in which an alkene inserts into a Pd—R group. The resulting alkyl then β-eliminates to give the product (Eq. 14.68). The initial R group must be stable to β elimination, of course, and this limits the reaction to aryls, vinyls, and allyls. Equation 14.69[35b] shows a typical example, the synthesis of a 2-quinolone. The role of the base is to make the reaction catalytic by removing the hydrogen halide from the Pd(II) product and so regenerate the Pd(0) catalyst. This reaction has also been applied by Hegedus et al.[35c] to the syntheses of *N*-acetyl claviciptic acid. Bromides and iodides are the most suitable as substrates.

Hydroformylation

Leighton[36a] has applied Rh/PR_3-catalyzed hydroformylation (Section 9.3) to a polyol synthesis (Eq. 14.71) as a useful alternative to the aldol strategy (Eq. 14.70).

In the cyclic enol acetal of Eq. 14.72, this has the great advantage of being highly diastereoselective, producing the 1,3-*syn*-diol derivative. The bulky phosphine, $P(o\text{-}t\text{-BuC}_6\text{H}_4\text{O})_3$ proved the most effective.

$$(14.68)$$

$$(14.69)$$

Silylformylation,[36b] in which the H_2 component of the H_2/CO mixture of hydroformylation is replaced by a silane to give net addition of R_3Si- and $-CHO$ across an unsaturated bond, is normally effective for alkynes but not alkenes. Leighton and co-workers have used a chelation strategy (Eq. 14.73) to produce products that can be further functionalized by oxidation of the C–Si bond.

$$(14.70)$$

$$(14.71)$$

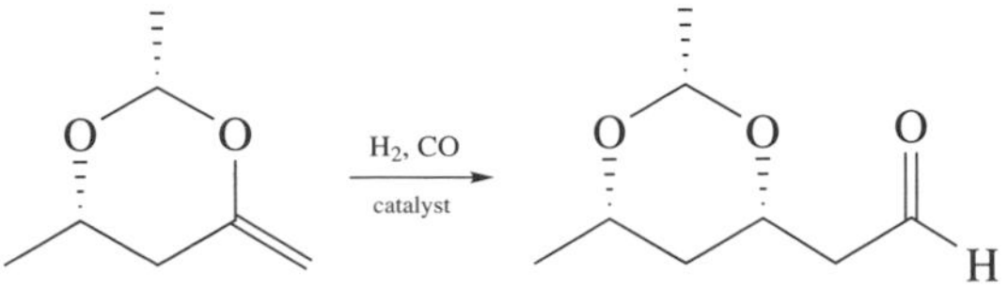

$$(14.72)$$

(80%, 93% *syn*)

$$(14.73)$$

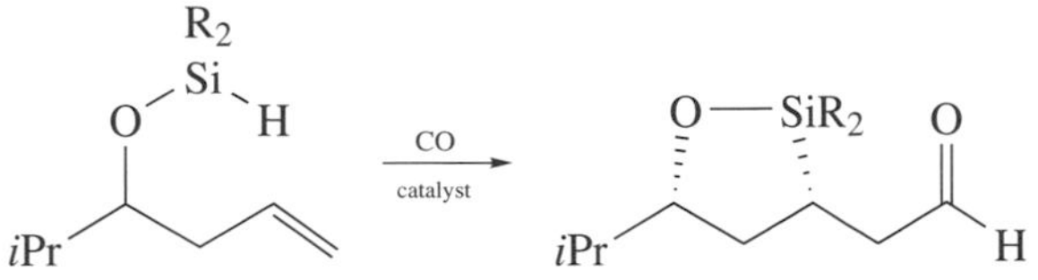

(71%, 90% *syn*)

Many other carbonylation reactions are also useful in total synthesis, including that shown in Eq. 14.74.[36c] The last step in the likely mechanism is nucleophilic attack on the metal acyl with abstraction of the acyl group (Section 8.3).

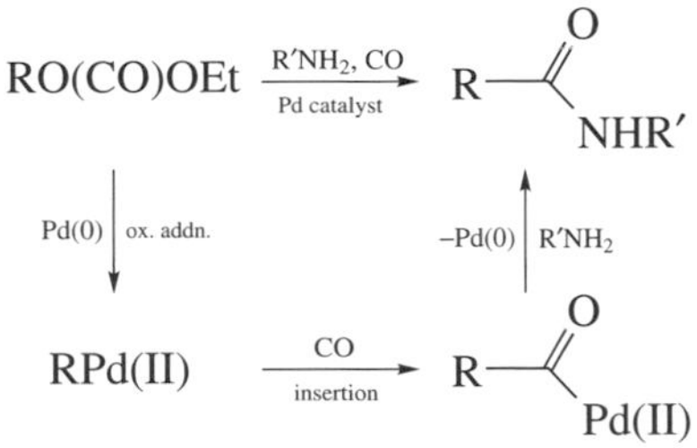

$$(14.74)$$

Cascade Carbometallation

This can be used to construct multiple rings as shown below.[37] The reaction starts with the oxidative addition of the vinyl iodide and the resulting alkyl undergoes insertion with two alkynes and two alkenes to give the tetracyclic Pd alkyl shown, which then β-eliminates to regenerate the Pd(0) catalyst. Many interesting variations of this reaction have been investigated.

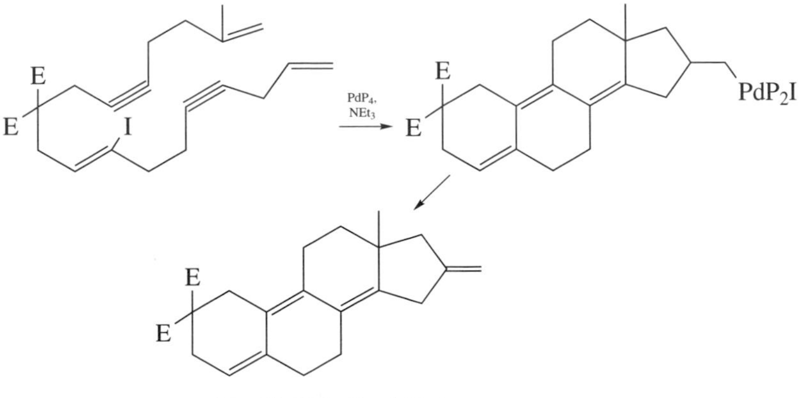

$$(14.75)$$

(E = COOMe, P = PPh₃)

Decarbonylation

The reverse of CO insertion can be mediated by transition metal reagents in the case of aldehydes. For example, $RhCl(PPh_3)_3$ reacts with RCHO to give $RhCl(CO)(PPh_3)_2$ and RH. Oxidative addition of the aldehyde C$-$H bond to rhodium is followed by a retromigratory insertion to give $Rh(R)(H)Cl(CO)(PPh_3)_2$. This loses RH by reductive elimination, and the net reaction goes with retention of configuration at carbon. It is also intramolecular as shown by crossover studies on a mixture of RCHO and R'CDO. Unfortunately, the $RhCl(CO)(PPh_3)_2$ product is no longer sufficiently reactive to add to a new aldehyde C$-$H bond, and so the reaction is not catalytic. $[RhCO\{PhP(CH_2CH_2PPh_2)_2\}]SbF_6$, shown below, is catalytic,[38] however, perhaps because this less basic, cationic system favors carbonyl loss from the metal to regenerate the active catalyst, $[Rh(triphos)]^+$.

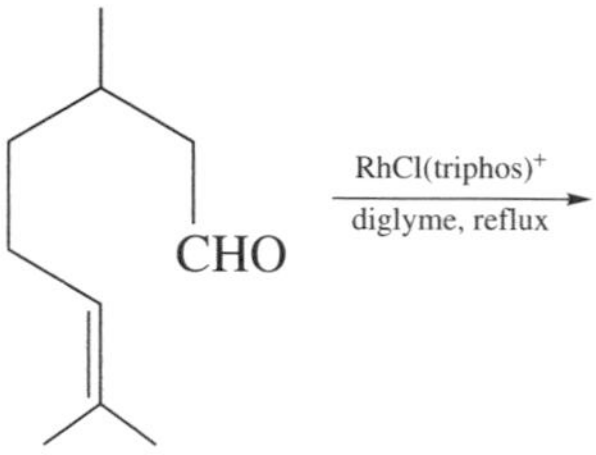

$$(14.76)$$

Intramolecular hydroacylation of 4-alkenals is a well-established method for producing cyclopentanones in which a C=C bond inserts into the Rh hydride formed by C$-$H oxidative addition to Rh(I); with chiral ligands and suitable alkenal substrates, useful asymmetric induction is possible.[39a] Extension to cyclopentenone synthesis requires a trans addition of Rh$-$H across the alkyne (see Section 7.2).[39b]

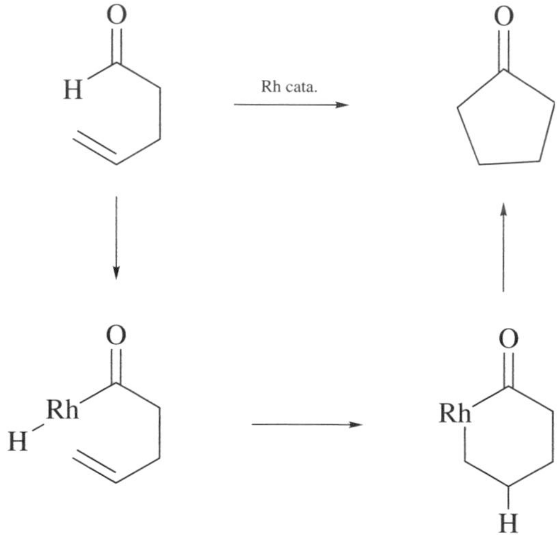

$$(14.77)$$

$$(14.78)$$

$$(14.79)$$

Hydroamination of alkenes can be catalyzed by lanthanide alkyls via insertion, as shown in Eq. 14.79.[39c]

14.6 NUCLEOPHILIC ATTACK ON A LIGAND

As we saw in Section 8.2, the binding of a polyene or polyenyl ligand to a metal can suppress the reactivity toward electrophiles usually seen for the free polyene and encourages attack by nucleophiles instead. This reversal of the normal reactivity pattern (umpolung) has been very widely used in organic synthesis.

Palladium Allyls

Of all the applications of nucleophilic attack, that on an allyl group coordinated to palladium is perhaps the one that has been most widely applied to organic synthesis.[40,41] The allyl group is usually formed either from $PdCl_2$ and an alkene by C−H activation or from Pd(0) and an allylic acetate by oxidative addition. Where the substrate is an alkene, a mixture of $[(\pi\text{-allyl})\text{PdCl}]_2$ complexes is sometimes formed because there may be a choice of C=C groups or of C−H bonds to attack, but in general the more substituted alkene is more reactive and the regiochemistry of the C−H activation step can be moderately selective. The allylic acetate route is useful in that the Pd ends up attached to the allyl group of the substrate in a defined regio- and stereochemistry. Subsequent rearrangement can degrade the stereochemistry of the allyl, however, and so the nucleophilic attack step should be carried out without delay. In addition, the product of oxidative addition to $\text{Pd}(\text{PPh}_3)_4$ is the cationic $[(\text{allyl})\text{PdL}_2]^+\text{OAc}^-$, rather than the neutral halo complex formed from the halide. This cationic charge helps activate the allyl group for subsequent nucleophilic attack. In addition, the reactions are often catalytic with the acetates, and with suitable asymmetric ligands, can give useful levels of asymmetric induction;[40] these catalysts have been used in numerous total syntheses.[40a]

$$\text{CH}_2\text{=CHCH}_2\text{OAc} \xrightarrow{\text{Pd}(0)} (\pi\text{-allyl})\text{PdL}_2{}^+$$

$$\xrightarrow{\text{Li[Nu]}} \text{CH}_2\text{=CHCH}_2\text{Nu} + \text{Pd}(0) + \text{LiOAc} \quad (14.80)$$

The palladium selectively attacks an allylic acetate with inversion, even in the presence of other reactive groups, such as a C−Hal bond; nucleophilic attack then occurs exclusively at the allyl group, showing the strongly activating effect of the metal (Eq. 14.81):[40a]

$$(14.81)$$

The nucleophile usually attacks the exo face of the allyl group (the one opposite the metal), and at the least hindered terminus of the allyl group (although this preference can be reversed with suitable ligands).[40c] The stereochemical consequences of this sequence have been used to define the relative stereochemistries of two chiral centers five carbons apart in an acyclic system, during the synthesis of the side chain (**14.11**) of vitamin E (Eq. 14.82).[42] Unfortunately, only stabilized carbanions, such as malonates, have proved effective carbon nucleophiles in most cases.

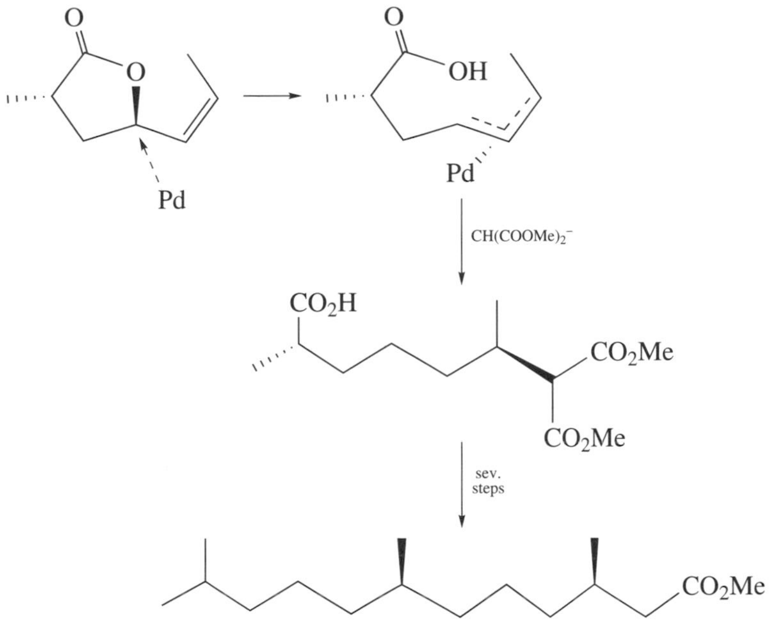

14.11

(14.82)

Rather than give direct attack at the exo face of the ligand, the nucleophile may bind to the metal first, in which case it can be transferred to the endo face of the allyl group; this changeover of stereochemistry can occur as a result of relatively small changes in the conditions (Eq. 14.83).[41] In the presence of excess LiCl, the acetate is prevented from coordinating to the metal and the cis product is formed; conversely, the presence of LiOAc encourages coordination of the OAc⁻ anion to the metal, and therefore, the production of the cis product.

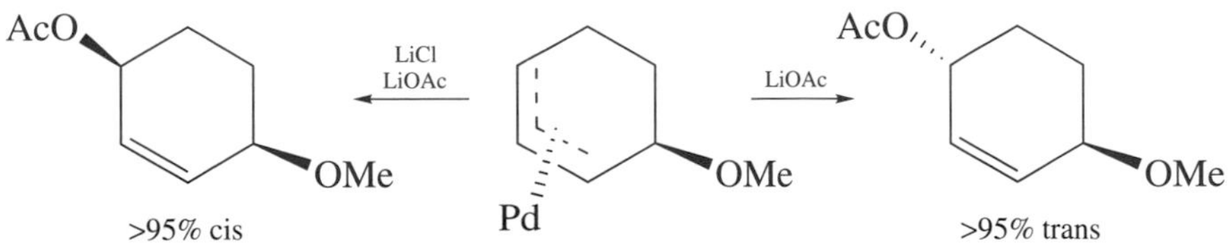

(14.83)

By starting from the diene, a 1,4-bis acetoxylation can be carried out to give cis or trans product, according to the exact conditions. The intermediates **14.12a** and **14.12b** are invoked to explain these products. The benzoquinone serves to reoxidize Pd(0) and make the reaction catalytic (Eq. 14.84):

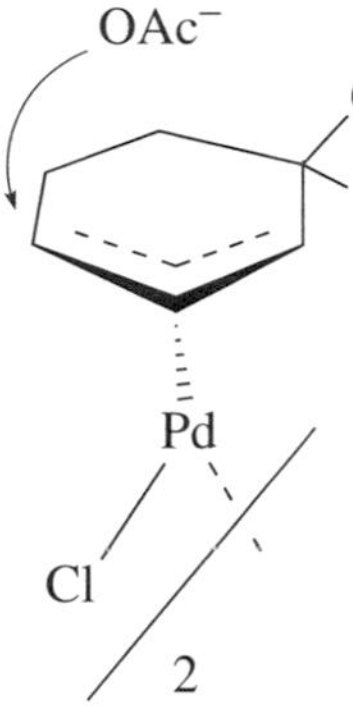

14.12a 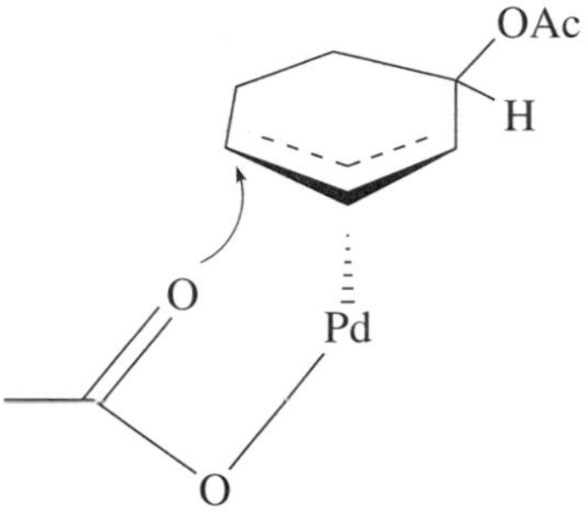   **14.12b**

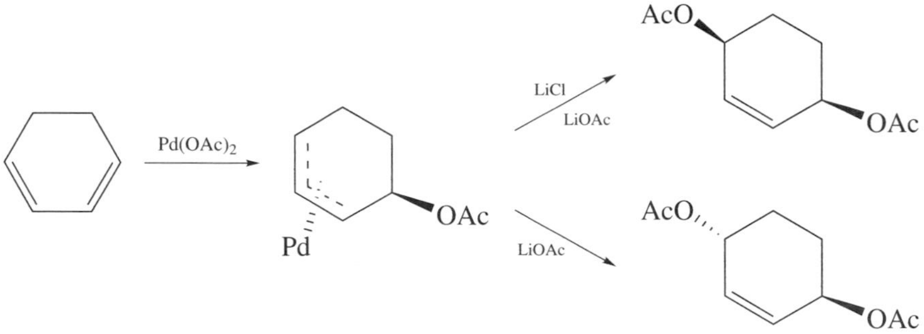

$$(14.84)$$

The element most widely used today in organic synthesis is palladium. Not only do we have the catalytic palladium allyl chemistry mentioned above but there is also a wide range of Pd coupling reactions available (Section 9.6); Pd reactions are very tolerant of functionality and give predictable products.[43]

Rhodium Acetate–Catalyzed Carbene Reactions

This reaction involving a metal carbene is illustrated by Eq. 14.85, where $Rh_2(OAc)_4$ is the catalyst. A diazoketone acts as a source of a carbene that inserts into an activated C–H[44a] or O–H[44b] bond. The presumed intermediate rhodium carbene complex is too unstable to isolate.

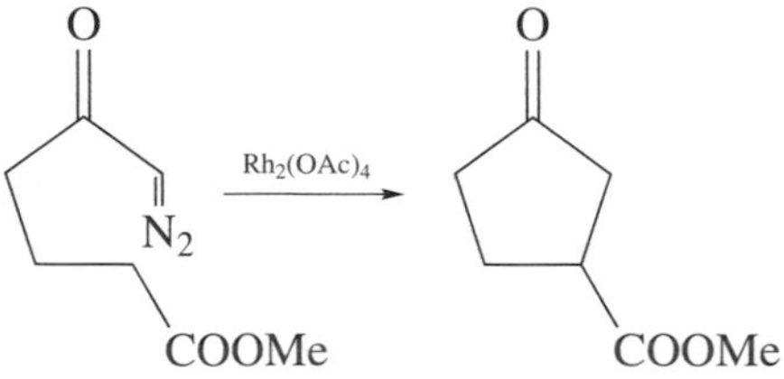

$$(14.85)$$

Wood et al.[44b] have shown how the product from such a reaction can set up the system to undergo a subsequent Claisen rearrangement (Eq. 14.86). Instead of the expected ketone **14.13**, however, the nature of the final product suggests that the intermediate is enol **14.14**, formed by transfer of the alcohol proton to O rather than C. Starting with one enantiomer of the chiral alcohol leads to good transfer of chirality to the final product.

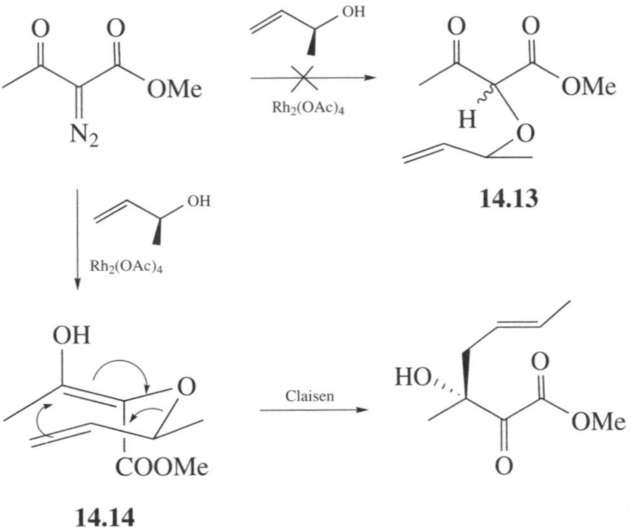

14.13

14.14

(14.86)

Alkene Metathesis[45]

This reaction has been increasingly used in organic synthesis since catalysts tolerant of functionality (particularly Grubbs's catalyst, Section 12.1) have become available. The four major variants of the reaction common in organic synthesis are illustrated: ring-closing metathesis (RCM), cross-metathesis (CM), ring-opening metathesis (ROM), and enyne metathesis (EYM). RCM, CM, and ROM follow the standard metathesis pattern (Eq. 12.1–12.7), but EYM[45d] is a little different in that the CR_2 alkylidene fragment of the alkene is transferred to the alkyne during the reaction, by the route shown below.

The second example of ring-closing metathesis (Eq. 14.92)[46] shows that in favorable situations even a large ring can be formed by eliminating C_2H_4 from a diene.

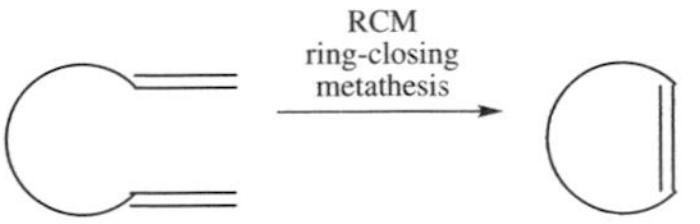

(14.87)

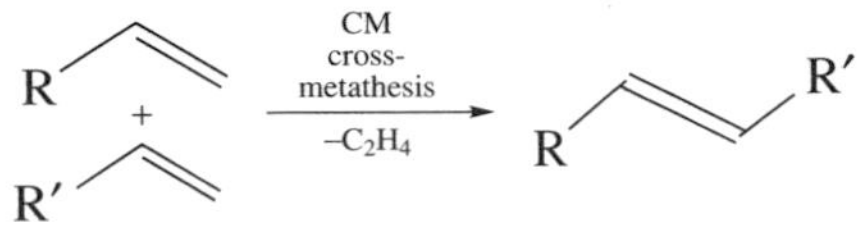

(14.88)

$$(14.89)$$

$$(14.90)$$

RCM was used by Nicolaou et al.[45c] in the key step of his Epothilone A synthesis to give a 16-membered ring. Whenever a diene is used as substrate, RCM ring closure and undesired ADMET polymerization (Section 12.1) are in principle competitive (Eq. 14.93), high dilution favoring the former by suppressing intermolecular pathways. Enynes can give an interesting cascade cyclization (Eq. 14.94).

$$(14.91)$$

$$(14.92)$$

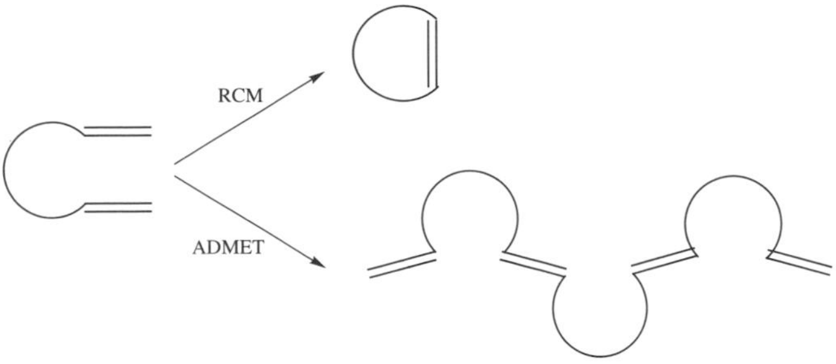

(14.93)

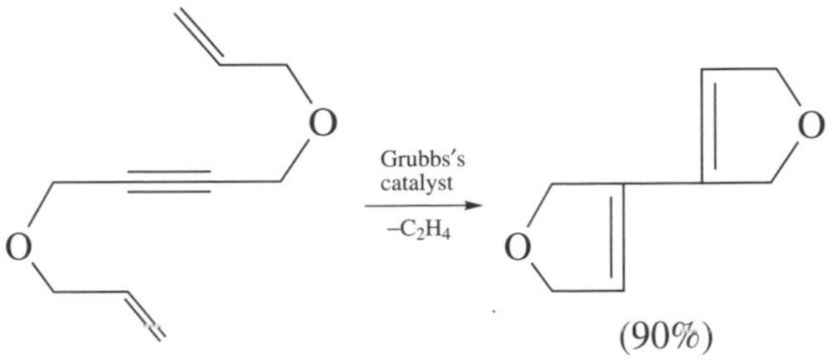

(14.94)

(90%)

Crowe and Goldberg[47a] and Grubbs et al.[47b] have found conditions under which cross-metathesis (CM) of two different monoalkenes can give good selectivity for the desired cross-dimer over other products. Equation 14.95 shows an example that relies on the fact that acrylonitrile is inactive for self-metathesis but takes part in the cross-reaction. Differential reactivity of different carbene intermediates is also responsible for the selectivity of ring-opening metathesis; Eq. 14.96 shows a ROM example by Snapper et al.[48] Asymmetric ROM catalysis is also possible.[49]

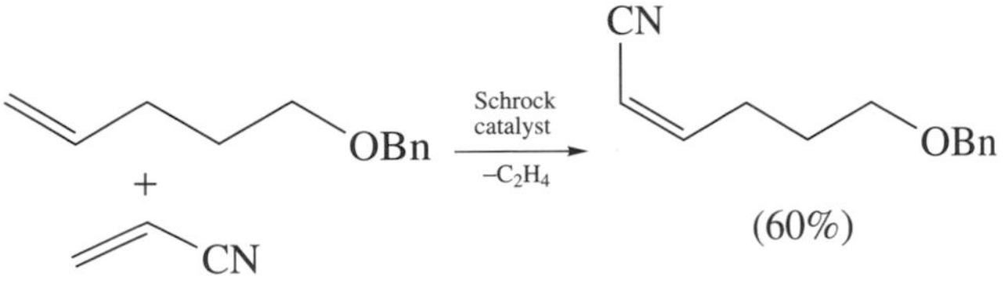

(14.95)

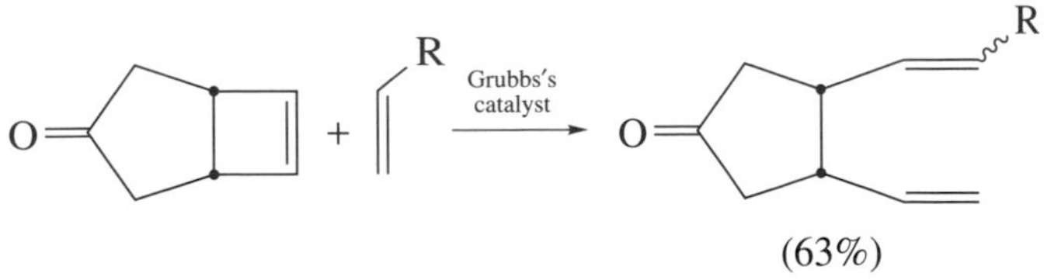

(60%)

(14.96)

(63%)

Alkyne Metathesis

Bunz has made a very convenient and robust catalyst[50] for alkyne metathesis available simply from combining $Mo(CO)_6$ with $p\text{-}XC_6H_4OH$ (X = Cl or CF_3) and heating to 125–150°C; the active catalyst is believed to be $(ArO)_3Mo\equiv CR$.

The system has been used for alkyne polymerization to give useful fluorescent polymers and oligomerization to give molecular triangles and squares (Eq. 14.97).

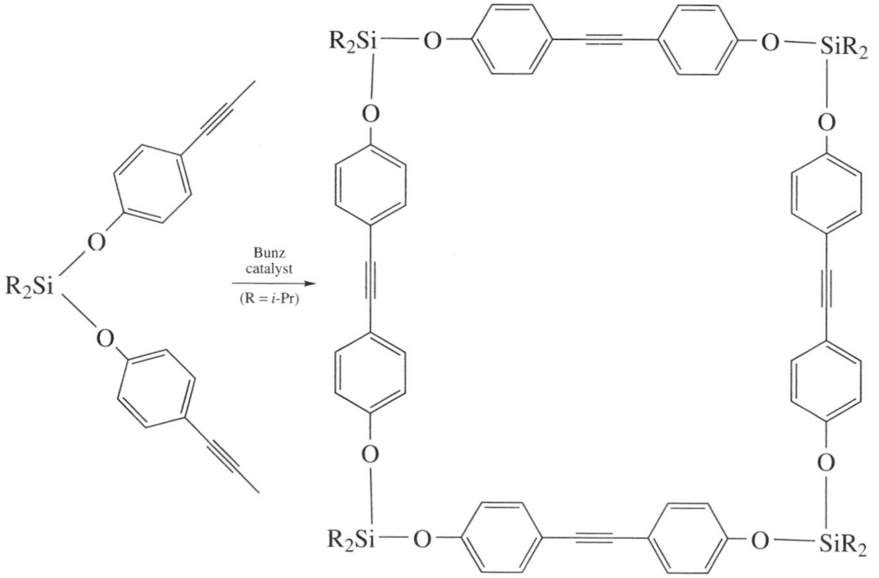

$$(14.97)$$

Metathesis seems to be exceptionally widely applicable and is rapidly becoming a standard reaction in synthesis.

We can confidently predict that the whole area of organometallic chemistry in organic synthesis will continue to grow strongly. It is likely that transition metal reagents will be involved in many of the new organic synthetic methods to be developed in the near future.

- Metathesis allows a wide variety of unusual reaction pathways and provides entirely new strategies for organic and polymer synthesis.

14.7   HETEROCYCLES

Sharpless has called for the need to develop a series of highly reproducible, high-yield reactions, *click chemistry*, that can be relied on to couple molecules in a variety of media. Catalysis plays its usual role in mediating such reactions. In one of the best examples, a 1-alkyne can be coupled with an organic azide to give a 4-substituted 1,2,3-triazole. A Cu(II)/ascorbate or [Cu(I)(NCMe)$_4$]BF$_4$ catalyst ensures that 1,4-selectivity is achieved (the 1,5-substituted product also forms in uncatalyzed reactions). The mechanism goes via the Cu(I) acetylide, and a tris(triazolyl)amine ligand helps stabilize the Cu(I) state.[51a]

$$(14.98)$$

This reaction has been used to make rapid and reliable covalent connections to micromolar concentrations of protein in a bioconjugation reaction in water, as well as construct a dendrimer with exceptionally high yields.[50b]

14.8 MORE COMPLEX MOLECULES

The synthesis of complex organic molecules can be considered an art[52] in that elegance of design is a major criterion of excellence. Today, many syntheses are designed to give homochiral rather than racemic products, so asymmetric catalysis often plays a large role. In addition, the strategy must avoid incompatibilities—a reagent must be selective enough to react only in the desired way at the desired part of the molecule.

Equation 14.99 shows how the Mizoroki–Heck reaction can be used in the last step of a synthesis of (−)-tubifolene.[53] The conformation of the molecule sets the regiochemistry—CC bond formation at C7—and the geometry—attack from the front as written. Subsequent β elimination, followed by isomerization provides the final product.

In Eq. 14.100, the Grubbs catalyst is used to construct three rings selectively in a synthesis of halicholactone.[54] Equation 14.101 shows a Pd allyl coupling to give a bicyclic ketone.[55]

Finally, Eq. 14.102 shows an asymmetric reaction[56] with a possible mechanism.

$$(14.99)$$

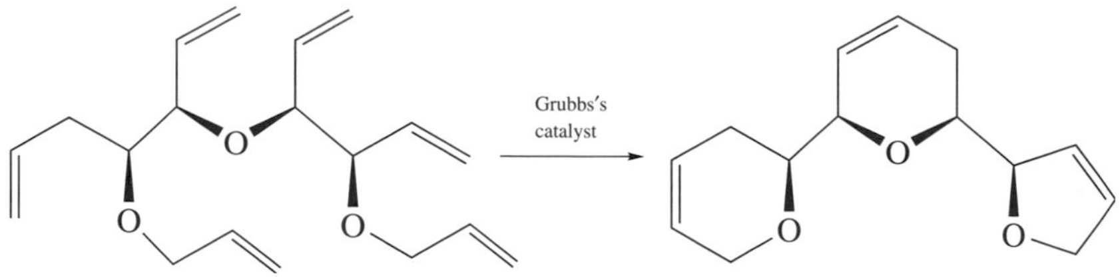

(14.100)

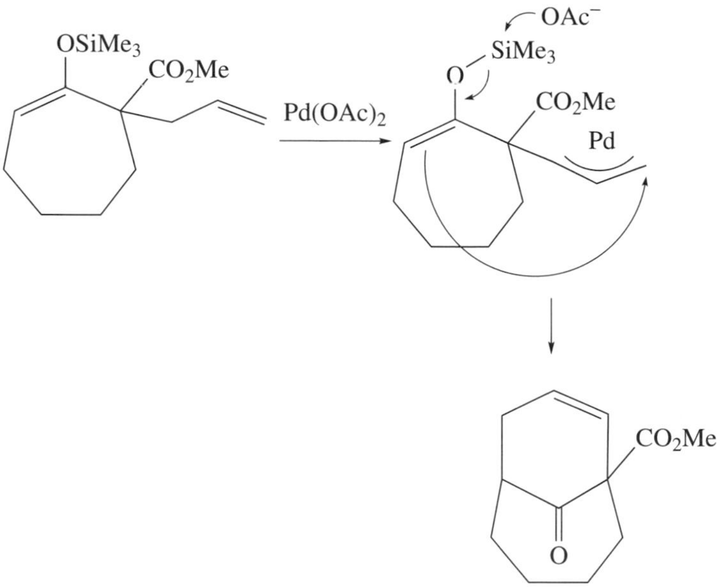

(14.101)

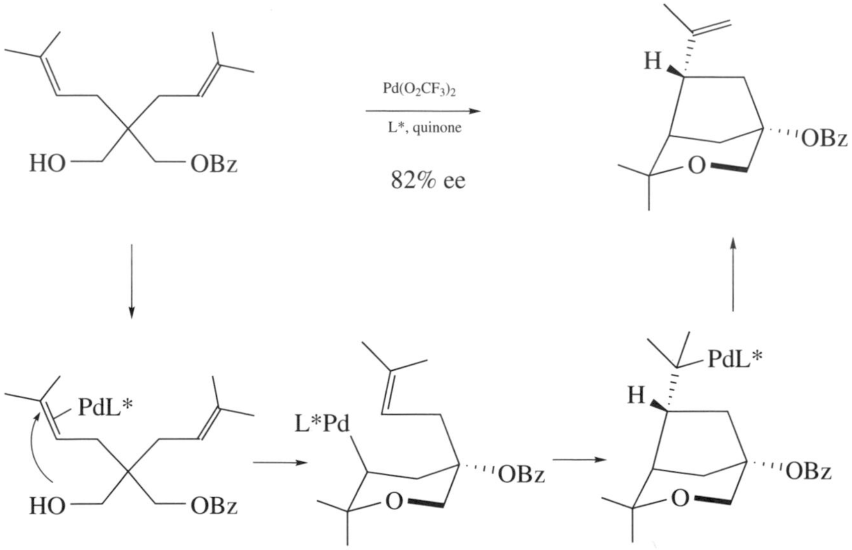

(14.102)

REFERENCES

1. (a) D. Seebach, *Angew. Chem., Int. Ed.* **29**, 1320, 1990; (b) P. J. Harrington, *Transition Metals in Total Synthesis*, Wiley, New York, 1990; S. V. Ley, ed., *Comprehensive Organic Synthesis*, Pergamom, New York, 1991; (c) L. S. Hegedus, *Transition Metals in Organic Synthesis*, University Science Books, Mill Valley, CA, 1994.

2. (a) R. A. Sheldon, *Pure Appl. Chem.* **72**, 1233, 2000.

3. M. Schlosser, *Organometallics in Synthesis*, Wiley, New York, 2002.

4. (a) R. G. Jones and H. Gilman, *Organic Reactions* **6**, 339, 195l; (b) R. D. Rieke, *Accts. Chem. Res.* **10**, 301, 1977; (c) H. M. Walborsky, *Accts. Chem. Res.* **23**, 286, 1990; (d) E. C. Ashby, *Pure Appl. Chem.* **52**, 545, 1980; (e) R. Noyori and M. Kitamura, *Angew. Chem., Int. Ed.* **30**, 49, 1991.

5. (a) D. S. Matteson, *Stereodirected Synthesis with Organoboranes*, Springer, Berlin, 1995; (b) M. E. Brook, *Silicon in Organic, Organometallic and Polymer Chemistry*, Wiley, New York, 2000.

6. (a) I. Fleming et al., *Chem. Comm.* 176, 1978; (b) D. P. Curran et al, *J. Am. Chem. Soc.* **121**, 6607, 1999; A. G. Davies, *Organotin Chemistry*, Wiley, New York, 2004.

7. (a) C. Lutz and P. Knochel, *J. Org. Chem.*, **62**, 7895, 1997; (b) H. Lebel, J. F. Marcoux, C. Molinaro, and A. B. Charette, *Chem. Rev.* **103**, 977, 2003; (c) A. Boudier, L. O. Bromm, M. Lotz, and P. Knochel, *Angew. Chem. Int. Ed.* **39**, 4414, 2000; (d) R. C. Larock, *Organomercury Compounds in Organic Synthesis*, Springer, New York, 1985; (e) R. H. Crabtree, *Pure Appl. Chem.* **67**, 39, 1995.

8. S. Woodward, *Chem. Soc. Rev.* **29**, 393, 2000.

9. G. H. Posner, *Org. React.* **19**, 1, 1972.

10. J. F. Normant and A. Alexakis, *Synthesis* 841, 1981.

11. T. Imamoto, *Lanthanide Reagents in Organic Synthesis*, Academic, London, 1994.

12. (a) P. Wipf et al., *Pure Appl. Chem.* **69**, 639, 1997; (b) R. J. Whitby et al., *Pure Appl. Chem.* **69**, 633, 1997; (c) S. Murai et al., *Pure Appl. Chem.* **69**, 589, 1997; Y Guari, A. Castellanos, S. Sabo-Etienne, and B. Chaudret, *J. Mol. Catal. A* **212**, 77, 2004; (d) A. Fürstner et al., *Pure Appl. Chem.* **70**, 1071, 1998.

13. (a) K. B. Sharpless, *J. Am. Chem. Soc.* **111**, 1123, 1989; (b) K. B. Sharpless et al., *Chem. Brit.* **22**, 38, 1986; (c) K. B. Sharpless and R. B. Michaelson, *J. Am. Chem. Soc.* **95**, 6136, 1973; G. R. Cook, *Curr. Org. Chem.* **4**, 869, 2000.

14. R. M. Hanson and K. B. Sharpless, *J. Org. Chem.* **51**, 1922, 1986.

15. (a) E. N. Jacobsen et al., *Angew. Chem. Int. Ed*, **36**, 1720, 1997; (b) *Acct. Chem. Res.*, **33**, 421, 2000.

16. M. J. Burk et al., *J. Org. Chem.* **68**, 5731, 2003.

17. R. Noyori et al., *Adv. Synth. Catal.* **345**, 15, 2003.

18. (a) K. Burgess et al., *Chem. Rev.* **91**, 1179, 1991; (b) T. Hayashi, Y. Matsumoto, and Y. Ito, *J. Am. Chem. Soc.* **111**, 3426, 1989; (c) D. A. Evans, G. C. Fu, and A. H. Hoveyda, *J. Am. Chem. Soc.* **110**, 6917, 1988; (d) H. B. Kagan, *Tetrahedron* **59**, 10, 351, 2003.

19. M. Rosenblum et al., *J. Org. Chem.* **45**, 1984, 1980.

20. (a) Y. K. Chung, *Coord. Chem. Rev.* **188**, 297, 1999. (b) B. J. Teobald, *Tetrahedron*, **58**, 4133, 2002.

21. H. J. Knölker, *Chem. Rev.* **100**, 241, 2000.

22. E. P. Kündig, *Pure Appl. Chem.* **69**, 543, 1997.

23. (a) J. Rebek et al., *J. Am. Chem. Soc.* **97**, 3453, 1975; (b) K. B. Wiberg et al., *J. Am. Chem. Soc.* **96**, 6531, 1974.

24. (a) D. H. Valentine and D. H. Hillhouse, *Synthesis* 2437, 2003; (b) J. F. Hartwig et al., *J. Am. Chem. Soc.* **121**, 3224, 1999; *Acct. Chem. Res.* **36**, 234, 2003; S. L. Buchwald et al., *Angew. Chem., Int. Ed.* **38**, 2411, 1999; *Top. Curr. Chem.* **219**, 131, 2002.

25. G. C. Fu et al., *J. Org. Chem.* **64**, 10, 1999.

26. (a) C. Kibayashi et al., *J. Am. Chem. Soc.* **121**, 9873, 1999; (b) E. Negishi et al., *J. Am. Chem. Soc.* **102**, 3298, 1980.

27. K. C. Nicolaou et al., *J. Am. Chem. Soc.* **119**, 5467, 1997.

28. M. Nakata et al., *Tetrahedron Lett.* **40**, 7501, 1999.

29. (a) K. P. C. Vollhardt, *Angew. Chem., Int. Ed.* **23**, 539, 1984; (b) K. R. Roesch and R. C. Larock, *Org. Lett.* **1**, 553, 1999; M. Rubin, A. W. Sromek, and V. Gevorgyan, *Synlett* 2265, 2003.

30. K. P. C. Vollhardt et al., *Tetrahedron* **39**, 905, 1983.

31. (a) R. L. Funk and K. P. C. Vollhardt, *J. Am. Chem. Soc.* **101**, 215, 1979; **102**, 5253, 1980; (b) K. P. C. Vollhardt et al., *Tetrahedron* **41**, 5791, 1985; (c) K. P. C. Vollhardt et al., *J. Org. Chem.* **49**, 4786, 1984.

32. (a) S. E. Gibson and A. Stevenazzi, *Angew. Chem. Int. Ed.* **42**, 1800, 2003; (b) S. L. Schreiber, T. Sammakia, and W. E. Crowe, *J. Am. Chem. Soc.* **108**, 3128, 1986.

33. J. E. McMurry, *Acc. Chem. Res.* **16**, 405, 1983; A. Gansäuer and H. Bluhm, *Chem. Rev.* **100**, 2771, 2000.

34. (a) B. Bosnich et al., *Organometallics* **11**, 2745, 1992; (b) J. W. Faller et al., *J. Am. Chem. Soc.* **113**, 1579, 1991.

35. (a) I. P. Beletskaya and A. V. Cheprakov, *Chem. Rev.* **100**, 3009, 2000; G. T. Crisp, *Chem. Soc. Rev.* **27**, 427, 1998; M. Ikeda et al., *Heterocycles* **51**, 1957, 1999; (b) R. F. Heck et al., *J. Org. Chem.* **43**, 2952, 1978; (c) L. S. Hegedus et al., *J. Am. Chem. Soc.* **109**, 4335, 1987.

36. (a) J. L. Leighton et al., *J. Am. Chem. Soc.* **119**, 11118, 12416, 1997; (b) S. Murai et al., *Synlett* 414, 1996; I. Ojima et al., *Organometallics* **10**, 38, 1991; (c) T. P. Loh et al., *Tetrahedron Lett.* **40**, 2649, 1999.

37. E. Negishi, *Pure Appl. Chem.* **64**, 323, 1992.

38. C. M. Beck, S. E. Rathmill, Y. J. Park, J. Chen, and R. H. Crabtree, *Organometallics* **18**, 5311, 1999.

39. (a) Bosnich, B. *Acc. Chem. Res.,* **31**, 667 1998 and references cited. (b) Tanaka, K, Fu, G, *J. Am. Chem. Soc.,* **123**, 11492, 2001. (c) Li YW, Marks TJ, *J. Am. Chem. Soc.,* **118**, 9295, 1996.

40. (a) B. M. Trost, *Pure Appl. Chem.* **68**, 779, 1996; (b) J. W. Faller et al., *Organometallics* **3**, 927, 1231, 1984; (c) G. Helchen et al., *Pure Appl. Chem.* **69**, 513, 1997.

41. J.-E. Backvall, *Acc. Chem. Res.* **16**, 335, 1983; *Pure Appl. Chem.* **64**, 429, 1992; *Chem. Eur. J.* **4**, 1083, 1998.

42. J. E. McMurry et al., *Chem. Rev.* **89**, 1513, 1989.

43. B. Söderberg, *Coord. Chem. Rev.* **247**, 79, 2003.

44. (a) M. P. Doyle et al., *Prog. Inorg. Chem.* **49**, 113, 2001; (b) J. L. Wood et al. **121**, 1748, 1999.

45. (a) R. H. Grubbs and S. Chang, *Tetrahedron* **54**, 4413, 1998; (b) A. Fürstner ed., *Alkene Metathesis in Organic Synthesis*, Springer, Berlin, 1998; (c) K. C. Nicolaou et al., p. 73 in Ref. 44b, 1998; (d) S. T. Diver and A. J. Giessert, *Chem. Rev.* **104**, 2667, 2004.

46. A. Fürstner, P. H. Dixneuf et al., *Chem. Commun.* 1315, 1998.

47. (a) W. E. Crowe and D. R. Goldberg, *J. Am. Chem. Soc.* **117**, 5162, 1995; (b) R. H. Grubbs et al., *J. Am. Chem. Soc.* **122**, 58, 2000.

48. M. L. Snapper et al., *J. Am. Chem. Soc.* **117**, 9610, 1995.

49. A. H. Hoveyda et al., *J. Am. Chem. Soc.* **121**, 11603, 1999.

50. U. H. F. Bunz, in *Handbook of metathesis*, R. H. Grubbs, ed., Wiley-VCH, New York, 2003, Chapter 3.10; A. Fürstner, Chapter 2.12 in the same work.

51. (a) V. V. Rostovtsev, L. G. Green, V. V. Fokin, and K. B. Sharpless, *Angew. Chem. Int. Ed.* **41**, 2596, 2002; (b) Q. Wang, T. R. Chan, R. Hilgraf, V. V. Fokin, K. B. Sharpless, and M. G. Finn, *J. Am. Chem. Soc.* **125**, 3192, 2003.

52. C. Rucker, G. Rucker, and S. H. Bertz, *J. Chem. Inform. Comput. Sci.* **44**, 378, 2004; *New J. Chem.* **27**, 860, 2003; W. A. Smit, A. F. Bochkov, and R. Caple, *Organic Synthesis: The Science behind the Art*, RSC, Cambridge, 1998; G. Stork, *CR Acad. Sci. Chim.* **323**, 441, 2000.

53. M. Mori, M. Nakanishi, D. Kajishima, and Y. Sato, *Org. Lett.* **3**, 1913, 2001.

54. Y. Baba, G. Saha, S. Nakao, C. Iwata, T. Tanaka, T. Ibuka, H. Ohishi, and Y. Takemoto, *J. Org. Chem.* **66**, 81, 2001.

55. M. Toyota, V. J. Majo, and M. Ihara, *Tetrahedron Lett.* **42**, 1555, 2001.

56. M. A. Arai, M. Kuraishi, T. Arai, and H. Sasaki, *J. Am. Chem. Soc.* **123**, 2907, 2001.

PROBLEMS

1. *o*-Iodoaniline and CH_2=CHCH(OMe)$_2$ give quinoline (1-azanaphthaline) with Pd(PPh$_3$)$_4$. Suggest a mechanism.

2. The epoxides from cis and trans 2-hexene are reduced to the parent alkenes with retention of stereochemistry by treatment with (i) Fp$^-$, (ii) H$^+$, and (iii) NaI/acetone. Suggest a mechanism.

3. [CpFe(CO)$_2$]$_2$ catalyzes the addition of CCl$_4$ to an alkene as shown below:

$$CCl_4 + RHC=CH_2 = RHCCl-CH_2CCl_3$$

The reaction is not affected by light, and running the reaction with mixed CCl$_4$ and CBr$_4$ gave no crossover products such as RHCBr−CH$_2$CCl$_3$, but only RHCCl−CH$_2$CCl$_3$ and RHCBr−CH$_2$CBr$_3$. Suggest a mechanism. (R. Davis et al., *Chem. Commun.* 1387, 1986.)

4. On treating compound **A** with RhCl(CO)(PPh$_3$)$_2$ an acidic compound is obtained. Treatment of **B** with Fe(CO)$_5$ gives a diene complex. What do you think these new species are?

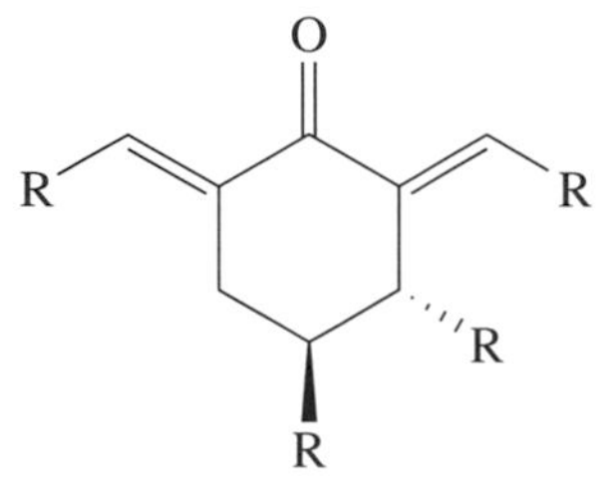

A

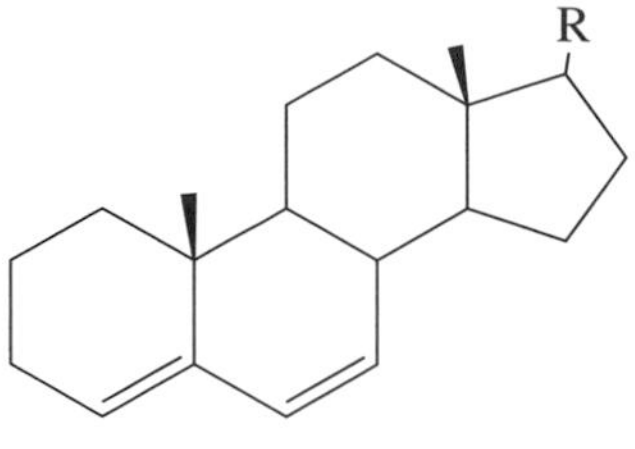

B

5. Compound **C** gives **D** on treatment with PdCl$_2$ and PPh$_3$ in methanol, followed by CO and then MeLi. Account for the stereochemistry of the product and explain the role of the PPh$_3$. In a related reaction, (cod)PdCl$_2$ is first treated with aqueous base, and then CO. The final product has the formula C$_9$H$_{14}$O$_3$. What is its structure and stereochemistry?

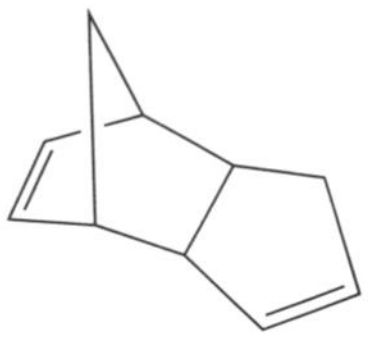

C 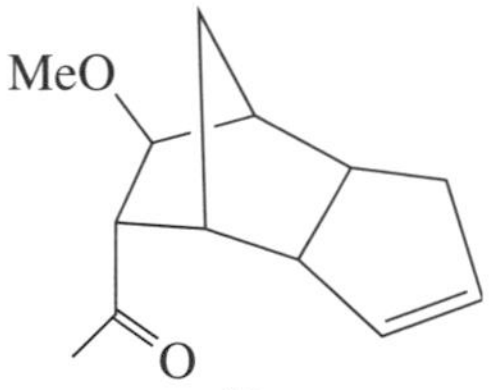   **D**

6. Compound **E** reacts with PdCl$_4{}^{2-}$ to give a complex. This, in turn, reacts with NaCH(COOEt)$_2$ and base to give **F**. Account for the formation of this product. In particular, why did the nucleophile attack where it did, and why is the double bond where it is in **F**? Compound **F** reacts with PdCl$_4{}^{2-}$ to give a new complex, which in turn reacts with (i) ClCH$_2$CH$_2$OH/base and (ii) CH$_2$=CHCOR, to give **G**, which can be converted to a number of prostaglandins. Account for the transformation of **F** to **G**.

E 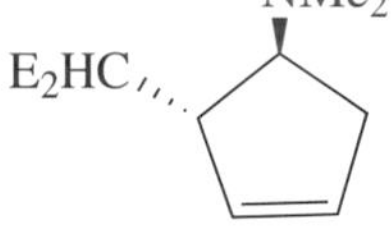   **F** 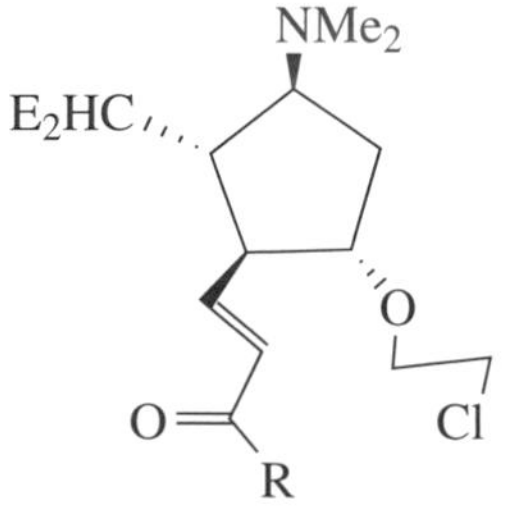   **G**

7. Fp^- reacts with $ClCH_2SMe$ to give a product that can be methylated with Me_3O^+. The methylation product reacts with cyclooctene to give **H**, shown below. Account for the formation of H.

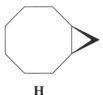

H

8. Although aldehydes can be decarbonylated with $RhCl(PPh_3)_3$, ketones are unaffected. Why do you think this is so? What products, organic and inorganic, do you think would be formed from $RCOCl$ and $RhCl(PPh_3)_3$?

9. Although decarbonylation of RCHO is not catalytic with $RhCl(PPh_3)_3$, both compounds become catalytic using $RhCl(dpe)_2 (dpe = 1, 2$-diphenylphosphinoethane) at 120°C or above. What is the origin of the difference in properties between $RhCl(dpe)_2$ and $RhCl(PPh_3)_3$?

10. Cyclohexene reacts with $HgCl_2$ and MeOH, followed by $PdCl_4^{2-}$ and CO, also in MeOH to give a compound $C_7H_{16}O_3$. What is this compound, what stereochemistry does it have, and how was it formed? Propargyl alcohol, $HC\equiv CCH_2OH$, gives a compound $C_5H_4O_3$ under similar conditions. What is the structure of this species?

11. In the Heck arylation of 1-methylcyclohexene by an aryl bromide, what would you expect the stereochemistry of the insertion and β-elimination steps (syn or anti) to be? Given this stereochemistry, what regiochemistry would you expect for the C=C double bond in the final product (i.e., formation of the 1- 2- or 3-alkene)? (R. Semmelhack, *Pure Appl. Chem.* **53**, 2379, 1981.)

12. Maleic anhydride (MA) reacts with $CoCl(PPh_3)_3$ to give an adduct Co(MA) $Cl(PPh_3)_2$. This adduct, in turn, reacts with 2-butyne to give 2,3-dimethylbenzoquinone. What structure do you propose for the adduct, and what methods might you use to test your suggestion? (L. N. Liebeskind et al., *Organometallics* **5**, 1086, 1986; **1**, 771, 1982; *J. Am. Chem. Soc.* **102**, 7397, 1980.)

13. Propose a mechanism for Zhang's cyclocarbonylation reaction. (X. Zhang et al., *J. Am. Chem. Soc.* **121**, 7708, 1999.)

14. Propose a mechanism for the cascade cyclization of Eq. 14.94. What ynedi-
    ene precursor would you choose as a starting material to make the bicyclic
    diene shown below? (R. H. Grubbs et al., *J. Am. Chem. Soc.* **116**, 10801,
    1994; *J. Org. Chem.* **61**, 1073, 1996.)

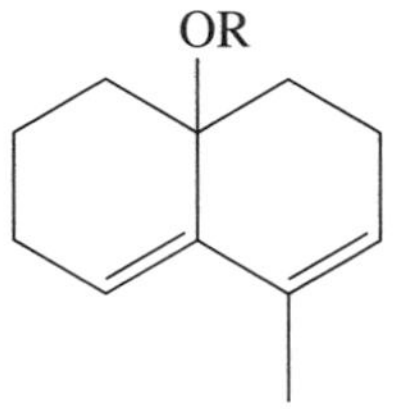

OR

15

PARAMAGNETIC, HIGH-OXIDATION-STATE, AND HIGH-COORDINATION-NUMBER COMPLEXES

Diamagnetic complexes have dominated the discussion up to this point because they are the easiest to study. When we move to first-row transition metals, as for the majority of metals in biology (Chapter 16), for example, paramagnetism (Section 15.1) is much more common. One reason is the propensity of these metals to undergo one-electron redox processes to give odd-electron monomeric d^n configurations that are necessarily paramagnetic. Another is the lower ligand field splitting for the first-row transition metals, which makes high-spin paramagnetic complexes possible for even-electron d^n configurations. The f-block metals (Section 15.5) are also normally paramagnetic because of the partial occupation of f orbitals. In all of these complexes, we move away from 18e "closed-shell" configurations into "open-shell" territory where at least one orbital has only one electron.

Low oxidation states have also dominated the discussion because they are most capable of binding soft, π-acceptor ligands (CO, C_2H_4, etc.) that are so typical of organometallic chemistry. When we avoid these ligands by restricting the coordination sphere to ligands like alkyl, aryl, H, and Cp, however, much higher oxidation states can be achieved; these are typically d^0 diamagnetic compounds. We look at polyalkyls like WMe_6 in Section 15.2 and cyclopentadienyls like $Cp*ReMe_4$ in Section 15.4. Finally, the highest coordination numbers are attained with the smallest ligand, hydride: polyhydrides like $[ReH_9]^{2-}$ appear in Section 15.3; like polyalkyls, these are also often d^0 and diamagnetic.

The maximum oxidation state possible for any transition element is the group number, N, because only N valence electrons are available for ionization or for forming covalent bonds. These d^0 compounds are normally diamagnetic. Re

The Organometallic Chemistry of the Transition Metals, Fourth Edition, by Robert H. Crabtree
Copyright © 2005 John Wiley & Sons, Inc.

in group 7 and Os in group 8 are the last elements that are able to attain their theoretical maximum oxidation states (e.g., ReF_7 and OsO_4); Ir and Pt only reach M(VI) in MF_6, and gold shows its highest oxidation state, Au(V), in $[AuF_6]^-$. It is therefore not surprising that most of the organometallic complexes having an oxidation state in excess of 4 come from the elements Ta, W, Re, Os, and Ir. While high oxidation states are usual for the earlier elements [e.g., Ti(IV), Ta(V)], high oxidation states are rare for the later elements, and it is here that we might expect to see interesting oxidizing properties. Just as the study of low-valent organotransition metal complexes led to the development of methods for the selective reduction of organic compounds, we can anticipate that high-oxidation-state chemistry will lead to better methods of oxidation. We already looked at OsO_4 in Section 14.2. The higher oxidation states in general are more stable for the third-row transition metals (Section 2.7). We will see that this is also true for organometallic compounds.

As we saw in Section 2.2, the 18e rule is most likely to be obeyed by low-valent diamagnetic complexes. In this chapter, we will find many examples of stable species with electron counts less than 18e, but this is especially true of polyalkyls, some of which are paramagnetic. One reason is that an alkyl ligand occupies much space around the metal in exchange for a modest contribution to the electron count. Second, the high ∂^+ character of the metal leads to a contraction in its covalent radius because the metal electrons are contracted by the positive charge. Note that this only leads to a slight decrease in the M−L bond lengths because the ligands acquire ∂^- character and so their covalent radii increase. An increase in the ligand size and a decrease in the metal size makes it more difficult to fit a given number of ligands around a metal in the high-valent case. The low apparent electron count in such species as $MeReO_3$ may be augmented somewhat by contributions from the ligand (O, Cl, NR, etc.) lone pairs. Agostic interactions with the alkyl C−H bonds are probably not widespread in d^0 and high-valent complexes because this interaction needs back donation from the metal (Chapter 3). This means that electron counting in these species is not completely unambiguous. High-valent Cp complexes are more likely to be conventional 18e species because Cp contributes many more electrons to the metal in proportion to the space it occupies than do alkyl groups. Polyhydrides are almost always 18e, as we might expect for what is one of the smallest, and one of the least electronegative, ligands present in the complexes discussed in this section.

- Paramagnetic and high-oxidation-state organometallics have been relatively neglected because they are harder to study.

15.1 MAGNETISM AND SPIN STATES

Diamagnetic materials are weakly repelled by a magnetic field gradient while paramagnetic ones are attracted. From the weight change of a sample in the presence or absence of a magnetic field gradient, or by an NMR method (Evans method; Section 10.11), one can measure the magnetic moment of a complex. This gives the number of unpaired electrons on the central metal. Specialist texts[1] cover a number of possible factors that can complicate the interpretation, such as spin coupling in metal clusters and orbital contributions in third-row ($5d$) transition metals. Table 15.1 shows the situation in the absence of such complications, where the measured magnetic moment in Bohr magnetons gives the number of unpaired electrons. This number is often indicated by the spin quantum number, S, which is simply half the number of unpaired electrons. The multiplicity (singlet, doublet, triplet, etc.) is also used as shown in the table.

The S value of a complex depends first on the d^n configuration. The d^0 and d^{10} cases are necessarily diamagnetic ($S = 0$), having no possibility for unpaired electrons. In contrast, d^1 and d^9 are necessarily paramagnetic with one unpaired electron ($\uparrow$, $S = \frac{1}{2}$). The d^3, d^5, and d^7 odd-electron configurations are necessarily paramagnetic but may have different accessible spin states depending on how the spins are paired [e.g., ($\uparrow\uparrow\uparrow$, $S = \frac{3}{2}$) or ($\uparrow\uparrow\downarrow$, $S = \frac{1}{2}$) for d^3]. Even-electron d^2, d^4, d^6, and d^8 may be diamagnetic or paramagnetic with the spin states depending on spin pairing [e.g., ($\uparrow\uparrow$, $S = 1$) or ($\uparrow\downarrow$, $S = 0$) for d^2].

Spin States

Spin states are isomeric forms with distinct energies, structures, and reactivities. Which spin state is stablest for a given metal and oxidation state depends on the geometry and ligand set that lead to a splitting pattern for the d orbitals. As we fill these orbitals, we have alternative spin states whenever we have choices in the electron filling pattern. Instead of the idealized octahedral splitting pattern of three d_π orbitals below two d_σ orbitals that we considered in Chapter 1, which gives the high-spin/low-spin alternative spin states of Fig. 1.2, we deal instead with more realistic splitting patterns of low-symmetry organometallic complexes.

TABLE 15.1 Terms Used in Discussing Magnetism

Spin Quantum Number, S	Number of Unpaired Electrons	Multiplicity	Magnetic Moment (bohr magnetons)[a]
0	0	Singlet	0
$\frac{1}{2}$	1	Doublet	1.73
1	2	Triplet	2.83
$\frac{3}{2}$	3	Quartet	3.87
2	4	Pentet	4.90
$\frac{5}{2}$	5	Sextet	5.92

[a] The magnetic moment can also be affected by orbital contributions and magnetic coupling in metal clusters, effects that we ignore here.

As discussed by Poli,[2a] a simple picture, based on the ionic model, starts from the coordination number, represented in what follows by the symbol m, a value determined by Eq. 15.1 for the complex $[MX_aL_b]^{c+}$. Of the nine valence orbitals of the metal, we expect to find m orbitals in the M$-$L σ^* group (Fig. 15.1a). Of these m orbitals, four are the single s and the three p orbitals, so $(m-4)$ is the number of d orbitals in this M$-$L σ^* group. For the octahedral case, we have $(6-4)$, or two d orbitals, in agreement with the presence of just two d_σ orbitals in the familiar octahedral crystal field pattern. We can usually avoid further consideration of these $(m-4)$ orbitals because electrons rarely go into M$-$L σ^* antibonding orbitals in organometallic complexes, although this is not uncommon in Werner complexes with their generally lower Δ values. In the middle set of orbitals, in a dotted box in Fig. 15.1, we find $(9-m)$ d orbitals, which are either nonbonding or involved in π back bonding. For the familiar octahedral case, we have $(9-6)$ or three orbitals, corresponding with the familiar d_π set. Below these orbitals, we have m M$-$L σ-bonding levels. The electron count of the complex will be $(2m+n)$; for the familiar d^6 octahedral case, this will be $(2 \times 6 + 6)$, or 18 electrons.

$$CN = m = a + b \tag{15.1}$$

$$\text{Number of M}-\text{L antibonding } d \text{ orbitals} = (m-4) \tag{15.2}$$

$$\text{Number of M}-\text{L nonbonding } d \text{ orbitals} = (9-m) \tag{15.3}$$

To find the possible spin states for any system, we first find the d^n configuration, then we see what choices are available to distribute n electrons among $(9-m)$ orbitals. To take the d^2 case, typical coordination numbers seen in real complexes are 6 and 7. The examples of Fig. 15.2b and 15.2c show how the LX_2-type Cp ligand contributes three to the coordination number. Small changes in the ligand set can be sufficient to alter the energies of the d orbitals so that the magnetism changes from one spin state to the other. If the energies of the two states are close enough together, there can even be a spin equilibrium between the two forms, as for $S = 0$ and $S = 1$ spin isomers of $[(C_5H_4Me)NbCl_2(PEt_3)_2]$.

The relative energies of the spin states in such a case is decided by the relative magnitudes of the electron pairing energy and the HOMO$-$LUMO splitting, Δ. A large electron pairing energy (PE) favors the $S = 1$ state because it makes it difficult to put two electrons in the same orbital where they repel each other more strongly than when they are in different orbitals. A large Δ favors the $S = 0$ state because it makes it difficult to promote the electron because this now requires more energy. In Fig. 15.2b and 15.2c, Δ_1 is larger than Δ_2 and Δ_3 is larger than Δ_4, as expected on the basis of this argument.

The value of Δ depends on the geometry, ligands, and metal. The geometry therefore often changes to some extent when the spin state changes. An example where a large change occurs is d^8 16e $NiX_2(PR_3)_2$: the $S = 0$ complexes are square planar and the $S = 1$ species are tetrahedral. The Δ often increases as we move from $3d$ to $4d$ and $5d$ metals; for example, $PdX_2(PR_3)_2$ and $PtX_2(PR_3)_2$

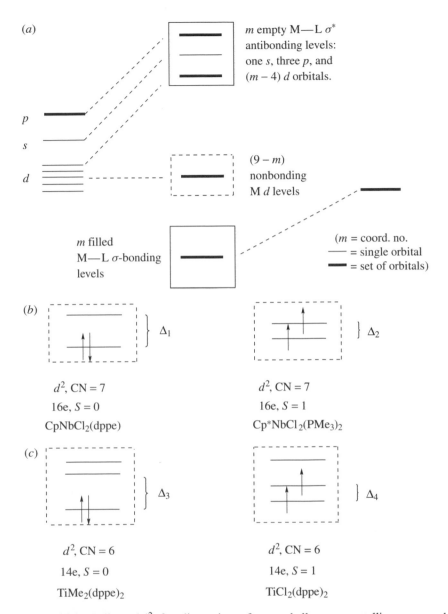

FIGURE 15.1 Poli model[2] for discussion of open-shell organometallic compounds (dppe = $Ph_2CH_2CH_2PPh_2$). (*a*) The number of nonbonding levels (dotted box) depends on the coordination number, *m*. The number of electrons, *n*, available to fill these levels depends on the d^n configuration. (*b, c*) For 6- and 7-coordinate species, such as the ones shown, two spin states are possible, $S = 0$ and $S = 1$. Thick lines denote sets of orbitals.

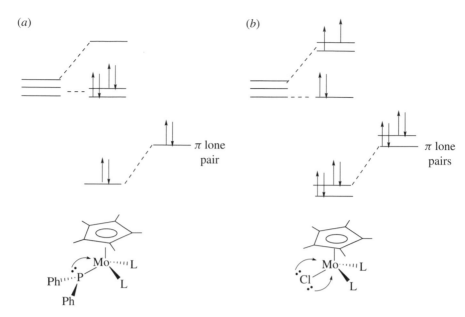

FIGURE 15.2 (*a*) The single π-donor lone pair of PPh$_2$ splits the *d* orbitals so that the four *d* electrons prefer to occupy the two lower levels leading to an $S = 0$ state. (*a*) The pair of π-donor lone pairs of Cl split the *d* orbitals so that the four *d* electrons now prefer to occupy the three lower levels as shown, leading to an $S = 1$ state. The two unpaired electrons are parallel according to Hund's rule.

are always square planar with $S = 0$ as a result of Pd and Pt having higher Δ values than Ni.

The π bonding also strongly alters Δ by the mechanism of Figs. 1.7 and 1.8 if different orbitals are differently affected. In [Cp*Mo(PMe$_3$)$_2$(PPh$_2$)], for example (Fig. 15.2), there is one π-bonding lone pair on the phosphide ligand that raises one of the three nonbonding *d* levels appropriate for this 6-coordinate system. The result is a diamagnetic $S = 0$ state for this d^4 case. If the ligand has two π-bonding lone pairs, as in the chloro analog [Cp*Mo(PMe$_3$)$_2$Cl], however, the two *d*-orbitals now affected by π bonding are both raised in energy, resulting in an $S = 1$ state.

Influence of Spin State Changes on Kinetics and Thermodynamics

Often, one spin state may be very reactive, the other not. Where alternate spin states are possible, there may be a change of spin state in a reaction, as has been discussed by Shaik et al.[2b] and by Harvey et al.[2c]. A molecule in one spin state could undergo a spin change to give a reactive form if the latter is close enough in energy; the energy cost of the spin state change would merely contribute to the reaction barrier. Such a case is illustrated in Fig. 15.3*a* for the reaction of **A** to give **B** in a case where we have a ground spin state with a high reaction barrier and an excited spin state with a low barrier. If the spin state change were very fast, the system could take the path $\mathbf{A} \rightarrow 1 \rightarrow 2 \rightarrow 3 \rightarrow \mathbf{B}$. If the spin change could not occur rapidly enough to happen during the reaction, however, we would have to go via the pathway $\mathbf{A} \rightarrow \mathbf{A}^* \rightarrow 2 \rightarrow \mathbf{B}^* \rightarrow \mathbf{B}$ (where $\mathbf{A}^*$ and $\mathbf{B}^*$ are the excited spin states of reactant and product). In either case, the reaction would still be faster than going via point 4, which would be the case if there were no alternate spin states available (as is often the case in conventional low-valent organometallic chemistry). This implies that organometallic species with alternate spin states could be more kinetically labile than other cases, but good data are still lacking.

Another situation, discussed by Poli,[2a] also involves a system with alternate spin states but with a change of spin state occurring during the reaction. As shown in Fig. 15.3b, this can alter the thermodynamics of the reaction. Assume

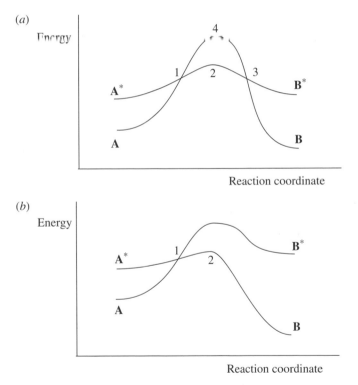

FIGURE 15.3 Reactivity patterns for species with alternate spin states. (a) The kinetics of a reaction can be accelerated if a more reactive accessible excited spin state exists with a lower net barrier for the reaction. We assume that spin change is fast. (b) The thermodynamics of a reaction can be affected if the product has a spin state different from that of the reagent. In this case, a reaction is unfavorable in the starting spin state but favorable if the system crosses to the other spin state. The star refers to the excited (less stable) spin state in each case.

the reagent spin state, **A**, leads to an excited spin state of the product, **B***; this can even be an endothermic, unfavorable process, as shown here. If this reaction pathway intersects the corresponding curve for the other spin state, crossover is expected to give not **B*** but **B**. The path is now **A** $\rightarrow$ 1 $\rightarrow$ 2 $\rightarrow$ **B** and the reaction now becomes thermodynamically favorable thanks to the accessibility of the alternate spin state.

If the unsaturated product of ligand loss is stabilized by this mechanism, the M−L bond strength will be lower than if no such stabilization occurred. This is because the bond strength is defined as the difference in energy between L_nM-L and ground state $L_nM + L$. Indeed, exceptionally low M−CO bond energies of 10–15 kcal/mol have been reported for a series of compounds where spin state changes of this sort occur.[3]

Examples of spin state control of reaction rates have been given by Harvey et al.[2c] For example, the slow addition of H_2 to Schrock's $[W\{N(CH_2CH_2NSiMe_3)_3\}H]$ is "spin-blocked" with a high barrier due to the crossing between reactant triplet and product singlet surfaces. In contrast, addition of CO to Theopold's $[TpCo(CO)]$ is fast because the triplet and singlet surfaces cross at an early stage of reaction and therefore at low energy.

3d Versus 4d and 5d Metals

First-row ($3d$) transition metals are the most likely to be paramagnetic with a <18e structure. Later metal analogs often adopt a different, often 18e, structure. For example, the $CpMCl_2$ series (M = Cr, Mo, W), shown below, starts with **15.1** without M−M bonds, where each Cr is $S = \frac{3}{2}$ 15e Cr. In contrast, the Mo and W analogs **15.2 and 15.3** are both 18e, $S = 0$ with M−M bonds. Similarly, the $3d$ metals may have a lower coordination number in their compounds. For example, **15.1** reacts with dppe to give $S = \frac{3}{2}$, 15e **15.4** having a monodentate dppe, but with **15.2** to give $S = \frac{1}{2}$, 17e **15.5**.[4,5]

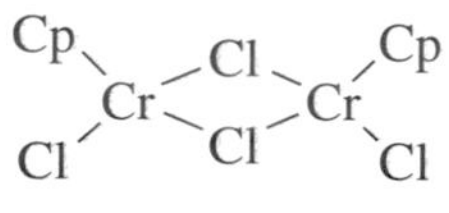

15.1 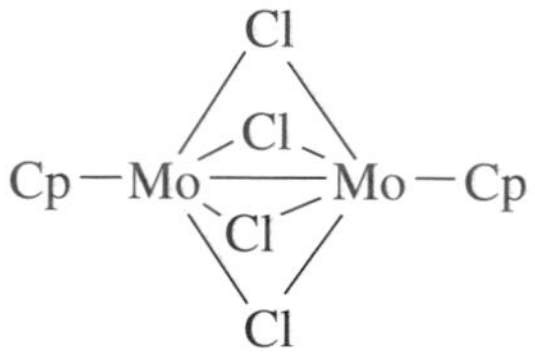   15.2 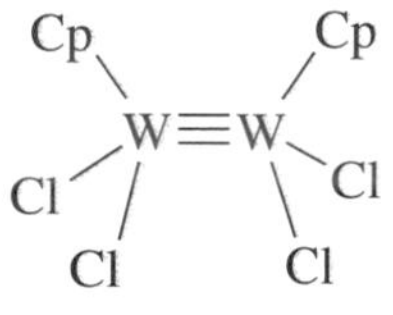   15.3

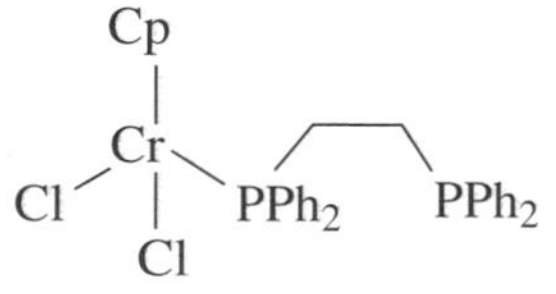

15.4 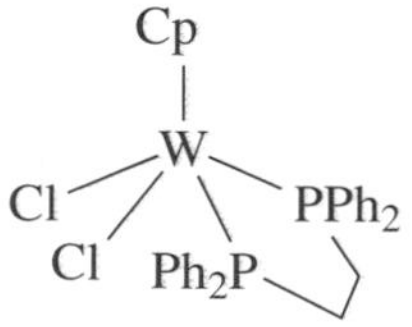   15.5

> - Simple models are available to predict the magnetism of organometallics.
> - The reactions may involve crossing between one potential energy surface and another, which can lead to faster reaction (Fig. 15.3)

15.2 POLYALKYLS

Group 4

We saw in Section 14.1 how $MeTiCl_3$ is used in organic synthesis. The homoleptic $TiMe_4$ (a homoleptic complex contains only one type of ligand) was reported as early as 1959.[6] The bright yellow crystalline material decomposes above $\sim 0°C$ to methane and a black powder containing Ti, C, and H. Adducts with such ligands as NMe_3, tmeda, or PMe_3 are thermally more stable. Note the hard character of the ligands that bind to $TiMe_4$; this suggests that the high formal oxidation state is real and that the electrophilic metal requires good σ-donor ligands but is incapable of significant back donation. Another clue that points in the same direction is the Grignard-like reactivity of the Ti(IV) alkyls (Section 14.1), which implies the presence of a ∂^- carbon. Since the electronegativity difference between C (2.5) and Ti (1.5) is considerable, the real charge on Ti must be quite positive. As we go to the right and down in the periodic table from Ti, we find that the electronegativity increases from 1.5 to about 2.2 for the heavy platinum metals, and so the M−C bond becomes less polar for these elements. This means the metal will be less positive and the alkyl groups less negatively charged in homoleptic alkyls of the later metals in a given oxidation state.

The red $Ti(CH_2Ph)_4$ has been studied crystallographically,[7] and it has been found that the $Ti-C_a-C_b$ angle is only $84°-86°$ (Fig. 15.4). The C_b carbon of the aromatic ring interacts to some extent with the metal and the structure is reminiscent of the η^2-allyl (Section 5.2). The soft ligand CO does react with $Ti(CH_2Ph)_4$, but although initial formation of a CO adduct has been proposed, the

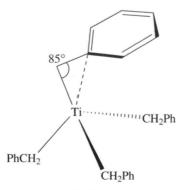

FIGURE 15.4 The structure of $Ti(CH_2Ph)_4$ showing the unusual distortion of the $T-C_\alpha-C_\beta$ bond.

final product is Ti(COCH$_2$Ph)$_2$(CH$_2$Ph)$_2$.[8] In contrast to the low thermal stability and high air and acid sensitivity of these alkyls, the bulky complexes **15.6** and **15.7** are unusually stable, thanks to steric protection of the metal. Complex **15.6** decomposes only over several days at 100°C, is stable to air even in solution, and decomposes only slowly in dilute H$_2$SO$_4$,[9] and **15.7** is stable enough to melt at 234°C.[10] The Zr and Hf alkyls are less well studied but behave rather similarly to their Ti analogs.

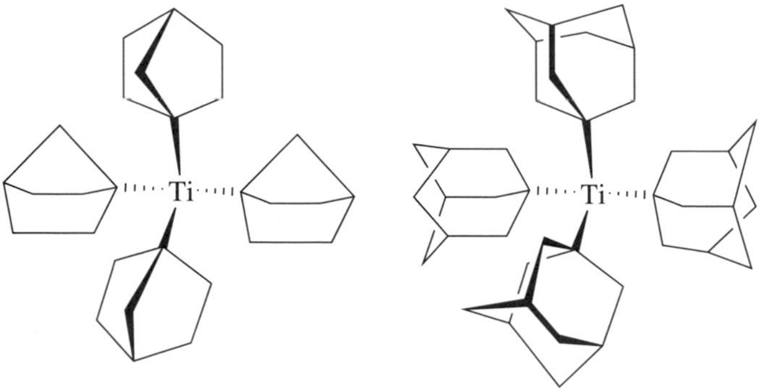

15.6   **15.7**

Group 5

Even though vanadium has a stable (V) oxidation state, the only alkyls so far discovered are the dark paramagnetic d^1 VR$_4$ species, such as the green-black benzyl complex. The 1-norbornyl is the most stable, decomposing only slowly at 100°. Tantalum, the third-row element gives stable alkyls, such as TaMe$_5$, which forms a dmpe adduct.[11] As we go to the right in the transition series, the differences between the first-, second-, and third-row elements become more marked. An example is the increasing reluctance of the first- and even second-row elements to give alkyls having the highest possible oxidation state, a feature that first appears in group 5 and becomes dominant in groups 6 and 7. TaMe$_5$ is trigonal bipyramidal, but attempts to make bulkier TaR$_5$ complexes always lead to α elimination to carbenes.

Group 6

A dark red Cr(IV) alkyl [Cr(CH$_2$SiMe$_3$)$_4$] is known, but Cr(III) is the common oxidation state, as in the orange Li$_3$[CrPh$_6$]. WMe$_6$ was the first homoleptic alkyl of group 6 having maximum oxidation state allowed for the group. It can decompose explosively at room temperature, but the reactions shown below have been identified.[12]

$$WMe_6 \xrightarrow{O_2} W(OMe)_6$$

$$WMe_6 \xrightarrow{CO} W(CO)_6 + Me_2CO$$

$$WMe_6 \xrightarrow{heat} 3MeH + C_2H_6 (traces)$$

$$WMe_6 \xrightarrow{Hal_2} WHal_6 + MeHal \tag{15.4}$$

$$WMe_6 \xrightarrow{NO} \tag{15.5}$$

The reaction with CO may go by migratory insertion, then reductive elimination of species containing the W(COMe)Me unit. The reaction with NO may go via insertion to give W−O−N•−Me, the N-centered radical center may then bind a further NO to give the final product.

Schrock and co-workers[13] have found that the hydrolysis of some of their alkylidyne complexes lead to oxoalkyls, such as neopentyl tungsten trioxide, which is air stable and is hydrolyzed further only by strong acid or base. The S(TMS)$_2$ reagent (Eq. 15.6)[13] is useful for replacing oxygen with sulfur because the formation of Si−O bonds provides a strong driving force. The mechanistic scheme proposed for the hydrolysis is also shown (Eq. 15.7). Note in Eq. 15.8 how the alkyl groups resist hydrolysis under conditions that would lead to cleavage of Ti−C bonds, a sign of the greater electronegativity of W compared to Ti.

$$t\text{-BuC}{\equiv}W(Ot\text{-Bu})_3 \xrightarrow{OH^-} t\text{-BuCH}_2{-}WO_3 \xrightarrow{S(TMS)_2} t\text{-BuCH}_2{-}WS_3 \tag{15.6}$$

$$t\text{-BuC}{\equiv}W(Ot\text{-Bu})_3 \xrightarrow{OH^-} \{t\text{-BuC}{\equiv}W(OH)(Ot\text{-Bu})_3\}^-$$

$$\xrightarrow{-t\text{-BuOH}} \{t\text{-BuCH}{=}W({=}O)(Ot\text{-Bu})_2\}^-$$

$$\xrightarrow{H_2O} \{t\text{-BuCH}_2{-}W({=}O)(OH)(Ot\text{-Bu})_2\}^-$$

$$\xrightarrow{H_2O} t\text{-BuCH}_2{-}WO_3 + 3t\text{-BuOH} \tag{15.7}$$

$$t\text{-BuC}{\equiv}W(CH_2t\text{-Bu})_3 \xrightarrow{H_2O} \{(CH_2t\text{-Bu})_3W({=}O)\}_2(\mu\text{-}O) \tag{15.8}$$

Wilkinson et al.[14] have made an analogous series of M(VI) complexes of the type M(=Nt-Bu)$_2$(2,4,6-Me$_3$C$_6$H$_2$)$_2$ for Cr, Mo, and W. The Cr complex is deep red and air stable.

Group 7

Only one Mn(IV) alkyl is known, the green Mn(1-norbornyl)$_4$, but rhenium has one of the most extensive series of high-oxidation-state alkyls, some of which are illustrated in Eq. 15.9.[15]

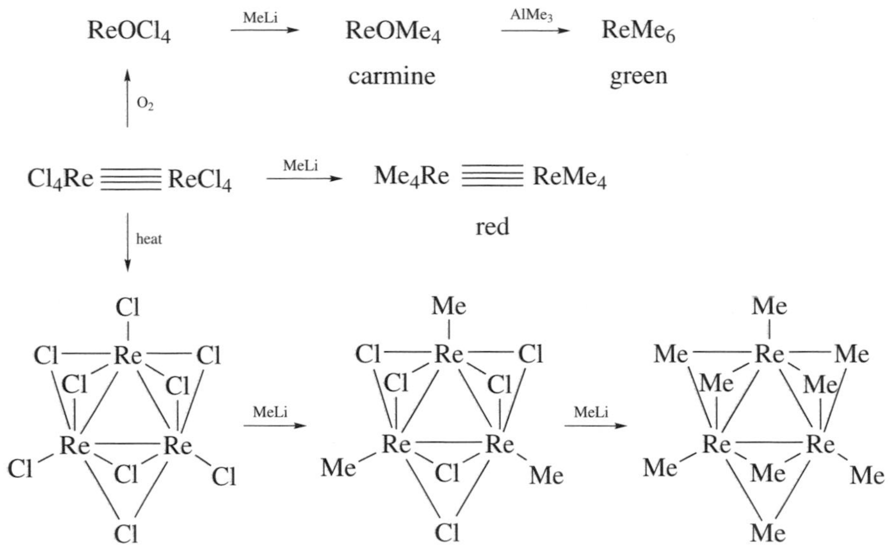

$$ReOCl_4 \xrightarrow{MeLi} ReOMe_4 \xrightarrow{AlMe_3} ReMe_6$$

$$\text{carmine} \qquad \text{green}$$

O_2 ↑

$$Cl_4Re \equiv ReCl_4 \xrightarrow{MeLi} Me_4Re \equiv ReMe_4$$

$$\text{red}$$

heat ↓

(15.9)

In contrast to the reactions of O_2 and NO with WMe_6 (Eqs. 15.4 and 15.5), interesting oxo-alkyls can be obtained by oxidation of $ReMe_6$ with these oxidants. The higher electronegativity of Re compared to W may make the Re alkyls generally more stable to air, acids, and attack by nucleophiles. $ReOMe_4$ fails to react with the Lewis bases that usually give complexes with the polyalkyls of the earlier metals. The dirhenium alkyls probably have the eclipsed structure characteristic of quadruply bonded metals (Section 13.1), and the trirhenium complexes are triangular clusters with Re—Re bonds and bridging halide or alkyl groups.[15]

$$ReMe_6 \xrightarrow{O_2} ReOMe_4 \xrightarrow{NO} cis\text{-}ReO_2Me_3 \xrightarrow{O_2} ReO_3Me \qquad (15.10)$$

The NO reactions are said to go as follows:

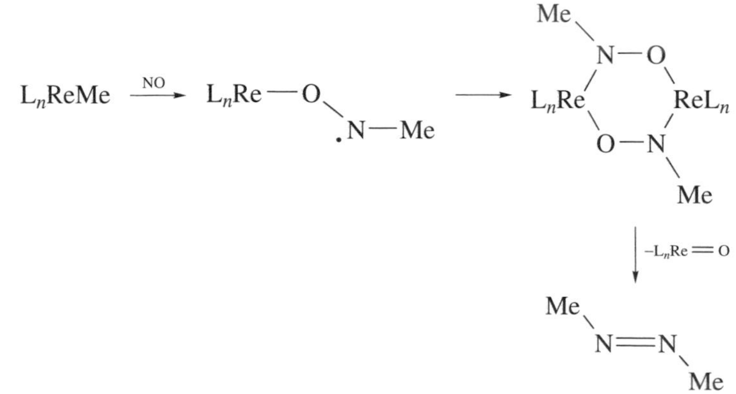

(15.11)

$ZnNp_2 (Np = t\text{-}BuCH_2)$ and $ReOCl_3(PPh_3)_2$ give the unusual dirhenium tetraalkyl shown in Eq. 15.12.[16] The presence of an Re—Re bond is believed to account for the short intermetallic distance of 2.6 Å.

$$ReOCl_3(PPh_3)_2 \xrightarrow{ZnNp_2} \begin{array}{c} Np\ \ \ \ O\ \ \ \ Np \\ \diagdown \diagup \diagdown \diagup \\ O{=}Re{-}Re{=}O \\ \diagup \diagdown \diagup \diagdown \\ Np\ \ \ \ O\ \ \ \ Np \end{array} \qquad (15.12)$$

Groups 8–10

Purple Fe(IV) and brown Co(IV) norbornyls are known, but most alkyls of these groups are M(II) or M(III) such as the yellow $Li_2[FeMe_4]$ or *fac*-$[RhMe_3(PMe_3)_3]$. Co(III) alkyls have been studied in connection with coenzyme B_{12} chemistry (Section 16.2). Ir(IV) aryls have been made by electrochemical oxidation of Ir(III) precursors.[17] The biphenyl-1,2-diyl ligand seems to be especially stabilizing for high oxidation states and is the C analog of the bipyridyl ligand that has proved so useful in coordination chemistry. Note how the strained ring in the biphenylene starting material helps drive the C—C bond cleavage reaction:

$$(15.13)$$

Nickel alkyls are always and Pd alkyls often M(II), such as the golden-yellow $Li_2[NiMe_4]$ or $PdMe_2(bipy)$. In many organic synthetic applications of Pd, formation of a Pd(IV) alkyl had to be postulated, but for many years no isolable example was found.[18a] The first aryl, $PdCl_3(C_6F_5)(bipy)$ (1975),[18b] and the first alkyl, $PdIMe_3(bipy)$ (1986)[18c] (Eq. 15.14), both made use of both of the stabilizing N-donor bipy group and the exceptionally strong $M\text{-}C_6F_5$ and M—Me bonds.

$$Pd(bipy)Me_2 \xrightarrow{MeI} Pd(bipy)Me_3I \qquad (15.14)$$

Of all polyalkyls, the longest known are the Pt(IV) species. The orange complex $[Me_3Pt(\mu^3\text{-}I)]_4$, which has a cubane structure with octahedral platinum, was described by Pope and Peachey in 1907–1909.[18d] Some of its reactions (Eqs. 15.15–15.17; L = NH$_3$, en, py, PMe$_3$) illustrate how the chemistry resembles that for aqueous high-valent metal ions, such as the Co(III) Werner compounds that we looked at in Chapter 1.

$$[Me_3Pt(\mu^3\text{-}I)]_4 \xrightarrow{\text{L}} Me_3PtIL_2 \qquad (15.15)$$

$$\xrightarrow{\text{acacH}} [PtMe_3(acac)]_2 \qquad (15.16)$$

$$\xrightarrow{\text{water}} [PtMe_3(H_2O)_3]^+ \qquad (15.17)$$

Group 11

Cu and Ag give only M(I) alkyls, such as the bright yellow and explosive $[CuMe]_n$, but Au forms compounds from Au(I) to (III) such as $[Au(C_6F_5)_4]^-$. With many examples recently known, the reactions of high-valent alkyls now need to be investigated in more detail.

15.3 POLYHYDRIDES

Polyhydrides[19a] are complexes such as FeH$_4$(PR$_3$)$_3$, with a H : M ratio exceeding 3. Hydrogen is not as electronegative as carbon, and so the metal in a polyhydride is not as oxidized as in a polyalkyl. Polyhydrides therefore retain more of the properties of low-valent complexes than do polyalkyls. For example, many of them are 18e, and relatively soft ligands (in the vast majority of cases a phosphine or a cyclopentadienyl) are required to stabilize them. Rare examples of N-donor-stabilized polyhydrides are [TpReH$_6$] and [BpReH$_7$] (Tp = tris-pyrazolylborate (**5.37**) Bp = bis-pyrazolylmethane).[19b]

A second reason why the metal may not be as highly oxidized as is suggested by the high formal oxidation state is that not all polyhydrides have a *classical* structure, with all-terminal M–H bonds. Some are really dihydrogen complexes.[20] For example, IrH$_5$(P{C$_6$H$_{11}$}$_3$)$_2$ is classical and so authentically Ir(V), but [IrH$_6$(P{C$_6$H$_{11}$}$_3$)$_2$]$^+$ is in fact[21] [Ir^{III}H$_2$(H$_2$)$_2$(P{C$_6$H$_{11}$}$_3$)$_2$]$^+$, and so is Ir(III) not Ir(VII) because the dihydrogen ligand must be regarded as a 2e L-type ligand, contributing nothing to the oxidation state (Eq. 15.18).

$$(15.8)$$

$ReH_7(P\{p\text{-tolyl}\}_3)_2$ has the structure $ReH_5(H_2)L_2$ with a stretched H−H distance (1.357 Å[22a] instead of 0.8−1.0 Å in normal or unstretched H_2 complexes) and so the oxidation state is difficult to define because the structure is half way between the Re(V) and Re(VII) extreme formulations. $Re^{VII}H_7(dppe)$ is classical, however.[22b] A related Re tetrahydride exists in a tautomeric equilibrium (Eq. 15.19).[23a]

$$[ReH_4(CO)(PMe_2Ph)_3]^+ \rightleftharpoons [ReH_2(H_2)(CO)(PMe_2Ph)_3]^+ \qquad (15.19)$$

There is still doubt about the structures of some other polyhydrides, and this is an area in which X-ray crystallography is of limited use because of the small X-ray scattering factor for H. Crystals of the size appropriate for neutron work can be difficult to grow (Section 10.10), and NMR spectroscopic data (Section 10.7) are not always definitive.

Polyhydrides often have coordination numbers in excess of 6, a consequence of the small size of the hydride ligand. Nine is the normal limit on the number of ligands imposed by the availability of nine orbitals, but if a polyhydride can adopt a nonclassical structure with an H_2 molecule bound via a single metal orbital, this limit can be exceeded. A rare example of such a complex is "$[WH_7(PPh(CH_2CH_2PPh_2)_2)]^+$" (Eq. 15.20), which is stable up to −20°C in solution.[23b] Since **15.8** is classical with terminal M−H bonds, and therefore d^0, there are no metal lone pairs and so protonation must occur at the M−H bond to give an H_2 complex directly. If it were classical, **15.9** would exceed the maximum allowed oxidation state and coordination number for a transition metal.

$$WH_6(triphos) + H^+ \longrightarrow \text{``}[WH_7(triphos)]^+\text{''} \qquad (15.20)$$
$$\mathbf{15.8} \qquad\qquad\qquad\qquad \mathbf{15.9}$$

$$(triphos = PPh(CH_2CH_2PPh_2)_2)$$

Compound **15.9** must therefore have at least one H_2 ligand present but is probably $[WH_3(H_2)_2(triphos)]^+$. This d^2 formulation would allow for some back bonding to the H_2 ligands to help stabilize the M−(H_2) bond; d^0H_2 complexes are unknown. Spectroscopic methods show that some H_2 ligands are present but do not tell the number. The 7-coordinate polyhydrides, such as $IrH_5(PEt_2Ph)_2$, have a pentagonal bipyramidal structure, rather than the much more usual capped octahedron. This is also a consequence of the small size of the hydride ligand, five of which can bind in the equatorial plane of the complex. The 8-coordinate examples [e.g., $MoH_4(PMePh_2)_4$, **15.10**] tend to be dodecahedral, with the H ligands in the more hindered A sites (see Table 2.5). Nine coordinate hydrides are always found in the tricapped trigonal prismatic geometry first seen for $[ReH_9]^{2-}$ (**15.11**), an unusual example of a homoleptic hydride.

Almost all polyhydrides are fluxional in the ^1H NMR and the hydrides show coupling to any phosphines present. The number of hydrides present (n) can be predicted with some confidence from the 18e rule, but a useful experimental method involves counting the multiplicity ($n + 1$) of the ^{31}P NMR peak, after the phosphine ligand protons have been selectively decoupled (Section 10.4).

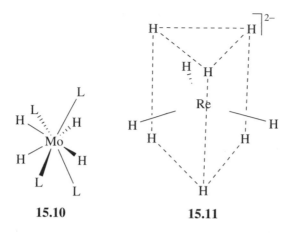

15.10   **15.11**

The basic character of polyhydrides is shown by the fact that many of them protonate, either to give stable cationic polyhydrides or to lose H_2 to give coordinatively unsaturated species, which can bind any ligand available, such as the solvent (Eq. 15.21).[24] Other polyhydrides can lose H_2 and bind N_2 or CO (= L in Eq. 15.23);[25] for nonclassical species this is especially easy. $ReH_7(PPh_3)_2$ is particularly interesting in that it can bind ligands such as pyridine,[26] phosphines,[27] and polyenes[28] to give substituted polyhydrides (Eq. 15.24).

$$MoH_4(PMePh_2)_4 \xrightarrow{\text{HBF}_4,\text{ MeCN}} [MoH_2(MeCN)_3(PMePh_2)_3](BF_4)_2 \quad (15.21)$$

$$WH_4(PMePh_2)_4 \xrightarrow{\text{HBF}_4,\text{ thf}} [WH_5(PMePh_2)_4]^+ \quad (15.22)$$

$$Ru(H_2)H_2(PPh_3)_2 \xrightarrow{\text{L}} RuLH_2(PPh_3)_2 \quad (15.23)$$

$$Re(H_2)H_5(PPh_3)_2 \xrightarrow{\text{diene}} (diene)ReH_3(PPh_3)_2 \quad (15.24)$$

Photochemical substitution is useful because it usually expels H_2 to generate one or more 2e sites at the metal (Eq. 15.25).[28]

$$MoH_4(PMePh_2)_4 \xrightarrow{h\nu,\text{ C}_2\text{H}_4} Mo(C_2H_4)_2(PMePh_2)_4 \quad (15.25)$$

The importance of polyhydrides in the activation of alkanes has already been discussed (Eq. 12.34).

> • Hydrides can give very high coordination
>   numbers, but polyhydrides are otherwise
>   rather conventional in their behavior.

15.4 CYCLOPENTADIENYL COMPLEXES

The Cp and especially the Cp* ligands are very effective at stabilizing high oxidation states. While the Cp complexes can be polymeric and difficult to characterize, the Cp* species are often well-behaved, soluble complexes. Several high-oxidation-state halo complexes have been known for many years, for example, Cp_2TiCl_2, Cp_2NbCl_3, Cp_2TaCl_3, and $[Cp_2MoCl_2]^+$. A well-known route to oxo and halo species is oxidation of the cyclopentadienyl carbonyls or the metallocenes.[29–32] The $[CpMO]_4$ complexes, of which the earliest (1960) was Fischer's $[CpCrO]_4$, have the cubane structure (**15.12**).[30]

$$CpV(CO)_4 \xrightarrow{\text{HBr/O}_2,\ \text{or Cl}_2} CpVOX_2 \qquad\qquad (15.26)^{29}$$

$$Cp_2Cr \xrightarrow{\text{O}_2}$$

15.12

$$[CpMo(CO)_3]_2 \xrightarrow{\text{O}_2,\ h\nu} CpMoO_2Cl + (CpMoO)_2(\mu\text{-O})_2$$

$$+[(CpMoO_2)_2(\mu\text{-O})] \xrightarrow{\text{hydrolysis}} [CpMoO]_4 \qquad\qquad (15.28)^{31}$$

$$CpMo(CO)_3Me \xrightarrow{\text{PCl}_5} CpMoCl_4 \qquad\qquad (15.29)^{32a}$$

Reaction of carbonyls with air or with PCl_5 seems to be a general method for preparing oxo and chloro complexes (Eqs. 15.28 and 15.29). These compounds can also react with organic peroxides; for example, $Cp^*W(=O)_2Me$ gives the very unusual η^2-peroxo complex, $Cp^*W(=O)(\eta^2\text{-O}_2)Me$.[32b]

Rhenium

As one might perhaps expect, rhenium seems to have the most extensive oxo chemistry of this type. The early elements are so oxophilic that organometallic groups are unlikely to survive, when lower valent species are oxidized or hydrolyzed. Re is the last element, as we go to the right in the periodic table, for which the M=O bond is still reasonably stable. Herrmann et al.[33] have shown how to make a whole series of oxo complexes of Cp*Re. The Re=O vibrations show up very strongly in the IR spectrum, as for the yellow Cp^*ReO_3 at 878 and 909 cm^{-1}, and the IR provides useful data for the characterization of all the complexes shown.

$$Cp^*Re(CO)_3 \xrightarrow{O_2, h\nu, \text{ or } H_2O_2} Cp^*ReO_3 \qquad (15.30)^{33}$$

Partial reduction of Cp^*ReO_3 under various conditions can lead to the species shown in Fig. 15.5.[33] Note the selectivity of SnMe$_4$ alkylation versus MeMgBr. The binuclear species are interesting because the short M$-$M distances found indicate that M$-$M bonds are present, a somewhat unexpected feature for such high-valent metals. CO reduction gives an unusual oxocarbonyl; CO is characteristic of low-valent, and M=O of high-valent, metals, and the two ligands are rarely seen in the same complex. Compound **15.14** is interesting in being an unusual cluster hydride. Compound **15.15** is a mixed-valent species, the metal bearing the terminal CO being Re(II), and the one bearing the terminal oxo group

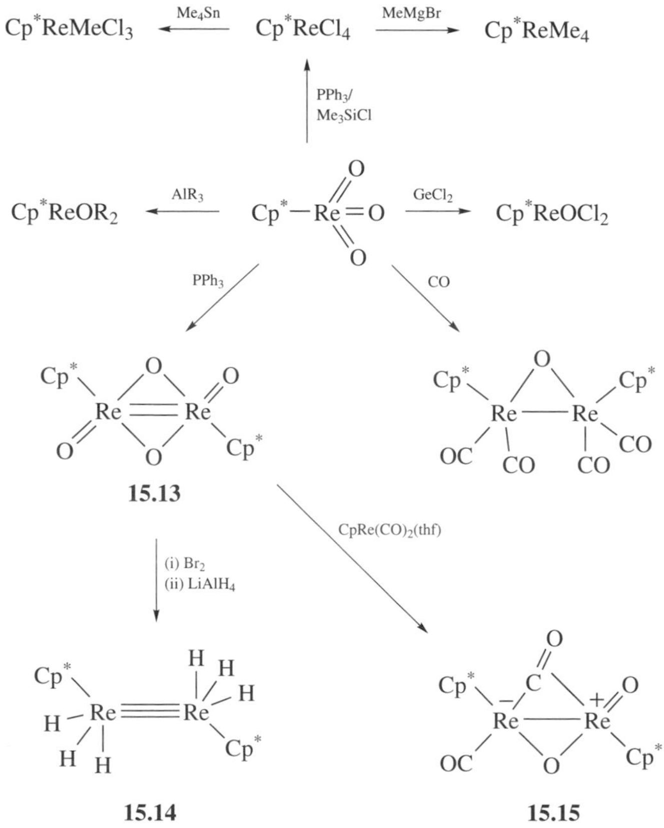

FIGURE 15.5 Some high-oxidation-state organometallic chemistry of the Cp*Re fragment.

being Re(IV); the semibridging CO is also a striking feature of the complex. The Cp^*ReX_4 systems in Fig. 15.5 all have low- and high-spin forms in equilibrium leading to unusual temperature-dependent shifts in the ^1H NMR spectra, for example, the Re*Me* signal in Cp^*ReCl_3Me is broad and shifts from 13.5δ at $-50°C$ to 36.5δ at $+50°C$ in $CDCl_3$.

Other Metals

Maitlis et al.[34] has described a number of Ir(V) and Rh(V) alkyls, such as Cp^*IrMe_4. $M(\eta^3$-allyl$)_4$ complexes also exist for Zr, Nb, Ta, Mo, and W.[35] Some complexes with agostic bonding have occasionally been considered as having a high oxidation state, as if oxidative addition of the agostic bonds had occurred; even Pd(VI) has been proposed in a doubly agostic case that is unusual in that the agostic bond is a Si−Si and not a C−H or Si−H.[36a] These are probably better interpreted in more conventional oxidation states by counting the agostic bond as a neutral ligand (Section 3.4); if so, doubly agostic Pd(VI) becomes Pd(II), for example.[36b]

15.5   *f*-BLOCK COMPLEXES

The *f*-block[37] consists of the $4f$ metals, La–Lu, and the $5f$ metals, Ac–Lr. The common terms *lanthanide* and *actinide* derive from the names of the first elements of each series, and the symbol Ln, not assigned to any particular element, is a useful way to designate the lanthanides as a class. The older term for lanthanides, *rare earths*, is sometimes encountered. The actinides are radioactive, and only Th and U are sufficiently stable to be readily handled outside high-level radiochemical facilities (^{238}U, $t_{1/2} = 4.5 \times 10^9$ years; ^{232}Th, $t_{1/2} = 1.4 \times 10^{10}$ years). Even though they have no *f* electrons, scandium (Sc) and yttrium (Y) in group 3 are also typically considered with the *f*-block elements because of their rather similar chemistry.

Unlike *d* electrons in the *d* block, $4f$ electrons are unavailable for bonding, and there is no equivalent of ligand field effects or of the 18e rule. Instead, the complexes tend to be predominantly ionic with no electronic preferences for particular geometries—indeed, irregular geometries are common. They become sterically saturated as a result of ligands continuing to add until the available space around the metal is filled. If the ligand set chosen does not saturate the metal sterically, oligomeric or polymeric structures are seen in which bridging occurs to adjacent molecules; a larger ligand would be needed to prevent bridging and provide a monomeric structure. This accounts for the key role of carefully adjusting steric effects in designing ligands for these elements. The high tendency to bridge also means that ligand redistribution and exchange is often fast.

The absence of ligand field effects has the further consequence that the magnetism of *f* block complexes is the same as that of the parent ion. In the *d* block, a d^2 complex such as Cp_2WCl_2 is typically diamagnetic as a result of *d* orbital splitting (Fig 5.5); in contrast, $5f^2$ Cp_2UCl_2 has two unpaired electrons.

Variable valence is a key feature of the d-block elements—in contrast, the $4f$ elements generally prefer the tripositive state. Table 15.2 shows the atomic electron configurations of the $4f$ elements, together with the configurations of their common oxidation states. There is an evident tendency to prefer an unfilled, a half-filled, or a filled f shell, accounting for the existence of some of the non-M(III) states, Ce(IV), Eu(II), Tb(IV), and Yb(II). The oxidation states with unfilled and fully filled f shells are also important because they are diamagnetic, allowing easy study of the complexes by conventional proton and carbon NMR. Line broadening is relatively small in many cases, with the paramagnetic Pr(III), Sm(II), Sm(III), and Eu(III) giving the most easily observable spectra. No doubt for this reason La(III), Ce(IV), Yb(II), and Lu(III)—together with diamagnetic Sc(III) and Y(III) from group 3—are among the most intensively studied states.

The trend in radius, shown for the M(III) ion in Table 15.2, is the result of the increasing number of protons in the nucleus causing the electron shells to contract; the f electrons added are deep-lying and inefficient at screening the nuclear charge. In most of chemistry, when we move from one element to the next, the changes in atomic size and preferred valency are abrupt. Here, in contrast, the radius varies smoothly and the M(III) valence state remains preferred, so we have a nice control over the M—L bond length. As this varies, the effective steric size of the ligands gradually varies because the ligand cone angle (Section 4.2) increases as the ligand gets closer to the metal. This *lanthanide contraction* from La–Lu helps account for the fact that the third-row d-block metals, Hf–Hg, which come just after the lanthanides in the periodic table, have a smaller increment in atomic radius over the second row than would be expected by extrapolation of the radius

TABLE 15.2 Lanthanide Electronic Configurations and Ion Radii[a]

Element	Atom Config.	M(II) Config.	M(III) Config.	M(IV) Config.	Radius M(III) (Å)
Lanthanum, La	$4f^05d^16s^2$		$4f^0$		1.16
Cerium, La	$4f^25d^06s^2$		$4f^1$	$4f^0$	1.14
Praseodymium, Pr	$4f^35d^06s^2$		$4f^2$	$4f^1$	1.13
Neodymium, Nd	$4f^45d^06s^2$	$4f^4$	$4f^3$		1.11
Promethium, Pm	$4f^55d^06s^2$		$4f^4$		1.09
Samarium, Sm	$4f^65d^06s^2$	$4f^6$	$4f^5$		1.08
Europium, Eu	$4f^75d^06s^2$	$4f^7$	$4f^6$		1.07
Gadolinium, Gd	$4f^75d^16s^2$		$4f^7$		1.05
Terbium, Tb	$4f^85d^06s^2$		$4f^8$	$4f^7$	1.04
Dysprosium, Dy	$4f^95d^06s^2$	$4f^{10}$	$4f^9$		1.03
Holmium, Ho	$4f^{10}5d^06s^2$		$4f^{10}$		1.02
Erbium, Er	$4f^{11}5d^06s^2$		$4f^{11}$		1.00
Thulium, Tm	$4f^{12}5d^06s^2$	$4f^{13}$	$4f^{12}$		0.99
Ytterbium, Yb	$4f^{13}5d^06s^2$	$4f^{14}$	$4f^{13}$		0.99
Lutetium, Lu	$4f^{14}5d^16s^2$		$4f^{14}$		0.98

[a]Oxidation state exists whenever configuration is shown.

change between the first- and second-row *d*-block metals. This is illustrated by the metallic radius trends for some triads: Ti, 1.47 Å; Zr, 1.60 Å; Hf, 1.59 Å; Cr, 1.29 Å; Mo, 1.40 Å; W, 1.41 Å; Ni, 1.25 Å; Pd, 1.37 Å; Pt, 1.39 Å.

As the ionic radius changes, the preferred coordination number can change. For the aqua ions $[Ln(H_2O)_n]^{3+}$, n is 9 for the larger ions, La–Eu, and 8 for the smaller ions, Tb–Lu. For Gd^{3+}, $n = 8$ and $n = 9$ ions have about the same energies. The later lanthanide ions, being smaller, have a slightly greater Lewis acidity.

Ionic bonding plays a greater, although not exclusive, role in their chemistry, compared to the *d* block, consistent with the low Pauling electronegativities of these elements (1.0–1.25). The *f* electrons are low-lying in the ions and complexes and do not participate to any great extent in bonding, as shown by the fact that both the magnetic moments and the color are practically the same in the free ion and in any of the complexes. The *f*-*f* transitions in the UV–visible spectrum, responsible for the color of the ions, are very sharp because the deep-lying *f* electrons are isolated from the effects of ligand binding or solvation. These transitions are also involved in the strong luminescence often seen for lanthanide compounds, as in the red Eu based phosphor in color TVs and Nd-based YAG lasers. It does not take very much energy to promote an *f* electron to the *d* level, however, and the resulting $4f^n$ to $4f^{n-1}5d^1$ transition can often be detected in the UV region. Since the $5d$ levels *are* affected by the ligands, this $f \rightarrow d$ band is broad, and the wavelength does depend on the nature of the complex. For example, in $[\{\eta^5\text{-}C_5H_3(SiMe_3)_2\}_3Ce]$ the $f \rightarrow d$ band is reduced to such an extent in energy that it appears in the visible range at 17,650 cm^{-1} compared to 49,740 cm^{-1} in the gas-phase UV spectrum of the bare Ce^{3+} ion.

Among the $5f$ elements, we look at Th, with its strongly preferred $5f^0$ Th(IV) state, and U with $5f^3$ U(III), $5f^2$ (IV), $5f^1$ (V) and $5f^0$ (VI) states all accessible. In the actinides, the complexes have somewhat more covalency in their bonding than do the $4f$ elements, in line with the higher electronegativities (U, 1.38), and in the case of reduced states of U, a significant tendency to back bond. The $5f$ level is somewhat more available for bonding than is $4f$ in the lanthanides.

Lanthanide Organometallic Chemistry

In the Ln(III) state, this broadly resembles the chemistry of the early *d*-block elements in their highest oxidation states. The lanthanide complexes are, of course, paramagnetic for all configurations from $4f^1$ to $4f^{13}$. Another difference is that the larger size of the Ln(III) ions versus Ti(IV)–Hf(IV) make the preferred coordination number higher for the *f* block; 8-coordination is typical for lanthanide complexes.

As hard Lewis acids, Ln^{3+} tend to prefer hard ligands, typically O donors, hence the term *oxophilic* often applied to these ions. Marks' series of bond energies for Cp*$_2$Sm-X compounds illustrates the bonding preferences are not quite as clear-cut as hard/soft ideas would have it: Cl > C≡CPh > Br > O(*t*-Bu) > S(*n*Pr) > I > H > NMe$_2$ > PEt$_2$.[38]

Simple alkyls, typically formed from LiR and $LnCl_3$, are possible when R is β elimination resistant, such as in the $[LnMe_6]^{3-}$ series of -*ate* (anionic) complexes. Bulky alkyls are necessary if bridging is to be avoided, as in the triangular 3-coordinate series $[Ln\{CH(SiMe_3)_2\}_3]$.[39] β Elimination has a lower driving force in the f than in the d block because the M—H/M—C bond energy difference is less favorable to M—H. Indeed, β-alkyl elimination, not generally seen in the d block, is common here for the same reason.[40]

Cyclopentadienyl ligands[41] have attracted most attention because they are ideally suited to these metals since they are capable of ionic bonding and can be readily sterically tuned with a variety of substituents. The first $LnCp_3$ complexes were prepared by Wilkinson and Birmingham as early as 1954. The 18e rule is entirely inapplicable: The formal electron count of $LnCp_3$ is 18 from the Cp^- ligands (the ionic model is most appropriate for this case) plus n from the ion, $4f^n$ being the ion configuration. In reality the Cp electrons stay largely on the ligand, but the metal–ligand bond strength can still be very high as a result of the 3+ charge on the metal. The pronounced oxophilicity leads to the formation of a THF complex that requires a temperature of >200°C to desolvate (Eq. 15.31).

$$LnCl_3 + NaCp \xrightarrow{\text{THF}} Cp_3Ln{\cdot}THF \xrightarrow[>200°C]{\text{sublime}} Cp_3Ln \qquad (15.31)$$

The solid-state structures adopted form a delicately ordered set. A strictly monomeric structure is only seen for $(\eta^5\text{-}Cp)_3Yb$, where steric saturation is precisely attained without the need for bridging. All the other cases involve some degree of Cp bridging between metals. The ions smaller than Yb, Lu, and Sc have $[(\eta^5\text{-}Cp)_2M]^+$ units bridged in an infinite chain by $\eta^1\text{-}Cp^-$ groups. The ions larger than Yb have a $(\eta^5\text{-}Cp)_3M$ structure with space available for bridges to adjacent Cp_3M units.

Bis-cyclopentadienyl complexes are also seen; the example of Y (Eq. 15.32) shows how LiCl can form an adduct with the product—an adduct that is only cleaved at 285°. Note how the monobridged structure of the product in Eq. 15.32 contrasts with the bis-bridged $[Cp_2Y(\mu\text{-}Cl)_2YCp_2]$, again resulting from steric differences, this time between Cp and Cp*.

$$YCl_3 + LiCp^* \xrightarrow{\text{THF}} Cp_2^*Y\underset{Cl}{\overset{Cl}{\diagdown}}Li(THF)_2 \xrightarrow[285°C]{\text{sublime}} Cp_2^*Y\overset{Cl}{\diagdown}\underset{\underset{Cl}{|}}{YCp_2^*}$$

$$(15.32)$$

Just as the lanthanides are oxophilic, they are also very fluorophilic, so BF_4^- is far from being the relatively noncoordination anion it is in late d-block chemistry: An example is the chelating fluoroborate, **15.16**. Methyl groups are also able to bridge, as in $[Cp_2Lu(\mu\text{-}Me)_2AlMe_2]$. Their oxophilicity also makes $4f$ and $5f$ organometallics very water and air unstable, resembling early d-block metals in this respect.

Cp rings can be connected to give an *ansa* system (Latin = handle), of which two examples are shown in **15.17** and **15.18**.

15.16   **15.17**   **15.18**

One of the most striking early discoveries (1982) in alkane activation chemistry was Watson's exchange reaction between a coordinated methyl group and free methane, via σ bond metathesis, discovered by ^{13}C isotope labeling of the methane carbon.

The same alkyl also undergoes hydrogenolysis with H_2 (Eq. 15.33), as well as insertion with alkenes, with β-alkyl elimination also being possible (Eq. 15.34).

$$(15.33)$$

$$(15.34)$$

For many years Cp* was never seen to form Cp_3^*Ln compounds and it was assumed that it was just too large. The reaction of Eq. 15.35, with its large driving force, permits the formation of the tris species. The tetraene takes one electron from each of two Sm(II) units to give two Sm(III) complexes. Detailed study of the tris complex showed that the Sm—C bond lengths are longer (av. 2.82 Å) than usual (2.75 Å) as a result of steric crowding forcing the Cp* ligands to move away from the metal. As might be expected, one of the Cp* groups easily departs, as in Eq. 15.36.[42]

$$(15.35)$$

$$2Cp_3^*Sm \ + \ H_2 \ \longrightarrow \ 2Cp^*H \ + \ Cp_2^*Sm\underset{H}{\overset{H}{\diagup\!\!\diagdown}}SmCp_2^* \qquad (15.36)$$

Cp_2^*Sm is an example of a reduced organolanthanide [Sm(II)]. Its special feature is a strongly bent structure quite unlike that of ferrocene. One possible reason is that this predominantly ionic system has no special geometric preference, and the bent arrangement generates a dipole that interacts favorably with neighboring Cp_2^*Sm dipoles.

Soft ligands like CO bind very weakly to $4f$ elements: for example, Cp_2^* Eu and CO are in equilibrium with $Cp_2^*Eu(CO)$.[43] For Cp_2^*Yb, the equilibrium includes both $Cp_2^*Yb(CO)$ and $Cp_2^*Yb(CO)_2$. Crystal structures not being useful here, the IR spectral data for $\nu(CO)$ were interpreted by comparison with the spectra predicted from DFT calculations.[44] These suggest that CO in $Cp_2^*Eu(CO)$ is conventionally C bound, but that for Yb, the adducts contain O-bound *isocarbonyls*: $Cp_2^*Yb(OC)$ and $Cp_2^*Yb(OC)_2$. This shows both the power of modern computational chemistry as well as the very high oxophilicity of the $4f$ metal. The bonding between Cp_2^*Ln and CO is largely dipole–dipole in character and the change from carbonyl to isocarbonyl from Eu to Yb is attributed to larger electron–electron repulsions with the more electron-rich carbon end of the CO in $4f^{14}$ Yb(II) versus $4f^7$ Eu(II). The weak adduct between Cp_2^*Yb and another soft ligand, $MeC\equiv CMe$ has been isolated and even characterized by X-ray crystallography, but the resulting Yb–C distance, 2.85 Å, is rather long compared to 2.66 Åfor the Yb–C distances to the Cp carbons.[45] Isonitriles, RNC, do bind well to Ln(III), as in $Cp_3Ln(CNPh)$, but only because RNC is a substantial σ donor; back donation is minimal, as shown by the increase in $\nu(NC)$ of 60–70 cm^{-1} on binding to Ln(III), compared to the decrease seen in complexes like Cp_2 W(NCPh).

Rare examples of Ln(0) species are known, such as the bis-arene [Ln{η^6-$C_6H_3(CMe_3)_3$}$_2$] made by metal vapor synthesis.[46]

As we saw in Chapter 14, lanthanides have some applications in organic synthesis.

Actinide Organometallic Chemistry

Uranocene[47] (**15.19**), a key discovery from 1968, showed that the higher radius and charge of U^{4+} relative to the lanthanides allows stabilization of the planar, aromatic, 10π-electron cyclooctatetraene dianion (cot^{2-}) in U(cot)$_2$. This pyrophoric 22e compound also shows the failure of the 18e rule in the $5f$ elements.

Cyclopentadienyls are again widely used as spectator ligands. Equation 15.37 shows how a thorium alkyl is hydrogenolyzed by H_2

$$[Cp_2^*ThCH_2SiMe_3]^+ + H_2 = [Cp_2^*ThH]^+ + SiMe_4 \qquad (15.37)$$

Carbonyls are somewhat more stable in the $5f$ series. $(Me_3SiC_5H_4)_3U(CO)$ has a relatively low $\nu(CO)$ value of 1976 cm^{-1}, but it easily loses CO. The more basic $(C_5Me_4H)_3U$ gave $(C_5Me_4H)_3U(CO)$ quantitatively with the surprisingly low $\nu(CO)$ of 1880 cm^{-1}, suggesting strong U–CO π backbonding.[48]

- Lanthanides are of growing importance because of their distinctly different properties from the d block.
- Ln(III) dominates but Ln(II) and Ln(IV) are known, specially when this results in an f^0, f^7, or f^{14} configuration.
- The smooth variation in ionic radius means that steric effects can be finely tuned by variation of the lanthanide.
- Steric saturation, not electron count, decides structure, so steric considerations dominate ligand choices.
- The f electrons are in the core, ligand field effects are absent, and distorted geometries are common.
- In actinides, the 5f, electrons do contribute somewhat to the bonding

---

REFERENCES

1. B. N. Figgis and M. A. Hitchman, *Ligand Field Theory and Its Applications*, Wiley, New York, 2000.

2. (a) R. Poli, *Chem. Rev.* **96**, 2135, 1996; J. N. Harvey, R. Poli, and K. M. Smith, *Coord. Chem. Rev.* **238**, 347, 2003; (b) D. Schroder, S. Shaik, and H. Schwartz, *Acc. Chem. Res.* **33**, 139, 2000; (c) J. L. Carredon-Macedo and J. N. Harvey *J. Am. Chem. Soc.* **126**, 5789, 2004.

3. A. L. Rheingold, K. Theopold et al., *J. Am. Chem. Soc.* **117**, 11745, 1995.

4. F. H. Köhler et al., *J. Organomet. Chem.* **365**, C15, 1989.

5. R. Poli, *J. Coord. Chem. B* **29**, 121, 1993.

6. K. Clauss and C. Beermann, *Angew. Chem.* **71**, 627, 1959.

7. W. Bassi et al., *J. Am. Chem. Soc.* **93**, 3788, 1971.

8. L. Marko et al., *J. Organomet. Chem.* **199**, C31, 1980.

9. B. K. Bower and H. G. Tennent, *J. Am. Chem. Soc.* **94**, 2512, 1972.

10. R. M. G. Roberts, *J. Organomet. Chem.* **63**, 159, 1973.

11. G. Wilkinson et al., *Chem. Commun.* 159, 1976; R. R. Schrock and P. Meakin, *J. Am. Chem. Soc.* **96**, 159, 1974.

12. G. Wilkinson et al., *J. Chem. Soc., Dalton* 872, 1973.

13. I. Feinstein-Jaffé, J. C. Dewan, and R. R. Schrock, *Organometallics* **4**, 1189, 1985.

14. G. Wilkinson et al., *Chem. Commun.* 1398, 1986.

15. (a) G. Wilkinson et al., *J. Chem. Soc., Dalton* 607, 1975; 1488, 1976; 334, 1980; *Nouv. J. Chim.* **1**, 389, 1977; I. R. Beattie and P. J. Jones, *Inorg. Chem.* **18**, 2318, 1979; (b) W. A. Herrmann, *Angew. Chem., Int. Ed.* **27**, 1297, 1988.

16. J. M. Huggins et al., *J. Organomet. Chem.* **312**, C15, 1986.

17. Z. Lu and R. H. Crabtree, *J. Am. Chem. Soc.* **117**, 3994, 1995.

18. (a) A. J. Canty, *Acc. Chem. Res.* **25**, 83, 1992; (b) R. Usón et al., *J. Organomet. Chem.* **96**, 307, 1975; (c) A. J. Canty et al., *Chem. Commun.* 1722, 1986; (d) W. J. Pope and S. J. Peachey, *J. Chem. Soc.* 571, 1909.

19. (a) F. Maseras, A. Lledos, E. Clot, and O. Eisenstein, *Chem. Rev.* **100** 601, 2000; (b) D. G. Hamilton, X.-L. Luo, and R. H. Crabtree, *Inorg. Chem.* **28**, 3198–3203, 1989.

20. G. J. Kubas *Metal Dihydrogen and Sigma Bond Complexes*, Kluwer, New York, 2001; R. H. Crabtree, *Angew. Chem., Int. Ed.* **32**, 789, 1993.

21. R. H. Crabtree and M. Lavin, *Chem. Commun.* 1661, 1985; *J. Am. Chem. Soc.* **108**, 4032, 1986; R. H. Crabtree and D. Hamilton, *J. Am. Chem. Soc.* **108**, 3124, 1986.

22. J. A. K. Howard et al., *Chem. Commun.* (a) 241, 1991; (b) 1502, 1988.

23. (a) X. L. Luo and R. H. Crabtree, *J. Am. Chem. Soc.* **112**, 6912, 1990; (b) D. Michos, X.-L. Luo, J. W. Faller, and R. H. Crabtree, *Inorg. Chem.* **32**, 1370, 1993.

24. K. G. Caulton et al., *Inorg. Chem.* **21**, 4185, 1982.

25. G. Wilkinson et al., *J. Chem. Soc., Dalton* 1716, 1977; V. D. Makhaev, A. P. Borisov et al., *Koord. Khim.* **4**, 1274, 1978; **8**, 963, 1982.

26. R. O. Harris et al., *J. Organomet. Chem.* **54**, 259, 1973.

27. D. Baudry, M. Ephritikine, H. Felkin et al., *J. Organomet. Chem.* **224**, 363, 1982.

28. M. S. Wrighton, G. L. Geoffroy et al., *J. Am. Chem. Soc.* **104**, 7526, 1982.

29. H. J. Leifde-Meijer et al., *Rec. Trav. Chem.* **80**, 831, 1961; *Chem. Ind. (Lond.)* 119, 1960.

30. (a) E. O. Fischer et al., *Chem. Ber.* **93**, 2167, 1960; (b) F. Bottomley et al., *J. Am. Chem. Soc.* **104**, 5651, 1982.

31. M. L. H. Green et al., *J. Chem. Soc.* 1567, 1964.

32. (a) R. R. Schrock et al., *Organometallics* **4**, 953, 1985; (b) J. W. Faller and Y. Ma, *J. Organomet. Chem.* **368**, 45, 1989.

33. W. A. Herrmann et al., *Angew. Chem., Int. Ed.* **23**, 383, 515, 1983; **24**, 50, 860, 1984; *J. Organomet. Chem.* **272**, 55, 287, 329, 1985; **300**, 111, 1986.

34. P. Maitlis et al., *Chem. Commun.* 310, 1982.

35. G. Wilke, *Angew. Chem., Int. Ed.* **2**, 105, 1963; **5**, 151, 1966.

36. (a) S. Shimada, M. Tanaka et al., *Science* **295**, 308, 2002; (b) R. H Crabtree, *Science* **295**, 288, 2002.

37. H. C. Aspinall, *Chemistry of the f-Block Elements*, Gordon & Breach, London, 2001.

38. S. P. Nolan, D. Stern, and T. J. Marks *J. Am. Chem. Soc.* **111**, 7844, 1989.

39. P. B. Hitchcock, M. F. Lappert, R. G. Smith, R. A. Bartlett, and P. P. Power, *Chem. Commun.* 1007, 1988.

40. E. Bunel, B. J. Burger, and J. E. Bercaw, *J. Am. Chem. Soc.* **110**, 976, 1988.

41. S. Arndt and J. Okuda, *Chem. Rev.* **102**, 1953, 2002.

42. W. J. Evans and B. L. Davis, *Chem. Rev.* **102**, 2119, 2002.

43. M. Schultz, C. J. Burns, D. J. Schwartz, and R. A. Andersen, *Organometallics* **20**, 5690, 2001.

44. L. Perrin, L. Maron, O. Eisenstein, and R. A. Andersen, *J. Am. Chem. Soc.* **124**, 5614, 2002.

45. C. J. Burns and R. A. Andersen, *J. Am. Chem. Soc.* **109**, 941, 1987.

46. J. G. Brennan, F. G. N. Cloke, A. A. Sameh, and A. Zalkin, *Chem. Comm.* 1668, 1987.

47. D. Seyferth, *Organometallics* **23**, 3562, 2004.

48. R. A. Andersen et al., *J. Am. Chem. Soc.* **108**, 335, 1986; E. Carmona et al., *J. Am. Chem. Soc.* **117**, 2649, 1995.

PROBLEMS

1. Suggest reasons why $Ti(CH_2Ph)_4$ does not form a stable CO adduct.

2. Given that an unstable CO adduct of $Ti(CH_2Ph)_4$ is an intermediate on the way to forming $Ti(COCH_2Ph)_2(CH_2Ph)_2$, suggest reasons why this adduct might be especially reactive.

3. Why do you think V only gives VR_4 as the highest-oxidation-state alkyl, but Ta can give TaR_5?

4. What mechanism is likely for Eq. 15.4 (reaction with Br_2), and would **15.6** and **15.7** be likely to give the same type of reaction?

5. The ethylenes in $Mo(C_2H_4)_2(PR_3)_4$ are mutually trans. What do you think the orientation of their C−C bonds would be with respect to one another? (Draw this looking down the principal axis of the molecule.)

6. Why are alkene polyhydrides so rare? Why is $Re(cod)H_3(PR_3)_2$ an exception, given that its stereochemistry is pentagonal bipyramidal, with the phosphines axial?

7. What values of the spin quantum number S are theoretically possible for: $CpCrLX_2$, $CpMnL_2X_2$, $CpFeLX_2$, $CpCoLX_2$?

8. $Cp_2^* Lu H$ reacts with C_6H_6 to give $[(Cp_2^* Lu)_2C_6H_4]$. What structure do you predict for this compound?

9. What spin states are in principle possible for: (a) d^6 octahedral; (b) f^2 8-coordinate; and (c) d^3 octahedral.

16

BIOORGANOMETALLIC CHEMISTRY

In the future, chemistry will be increasingly influenced by biology as a result of the dramatic advances in our understanding of the chemical basis of life.[1a] Both organic and inorganic[1b] motifs have long been known to be present in living things. Only with coenzyme B_{12} (Section 16.2) did it become clear that organometallic species also occur in biology, both as stable species and as reaction intermediates. Nature uses organometallic chemistry sparingly, but it has been suggested[2] that the examples we see today are relics of early life forms, which had to live on simple molecules, such as H_2, CO, and CH_4, that may have used organometallic chemistry more extensively. The elements Co and Ni are rather unusual in biology, but when they are found, it is often in the context of organometallic chemistry. The term *bioorganometallic chemistry* dates from 1985.[1c] We will first review the basic aspects of biochemistry as they apply to enzymes.[1a]

All the systems described in this chapter are organometallic in character. Coenzyme B_{12} has several forms with M—C or M—H bonds. In nitrogen fixation, CO binds competitively at the active site. The nickel enzymes are believed to operate via intermediates with M—H (H_2ase) or M—C bonds (CODH and MeCoM reductase).

- Nature sometimes uses organometallic chemistry, but much less often than she uses coordination chemistry.

The Organometallic Chemistry of the Transition Metals, Fourth Edition, by Robert H. Crabtree
Copyright © 2005 John Wiley & Sons, Inc.

16.1 INTRODUCTION

One of the most important features of the chemistry of life is that biochemical reactions have to be kept under strict control. They must only happen as they are required, where they are required. One way of doing this is to employ reactions that can only proceed when catalyzed. The organism now only has to turn the appropriate catalysts on and off to control its biochemistry. The catalysts of biology are called *enzymes*, and they can be soluble, or bound to a membrane, or even part of an enzyme complex, in which case they act as a cog in a larger piece of biochemical machinery.

Proteins

Essentially all enzymes are *proteins*; that is, they are made up of one or more polypeptide chains having the structure shown in **16.1**. The value of *n* usually ranges from 20 to 100, and there may be several separate polypeptide chains or *subunits* in each enzyme. Sometimes two or more proteins must bind together to give the active enzyme. The monomers from which protein polymers are built up are the amino acids, $RCH(NH_2)COOH$, which always have the L configuration. There are more than 20 different amino acids commonly found in proteins, each having a different R group (see Table 16.1). The ordering of the R groups along the protein chain is its *primary structure* and is of great significance. Each enzyme has its own specific ordering, which often differs in minor ways if we isolate the same enzyme from one species rather than another. Chains that have similar sequences are said to be *homologous*. In spite of minor sequence differences, the chains can fold in the same way in all cases to give an active enzyme. The sequence of the R groups is believed to decide the way in which the chain will fold, and the R groups also provide the chemical functional groups that enable the protein to perform its function. The problem of predicting the folding pattern of a polypeptide (usually found by X-ray diffraction or NMR) from its primary structure is still unsolved. Two types of *secondary structure* are common, the rodlike α helix and the flat β sheet. In each case the folding is decided by the patterns of many hydrogen bonds formed between N–H groups of one peptide bond and CO groups of another. *Tertiary structure* refers to the finer details resulting from H-bonding or other interactions between the R groups of the residues. Finally, *quaternary structure* refers to the way the subunits pack together. Greek letters are used to designate subunit structure; for example, an $(\alpha\beta)_6$ structure is one in which two different chains α and β form a heterodimer, which, in turn, associates into a hexamer in the native form of the protein.

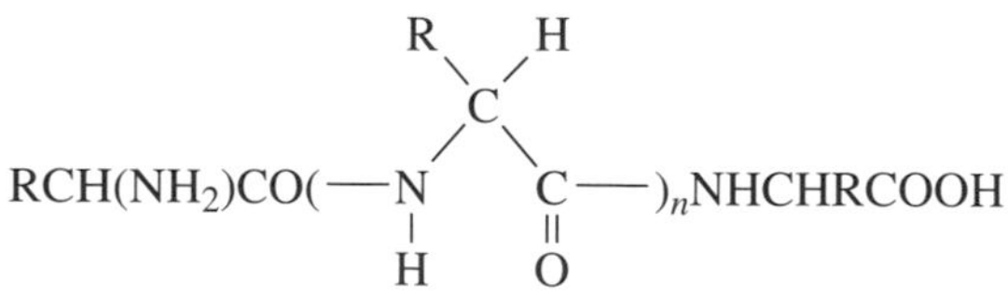

$$
RCH(NH_2)CO(\!\!-\!N\overset{\displaystyle \underset{H}{|}}{}\quad \overset{\displaystyle \underset{\parallel}{C}}{\underset{O}{}}\!\!-\!)_n NHCHRCOOH
$$

R, H on C; C–N, C=C structure

16.1

TABLE 16.1 Some Common Amino Acids

Name	Symbol	R	Remarks
Glycine	Gly	H	Nonpolar R group
Alanine	Ala	Me	"
Valine	Val	i-Pr	"
Leucine	Leu	i-PrCH$_2$	"
Phenylalanine	Phe	PhCH$_2$	"
Glutamic acid	Glu	$^-$O$_2$CCH$_2$CH$_2$	Anionic R group[a]
Aspartic acid	Asp	$^-$O$_2$CCH$_2$	"
Lysine	Lys	$^+$H$_3$N(CH$_2$)$_4$	Cationic R group[a]
Arginine	Arg	$^+$H$_2$N=C(NH$_2$)NH(CH$_2$)$_3$	"
Tyrosine	Tyr	HO(C$_6$H$_4$)CH$_2$	Polar but not ionized
Serine	Ser	HOCH$_2$	"
Threonine	Thr	MeCH(OH)	"
Asparagine	Asn	H$_2$NOCCH$_2$	"
Methionine	Met	MeSCH$_2$CH$_2$	Soft nucleophile
Cysteine	Cys	HSCH$_2$	—[b]
Histidine	His	C$_3$N$_2$H$_4$CH$_2$	—[c]

[a]Predominant protonation state at pH 7.
[b]Binds metal ions and links polypeptide chains via an −CH$_2$S−SCH$_2$− group.
[c]Heterocyclic amine base that acts as a nucleophile or binds metal ions.

Certain R groups are "greasy" and will tend to be found in the interior of the structure. Others are hydrophilic and are likely to be found at the surface. Some are sufficiently acidic or basic so as to be deprotonated or protonated at physiological pH (generally close to 7); these provide a positive or negative charge at the surface of the protein. When histidine is present, it usually serves one of two special functions: either as a nucleophile to attack the substrate, or to ligate any metal ions present. Similarly, cysteine either holds chains together by formation of a disulfide link (RS−SR) with a cysteine in another chain or binds a metal ion as a thiolate complex (RS−ML$_n$). Any nonpolypeptide component of the protein required for activity (e.g., a metal ion, or an organic molecule) is called a *cofactor*. Sometimes two or more closely related protein conformations are possible. Which is adopted may depend on whether the substrate for the protein or the required cofactors are bound. Such a "conformational change" may turn the enzyme on or off or otherwise modify its properties. Proteins can lose the conformation required for activity if we heat, add urea (which breaks up the H-bond network) or salts, or move out of the pH range in which the native conformation is stable. This leads to denatured, inactive protein, which in certain cases can refold correctly when the favorable conditions of temperature, ionic strength, and pH are reestablished.

Metalloenzymes

More than half of all enzymes have metal ions in their structure; these are called *metalloenzymes*. In most cases, the metals are essential to the action of the enzyme and are often at the active site where the substrate for the biochemical reaction is bound. Most organisms require certain "trace elements" for growth. Some of these trace elements are the metal ions that the organism incorporates into its metalloenzymes. Of the inorganic elements, the following have been found to be essential for some species of plant or animal: Mg, V, Cr, Mn, Fe, Co, Ni, Cu, Zn, Mo, B, Si, Se, F, Br(?), and I. New elements are added to the list from time to time, and the role of the established trace elements is gradually becoming more clear. In addition, Na, K, Ca, phosphate, sulfate, and chloride are required in bulk rather than trace amounts. Metal ions also play an important role in nucleic acid chemistry. The biochemistry of these elements has been termed *bioinorganic* chemistry.[1b]

Modeling

In addition to purely biochemical studies, bioinorganic chemistry also includes studies that try to elicit the chemical principles that are at work in biological systems. Two such areas are structural and functional modeling. In structural modeling, the goal is to prepare a small molecule, such as a metal complex, that can be structurally and spectroscopically characterized in order to compare the results with physical measurements on the biological system. This can help determine the structure, oxidation state, or spin state of a metal cofactor. It is often the case that a small molecule complex can reproduce many important physical properties of the target. Less common is functional modeling, where the goal is to reproduce some chemical property of the target in a small molecule complex and so try to understand what features of the structure promote the chemistry. Typical properties include the redox potential of a metal center or its catalytic activity. Functional models *with the correct metal and ligand set* that reproduce the catalytic activity of the target system are still rare. Many so-called models use the "wrong" metal or ligands and so provide less relevant information.

Molecular Recognition

A key principle of biochemistry is the recognition of one molecule or fragment of a biochemical structure by another. One entity will bind strongly to another, whether it is binding of the substrate with its specific enzyme, or of a hormone with its receptor protein, or of a drug with its receptor. This happens as a result of complementarity between the two fragments with regard to shape, surface charges, and the possibility of forming hydrogen bonds. It is this chemical recognition that accounts for the astonishing specificity of biology; for example, only one enantiomer of a compound may be accepted by an enzyme, and only the human, but not the monkey, version of a given protein may be recognized by a suitable antibody (specific binding protein). It is largely the three-dimensional rigidity and the rich pattern of possible chemical functional groups in proteins that allows this to happen.

If a protein selectively recognizes and binds the transition state for a reaction, then that reaction will be accelerated by catalysis. This is because a reaction will

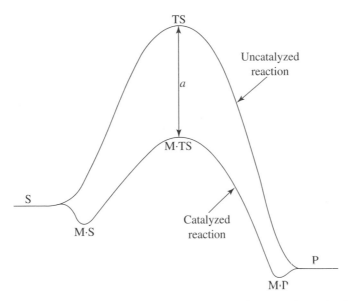

FIGURE 16.1 An enzyme lowers the activation energy for a reaction, often by binding the transition state (TS) for the reaction more tightly than the substrate (S) or product (P). The binding energy for the TS is represented as *a* in this plot of energy versus reaction coordinate.

go faster if it becomes easier to reach the transition state, which will be the case if the transition state is stabilized more than is the substrate (note how TS, but not M·S on M·P, is stabilized in this way in Fig. 16.1). An enzyme that hydrolyzes an ester RCOOMe as substrate should recognize the transition state **16.2** for the attack of water on the ester. Such an enzyme may bind a transition state analog, such as the phosphate **16.3** much more tightly than it binds the starting ester RCOOMe and inhibit the enzyme (poison the catalyst). Drugs are often selective inhibitors of certain target enzymes.

16.2   **16.3**

Coenzymes

Just as a whole set of reactions may require a given reagent, sometimes a whole set of enzymes require a given *coenzyme*. The first organometallic system we shall study is coenzyme B_{12}, a small molecule containing Co, which is required for the activity of a number of enzymes, which are therefore said to be "B_{12}

dependent." Only when the coenzyme binds, does the enzyme become functional. The alternative strategy of incorporating a Co into each mole of enzyme would make less efficient use of this rare element.

Protein Structure

The structures of proteins are generally studied by crystallography,[3a] by no means a straightforward procedure for such large molecules. The structural data cannot reveal the oxidation state of any metal present, and for this we normally need to compare the UV–visible or EPR spectra of the protein with those of model compounds.[3b] If the natural enzyme has a metal such as Zn^{2+} that gives uninformative electronic spectra or is EPR silent, it is sometimes possible to replace it with an unnatural but more informative metal, such as Co^{2+}.

Many interesting metalloproteins are not yet crystallographically characterized, but it is always possible to use X-ray spectroscopy even in the absence of suitable crystals. For example, the fine structure on the X-ray absorbtion edge (EXAFS)[4] for the metal may reveal the number of ligand atoms, their distance, and whether they are first (N,O) or second row (S). The X-ray photon expels a photoelectron from the metal, if it has a certain minimum photon energy required to ionize electrons from a given shell (say, the 2s); an absorbtion edge appears at this energy in the X-ray absorbtion spectrum. As we go to slightly higher X-ray photon energies, the photoelectron leaves the metal atom with a certain small translational energy because of the slight excess energy of the X-ray photon relative to the absorbtion edge of the metal. The wavelength of the photoelectron will depend on the amount of excess energy of the X-ray photon. The backscattering of the electron from the ligands around the metal will also be wavelength-dependent and will affect the probability for absorbtion of the X-ray. Crudely speaking, the ligand atom may backscatter the photoelectron wave in such a way as to give a constructive or destructive interference and so raise or lower the probability of the electron leaving the vicinity of the metal; the probability of absorbtion of the X-ray photon will be raised or lowered in consequence. Interpretation of EXAFS data is not entirely straightforward and is considerably helped by making measurements on model complexes. Normally the M−L distance(s) can be extracted to an accuracy of ±0.002 Å, but the number of ligands of a given type is much less well determined (e.g., ±1). The energy of the X-ray absorbtion edge is related to the charge on the metal. Unfortunately, this is not related directly to the formal oxidation state for the reasons we considered in Section 2.4.

Another useful physical method is resonance Raman spectroscopy.[5a] It is found that if the exciting radiation in a Raman experiment is near an absorbtion feature of the metal ion in the electronic spectrum, then Raman scattering involving bonds in the immediate vicinity of the metal is greatly enhanced. This selectivity for the vicinity of the active site is very useful in bioinorganic studies because the absorbtions from the active site would otherwise be buried under the multitude of absorbtions from the rest of the protein. For iron proteins, Mössbauer measurements[5b] can help determine oxidation state and help distinguish 4- from 5- and 6-coordinate metals.

16.2   COENZYME B$_{12}$[6]

The story begins with the observation, made early in the twentieth century, that raw beef liver is a cure for the otherwise uniformly fatal disease, pernicious anemia. The active component of liver extract was first separated and finally crystallized in 1948. In 1965 Dorothy Hodgkin[7] determined the structure **16.4** crystallographically. This showed that the molecule is an octahedral cobalt complex with a 15-membered 4-nitrogen ring L$_3$X ligand, called a *corrin*, occupying the equatorial plane. Connected to the corrin is a side chain containing a benzimidazole, which can bind as an axial ligand. The sixth site of the octahedron can be occupied by a number of different ligands. As a result of the isolation procedure commonly used, cyanide binds at the sixth site, and the final product is cyanocobalamin, the species studied by Hodgkin. In nature, several other ligands can be present including water (aquacobalamin or B$_{12a}$), or methyl (methylcobalamin) or adenosyl groups, **16.5** (the vitamin B$_{12}$ coenzyme). Other than B$_{12a}$, all these species have a Co—C bond, the first M—C bonds of any sort to be recognized in biology.

16.4

16.5

The coenzyme acts in concert with a variety of enzymes that catalyze reactions of three main types. In the first, two substituents on adjacent carbon atoms, -X and -H, are permuted; this is called the *isomerase reaction*. The generalized process is shown in Eq. 16.1 and specific examples are given in Eq. 16.2–16.4. Note that CoA has nothing to do with cobalt, but is the biochemical symbol for coenzyme A, a thiol that activates carboxylic acids by forming a thioester.

$$R'CHX-CH_2R \xrightarrow{\text{enzyme, B}_{12}\text{ coenzyme}} R'CH_2-CHXR \qquad (16.1)$$

$$HOOCCH_2-CH_2COS(CoA) \xrightarrow{\substack{\text{methylmalonyl CoA mutase,} \\ \text{B}_{12}\text{ coenzyme}}} HOOCCHCOS(CoA)-CH_3$$

succinyl coenzyme A   methylmalonyl coenzyme A

$$(16.2)$$

In the second general type, methylcobalamin methylates a substrate, as in the conversion of homocysteine to methionine, for example.

$$HSCH_2CH_2CH(COOH)NH_2 \xrightarrow{\substack{\text{methionine synthetase,} \\ \text{methylcobalamin}}} MeSCH_2CH_2CH(COOH)NH_2$$

homocysteine   methionine

$$(16.3)$$

Finally, B_{12} is also involved in the conversion of the ribose ring of the ribonucleotides that go to make RNA to the deoxyribose ring of the deoxyribonucleotides that go to make DNA. The schematic reaction is shown in Eq. 16.4.

$$-CHOH-CHOH- \xrightarrow{\substack{\text{deoxyribose synthetase,} \\ \text{B}_{12}\text{ coenzyme}}} -CHOH-CH_2- \qquad (16.4)$$

The coenzyme is required only in small amounts; 2–5 mg is present in the average human, for example, and one of the first signs of deficiency is the failure to form red blood cells. Hence the anemia, but the disease is not treated successfully by the methods that work for the usual iron-deficiency anemia, which explains the term "pernicious" anemia.

The B_{12a} system was found to be easily reducible, first to B_{12r} and then to B_{12s} (r stands for reduced and s for superreduced). Physical studies showed that B_{12r} contains Co(II), and by comparison with model compounds, B_{12s} was shown to contain Co(I), probably in a 4- or 5-coordinate form. The B$_{12s}$ state turns out to be one of the most powerful nucleophiles known, and it reacts rapidly with MeI, or the natural Me$^+$ donor, N^5-methyl tetrahydrofolate, to give methylcobalamin.

Model Studies

Is this chemistry unique to the natural system, or is it a general property of cobalt in a 5-nitrogen ligand environment? At the time that the original model studies were carried out (1960s), it was believed that transition metal alkyls were stable only with very strong ligand field ligands, such as CO or PPh$_3$. This problem was better understood by studying model systems. Early studies revealed that the simple ligand dimethylglyoxime (dmgH) **16.6** gives a series of Co(III) complexes (called *cobaloximes*) **16.7** that have much in common with the natural system. Two dmg ligands model the corrin, a pyridine models the axial base, and the sixth position can be an alkyl group or water. It was found that these alkyls are stable when the equatorial ligand had some, but not too much, electron delocalization. Neither fully saturated ligands nor the more extensively delocalized porphyrin system, common in other metalloenzymes, allow cobalt to form alkyls easily. The second interesting point is that the longer-chain alkyls, such as -Et or -adenosyl, do not β-eliminate easily. We can now see that this is because the equatorial ligand prevents a vacant site from being formed cis to the alkyl in this 18e system. Such a site would be needed for β elimination to take place by a concerted mechanism (Section 6.5).

16.6   **16.7**

The nature of the B$_{12r}$ and B$_{12s}$ states was made clearer when it was found that cobaloxime could be reduced to Co(II) and Co(I) oxidation states. The Co(I) form, [Co(dmg)$_2$py]$^-$, proved to be a supernucleophile, reacting very fast with MeI to give [MeCo(dmg)$_2$py] (Eq. 16.5).

$$[\text{Co(dmg)}_2\text{py}]^- + \text{MeI} \longrightarrow [\text{MeCo(dmg)}_2\text{py}] + \text{I}^- \qquad (16.5)$$

Homolytic Mechanisms

The mechanism of the isomerase reactions involving the coenzyme is believed to start with reversible homolysis of the Co(III)–C bond to generate the Co(II) "radical," B_{12r}, and the adenosyl radical, $RCH_2\cdot$. This carbon radical abstracts a hydrogen atom from the substrate, QH, to give RCH_3, and the substrate radical, Q$\cdot$. This radical is believed to undergo a 1,2 shift of the X group (see Eq. 16.8), to give the product radical. Hydrogen atom transfer from RCH_3 to the product radical gives the final product:

$$L_nCo-CH_2R \longrightarrow L_nCo\cdot + \cdot CH_2R \qquad (16.6)$$

$$R'HCX-CH_2R + \cdot CH_2R \longrightarrow R'HCX-\cdot CHR + CH_3R \qquad (16.7)$$

$$R'HCX-\cdot CHR \longrightarrow R'HC\cdot -CXHR \qquad (16.8)$$

$$R'HC\cdot -CXHR + CH_3R \longrightarrow R'CH_2-CXHR + \cdot CH_2R \qquad (16.9)$$

This mechanism implies that the Co$-$C bond in the coenzyme is not particularly strong because Eq. 16.6 requires that it must be spontaneously hemolyzing at ambient temperatures at a rate fast enough to account for the rapid turnover observed for the B_{12}-dependent enzymes ($\sim 10^2$ s^{-1}). Halpern[8] has estimated Co$-$C bond strengths in B_{12} models by two methods. The first involves measuring the equilibrium constant for Eq. 16.10. From the ΔH and ΔS values, and given the known heats of formation of $PhCH=CH_2$ and $PhCH\cdot -CH_3$, the ΔH and ΔS for Eq. 16.6 can be deduced.

$$(py)dmg_2Co-CHMePh \rightleftharpoons (py)dmg_2Co\cdot + PhCH=CH_2 + \tfrac{1}{2}H_2$$

$$(\Delta H = 22.1 \text{ kcal/mol\{measured\}}) \qquad (16.10)$$

$$PhCH=CH_2 + \tfrac{1}{2}H_2 \rightleftharpoons PhCH\cdot -CH_3$$

$$(\Delta H = -2.2 \text{ kcal/mol \{calculated\}}) \qquad (16.11)$$

$$(py)dmg_2Co-CHMePh \rightleftharpoons (py)dmg_2Co\cdot + PhCH\cdot -CH_3$$

$$(\Delta H = 19.9 \text{ kcal/mol \{calculated\}}) \qquad (16.12)$$

Note that Eq. 16.10 looks like a β elimination of the sort that we said should be prevented by the unavailability of a 2e vacant site at the metal. In fact, the reaction probably goes by Co$-$C bond homolysis, followed by abstraction of a hydrogen atom from the carbon radical by the Co(II) (Eqs. 16.13$-$16.15), not by a concerted mechanism at all.

$$(py)dmg_2Co-CHMePh \rightleftharpoons (py)dmg_2Co\cdot + PhCH\cdot -CH_3 \qquad (16.13)$$

$$(py)dmg_2Co\cdot + PhCH\cdot -CH_3 \rightleftharpoons (py)dmg_2CoH + PhCH=CH_2 \qquad (16.14)$$

$$2(py)dmg_2CoH \rightleftharpoons 2(py)dmg_2Co\cdot + H_2 \qquad (16.15)$$

Halpern's second method of determining the Co$-$C bond strength is to trap the R$\cdot$ intermediate using a second Co(II) complex as the trap. The $\Delta H^{\ddagger}$ found for this process should be a measure of the Co$-$C bond strength. In the case above where we already know the Co—C bond strength is approximately 20 kcal/mol, the answer by the kinetic method comes out to be 22 kcal/mol. The extra 2 kcal probably represents the activation energy for the homolysis. Applying the method to coenzyme B$_{12}$ itself gives a figure of 28.6 kcal/mol for the Co—CH$_2$R bond strength. This figure is too high to account for the rate of turnover of the B$_{12}$-dependent enzymes because the rate of the homolysis of such a strong bond would be much slower than 10^2 s^{-1}. On reflection, however, this Co—C bond strength is indeed reasonable because the coenzyme must be under control. It must not liberate a radical until required to do so. Very likely, when the coenzyme binds to the B$_{12}$-dependent enzyme, part of the binding energy of the B$_{12}$ to the enzyme is used to deform the coordination sphere around Co in such a way that the Co—C bond is made slightly weaker, and when the substrate also binds, the coenzyme may be further activated so that it is now able to hemolyze at the appropriate rate.

Halpern has also looked at the rearrangement step itself by making the proposed substrate-derived radical independently in the absence of metal by a standard method, the action of Bu$_3$SnH on the corresponding halide. He finds that for the methylmalonyl mutase reaction, the rate of rearrangement is 2.5 s^{-1}, only slightly slower than the 10^{-2} s^{-1} turnover rate for the enzyme. This difference is small enough to be accounted for by saying that the radical involved in the natural system is not free, but bound to the enzyme, which will hold it in the conformation most favorable for the rearrangement. All this does not prove that the substrate radical does not bond to cobalt in the course of the rearrangement, but at least we can say for the moment that a viable pathway exists without any such binding being necessary. The same goes for some of the other proposals that have been made for the rearrangement, notably redox reactions between the radical and the Co(II) to generate a putative carbonium ion or carbanion, either of which might also rearrange. B$_{12}$ generates a thiyl radical in the mechanism of DNA synthesis by a class of ribonucleotide reductases.[9]

Bioalkylation and Biodealkylation[10]

Methylcobalamin is important in biological methylation, itself of great importance in gene regulation and even in cancer.[10c] In some cases it has been found that Hg(II) in the sea can be methylated by these bacteria to give MeHg$^+$. This water-soluble organometallic species can be absorbed by shellfish, which can then become toxic to humans. Mercury is naturally present in small quantities in seawater, but the concentration can be dramatically increased by pollution. A notorious poisoning episode of this sort occurred at Minimata in Japan, where abnormally high amounts of mercury were found in the sea, as a result of industrial activity. Certain bacteria even have a pair of enzymes, organomercury lyase and mercuric ion reductase, that detoxify organomercury species via the processes shown in Eqs. 16.16 and 16.17. The lyase cleaves the R—Hg bond and the reductase reduces the resulting Hg(II) ion to the metallic state; in this form it evaporates from the organism. The mechanisms involved have been studied by

Walsh[10a] and O'Halloran.[10b] The retention of configuration observed in the lyase reduction of Z-2-butenylmercury chloride and the failure of radical probes such as **16.8** to give a radical rearrangement (to norbornadiene) led to the proposal that the reaction goes by an S_E2 mechanism in which a cysteine SH group of the protein cleaves the bond (Eq. 16.18; enz = lyase). The reduction of the Hg^{2+} to Hg(0) is believed to go via initial formation of a dithiolate that loses disulfide (Eq. 16.19; enz' = reductase).

16.8

$$R-Hg-Cl \xrightarrow{\text{organomercury lyase}} RH + Hg^{2+} + Ce^{-} \qquad (16.16)$$

$$Hg^{2+} \xrightarrow{Hg^{2+} \text{ reductase}} Hg(0) \qquad (16.17)$$

$$(16.18)$$

$$(16.19)$$

In the absence of Hg(II), the transcription and synthesis of these Hg detoxification enzymes is inhibited by a regulatory protein, called merR, that binds to a specific location in the *mer* operon, the section of DNA coding for mercury resistance. When Hg(II) is present, it binds to three Cys residues of the merR protein. This causes a conformational change in both the protein and in the DNA to which it is bound that leads to transcription of the lyase and reductase. In this way, these proteins are only produced when required.[10b]

In the early nineteenth century, certain green wallpapers contained copper arsenite (Scheele's green) as a dyestuff. In damp conditions, molds, such as *Scopulariopsis bevicaulis*, are able to convert the arsenic to the very toxic $AsMe_3$, by a B_{12}-dependent pathway, and many people died before the problem was recognized. It has even been argued[11a] that in 1821 Napoleon was accidentally poisoned in this way, when he was held at St. Helena by the British; others have blamed the British for deliberately poisoning him.[11b]

- Coenzyme B_{12}, the best-established organometallic cofactor in biology, provides a source of carbon-based radicals as well as a methylation reagent.

16.3 NITROGEN FIXATION[12]

It has been noticed by farming communities since antiquity that the presence of certain plants encourages the growth of crops. The presence of a fertility goddess in the plant was a colorful explanation developed in early times to account for this phenomenon. The truth is only slightly less remarkable: the roots of the plant in question are infected by various species of soil bacteria, which, provided in their new home with the necessary energy input by the plant, "fix" atmospheric N_2 to NH_3, by means of a metalloenzyme, nitrogenase. The resulting ammonia not only fertilizes the host plant but also escapes into the surroundings, where the growth of crop plants is stimulated. Before the advent of fertilizers, almost all the nitrogen required in human nutrition was obtained by biological nitrogen fixation; now, much of it comes from N_2 fixed by the Haber process (Eq. 16.20):

$$N_2 + 3H_2 \xrightarrow{\text{Fe catalyst}} 2NH_3 \qquad (16.20)$$

As early as 1930, it was realized that molybdenum was normally essential for biological nitrogen fixation; iron and magnesium are also required. More recently, alternative nitrogenases have been described that contain no Mo, but either V and Fe or Fe alone instead. The MoFe N_2ase is the best studied and this is the system referred to below, unless specifically stated. Ammonia is the first reduction product released by the enzyme, and there is no evidence for other species, such as hydrazine. The enzyme, like many organometallic complexes, is air sensitive, and CO and NO are strong inhibitors. It is believed that the CO or NO coordinate to the N_2 binding site, and that this site is a low-valent Fe—Mo cluster. Apart from N_2, the enzyme also reduces some other substrates very efficiently, such as C_2H_2 (but only to C_2H_4), MeNC (to MeH and MeNH$_2$), and N_3^-. Acetylene reduction is used as the standard assay for the enzyme. Since the VFe N_2ase reduces C_2H_2 to C_2H_6, its presence escaped detection by the classic assay.

The Mo enzyme consists of two components: (1) the Fe protein (molecular weight 57,000 daltons), which contains iron and sulfur (4 atoms of each per protein); and (2) the MoFe protein (220,000 daltons, $\alpha_2\beta_2$ subunits), which contains both metals (1 atom Mo, 32 atoms Fe). Each also contains S^{2-} ions (ca. one per iron), which act as bridging ligands for the metals. The protein contains special Fe—S clusters called "P clusters" that have EPR resonances like those of no other Fe—S cluster. A soluble protein-free molybdenum and iron-containing cluster can be separated from the enzyme. This iron–molybdenum cofactor, or FeMo-co, was known to have approximately 1 Mo, 7–8 Fe, 4–6 S^{2-}, and one molecule of homocitrate ion. As for the P cluster, there was no agreement on the structure of FeMo-co for many years. In purified form FeMo-co does crystallize, and it can restore N_2 reducing activity to samples of mutant N_2ase that are inactive because they lack FeMo-co.[13] On the other hand, no crystal structure of FeMo-co proved possible, and no synthetic model complex was found that could activate the mutant enzyme.

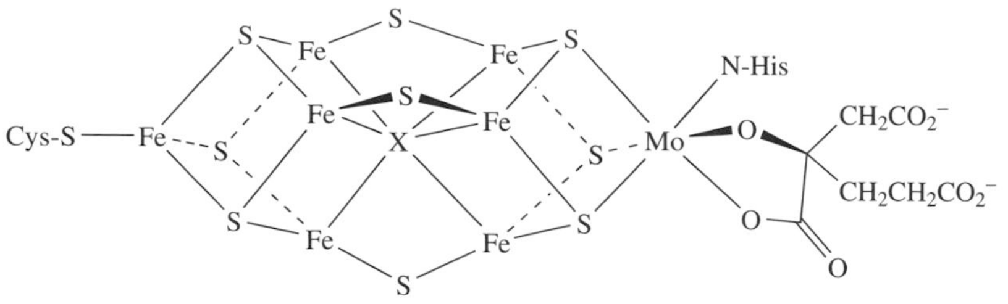

FeMo-co

FIGURE 16.2 Structure of the FeMo-co of *Azotobacter vinelandii* nitrogenase, as revealed by the crystallographic work of Rees et al.[14] X may be N^{3-}.

The crystal structure of the entire enzyme obtained by Rees et al.,[14] in 1992 has been very important in clearing up some of the mysteries surrounding the system. FeMo-co proves to have the structure shown in Fig. 16.2. One surprise is that the Mo is 6-coordinate, making it less likely to be the N_2 binding site. Model studies had for many years concentrated on this element. The possible noninvolvement of the Mo in binding N_2 illustrates one hazard of bioinorganic model chemistry: that the data on the biological system may undergo a reinterpretation that alters the significance or relevance of any model studies. An early state of the refinement suggests that six Fe atoms of the cofactor had the very low coordination number of 3, but the latest work puts a light atom, probably N, at the center of the cluster.

The isolated enzyme will reduce N_2 and the other substrates if a source of the electrons required by Eq. 16.21, such as $Na_2S_2O_4$, is provided. In addition, ATP is also consumed, even though the overall process of Eq. 16.21 is exergonic under physiological conditions, so the ATP must provide additional driving force to increase the rate. The Mo–Fe protein binds the N_2, and the Fe protein accepts electrons from the external reducing agent and passes them on to the MoFe protein. In the absence of N_2, N_2ase acts as a hydrogenase in reducing protons to H_2; indeed, H_2 is also formed even in the presence of N_2.

$$N_2 + 8H^+ + 8e^- \longrightarrow 2NH_3 + H_2 \qquad (16.21)$$

Dinitrogen and N_2 Complexes

Dinitrogen is very inert, and few systems are able to reduce it catalytically under the mild conditions employed by nitrogenase. N_2 will react with Li and Mg to give nitrides, but the only other nonbiological reaction of N_2 under mild conditions is the formation of N_2 complexes. More than 100 examples are now known, of which many contain Fe or Mo. In most cases, the N_2 is terminal and bound by one N atom, as in **16.9**. N_2 is isoelectronic with CO, so a comparison between the two ligands is useful. CO has a filled σ-lone pair orbital located on carbon, with which it forms a σ bond to the metal, and an empty π^* orbital for back bonding. N_2 also has a filled σ lone pair, but it lies at lower energy than the corresponding orbital in CO, probably because N is more electronegative than C, and so N_2 is the weaker σ donor. N_2 also has an empty π^* orbital. Although it is lower in energy, and so more accessible than the CO π^* orbital, it is equally distributed over N^1 and N^2 and therefore the $M-N$ π^* overlap is smaller than for $M-CO$, where the π^* is predominantly located on carbon. The result is that N_2 binds very much less efficiently than CO. Of the two $M-N$ interactions, the back donation is the more important for stability, and only strongly π-basic metals bind N_2. Because the two ends of N_2 are the same, the molecule can relatively easily act as a bridging ligand between two metals (**16.10**). If the back donation is large, the N_2 can be reduced to a hydrazide complex. The two forms **16.11** and **16.12**, shown below, are really resonance contributors to the real structure, which may more closely resemble **16.11** or **16.12**. The side-on bonding mode is rare.

$$M-N^1\equiv N^2 \qquad M-N\equiv N-M$$

terminal $\qquad\qquad$ bridging

16.9 $\qquad\qquad\qquad$ **16.10**

$$M-N\equiv N-M \qquad M=N-N=M$$

16.11 $\qquad\qquad\qquad$ **16.12**

The first dinitrogen complex to be recognized, $[Ru(NH_3)_5(N_2)]^{2+}$, was isolated in 1965 by Allen and Bottomley[15] during the attempted synthesis of $[Ru(NH_3)_6]^{2+}$ from $RuCl_3$ and hydrazine. The $N-N$ distance of this and related N_2 complexes is only slightly different (1.05–1.16 Å) from that of free N_2 (1.1 Å). An important property of the mononuclear complexes is the strong IR absorption due to the $N-N$ stretch at 1920–2150 cm^{-1}. Free N_2 is inactive in the IR, but binding to the metal causes a strong polarization of the molecule (see Section 2.6), with N^1 becoming positively charged and N^2 negatively charged. This contributes both to making the $N-N$ stretch IR active and to the chemical activation of the N_2 molecule.

Common preparative routes are reduction of a phosphine-substituted metal halide in the presence of N_2, degradation of a nitrogen-containing ligand, and displacement of a labile ligand by N_2.

$$MoCl_3(thf)_3 \xrightarrow{\text{Mg, dpe}} Mo(N_2)_2(dpe)_2 \qquad (16.22)$$

$$WCl_4(PMe_2Ph)_3 \xrightarrow{\text{Mg, PMe}_2\text{Ph}} W(N_2)_2(PMe_2Ph)_4 \quad (16.23)$$

$$ReCl_2(PPh_3)_2(N_2COPh) + PMe_2Ph \xrightarrow{\text{MeOH}} ReCl(N_2)(PMe_2Ph)_4$$

$$+ PhCOOMe + HCl \qquad (16.24)$$

$$FeH_2(H_2)(PEtPh_2)_3 \xrightarrow{N_2} FeH_2(N_2)(PEtPh_2)_3 \qquad (16.25)$$

Only on rare occasions is it possible to synthesize and purify a whole series of N_2 complexes with different ligands; the Mo, W, and Re systems shown above are perhaps the most versatile in this respect. N_2 can often displace η^2-H_2, as shown in Eq. 16.25; if this were the last step in the catalytic cycle, it would explain why N_2ase always produces at least one mole of H_2 per mole of N_2 reduced.

Some examples of complexes in which the N_2 bridges two metals are shown in Eqs. 16.26 and 16.27. In the ruthenium case, the system resembles **16.11**, and the μ-N_2 is little different in length from the terminal N_2 in $[Ru(NH_3)_5(N_2)]^{2+}$ itself. Some dinitrogen complexes are appreciably basic at N^2, showing once again the strong polarization of the N_2. These can bind Lewis acids at N^2 to give adducts, some of which have very low N$-$N stretching frequencies, and these seem to resemble **16.12**.

$$[Ru(NH_3)_5(N_2)]^{2+} + [Ru(NH_3)_5(H_2O)]^{2+} \longrightarrow [\{Ru(NH_3)_5\}_2(\mu\text{-}N_2)]^{4+}$$

$$(16.26)$$

$$\text{ReCl}(PMe_2Ph)_4(N_2) + MoCl_4(OEt_2)_2 \longrightarrow$$
yellow $[v(N_2) = 1925 \text{ cm}^{-1}]$

$$\text{Cl}(PMe_2Ph)_4Re(\mu\text{-}N_2)MoCl_4(\mu\text{-}N_2)ReCl(PMe_2Ph)_4 \qquad (16.27)$$
blue-black $[v(N_2) = 1680 \text{ cm}^{-1}]$

Reactions of N_2 Complexes

Only the most basic N_2 complexes, notably the bis-dinitrogen Mo and W complexes, can be protonated. According to the exact conditions, various N_2H_x complexes are obtained, and even, in some cases, free NH_3 and N_2H_4. As strongly reduced Mo(0) and W(0) complexes, the metal can apparently supply the six electrons required by Eq. 16.21, and so the metals are oxidized during the process. Note, too, that in Eq. 16.28, the loss of the very strong N$-$N triple bond is compensated by the formation of two N$-$H bonds and a metal nitrogen multiple bond.

$$W(N_2)_2(dpe)_2 \xrightarrow{\text{2HCl}} WCl_2(=N-NH_2)(dpe)_2$$

$$\xrightarrow{\text{base, }(-HCl)} WCl(=N-NH)(dpe)_2 \qquad (16.28)^{12a}$$

$$W(N_2)_2(PMe_2Ph)_4 \xrightarrow{\text{H}_2\text{SO}_4, \text{ MeOH}} N_2 + 2NH_3 + W(VI) \qquad (16.29)^{12a}$$

The mechanism shown in Eq. 16.30 (Chatt cycle) has been proposed for the N_2 reduction observed in these experiments. N_2 is a net electron acceptor from the metal, and so loss of the first N_2 leads to the metal acquiring a greater negative charge, and thus back donating more efficiently to the remaining N_2, which is therefore polarized and activated even further. Note that the final N$-$N bond breaking is again accompanied by the formation of a metal nitride; such species are known to hydrolyze easily to give ammonia. It is likely that the natural system may also go by similar intermediates.

$$L_nM(N_2)_2 \xrightarrow{-N_2} L_nM(N_2) \xrightarrow{H_2SO_4} L_{n-1}M(SO_4)(=N-NH_2) \xrightarrow{H+}$$

$$L_{n-1}M(SO_4)(=N-NH_3)^+ \longrightarrow L_{n-2}M(SO_4)(\equiv N)$$

$$+ NH_3 \xrightarrow{H+} M(VI) + 2NH_3 \qquad (16.30)$$

The greatest weakening of the N−N bond might be expected for early d^2 metals, which back-bond the most strongly to π-acceptor ligands. Cp_2^*Ti reacts with N_2 as shown in Eq. 16.31, where $Cp_2^*Ti(N_2)$ seems to have η^1 and η^2 forms and protonates with HCl to give N_2H_4. These show different $\nu(N_2)$ frequencies in the IR (2056 and 2023 cm^{-1}, respectively) and, most significantly, the ^{15}N NMR shows two mutually coupled ($J = 7$ Hz) resonances for the η^1 and a single resonance for the η^2 form.[16]

$$Cp_2^*Ti \xrightarrow{N_2} Cp_2^*Ti(\eta^1\text{-}N_2) \rightleftharpoons Cp_2^*Ti(\eta^2\text{-}N_2) \xrightarrow{Cp_2^*Ti} Cp_2^*Ti(NN)TiCp_2^*$$

$$(16.31)$$

Schrock et al.[17] have made $Cp^*Me_3M=N-N=MMe_3Cp^*$ (M = Mo or W), where the back donation is so strong that the N_2 is now effectively reduced to a hydrazide tetraanion, as shown by the N−N distance of 1.235 Å (Mo). Ammonia is formed with lutidine hydrochloride as proton source and Zn/Hg as reductant. Dinitrogen can also be reduced to ammonia at room temperature and 1 atm with the molybdenum catalyst $LMo(N_2)$, where L is the bulky trianionic tripodal tri-amide $[\{3,5\text{-}(2,4,6\text{-i-}Pr_3C_6H_2)_2C_6H_3NCH_2CH_2\}_3N]$. Addition of a lutidine salt as proton source, and decamethyl chromocene as reductant, gave four catalytic turnovers. The N_2 is reduced at a sterically protected, single molybdenum center that cycles from Mo(III) through Mo(VI).

In spite of much effort, no one has yet succeeded in making an N_2 complex using only thiolate and S^{2-}, ligands closer to those that are present in the enzyme. Indeed, the chemistry of sulfur ligands is plagued by their high tendency to bridge, and so soluble and characterizable materials can often be obtained only with thiolates having very bulky R groups. The binding site for N_2 in the enzyme may be one or more Fe atoms of the FeMo-co cluster.

Fe−S Clusters

The other surprise in the N_2ase structure, apart from the FeMo-co structure, is the nature of the P clusters. To understand this result, we must briefly look at iron−sulfur proteins, which have been known for many years, but the structures of the active sites having become clear only relatively recently. Structures **16.13–16.15** show some main cluster types that had been recognized.[18] There are also a number of triiron clusters.[19] In each case the R groups represent the cysteine residues by which the metal is bound to the protein chain. In the cases in which there is more than one iron atom, S^{2-} ions are also present and bridge the metals. The ferredoxin proteins contain Fe_4S_4 or Fe_2S_2 cores, and these have been extruded apparently intact from the enzyme by the addition of suitable thiols that can chelate the metal, to give a fully characterizable complex. The metal-free enzyme (the apoenzyme) can then be made active once again simply by adding Fe^{2+} and S^{2-}. These clusters are said to have the property of self-assembly; that is, they can form readily in solution on mixing the components (apoenzyme + metal ions or, for the model compounds, ligands + metal ions) under the correct conditions. This contrasts with FeMo-co, which as yet cannot be formed either from the apoenzyme and metal ions or in models from ligands and metal ions. At least three genes are present in nitrogen-fixing organisms whose role has been identified as the inorganic synthesis of the FeMo-co cluster.

16.13   **16.14**

16.15

It has been possible to synthesize model complexes with core geometries similar to those present in the natural Fe−S clusters. Some examples are shown in Eqs. 16.32 and 16.33. Normally, adding an oxidizing metal like Fe^{3+} to RSH simply leads to oxidation to RSSR, and so the choice of reaction conditions is critical. Millar and Koch have shown that metathesis from the phenoxide is very useful (Eq. 16.34), which allows synthesis of $Fe^{III}(SPh)_4^-$, an apparently very simple compound, but one that long resisted attempts to make it.[20]

$$FeCl_3 + RSH \xrightarrow{\text{NaSH, NaOMe}} (RS)_4Fe_4(\mu_2\text{-}S)_4 \qquad (16.32)$$

$$FeCl_3 + o\text{-}C_6H_4(CH_2SH)_2 \xrightarrow{\text{NaSH, NaOMe}}$$

$$C_6H_4(CH_2S)_2Fe(\mu\text{-}S)_2Fe(SCH_2)_2C_6H_4 \quad (16.33)$$

$$Fe^{III}(OPh)_4^- \xrightarrow{\text{PhSH}} Fe^{III}(SPh)_4^- \qquad (16.34)$$

The oxidation states present in the natural systems can be determined by comparison of the spectral properties of the natural system in its oxidized and reduced states with those of the synthetic models; the latter can be prepared in almost any desired oxidation state by electrochemical means. The results show that the monoiron systems indeed shuttle between Fe(II) and Fe(III) as expected. The diiron enzymes are Fe(III), Fe(III) in the oxidized state, and Fe(II), Fe(III) in the reduced state. The mixed-valence species are fully delocalized in all cases. There is also a superreduced state, Fe(II), Fe(II), which is probably not important in vivo. The 4-iron proteins shuttle between 3Fe(II), Fe(III) and 2Fe(II), 2Fe(III), such as in the ferredoxins (Fd). One class of 4-iron proteins have an unusually high oxidation potential (HIPIP, or high potential iron protein), because the system shuttles between 2Fe(II), 2Fe(III) and Fe(II), 3Fe(III).

$$
\begin{array}{ccc}
3Fe(II),\ Fe(III) \rightleftharpoons & 2Fe(II),\ 2Fe(III) \rightleftharpoons & Fe(II),\ 3Fe(III) \\
Fd_{red} & Fd_{ox} & Fe_{superox} \\
HIPIP_{superred} & HIPIP_{red} & HIPIP_{ox}
\end{array}
\qquad (16.35)
$$

The N_2ase crystal structure, apart from showing FeMo-co, also revealed the structure of the P clusters (**16.16**), which consist of a pair of Fe_4S_4 cubanes bridged by an S–S group.

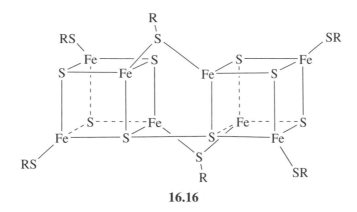

16.16

> • Nitrogen fixation is vital to life on Earth but is a very hard reaction to bring about.

16.4   NICKEL ENZYMES[21]

Urease is famous in enzymology for being the first enzyme to be purified and crystallized (1926). At the time enzymes were widely viewed as being too ill-defined for detailed chemical study. Sumner argued that its crystalline character meant that urease was a single defined substance and the fact that he could not find any cofactors led to the conclusion that polypeptides could have catalytic activity on their own. The existence of two essential Ni^{2+} ions per mole of urease was not proved until 1975. Sumner's conclusion that cofactors are not always required for catalytic activity is correct, but we now know that urease is not a valid example. Nickel was recognized as a significant catalytic element in a series of metalloenzymes only in the 1980s.[21] In three of these, hydrogenase (H_2ase), CO dehydrogenase (CODH), and MeCoM reductase (MCMR), organometallic Ni species are thought to be involved.

Archaea

This group of bacteria, including the methanogens, the thermoacidophiles, and the halobacteria, are sufficiently different from all other forms of life that it has been proposed that they be assigned to their own natural kingdom, the archaea.[22] The name indicates that they are proposed to be very early organisms in an evolutionary sense. One of the signs consistent with this antiquity is the fact that many archaea can live on the simple gases, such as H_2 and CO or CO_2, both as energy and carbon source, and on N_2 via nitrogen fixation as nitrogen source.[22] Higher organisms have more sophisticated nutritional requirements; humans, for example, must have such complex compounds as ascorbic acid (vitamin C) and vitamin B_{12} as part of the diet in order to survive; these compounds can only come from other life forms. Few, if any, other life forms must have existed when the earliest bacteria evolved, and they therefore had literally to live on air and water. A life form that can synthesize all its carbon constituents from CO_2 is called an *autotroph*; one that requires other C_1 compounds, such as methane or methanol, is called a *methylotroph*.

The archaea are very rich in nickel-containing enzymes and coenzymes, and Nature has clearly chosen this element to bring about the initial steps in the biochemical utilization of H_2, CO, CH_4, and other C_1 compounds, at least in an anaerobic environment. These steps almost certainly involve organonickel chemistry, although how this happens in detail is only just beginning to be understood.

CO Dehydrogenase

CODH[22] is unusual in that it can bring about two reactions (e.g., Eqs. 16.37 and 16.39) that are particularly interesting to the organometallic chemist: the reduction of atmospheric CO_2 to CO (CODH reaction, Eq. 16.37) and the synthesis of acetyl coenzyme A (ACS reaction, Eq. 16.39) from CO, a CH_3 group taken from a corrinoid iron–sulfur protein (denoted CoFeSP in the equation), and coenzyme A, a thiol. These are analogous to reactions we have seen earlier: the water–gas shift reaction (Eq. 16.36) and the Monsanto acetic acid process (Eq. 16.38).

The enzyme contains two metal clusters of special interest, denoted A and C. CODH activity occurs in the C cluster, which consists of a $NiFe_3S_4$ cubane unit

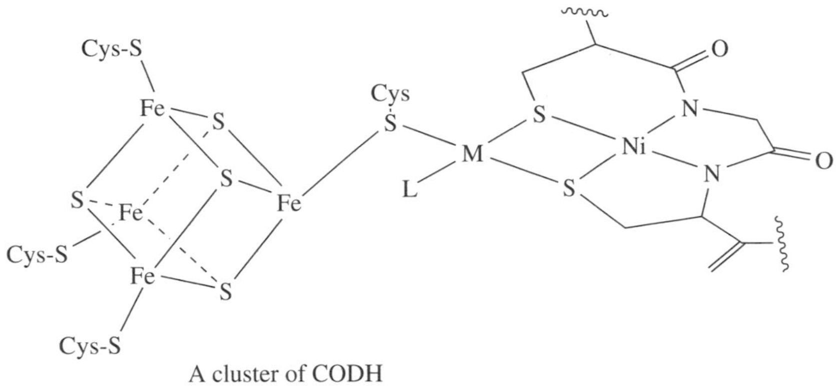

A cluster of CODH

FIGURE 16.3 The A cluster of ACS/CODH from *Moorella thermoacetica*. M is probably Ni, Cu, or Zn, Ni being the active form; L is an unknown nonprotein ligand.

capable of reversible CO_2 reduction. ACS activity occurs in the A cluster.[22b] The structure of the A cluster from *Moorella thermoacetica* showed a very unusual trinuclear active site (Fig. 16.3). An Fe_4S_4 cubane is bridged by a cysteine sulfur to a 4-coordinate metal, which may be Ni, Cu, or Zn. This is, in turn, bridged through two cysteine residues to a square-planar Ni(II) site, also ligated by two deprotonated peptide nitrogens from the peptide backbone. The square plane is completed by a fourth, still unidentified, but nonprotein ligand that is also bound to complete the coordination sphere. Current evidence suggests the Ni,Ni,Fe$_4$ form is responsible for the ACS activity[23], but the mechanism is still unknown.[22d]

$$CO + H_2O \rightleftharpoons CO_2 + H_2 \qquad (16.36)$$

$$CO + H_2O \rightleftharpoons CO_2 + 2H^+ + 2e^- \qquad (16.37)$$

$$MeOH + CO \longrightarrow MeCOOH \qquad (16.38)$$

$$Me\text{-}CoFeSP + CoA + CO \longrightarrow MeCOCoA + CoFeSP \qquad (16.39)$$

A fully functional model for the second Ni in CODH has been found: **16.17**.[24] This complex has the appropriate metal, Ni^{2+}, as well as an N-, O-, S-ligand environment and catalyzes the reaction shown in Eq. 16.37. The CO_2 is detected by precipitation with $Ca(OH)_2$, the H^+ production with a pH meter, and the electrons formed are transferred to the electron acceptor MV^{2+} (**16.18**) and gives the dark blue radical anion, $MV\cdot^+$. The reaction probably goes by CO splitting the Ni_2O_2 bridge to give LNiCO because the CO analog CN^- does so to give a stable complex $[LNiCN]^-$; CN^- is an inhibitor in both model and enzyme. Because Ni(II) is weakly back bonding, it would normally not bind CO at all (the S-ligand environment probably raises the basicity of the Ni d_π electrons in this case), but once bound, the CO should be very sensitive to nucleophilic attack because a CO bound to a weak π donor should be very ∂^+ at C. A possible

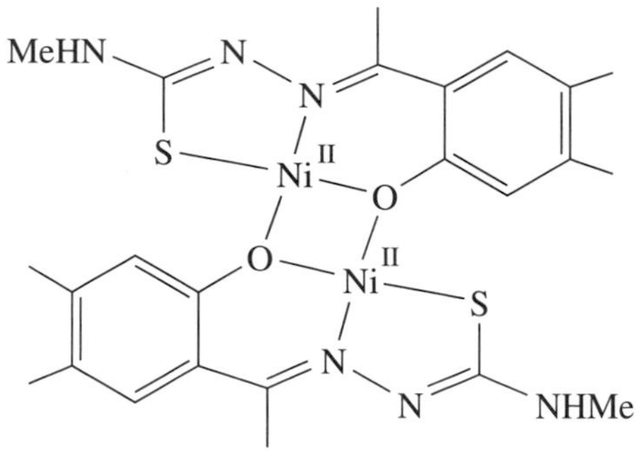

16.17

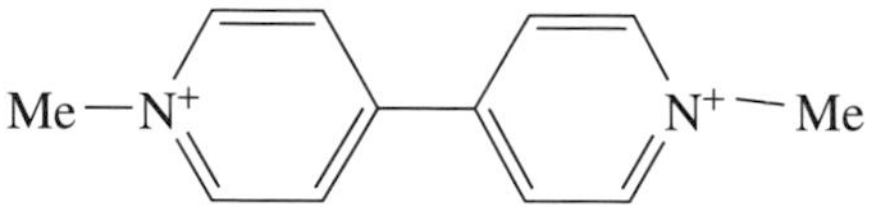

16.18

scheme based on analogy with the water–gas shift reaction is as follows:

$$LNi^{II}-CO + OH_2 \xrightarrow{-H^+} [LNi^{II}COOH]^- \xrightarrow{-2e,\ -H^+} LNi^{II} + CO_2 \qquad (16.40)$$

Note the iminothiolate S-donor group in **16.17**. An S-ligand environment is difficult to achieve while retaining an open site for catalytic activity because nickel thiolates have a very high tendency to bridge. This tends to remove any labile sites and prevents binding of the substrate CO. In **16.17** this problem is avoided by using an iminothiolate, which has two lone pairs on S, only one of which is strongly basic (see **16.19**). This is similar to the situation in acetate, where the lone pair syn to the C=O group is known to be less basic (see **16.20**). The other less basic S lone pair anti to the C=N group is only weakly basic, and so **16.17** prefers to bridge via phenolate O to give a weak bridge, easily opened up by ligands analogous to CO, like CN⁻.

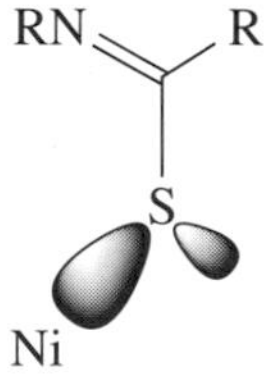

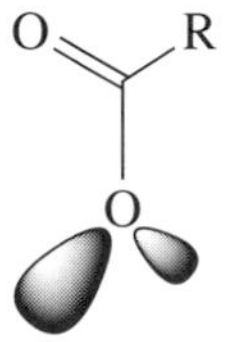

16.19   **16.20**

A stoichiometric model system by Holm et al.[25] for the acetylCoA synthase activity of CODH is shown in Eq. 16.41. This reaction is a property of the NiFeC cluster, of unknown structure, present in CODH. The enzyme brings about exchange between ^{14}CO and $Me^{12}COCoA$, which implies that formation of the C$-$S and Me$-$CO bonds is reversible.[26] This is consistent with CO insertion into a Ni$-$Me bond, and nucleophilic attack on the resulting Ni(COMe), both of which can be reversible.

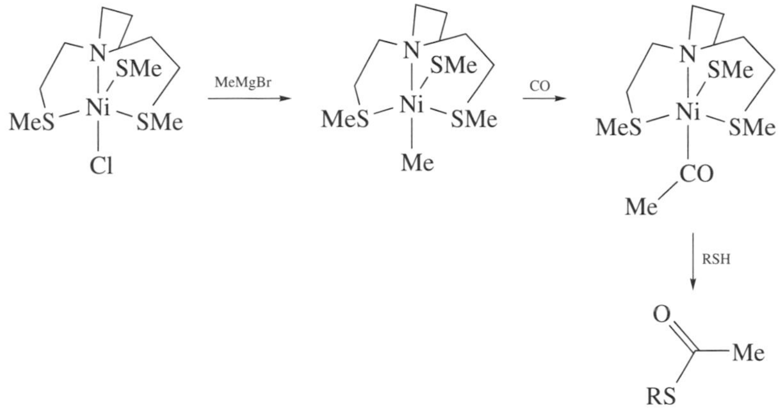

$$(16.41)$$

Methanogenesis

The methanogens reduce CO_2 to CH_4 and extract the resulting free energy; 10^9 tons of CH_4 are formed annually in this way. In the last step, methylcoenzyme M, **16.21**, is hydrogenolyzed to methane by a thiol cofactor HS$-$HTP ($=$ R$'$SH), catalyzed by the Ni enzyme, MCR.

$$CH_3SCH_2CH_2SO_3{}^- + R'SH \xrightarrow{\text{MCR}} CH_4 + R'S-SCH_2CH_2SO_3{}^- \qquad (16.42)$$

methylcoenzyme M   HS-HTP

16.21

A coenzyme, factor F_{430} (**16.22**), is bound to MCR and is believed to catalyze Eq. 16.42, perhaps via binding of methylcoenzyme M to the reduced form of F_{430}, leading to release of an incipient methyl radical by methyl CoM, that is immediately quenched by H atom transfer from the HS$-$HTP thiol to give methane. The resulting thiol radical abstracts the CoM thiolate from Ni to regenerate the Ni(I) form as well as give the observed heterodisulfide.[28,29]

$$Ni(I)-(Me-SCoM) + HS-HTP \Rightarrow Ni(II)-(SCoM) + MeH + \cdot S-HTP$$

$$\Rightarrow Ni(I) + CoM-S-S-HTP + MeH \qquad (16.43)$$

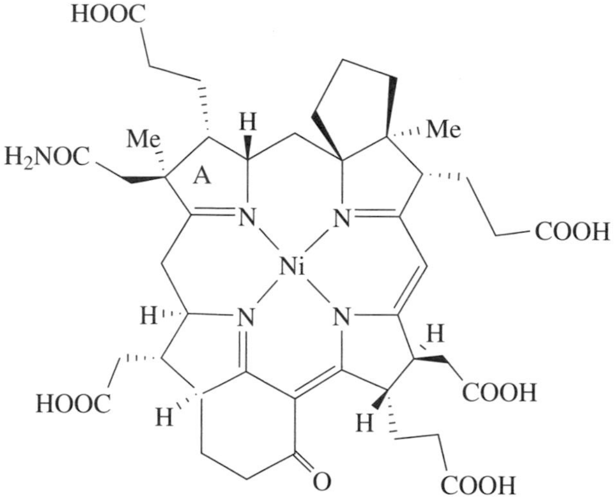

16.22

Factor F_{430}

Hydrogenase

The hydrogenases[30,31] bring about Eq. 16.44, which allows certain bacteria to live on H_2 as energy source, and others to get rid of excess electrons by combining them with protons for release as H_2. The nickel-containing [NiFe] hydrogenases are the largest class, but iron-only [FeFe] hydrogenases also exist. The number of metal ions present varies with the species studied, but the minimum cofactor composition is one Ni—Fe or Fe—Fe and one Fe_4S_4 cluster per enzyme.

$$H_2 \rightleftharpoons 2H^+ + 2e^- \qquad (16.44)$$

Both [NiFe] and [FeFe] hydrogenases have organometallic active-site clusters, as shown by X-ray crystallography and IR spectroscopy.[32] The [NiFe] protein active site cluster from *Desulfovibrio gigas* is shown as **16.23**, and the [FeFe] protein's H-cluster from *Clostridium pasteurianum* is shown as **16.24**. The active site **16.23** has a nickel tetrathiolate center bridged to a low-spin dicyanoiron(II) carbonyl group—the latter was then an unprecedented ligand set in biology. The bridging oxo or hydroxo group, X, is believed to be removed as H_2O when the enzyme converts to the active form by incubation under H_2 for some hours. Structure **16.24** has two monocyanoiron carbonyl groups bridged by a CO and a dithiolate, either propane-1,3-dithiolate or its aza analog (Y = NH). One iron has a labile ligand, thought to be water, where the H_2 presumably binds. Theoretical work[33] supports heterolytic splitting of such an intermediate, where the H^+ may move to an internal base (compare with Eq. 3.39), either a bridging sulfide or to the N lone pair of an azathiolate. As part of an interesting speculation on the origin of life,[2] iron sulfide, dissolved at deep-sea vents by CO, is proposed to give **16.25**, a complex that became incorporated into early proteins to give the first hydrogenases. In any event, **16.25** is a useful synthetic precursor to a series of complexes such as **16.26** that resemble the hydrogenase site.[34] The Fe−Fe distance of 2.5 Å in **16.26** is consistent with the metal−metal bonding required by the EAN rule.

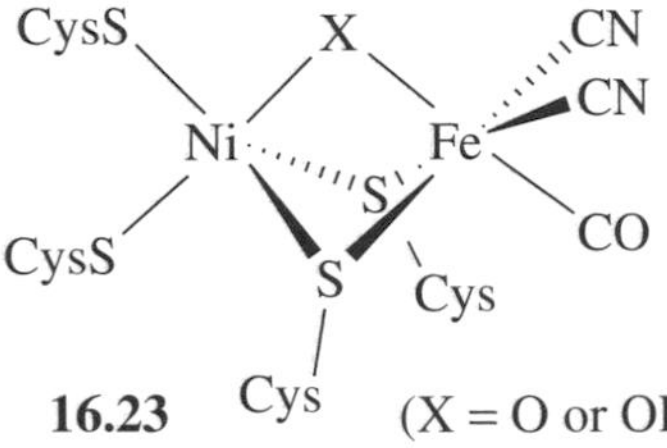

16.23 Cys (X = O or OH) 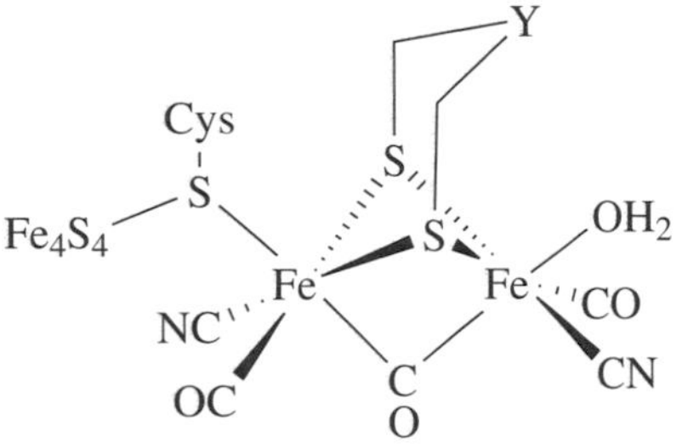   **16.24**   (Y = NH or CH$_2$)

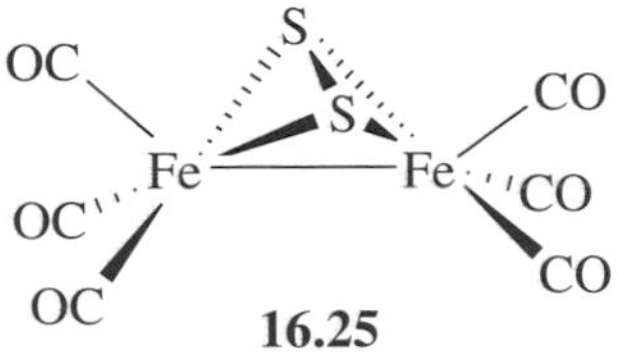

16.25 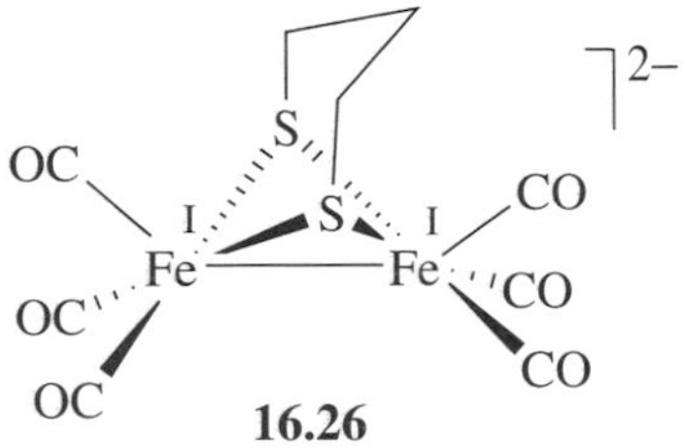   **16.26**

The nickel of the [NiFe] enzyme has an EPR active, odd-electron oxidized Ni(III) form but can be reduced to an EPR inactive Ni(II) form, and then to a more reduced, odd-electron Ni(I) form; an even more reduced even-electron form is also known.*

The Ni(III) state with a bridging X group (X = O or OH) seems to be formed as part of a mechanism for protecting the enzyme against exposure to air. The catalytically active form involves Ni(II) and more reduced states. Hydrogen activation by the enzyme is heterolytic because D$_2$ exchanges with solvent protons by Eq. 16.45;[31] dihydrogen complexes are known to catalyze similar reactions.[33b]

$$D_2 + ROH \rightleftharpoons HD + ROD \qquad (16.45)$$

Ni(III) is an unusual oxidation state, especially in an S environment, and so it is not surprising that a large amount of work has gone into looking for model compounds. The most easily oxidized Ni(II) species of this type is Millar's

---

*Ni(I) and (III) are convenient labels, implying that the oxidation or reduction are *at least in part* metal centered. The reader should be warned that inorganic chemists enjoy arguing about whether oxidation states such as these are valid descriptions of the species involved.

complex, **16.27**, for which the redox potential is -0.76 V.[35] Note the clever use of the cage structure to protect the metal and inhibit disulfide formation. Compound **16.28** is an interesting system in that all three oxidation states, I, II, and III, are all accessible without rearrangement or decomposition.[36] The g values seen in the EPR, also shown in the equation, are very different from those seen for organic radicals, which are always close to 2.0. This is the evidence that the reduction and oxidation are at least in large part metal centered, where g values different from 2 are common.

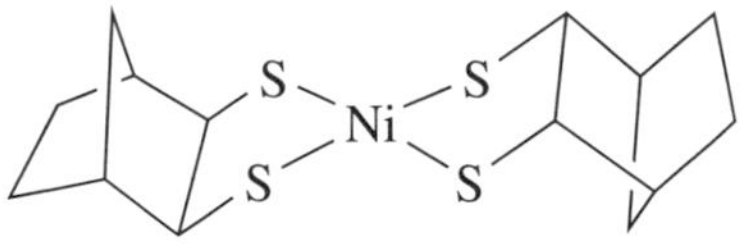

16.27

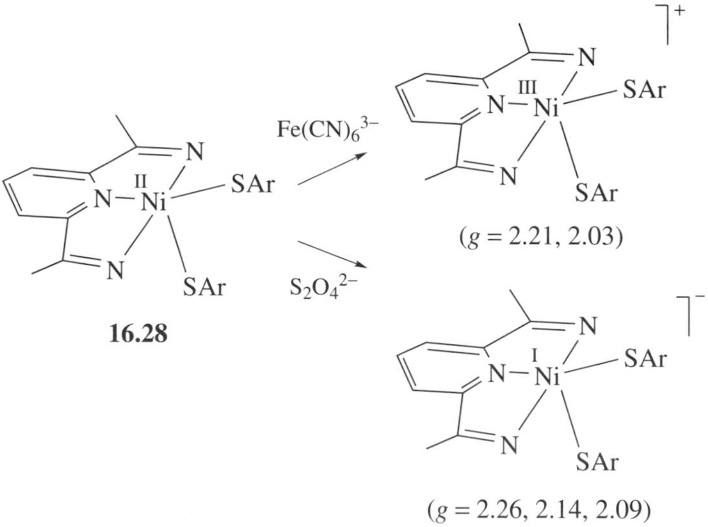

(16.46)

16.28

($g = 2.21, 2.03$)

($g = 2.26, 2.14, 2.09$)

Hemoglobin

As an O_2 transport protein, much of the chemistry of hemoglobin (Hb) falls outside the organometallic area. There is one exception, however: the structure of Hb-CO, the carbonyl complex.[37]

Hemoglobin contains four iron–porphyrin sites each capable of binding one O_2. Each site consists of an Fe(II) held within a porphyrin ring via coordination to the four N donors. The fifth Fe coordination position is occupied by a histidine, called the *proximal* His. The sixth site of the octahedral Fe(II) normally interacts with O_2, which binds reversibly in the bent form. One electron is transferred from Fe(II) to O_2 so that the O_2 adduct is best considered as an Fe(III) complex of superoxide ion (O_2^-). Unfortunately, the blood contains a small concentration of CO, formed in the biological degradation of the porphyrin ring. As a result, about

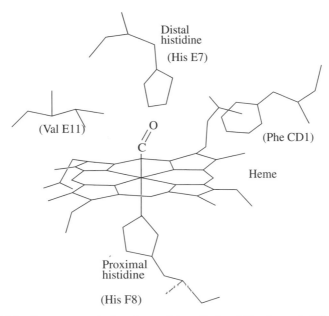

FIGURE 16.4 One proposal for the origin of destabilization of CO binding in hemoglobin. (Reproduced from Ref. 37 with permission.)

1% of the Hb in the blood is normally deactivated by irreversible binding of CO. This deactivation would be much more severe if it were not for the presence of the *distal* His in the vicinity of the CO binding site. The distal His is proposed to sterically disfavor binding of CO, which prefers linear binding, but not to interfere with O_2 binding in the bent superoxide form. When the CO binds, it is believed that it is forced to bend to some extent (Fig. 16.4) and this bending destabilizes the M-CO bond. This conclusion is controversial, however.[37]

- Nickel enzymes carry out the biological equivalent of the water–gas shift and the Monsanto acetic acid process.
- In hydrogenase, they also equilibrate H_2 with $2H^+ + 2e^-$.

16.5 BIOMEDICAL APPLICATIONS

Paul Ehrlich (1854–1915) is celebrated for his 1906 prediction that therapeutic compounds would be created "in the chemist's retort." As a physician strongly interested in chemistry, he had already been awarded the Nobel Prize (1908) for his work as the founder of chemotherapy, when he made his most important discovery—the application of the polymeric organoarsenical, Salvarsan, as the first antisyphilitic agent. This finding caused an international sensation and led

Salvarsan   16.29

to his being besieged by thousands of sufferers. In spite of this early success, organometallic compounds have received very little attention in pharmacology since that time.

The water-soluble carbonyl **16.29** has been used as a stable label for biomolecules where it is found to bind to histidine-repeat (His–His–His–) regions. An antibody has been labeled with the ^{99m}Tc complex so that the γ-ray image obtained as a result of the γ-emission of this Tc isotope shows the location of a tumor to which the antibody selectively binds.[38] Technetium imaging with various coordination complexes has become a standard procedure in medicine.[39]

It is likely that the few organometallic systems we currently recognize in biology represent a small fraction of the total, and that many others, including new organometallic pharmaceuticals, remain to be discovered. We can therefore anticipate growing interest in this new area.

- Salvarsan is very important in the history of medicine.
- Today the pharmaceutical industry tends not to use heteroatoms (B, Si, Fe,...) that might provide useful biologically active molecules.

REFERENCES

1. (a) J. Darnell, H. Lodish, and D. Baltimore, *Molecular Cell Biology*, 4th ed., Scientific American Books, New York, 2000; J. M. Berg, L. Stryer, and J. L. Tymoczko, *Biochemistry*, 5th ed, Freeman, San Francisco, 2002; (b) S. J. Lippard and J. Berg, *Principles of Bioinorganic Chemistry*, University Science Books, Mill Valley, CA, 1994; (c) G. Jasuen, M. J. McGlinchey, et al., *Organometallics*, **4**, 2143, 1985.

2. C. Huber and G. Wächtershäuser, *Science* **276**, 245, 1997.

3. (a) Chapter 4 in Ref. 1b; (b) A. J. Hoff, ed., *Advanced EPR*, Elsevier, Amsterdam, 1989.

4. S. P. Cramer and K. O. Hodgson, *Prog. Inorg. Chem.* **25**, 1, 1979; M. A. Newton, A. J. Dent, and J. Evans, *Chem. Soc. Rev.* **31**, 83, 2002.

5. (a) T. G. Spiro, ed., *Biological Applications of Raman Spectroscopy*, Wiley, New York, 1987; (b) T. C. Gibbs, *Principles of Mössbauer Spectroscopy*, Chapman, Hall, London, 1976.

6. R. Banerjee and S. W. Ragsdale, *Ann. Rev. Biochem.* **72**, 209, 2003; T. Toraya, *Chem. Rev.* **103**, 2095, 2003.

7. D. C. Hodgkin, *Proc. Roy. Soc. A* **288**, 294, 1965; *Chem. Br.* **13**, 138, 1977.

8. J. Halpern, *Pure Appl. Chem.* **55**, 1059, 1983; *Acc. Chem. Res.* **15**, 231, 1982; *Science* **227**, 869, 1985; *Inorg. Chem.* **38**, 2386, 1999.

9. M. D. Sintchak, G. Arjara, B. A. Kellogg, J. Stubbe, C. L. Drennan, *Nature Struct. Biol.*, **9**, 293, 2002.

10. (a) P. G. Schultz, K. G. Au, and C. T. Walsh, *Biochemistry* **24**, 6840, 1985; T. P. Begley and C. T. Walsh, *Biochemistry* **25**, 7186, 7192, 1986; (b) T. V. O'Halloran et al., *J. Am. Chem. Soc.* **120**, 12690 1999; *Adv. Inorg. Biochem.* **8**, 1, 1990; (c) M. Ehrlich, *J. Clin. Lig. Assay* **23**, 144, 2000.

11. (a) D. Jones, *New Sci.* 101, Oct. 14, 1982; (b) S. Forshufvud, *Who Killed Napoleon?* Hutchinson, London, 1962.

12. M. P. Shaver and M. D. Fryzuk, *Adv. Synth. Catal.* **345**, 1061, 2003; R. Y. Igarashi and L. C. Seefeldt, *Crit. Rev. Biochem. Mol. Biol.* **38**, 351, 2003; C. M. Kozak and P. Mountford, *Angew. Chem. Int. Ed.* **43**, 1186, 2004.

13. V. K. Shah and W. J. Brill, *Proc. Nat. Acad. Sci. (USA)* **74**, 3249, 1977.

14. D. C. Rees et al., *Nature* **360**, 553, 1992; *Science*, **297**, 1696, 2002.

15. See A. D. Allen and F. Bottomley, *Acc. Chem. Res.* **1**, 360, 1968.

16. J. E. Bercaw, *J. Am. Chem. Soc.* **96**, 5087, 1974.

17. R. R. Schrock et al., *Science*, **301**, 76, 2003.

18. R. H. Holm, *Prog. Inorg. Chem.* **38**, 1, 1990; *Pure Appl. Chem.* **70**, 931, 1998; *Science* **277**, 653, 1997; R. Cammack, *Adv. Inorg. Chem.* **38**, 281, 1992.

19. H. Beinert, *J. Biol. Inorg. Chem.* **5**, 2, 2000.

20. M. Millar and S. A. Koch, *Inorg. Chem.* **31**, 4594, 1992.

21. (a) J. A. Kovacs, Chap. 5 in Vol. 9 of *Advances in Inorganic Biochemistry*, G. L. Eichhorn, ed. Prentice-Hall, Englewood Cliffs, NJ, 1992; (b) J. R. Lancaster, ed., *The Bioinorganic Chemistry of Nickel*, VCH, Weinheim, 1988; A. C. Marr, D. J. E. Spencer, and M. Schröder, *Coord. Chem. Rev.* **219**, 1055, 2001.

22. (a) S. W. Ragsdale and M. Kumar, *Chem. Rev.* **96**, 2515, 1996; (b) T. I. Doukov, T. M. Iverson, J. Seravalli, S. W. Ragsdale, and C. L. Drennan, *Science* **298**, 567, 2002; C. Darnault, A. Volbeda, E. J. Kim, P. Legrand, X. Vernede, P. A. Lindahl, and J.-C. Fontecilla-Camps, *Nat. Struct. Biol.* **10**, 271, 2003; (c) S. W. Ragsdale, *Crit. Rev. Biochem. Mol. Biol.*, **39**, 165, 2004; (d) E. L. Hegg, *Accts. Chem. Res.*, **37**, 775, 2004.

23. J. Seravalli, Y. Xiao, W. Gu, S. P. Cramer, W. E. Antholine, V. Krymov, G. J. Gerfen, and S. W. Ragsdale, *Biochemistry*, **43**, 3944, 2004; C. E. Webster, M. Y. Darensbourg, P. A. Lindahl, and M. B. Hall, *J. Am. Chem. Soc.* **126**, 3410, 2004.

24. Z. Lu and R. H. Crabtree, *J. Am. Chem. Soc.* **117**, 3994, 1995.

25. R. H. Holm et al., *J. Am. Chem. Soc.* **112**, 5385, 1990.

26. S. W. Ragsdale and H. G. Wood, *J. Biol. Chem.* **260**, 3970, 1985.

27. U. Deppenmeier, *Prog. Nucl. Acid Res. Mol. Biol.* **71**, 223, 2002.

28. V. Pelmenschikov, M. R. A. Blomberg, P. E. M. Siegbahn, and R. H. Crabtree, *J. Am. Chem. Soc.* **124**, 4039, 2002.

29. V. Pelmenschikov and P. E. M. Siegbahn, *J. Biol. Inorg. Chem.* **8**, 653, 2003.

30. M. Frey, *Chembiochem* **3**, 153, 2002.

31. R. K. Thauer et al., *Chem. Rev.* **96**, 3031, 1996.

32. J. W. Peters et al., *Science* **282**, 1853, 1998; A. Volbeda et al., *J. Am. Chem. Soc.* **118**, 12989, 1996; S. P. A. Albracht et al., *Nature* **385**, 126, 1997.

33. (a) M. Pavlov, P. E. M. Siegbahn et al., *J. Am. Chem. Soc.* **120**, 548, 1998; I. Dance, *Chem. Commun.* 1655, 1999; R. H. Crabtree, *Inorg. Chim. Acta.* **125**, L7, 1986; (b) A. C. Albeniz, D. M. Heinekey, and R. H. Crabtree, *Inorg. Chem.* **30**, 3632, 1991.

34. M. Y. Darensbourg, E. J. Lyon, X. Zhao, and I. P. Georgakaki, *PNAS* **100**, 3683, 2003, and references cited.

35. M. Millar et al., *J. Am. Chem. Soc.* **112**, 3218, 1990.

36. P. K. Mascharak et al., *J. Am. Chem. Soc.* **114**, 9666, 1992.

37. J. P. Collman and L. Fu, *Acc. Chem. Res.* **32**, 455, 1999; T. G. Spiro and P. M. Kozlowski, *J. Biol. Inorg. Chem.* **2**, 516, 1997; J. A. Lukin and C. Ho, *Chem. Rev.* **104**, 1219, 2004.

38. A. Plückthun, *Nat. Biotechnol.* **17**, 897, 1999.

39. D. E. Reichert, J. S. Lewis, and C. J. Anderson, *Coord. Chem. Rev.* **184**, 3, 1999; R. Schibli and P. A. Schubiger, *Eur. J. Nucl. Med. Mol. Imag.* **29**, 1529, 2002.

PROBLEMS

1. Why do you think Nature uses first-row transition metals in most of the transition metalloenzymes?

2. The oxidation states found in the metal centers we have been discussing in this chapter, Fe(II), Fe(III), Ni(III), and Co(III), are often higher than those usually present in organometallic species we discussed in Chapters 1–14. Why do you think this is so?

3. Those mononuclear N_2 complexes, which have the lowest N–N stretching frequency in the IR, are in general also the complexes in which N_2 is most easily protonated. Explain.

4. Would you expect the following R groups to dissociate more or less readily as R• from cobaloxime than does •CH_2Ph: $-CH_3$, $-CF_3$, $-CPh_2H$? Explain.

5. Many N_2 complexes protonate. In the case of $ReCl(N_2)(PMe_2Ph)_4$, the protonated form $HReCl(N_2)(PMe_2Ph)_4^+$ (A) is relatively stable. What might happen to the N–N stretching frequency on protonation? Most N_2 complexes simply lose N_2 on protonation. Given that a complex of type A is the intermediate, explain why N_2 is lost.

USEFUL TEXTS ON ALLIED TOPICS

Bioinorganic Chemistry

S. J. Lippard and J. Berg, *Principles of Bioinorganic Chemistry*, University Science Books, Mill Valley, CA, 1994.

Homogeneous Catalysis

B. Cornils and W. A. Herrmann, *Applied Homogeneous Catalysis with Organometallic Compounds: A Comprehensive Handbook*, VCH, New York, 1996.

Encyclopedias

R. B. King, ed. *Encyclopedia of Inorganic Chemistry*, Wiley, New York., 2005.

G. Wilkinson et al., eds., *Comprehensive Organometallic Chemistry*, Pergamon, Oxford, 1982, 1987, 1995, 2006.

Group Theory

F. A. Cotton, *Chemical Applications of Group Theory*, Wiley, New York, 1990.

Inorganic Chemistry

F. A. Cotton, W. M. Bochmann, and C. A. Murillo, *Advanced Inorganic Chemistry*, 6th ed., Wiley, New York, 1998.

Kinetics and Mechanism

K. J. Laidler, *Chemical Kinetics*, 3rd. ed., Wiley, New York, 1987.

J. E. Espenson, *Chemical Kinetics and Reaction Mechanisms*, McGraw-Hill, New York, 1981.

R. B. Jordan, *Reaction Mechanisms of Inorganic and Organometallic Systems*, Oxford University Press, 1991.

R. G. Wilkins, *Kinetics and Mechanism of Reactions of Transition Metal Complexes*, 2nd ed., VCH, Weinheim, 1991.

NMR

W. Kemp, *NMR in Chemistry*, Macmillan, London, 1986.

K. A. McLaughlan, *Magnetic Resonance*, Clarendon, Oxford, 1982.

P. S. Pregosin and R. W. Kunz, *31P and 13C NMR Spectroscopy of Transition Metal Complexes*, Springer, Heidelberg, 1979.

G. N. Lamar, W. D. Horrocks, and R. H. Holm, *NMR of Paramagnetic Molecules*, Academic, New York, 1973.

Organic Chemistry, Organometallics in

M. Schlosser, *Organometallics in Synthesis*, Wiley, New York, 2002.

L. S. Hegedus, *Transition Metals in the Synthesis of Complex Organic Molecules*, University Science Books, Sausalito, CA, 1999.

Organometallic Chemistry

A. Yamamoto, *Organotransition Metal Chemistry*, Wiley, New York, 1990.

Photochemistry

G. L. Geoffroy and M. S. Wrighton, *Organometallic Photochemistry*, Academic, New York, 1979.

Preparative Techniques

D. F. Shriver, *The Handling of Air-Sensitive Compounds*, McGraw-Hill, New York, 1969.

Special Topics

W. A. Nugent and J. M. Mayer, *Metal–Ligand Multiple Bonds*, Wiley, New York, 1988.

Structure

E. A. V. Ebsworth, D. W. H. Rankin, and S. Cradock, *Structural Methods in Inorganic Chemistry*, Blackwell, Oxford, 1987.

MAJOR REACTION TYPES

Alphabetical List of Reaction Types and Where to Find Them in the Text

The Organometallic Chemistry of the Transition Metals, Fourth Edition, by Robert H. Crabtree
Copyright © 2005 John Wiley & Sons, Inc.

The major reaction types presented in this book are listed in Fig. A.1.

(OS) \ Δ(CN)	−2	−1	0	1	2
−2	Red. Elim. {−2} [6.5, 14.4] Deprotonatn. {0} [8.3]	Nucl. Abs. of X⁺ {0} [8.3]	Metalacyclo-butane Clvg. {2} [12.1] Red. Clvg. {2} [6.7]		
−1		Binucl. Red. Elim. {−1} [6.5]			
0	Dissoc of L. {−2} [4.3, photochem., 4.6] Dissoc or Abstrn of E⁺ {0} [6.5, 8.3]	Substn. of L. {0} [4.3–4.7] Insertn. & Elim. {0} [7.1–3, 9.1] SET {±1} [8.6] Ox. Cplg. {−2} [6.7]	Assoc of L. {2} [4.4] Alpha & Beta Elim. {+2} [7.4]		
1				Binucl. Ox Addn {1} [6.3] Assocn. of X• {1} [4.3, 6.3]	
2		Carbene–Alkene Cycloaddn. {−2} [12.1] Ox. Cplg. {−2} [6.7, 14.4]	Assoc of E⁺ incl. Protonation {0} [6.4, 8.4, 11.1]	Ox. Addn {2} [6.1–6.4, 12.4] Gamma, Delta Elim. {+2} [7.4]	

FIGURE A.1 Master list of reaction types. *Key:* Abs. = abstraction, Addn. = addition, Assoc. = association, Binucl. = binuclear, Cplg. = coupling, Dissoc. = dissociation, E⁺ = electrophile, Elim. = elimination, Fragtn. = fragmentation. L = 2e ligand, Ox. = oxidative, SET = single electron transfer, Substn. = substitution, X• = free radical: {encloses electron count change in the reaction} and [encloses section number for the topic.]

SOLUTIONS TO PROBLEMS

Chapter 1

1. 4 (if you thought 2, you forgot structures such as $[PtL_4]^{2+}[PtCl_4]^{2-}$).

2. A cubane with $PtMe_3$ and I at alternate corners to give the octahedral geometry required by Pt(IV).

3. The first diphosphine ligand gives a favorable five-membered ring on chelation, while the second gives an unfavorable four-membered ring. The second lone pair of water repels and destabilizes the d_π electrons.

4. (i) $[PtCl_4]^{2-}$ + tu, 1 equiv, which must give $[Pt(tu)Cl_3]^-$; (ii) NH_3, which replaces the Cl trans to the high trans effect tu ligand.

5. The Ti complex is a hard acid, so the order is N > P > C (hard base best); the W complex is a soft acid, so C > P > N (soft base best).

6. The tetrahedral structure with a two-below-three orbital pattern will be paramagnetic because in a d^8 ion the lower set of three orbitals will take six electrons, leaving two for the upper set of two orbitals; these must go in with parallel spin, so there will be two unpaired electrons.

7. Measure $\nu(CO)$, the better donors will cause the greater lowering because they will cause a greater charge buildup on the metal, which will lead to increased $M(d_\pi) \rightarrow CO(\pi^*)$ back donation and a lower C−O bond order.

8. The d orbitals are stabilized by the higher nuclear charge, and so back donation (required to form a strong M−CO bond) is reduced. Cu(I) rather than Cu(II) would be best because it would be a stronger π donor.

The Organometallic Chemistry of the Transition Metals, Fourth Edition, by Robert H. Crabtree
Copyright © 2005 John Wiley & Sons, Inc.

9. Reduced complexes will easily lose electrons to O_2 in an oxidation reaction but will not tend to bind a π donor such as H_2O.

10. Assume an octahedral three-below-two splitting pattern, then $MnCp_2$ has five unpaired electrons one in each of the five orbitals; $MnCp_2^*$ has 4e paired up in the lower pair of orbitals and one unpaired electron in the next higher orbital; Cp^* has the higher ligand field because it causes spin pairing.

Chapter 2

1. The first three are 16e, Pt(II), d^8, then 20e, Ni(II), d^8, 18e, Ru(II), d^6; 18e, Re(VII), d^0; 18e, Ir(V), d^4; 10e, Ta(V), d^0; 16e, Ti(IV), d^0, 14e, Re(VII), d^0.

2. $[\{(CO)_3Re\}(\mu_3\text{-}Cl)]_4$. A triply bridging Cl^- in a cubane structure allows each Cl^- to donate 5 electrons (6e ionic model).

3. $(\eta^6\text{-}PhC_6H_5)Cr(CO)_3$, with a π-bound arene ring.

4. Ti(0) if both ligands are considered as being 4e L_2, but Ti(II) if one is considered as being X_2 and bound via the two N atoms in the MeN$-$CH$=$CH$-$NMe form, and Ti(IV) if both are considered as being in the X_2 form.

5. The complex is 12e, 10e, and 8e in the Ti(0), (II), and (IV) forms.

6. M$-$M counts one for each metal. This rule allows the Os compound to reach 18e. The Rh compound has a tetrahedron of mutually bonded Rh atoms for a total of six Rh$-$Rh bonds is also 18e.

7. 8e C for $H_3C^+ \leftarrow :NH_3$ (three X ligands, one L, and a positive charge) and 8e for $H_2C \leftarrow :CO$ (two X ligands and one L).

8. Counting only one lone pair gives an 18e count in both cases.

9. 2e either way. A σ-acid metal favors the η^1 form in which the important bonding interaction is L $\rightarrow$ M σ donation and a π-basic metal favors the η^2 form where back donation into the C$=$O π^* is the most important interaction. η^1 binding should favor nucleophilic attack.

10. W, η^3, and η^5 gives an 18e count. If each triphos is η^2, we get a 16e count, which is appropriate for Pd(II), and this is the true structure; an $\eta^2 - \eta^3$ structure would be 18e and cannot be ruled out, but an $\eta^3 - \eta^3$ would be 20e and is unreasonable.

11. The left-hand complex has six L-type ligands, so we have 18e, d^6, W(0); the right-hand complex has five L and two X ligands, so we have 18e, d^4, W(II).

Chapter 3

1. Protonation of the Pt or oxidative addition give a Pt$-$H into which the acetylene inserts.

2. $M-CF_2-Me$ (σ-acceptor substituents, especially F strongly stabilize an alkyl).

3. Oxidative addition of MeCl, followed by reaction of the product with LiMe, which acts as a Me^- donor and replaces the $Ir-Cl$ by $Ir-Me$.

4. Bent, 18e, no π bonding.

5. 18e in all cases; both structures have the same electron count because (H_2) is a 2e L ligand and $(H)_2$ consists of two 1e X ligands, so no change. Both structures are in fact classical.

6. If X or Y have lone pairs, they may complete for binding. $Y-H-M$ is not competitive with lone-pair binding.

7. It is easier to reduce a more oxidized complex.

8. (a) To maximize $M \rightarrow Ph$ back bonding from the out-of-plane d_{z^2} orbital, the Ph will have to be in the square plane so the π cloud of the Ph ring is lined up with the d orbital. (b) To minimize steric repulsion, Ph will be out of the plane.

9. 17e, Ru(III), d^5; 18e, Cr(0), d^6; 12e, W(VI), d^0.

10. Initial formation of $Ir-(i\text{-Pr})$ with RMgX acting as source of R^- to replace the Cl^- initially bound to Ir. The alkyl then β eliminates to give propene as the other product.

11. Insertion of the alkene into the $M-H$ bond to give $M-CHMe(nPr)$, followed by β elimination to give $MeCH=CHMe$; insertion requires prior binding of the alkene and so does not happen in the 18e case.

Chapter 4

1. (a) Halide dissociation is bad for two reasons. The product is 16e and cationic, while for proton dissociation the product is 18e and anionic; 16e species are less favorable and cations are less well stabilized by the π-acceptor CO groups than anions. (b) Solvent likely to bind only to the 16e cation.

2. The NO can bend to accommodate the incoming ligand.

3. The more ∂^+ the CO carbon, the easier the reaction, so the order is: $Mn(CO)_6^+ > Mo(CO)_3(NO)_2 > Mo(CO)_6 > Mo(CO)_4(dpe) > Mo(CO)_2(dpe)_2 > Mo(CO)_5^{2-}$. [This order is decided by (1) cations > neutrals > anions, and (2) within each class, complexes with the better π-acceptor ligands > complexes with less good π-acceptor ligands.]

4. The $\nu(CO)$ in the IR or the ease of oxidation as measured electrochemically.

5. $CpWH_2$: 18e, W(IV), 8; $\{Cp_2W\}_2$: same; $ReCl(N_2)L_4$: 18e, Re(I), d^6; Re dimer: same; FeL_4: 16e, Fe(0), 4; cyclometalated form: 18e, Fe(II), 6; W compounds: 18e, W(0), 6.

6. NR_3 lacks significant π-acid character, but NF_3 should bind better thanks to its N$-$Fσ^* orbital, which should be polarized toward the metal and could act as π acceptor; this resembles the cases of CH_3 versus CF_3, where the same applies.

7. As a highly reduced metal, Ni(0) prefers π-acceptor ligands such as $P(OMe)_3$. PMe_3 as a poor π acceptor causes the electron density on the metal to rise so much that the NiL_3 fragment is a poor σ acceptor.

8. D, A, D, D, A, A. D for 18e, A for 16e species.

9. Eighteen electron structures (or 16e where appropriate) can be achieved as follows: η^6-Ph or BPh_4; η^3 and η^5-Ind groups; $[Me_3Pt(\mu\text{-}I)]_4$, cot must be η^4 to two $PtCl_2$ groups; μ-Cl required.

10. (a) $trans$-$L'_2Mo(CO)_4$$-L'$ labilizes the CO trans to itself; (b) cis-$L'_2Mo(CO)_4$$-$CO preferentially labilizes a CO trans to itself.

11. Six positive ionic charges on the complex rules it out because the metal would not retain enough π-donor power to bind NO. Very few complexes exceed a net ionic charge of ± 2.

12. Protonation at the metal (always allowed even for 18e complexes) should weaken M$-$CO and put a high-trans-effect ligand on the metal.

13. Extrapolation suggests a very high figure, 2270 cm^{-1} or above, implying the presence of a very weakly bound CO and that the compound would be very hard to make.

14. One factor must be the lack of back donation for NR_3, but the short M$-$N and N$-$R bonds relative to M$-$P and P$-$R may lead to a significant increase in steric size. For the pentacarbonyl, the lack of back donation is not a problem because there are so many good π-acceptor COs present and the steric problem is minimal because the COs are so small.

15. The hydride is 12e and the carbynes are 14e but the N lone pairs can act as π donors and raise both counts to 18e.

Chapter 5

1. The poorer π-back-bonding centers will have the highest alkene reactivity: Pd > Pt; cation > neutral; phosphite > phosphine.

2. Nucleophilic attack on a halide or tosylate (the latter may be better because the halide may dehydrohalogenate) $2L_nM^- + TsOCH_2CH_2OTs$. ^{13}C NMR should show two equivalent carbons with coupling to two directly attached H, and coupling to $2n$ L and 2 M nuclei (if these have I $\neq$ 0).

3. Oxidative coupling of two alkynes to give the metallole, followed by CO insertion and reductive elimination. The dienone should be a good ligand.

4. From $Cp_2MoClMe$ by abstraction of Cl^- with Ag^+ in the presence of ethylene. $C-C$ should be parallel to $Mo-Me$ for the best back donation because the back-bonding orbital lies in the plane shown in Fig. 5.6. NMR should show inequivalent CH_2 groups, one close to the methyl and one far from this group.

5. We expect more LX_2 character (see **5.16**) as L becomes more donor, so C_2C_3 should shorten.

6. The allyl mechanism of Fig. 9.2b to give $[(1, 5\text{-cod})IrCl]_2$ then removal of the cod with the phosphite. 1,5-Cod is less stable because it lacks the conjugated system of the 1,3-isomer. The formation of two strong $M-P$ bonds provides the driving force.

7. Two optical isomers are possible: the 2-carbon of propene has four different substituents: CH_3, H, CH_2 and Cl_3Pt.

8. There are three unpaired electrons for octahedral high spin d^7 Co(II).

9. The first complex is the 18o species, $[(\eta^6\text{-indane})IrL_2]^+$ formed by hydrogenation of the C=C bond by the IrH_2 group, and the second is $[(\eta^5\text{-indenyl})IrHL_2]^+$, formed by oxidative addition of an indane $C-H$ bond, β elimination, then loss of H_2 from the metal and oxidative addition of an indane $C-H$ bond. Substitution only of the arene complex by CO is possible because loss of arene is easier than loss of the Cp-like η^5-indenyl (see Section 5.7).

10. The d^0 carbonyls are rare (no back donation), but $M-H$ to CO π^* back donation may occur here.

Chapter 6

1. **A** reacts by S_N2, **B** by a radical route. i-PrI is an excellent substrate for radical reactions and $MeOSO_2Me$ for S_N2 (see Section 6.3).

2. Assuming steric effects are not important, only the bond strengths change, so these are in the order $Me-Me < M-Ph < M-H < M-SiR_3$, favoring silane addition and disfavoring methane addition.

3. True oxidative addition is more likely for electron-releasing ligands, good π-donor third-row elements, and better π-donor reduced forms. Dewar–Chatt binding is favored for a weak π-donor site that binds H_2 as a molecule.

4. For HCl the steps must be: (1) oxidative addition of HCl; (2a) a second oxidative addition of HCl followed by reductive elimination of H_2 and binding of Cl^- or (2b) electrophilic abstraction of H^- by H^+ and coordination of the second Cl^- to the empty site so formed. In either case H_2 is also formed. For t-BuCl: (1) SET to give •PtClL and t-Bu•. t-Bu• may abstract H from a second molecule of t-BuCl to give $Me_2C=CH_2$ and Cl•. In the final step, Cl• adds to $PtClL_2$• to give the product. A Pt(t-Bu) intermediate is also possible, but less likely ($M-t$-Bu is very rare).

5. Oxidative coupling to give the metallacycle followed by β elimination to give $L_nM(H)(CH_2CH_2CHCH_2)$, followed by reductive elimination of 1-butene.

6. C > D > B > A. The $\nu(CO)$ frequencies increase in the reverse order and lower $\nu(CO)$ correlates with a more reduced metal and so faster oxidative addition. After oxidative addition the frequencies should rise because oxidation of the metal should reduce its π basicity.

7. Reductive elimination of MeH and PhH are thermodynamically favored relative to reductive elimination of HCl.

8. Oxidative addition is not possible for d^0 species, so σ-bond metathesis must be implicated in the first step, probably via formation of H_2 complex, which is allowed in a 12e species. PMe_3 then displaces H_2 from intermediate MH_2 species to give the final product. The final H_2 is not lost because $W(PMe_3)_6$ is a rather unstable species, for the same reasons we saw for the Ni(0) analog in Problem 7 of Chapter 4.

9. The two Hs must be cis in the products. If we run the rearrangement under D_2, D incorporation into products will be seen if H_2 is lost.

10. PhCN has an unusually unhindered C−C bond, an intermediate η^2-arene complex is possible, and this may help bring the metal close to the C−C bond. Finally, M−CN is unusually strong for a C−C bond because of the π-bonding possible with this CO analog.

11. Insertion into $D_2C−O$ bond; then β elimination.

12. CO is axial but the bulky fullerene may oblige the bulky PR_3 ligands to fold back.

Chapter 7

1. (a) Migratory insertion should give the acyl $[CpRu(CO)(COMe)(PPh_3)]$; (b) insertion into M−H should give the allyl product; (c) attack at an 18e complex is allowed for SO_2 (see Section 7.3), so the $[CpFe(CO)_2(MeSO_2)]$ is formed; (d) no reaction is expected because the M−CF_3 bond is too strong.

2. Cyclometallation of the amine with loss of HCl gives **A**, followed by insertion of the cyclopropene to give **C** or oxidative addition of the strained C−C single bond of the cyclopropene followed by rearrangement to give **D**. Cyclometallation of the amine is not possible for $PhNMe_2$ because of the wrong ring size in this case.

3. α Elimination of CH_4 leaves $M=CH_2$ groups that couple to give $H_2C=CH_2$.

4. (1) RNC must bind, undergo migratory insertion, and the resulting imine undergo another insertion with the second hydride. (2) Migratory insertion twice over gives a bis-acyl that in its carbenoid canonical form (**7.2**) couples to give the new double bond. (3) Migratory insertion once, followed by alkyl

migration from the metal to the carbene carbon in the carbenoid resonance form of the cyclic acyl. (4) Insertion to give $MPh(O_2CPh)$ is probably followed by a cyclometallation by a σ-bond metathesis pathway with loss of PhH.

5. Oxidative addition of MeI is followed by reductive elimination. The possibility of binuclear reductive elimination is suggested from the label crossover data.

6. Ethylene displaces the agostic C$-$H to give $MEt(C_2H_4)$. Insertions of ethylene gives an agostic butyl with no β elimination of the growing chain. The process is repeated. The presence of an agostic C$-$H points to a weakly π-donor metal, which is unable to carry out a β elimination. In the Rh system, neutral Rh(I) is a better π donor and so β elimination is fast in the first formed butyl complex.

7. Possibilities are $-CH_2-CMe(OMe)_2$ or $-CH_2-CMePh_2$. For C$-$C bond breaking, we need a strained ring system such as $-CH_2-CMe(CH_2CH_2)$ or $-CH_2-CMe(CH_2CH_2CH_2)$.

8. More strongly ligating solvents, more electron-withdrawing ligands, and a poorer π-basic metal will all favor the reaction. The solvent stabilizes the product, and the ligands and metal make the CO more ∂^+ at carbon and so more reactive.

9. Cyclometallation should give $PtHClL_2$; the phosphine must cyclometallate in the $-CH_2Nb$ case, which would release CH_3Nb and leave a cyclometalated Pd complex.

10. The α-CH is β to the second metal, M_2, in a Me$-M_1-M_2$ cluster.

Chapter 8

1. The rules of Section 8.2 predict attack at (1) ethylene, (2) the terminal position, and (3) the butadiene.

2. (1) Protonation gives MeH and FpCl, (2) SET and nucleophilic abstraction gives MeCl, (3) electrophilic abstraction gives MeHgCl, and (4) protonation gives MeH and $CpL_2Fe(thf)^+$.

3. Reduction of Pd(II) to Pd(0) by nucleophilic attack of the amine on the diene complex is followed by oxidative addition of PhI and then insertion of the diene into the Pd$-$Ph bond to give a Pd(II) allyl. This can either β-eliminate to give the free diene or undergo nucleophilic attack by the amine to give the allylic amine.

4. The high $v(CO)$ and 2+ charge imply weak π back donation and means that the CO carbon is very δ^+ in character and very sensitive to nucleophilic attack.

5. The arene is activated for nucleophilic attack by coordination because the $Cr(CO)_3$ group is so electron withdrawing. The product should be $[(\eta^6\text{-}PhOMe)Cr(CO)_3]$.

6. The H⁻ group abstracted should be anti to the metal, but in β elimination, expected for a 16e complex, the metal abstracts the syn H.

7. We need to make the metal a better σ acid and π base, use a noncoordinating anion, sterically protect the site to prevent dimerization or binding of a solvent C—H bond, and use a poor donor solvent to prevent displacement.

8. Nucleophilic attack of MeOH to give the 2-methoxy-5-cyclooctene-1-yl complex is followed by a PR$_3$-induced β elimination to give **C** and the hydride. The 1,4-diene might also be formed. E and Z isomers of Me(I)C=C(Me)Et.

9. Nucleophilic attack of Me⁻ to give a vinyl complex is followed by electrophilic abstraction of the vinyl with I$_2$.

10. The P=O bond is too strong and the oxygen is less nucleophilic; dppe increases the back donation and so lowers the $\delta+$ charge at C making it less sensitive to nucleophilic attack; peroxysulfate or PhIO are more powerful reagents.

Chapter 9

1. Isomerization should bring all three double bonds together in the right-hand ring to give a phenol, a compound known to be acidic; the reaction is driven by the aromatic stabilization in the product.

2. Dissociation of L, required for activity, is unlikely for triphos because of chelation, but Cl⁻ abstraction by BF$_3$ or Tl⁺ opens the required site.

3. The initial terminal cyanation step should be followed by isomerization of the remaining internal C=C group to the terminal position and so should give the 1,5-dinitrile as the final product.

4. Successive H transfers to the ring are followed by oxidative addition of H$_2$ and further H transfers. The first H transfer to the arene will be difficult because the aromatic stabilization will be disrupted; this should be easier with naphthalene, where the aromatic stabilization is lower per ring and we only disrupt one ring.

5. Oxidative addition of the aldehyde C—H bond to Rh is followed by C=C insertion into the M—H to give a metallacycle; this gives the product shown after reductive elimination. Oxidative addition of the strained C—O bond is followed by β elimination and reductive elimination to give the enol that tautomerizes to acetone.

6. The first and second are thermodynamically unfavorable unless we find reagents to accept the H$_2$ or O$_2$, respectively. The third reaction is favorable, but it will be difficult to prevent overoxidation because the MeOH is usually much more reactive than MeH.

7. H$_2$[PtCl$_6$] (i.e., an acid, not a hydride).

8. Insertion into the M−Si rather than the M−H bond would give M−CR=CHSiR$_3$, and β elimination can now give the unsaturated product. This β elimination produces an MH$_2$ species that could hydrogenate some alkene to alkane, which is the third product.

9. Oxidative coupling, followed by β elimination and reductive elimination. If the β elimination were suppressed by avoiding β-H substituents, the metallacycle might be isolable. A 1,6-heptadiene is another possibility, where the bicyclic structure of the oxidative coupling product might make the metallacycle isolable.

10. Oxidative addition of H$_2$ is possible after the arene slips to the η^4 form. The substrate can displace the arene to give M(CO)$_3$(diene)H$_2$. If we consider that the diene adopts an LX$_2$ form, the observed product can be formed by two successive reductive eliminations. The cis product reflects the conformation of the bound diene, and the monoene is a much poorer ligand in this system and so does not bind and is therefore not reduced.

Chapter 10

1. The cis form has a doublet of quartets in the hydride region because of the presence of three P nuclei cis to each H and one P trans to H. The trans form has a quintet because of the presence of four P nuclei cis to each H. Using the HD complex will give a 1:1:1 triplet from H coupling to the I = 1 D nucleus and after dividing J(H,D) by six to adjust for the lower γ of the D isotope, we get the J(H,H), which is not observed in the dihydride because equivalent Hs do not couple.

2. MH$_3$ and MH(H$_2$) are the most likely. T$_1$(min) data and ^1J(H,D) in the H$_2$D complexes would be useful. The trihydride should have a long T$_1$ and a low J(H,D) (see Section 10.7).

3. One Ind could be η^3, in which case we should see two distinct sets of Ind resonances. If the two rings were rapidly fluxional, exchanging between η^3 and η^5 forms, one set of C resonances would be seen, but the presence of an IPR effect (see Section 10.8) in this case should make it distinguishable.

4. The presence of an IPR effect (see Section 10.8) would suggest the η^4 form.

5. 31 s^{-1}, 2500 × $\pi\sqrt{2}$ s^{-1}.

6. (1) c, a; (2), b, d; (3) d; (4) d; (5) d; (6) b.

7. Using Eq. 10.17 gives an angle close to 120°, consistent with a TBP structure with the COs equatorial.

8. The CO bond order falls when bridging as μ_2 and falls even further when bridged as μ_3.

9. 6-Coordination is expected in both cases, and so loss of Cl$^-$ is necessary to produce an η^2 form; the conductivity should be high for the ionic species and the IR of the two acetate binding modes are also different. Comparison of the IR with literature examples would be needed to distinguish the two cases.

10. If the plane of the pyridine ring is orthogonal to the square plane (as expected if steric effects dominate), we expect diastereotopy of the phosphine methyls because the methyl group of the pyridine breaks the plane of symmetry.

Chapter 11

1. Two moles of Tebbe's reagent should convert the ketone first to methylene cyclohexane and via that intermediate to product.

2. Initial intramolecular metalacycle formation, presumably with initial reversible CO loss, with metathesis-like cleavage leads to the product.

3. Initial oxidative coupling of the two ethylenes would have to be followed by β elimination and reductive elimination. The resulting 1-butene would have to resist displacement by ethylene (unlikely) but give an oxidative coupling of butene with ethylene, with the Et group always in the 1 position of the metalacycle and the β elimination would have to occur only at the former ethylene end of the metalacycle.

4. (a) $Ph_3P=CH_2$ has strong Schrock-like character, judging from the strongly nucleophilic character of the methylene group. This is consistent with Fig. 11.1 because C is more electronegative than P. (b) O is more electronegative than C, so $Re=O$ should be more nucleophilic than $Re=CH_2$.

5. Initial metathesis of the substrate $C=C$ bond gives $MeCH=CR(OR)$ and a $C=W$ carbene intermediate. This forms a metalacycle with the nearby alkyne and metathesis-like steps lead to product.

6. The CH_2 group lines up with the $Cp-M-Cp$ direction to benefit from back donation from W. The two extra electrons of the anion would have to go into the CH_2 p orbital. The CH_2 orientation would be at right angles to that in cation to minimize repulsion between the two filled orbitals.

7. The carbene is a neutral ligand with a lone pair while Ph is an anionic ligand with a lone pair.

Chapter 12

1. The reverse process should go by the reverse mechanism, which implies (see Fig. 12.9) that H_2 will oxidatively add to Pt(0) and then CO_2 will insert into the Pt−H bond.

2. The carbene formed on metathesis is stable.

3. Cyclometallation of a PMe group in preference to a PPh group is very unusual; perhaps the RLi deprotonates PMe, the CH_2^- group of which then binds to the metal.

4. As an 18e species, an η^1-CO_2 adduct is expected; for the indenyl case, slip could generate a site to allow η^2-OCO binding; the 18e complex could only plausibly react by H^- abstraction from the metal by CO_2, which would produce an η^1-OCHO complex. The Re anion is probably the best case because of the negative charge (after all, CO_2 reacts easily with OH^-).

5. Cyclometallation of the $ArCH_3$ group followed by CO insertion.

6. Loss of PhH by reductive elimination, binding of substrate via the isonitrile C, cyclometallation of the $ArCH_3$ group, migratory insertion involving the isonitrile, isomerization, and reductive elimination of the product.

7. Transfer of *endo*-Et to the metal, rotation of Cp, migration of Et back to a different point on the Cp ring, a 1,3 H shift on the exo face to bring an H into the endo position from which H transfer to the metal is possible.

8. Reductive elimination to form a cyclopropane that immediately oxidatively adds back to the metal.

9. Binding of formate as η^1-OCHO, followed by β elimination to deliver H^- to the metal and release CO_2. This can be a good synthetic route to hydrides.

10. CO_2 insertion into the terminal M−C bond to give an η^4-OCOCH$_2$CHCHCH$_2$ carboxylato-allyl complex. Oxidation then leads to the coupling of the allyls by binuclear reductive elimination.

11. Oxidative addition of Si−H, followed by coordination and insertion of the alkyne into M−H or M−Si, followed by reductive elimination.

12. The intermediate acyl could be hydrogenated; if so, with D_2 one would get $MeCD_2OH$. The methanol could undergo CH activation; if so, one would get $MeCH_2OH$.

Chapter 13

1. Any bridging CO complex with L_nM isolobal with CH, for example, $Cp_2Ni_2(CO)$. This might be formed from $NiCp_2$ and CO.

2–3. (1) 48e, 3 M−M bonds; (2) 50e, 2 M−M bonds; (3) 52e, 1 M−M bond. The S's are counted as vertex atoms—they retain their lone pair as shown by easy methylation.

4. This 60e cluster is 2e short of the 62e system expected; Wade's rules give 14 skeletal electrons appropriate for an octahedron counting each of the EtC carbons as vertices.

5. **B** is isolobal with tetrahedrane, **C** with cyclopropane.

6. The Fe$_4$ species is 60e and should be tetrahedral. Four Fe(CO)$_3$ groups are likely, which leaves a single CO, which might be bridging; but we cannot tell from counting electrons. The Ni$_5$ structure is 76e, and so a

square pyramid with one Ni—Ni bond opened up is most likely. The 36e Cr_2 system is expected to have no M—M bond but be held together by the bridging phosphine.

7. Two W≡C bonds bind to Pt in the cluster just as two alkynes should bind to Pt in the alkyne complex, so $n = 2$. On an 18e rule picture, the alkynes are 4e donors. The unsaturated ligands are orthogonal so that each X≡C bond (X = W or C) can back-bond to a different set of d_π orbitals.

8. The most symmetric structure is a square pyramid with Fe at the apex and four B's at the base; $(\eta^4\text{-}C_4H_4)Fe(CO)_3$ is the carbon analog.

9. Elements to the left of C are electron deficient; elements to the right are electron rich. As long as electron-deficient elements dominate a structure, a cluster product can be formed.

10. An $\eta^2\text{-}\mu\text{-}CH_2CO$ complex with the ligand bridging two O atoms that have lost their direct M—M bond.

Chapter 14

1. Oxidative addition of ArI is followed by insertion of the alkene; β elimination gives a new alkene, nucleophilic attack on which by the N lone pair is followed by loss of MeOH to aromatize the system.

2. Nucleophilic attack at C with displacement of the epoxide as an $-O^-$ group, is followed by protonation to give the alcohol, loss of water, formation of $Fp(\text{alkene})^+$, and displacement of the alkene with I^-.

3. Possibly an oxidative addition of Cl—CCl_3, insertion of C=C and reductive elimination, but this could also be a radical chain reaction initiated by the metal. In this case ·CCl_3 would add to the free alkene to give $RCH \cdot CH_2CCl_3$, which would abstract Cl from another mole of CCl_4. If the latter were true, however, we would see crossover, so we can rule out the radical pathway.

4. The phenol is formed by isomerization. Treatment with the iron carbonyl forms a diene complex in which the double bonds have been shifted by isomerization so that they are in the left-hand ring.

5. Chelation of the diene is followed by nucleophilic attack of MeOH on the exo face, then CO insertion and nucleophilic attack of MeLi on the resulting acyl. Net HO—CO_2H addition across C=C.

6. The NMe_2 group binds to the metal and so directs Pd to the front face, CHE_2^- attacks from the back to give a five-membered ring intermediate, which then β-eliminates. The second sequence is similar but includes a Heck reaction.

7. $FpCH_2SMe$ is formed first, then $FpCH_2SMe_2^+$. Loss of Me_2S gives the carbene, which cyclopropanates the alkene.

8. Ketones lack a reactive C—H bond. After oxidative addition of RCO—Cl, retromigratory insertion and reductive elimination of RCl, $RhCl(CO)L_2$ is formed.

9. The 16e $RhCl(CO)L_2$ does not lose CO easily, but the dpe complex gives $Rh(dpe)_2(CO)$, which, being 18e, loses CO more easily because Rh(I) prefers 16e to 18e.

10. Trans-2-methoxycyclohexane carboxylic ester is formed by trans methoxy-mercuration transfer of the alkyl to Pd, CO insertion, and hydrolysis.

11. Syn. If Ph is at position 2 and Pd at 1, β-elimination is expected at the methyl group.

12–14. See papers cited.

Chapter 15

1–2. The metal is d^0 and therefore CO does not bind well enough to give a stable complex, but weak binding is possible and the absence of back donation increases δ^+ character of CO carbon and speeds up migratory insertion in the weakly bound form.

3. The third-row element prefers the higher oxidation state and has longer M—C bonds, allowing a greater number of R groups to fit around the metal.

4. Electrophilic abstraction is likely for Eq. 15.4, but this is unlikely for **15.6** and **15.7** because the M—C carbons are sterically protected in these two compounds.

5. The two alkenes are orthogonal to allow the metal to back-donate efficiently to both alkenes by using different sets of d_π orbitals.

6. Alkene hydrogenation normally occurs in the presence of many hydride ligands. The stereochemistry of the Re compound makes the (C=C) groups of the bound alkene orthogonal to the M—H bonds and prevents insertion.

7. Cr, 1 and 3; Mn, 4, 2, 0; Fe, 5, 3, 1; 4, 2, 0.

8. Cp_2^* Lu groups at 1 and 4 positions on benzene ring.

Chapter 16

1. These are the most abundant metals in the biosphere.

2. Most organisms live in an oxidizing environment and proteins have mostly hard ligands.

3. A low $\nu(N_2)$ implies strong back donation, which also means that the terminal N will also have a large ∂^- charge and therefore be readily protonated.

4. The stability of radicals R• is measured by the R$-$H bond strength, which is the ΔH for splitting the bond into R• and H•. For these species this goes in the order HCN > CF$_3$H > CH$_4$ > PhCH$_3$ > Ph$_2$CH$_2$. C$-$H bonds to sp carbons are always unusually strong because of the high s character while Ph groups weaken C$-$H bonds by delocalizing the unpaired electron in the resulting radical. This is the reverse of the order of ease of loss of R•.

5. Protonation lowers the electron density on Re and reduces the back donation to N$_2$, resulting in an increase in $\nu(N_2)$ and weaker M$-$N$_2$ binding, making the N$_2$ more easily lost.

INDEX

Page numbers in **bold type** indicate the main entries.

The Organometallic Chemistry of the Transition Metals, Fourth Edition, by Robert H. Crabtree
Copyright © 2005 John Wiley & Sons, Inc.